Understanding Dying, Death,
and Bereavement

Harcourt
College Publishers

Where Learning Comes to Life

TECHNOLOGY
Technology is changing the learning experience, by increasing the power of your textbook and other learning materials; by allowing you to access more information, more quickly; and by bringing a wider array of choices in your course and content information sources.

Harcourt College Publishers has developed the most comprehensive Web sites, e-books, and electronic learning materials on the market to help you use technology to achieve your goals.

PARTNERS IN LEARNING
Harcourt partners with other companies to make technology work for you and to supply the learning resources you want and need. More importantly, Harcourt and its partners provide avenues to help you reduce your research time of numerous information sources.

Harcourt College Publishers and its partners offer increased opportunities to enhance your learning resources and address your learning style. With quick access to chapter-specific Web sites and e-books . . . from interactive study materials to quizzing, testing, and career advice . . . Harcourt and its partners bring learning to life.

Harcourt's partnership with Digital:Convergence™ brings :CRQ™ technology and the :CueCat™ reader to you and allows Harcourt to provide you with a complete and dynamic list of resources designed to help you achieve your learning goals. Just swipe the cue to view a list of Harcourt's partners and Harcourt's print and electronic learning solutions.

http://www.harcourtcollege.com/partners/

Understanding Dying, Death, and Bereavement

Fifth Edition

Michael R. Leming
St. Olaf College

George E. Dickinson
College of Charleston

Harcourt College Publishers

Fort Worth   Philadelphia   San Diego   New York   Orlando   Austin   San Antonio
Toronto   Montreal   London   Sydney   Tokyo

Publisher	Earl McPeek
Acquisitions Editor	Bryan Leake
Market Strategist	Laura Brennan
Project Manager	Andrea Archer

ISBN: 0-15-506618-8
Library of Congress Catalog Card Number: 2001090150

Address for Domestic Orders
Harcourt College Publishers, 6277 Sea Harbor Drive, Orlando, FL 32887-6777
800-782-4479

Address for International Orders
International Customer Service
Harcourt College Publishers, 6277 Sea Harbor Drive, Orlando, FL 32887-6777
407-345-3800
(fax) 407-345-4060
(e-mail) hbintl@harcourt.com

Address for Editorial Correspondence
Harcourt College Publishers, 301 Commerce Street, Suite 3700, Fort Worth, TX 76102
Web Site Address
http://www.harcourtcollege.com

Printed in the United States of America

1 2 3 4 5 6 7 8 9 066 9 8 7 6 5 4 3 2 1

Harcourt College Publishers

ABOUT THE AUTHORS

 **Michael R. Leming** is professor of sociology at St. Olaf College in Minnesota. He holds degrees from Westmont College (B.A.), Marquette University (M.A.), and the University of Utah (Ph.D.). He has completed additional graduate study at the University of California–Santa Barbara. He is the founder and former director of the St. Olaf College Social Research Center and is a former member of the board of directors of the Minnesota Coalition of Terminal Care. He has served as a steering committee member of the Northfield AIDS Response and as a hospice educator, volunteer, and grief counselor. Dr. Leming has lived in Thailand on several occasions and taught at Chiang Mai University there. He is the author of numerous articles on social thanatology and family issues and has taught courses on death and dying for over 25 years.

 **George E. Dickinson** is professor of sociology at the College of Charleston in South Carolina. He holds degrees from Baylor University (B.A. in biology and M.A. in sociology) and Louisiana State University (Ph.D. in sociology). He has completed postdoctoral studies in gerontology at Pennsylvania State University, in thanatology at the University of Kentucky School of Medicine, and in medical sociology at the University of Connecticut. He was Visiting Research Fellow in palliative medicine at the University of Sheffield School of Medicine in England in 1999. He has published over 50 articles in such professional journals as *Social Science and Medicine*, *Omega*, *Death Studies*, *Academic Medicine*, *Medical Care*, *Clinical Anatomy*, *Family Practice Research*, *Journal of the American Medial Women's Association*, *Loss, Grief, and Care*, *Journal of Dental Education*, *Journal of the South Carolina Medical Association*, *Archives of Internal Medicine*, *Journal of Nursing Education*, *American Journal of Pharmaceutical Education*, *Journal for the Scientific Study of Religion*, *Journal of Marriage and the Family*, and *Journal of Family Issues*. Dr. Dickinson has been teaching courses on death and dying for over 25 years and has been actively involved as a hospice educator. He is on the international editorial board of *Mortality* in the United Kingdom.

In addition to *Understanding Dying, Death, and Bereavement* (fifth edition), Leming and Dickinson are coauthors of *Understanding Families: Diversity, Continuity, and Change* (Allyn & Bacon, 1990, and Harcourt Brace, 1995), and they are coeditors with Alan C. Mermann of *Annual Editions: Dying, Death, and Bereavement* (Dushkin/McGrawHill, 1993, 1994, 1995, 1997, 2000, 2002).

PREFACE

Nearly twenty years ago, we began work on the first edition of *Understanding Dying, Death, and Bereavement*. Our goal in the fifth edition, as in the previous editions, was to create a book that was both informative and practical, yet theoretical. We visualized a humanistic text that was multidisciplinary in orientation and inclusive of the major foci of the interdisciplinary subject of social thanatology. Indeed, many changes have occurred in end-of-life issues since that first edition. The topic of dying, death, and bereavement is coming "out of the closet" and is less of a taboo topic than it was in the early 1980s. The current edition reflects these changes.

The fifth edition of *Understanding Dying, Death, and Bereavement* equips the reader with the necessary information to both understand and cope with the social aspects of dying, death, and bereavement. Because we have taught courses on thanatology for over twenty-five years, we are convinced that every student carries both academic and personal agendas when approaching the subject.

This textbook will make a significant contribution to your class because it is a proven text informed by over fifty years of combined experience. *Understanding Dying, Death, and Bereavement* is comprehensive and covers the wide range of topics in social thanatology. It is scholarly and academically sound and is practical for students because it addresses personal issues relating to an individual's ability to cope with the social and psychological processes of dying, death, and bereavement. This text appeals to a vast audience, not only because of its wide adaptability on college and university campuses, but also because of its practical implications for all persons. Although intended primarily for undergraduate students in sociology, psychology, anthropology, nursing, social work, kinesiology, religion, gerontology, health science, family studies, philosophy, and education courses, it is also appropriate for professional courses in medicine, nursing, mortuary science, social work, and pastoral counseling.

The current edition reflects major changes from previous editions. There are five new chapters: "Living with Dying," "Dying in the American Health-Care System," "The Business of Dying," "The Legal Aspects of Death," and "Grieving Throughout the Life Cycle." The other nine chapters have been significantly altered and updated to address issues of the 21st century: "Studying Dying, Death, and Bereavement," "The American Experience of Death," "Growing Up with Death," "Perspectives on Death and Life After Death," "The Dying Process," "Biomedical Issues and Euthanasia," "Suicide," "Diversity in Death Rituals," and "Coping with Loss." The boxed inserts in the chapters fall into four categories: (1) **Practical Matters** boxes basically offer practical advice; (2) **Listening to the Voices** boxes consist of excerpted material from people writing about their own experiences with dying and death; (3) **Words of Wisdom** boxes contain excerpted materials—poems, literature, and other words of wisdom; and (4) **Death Across Cultures** boxes examine cross-cultural examples of death practices and beliefs.

In addition to the boxes in each chapter, chapter **conclusions** and **summaries** serve as study aids to outline the important points of each chapter. **Discussion questions** probe students' understanding of the important issues. **Glossaries** and **suggested readings** should also enhance the students' understanding of thanatology. As in previous editions, pictures and cartoons are scattered throughout the book.

For the instructor, an instructor's manual and test bank for Understanding Dying, Death, and Bereavement (fifth edition) is available and can be obtained through your local Harcourt College representative or by writing to the Sociology Editor, Harcourt College Publishers, 301 Commerce Street, Suite 3700, Fort Worth, TX 76102.

Web Site

Rather than place Web sites in each chapter, we opted to refer the reader to Mike Leming's Web site and thanatology links, which he regularly updates, at **www.stolaf.edu/people/leming.** This Web site should be especially useful to you if you are doing research on a topic in thanatology or simply wish to do additional readings on dying, death, and bereavement. The links can open up numerous possibilities for you in your research or reading efforts. We encourage you to log on to this site and take advantage of Michael Leming's continuous updates on your behalf.

Acknowledgments

We gratefully acknowledge those who reviewed the fourth edition and made valuable suggestions, which are incorporated in this edition.

Our greatest gratitude goes to Sylvia Shepherd, our developmental editor. Sylvia's insight and her ability to see "the big picture" contributed significantly to the major reconstruction of the fifth edition. Her creative energies and editorial background combined to make her the editor that every author needs. Indeed, we were fortunate to have Sylvia assigned to our book.

We are also indebted to Bryan Leake, Acquisitions Editor; Laura Brennan, Market Strategist; Andrea Archer, Project Manager; Caroline Robbins; Charlotte Thomas; and Mike McConnell at Graphic World Publishing Services. We acknowedge our reviewers: Martha Shwayder, Metropolitan State College of Denver; James F. Paul, Kankakee Community College; Catherine Wright, Mitchell College; Dolores Mysliwec, Glenville State College; Gerry Cox, University of Wisconsin–La Crosse.

As in the past, we acknowledge those we have encountered in life. So much we have learned from significant others and our students who have shared their experiences with dying and death. Our own lives have indeed gained from these experiences. We hope our readers will experience an appreciation for life as they begin to understand, both intellectually and emotionally, the social-psychological processes of dying, death, and bereavement. This book is about dying and death, but it is really about living and life. May your reading of this book help to make your life more meaningful.

M.R.L.
G.E.D.

CONTENTS

Chapter 12
The Legal Aspects of Dying   425

1

STUDYING DYING, DEATH, AND BEREAVEMENT

The symbols of death say what life is and those of life define what death must be. The meanings of our fate are forever what we make them.
                    —Lloyd Warner, *The Living and the Dead*

This chapter opens with a photograph of Princess Diana and Mother Teresa, two of the world's most famous women of the latter half of the 20th century. Prior to September 1997, this picture would have evoked a considerably different response than it does today. At that time, these two women died within a week of each other. Why did the death of Princess Diana have such an impact on the world? Certainly her problems with the royal family made individuals identify with her, since she seemingly had been "wronged" in the eyes of many individuals. We tend to identify with the underdog. Her relatively young age made her death so unfair and illogical— death is not "supposed" to happen at that age. Princess Diana was a person frequently in the news; thus many individuals around the world felt that they "knew her." She came into our homes via television, and indeed was "often there." Her death to many was like a member of the family dying.

Media coverage of Princess Diana's death was almost unprecedented. The public reaction, along with media coverage, fed on each other. The whole event simply swelled out of proportion. The public, fanned by the media, felt a need to express their sympathy for this "unfortunate death." The death had all the makings of a suspense movie—Was the driver drunk? Why were they leaving in such a hurry? Did the princess really love this man? Had he planned to give her an engagement ring later that night? The drama built out of sight, with public demands and the media pushing each other, until much of the world became saturated with "the story." Cinderella was not mistreated by her "wicked old step-sisters" but by the press and the royal family. The clock struck midnight, and the story did not have a happy ending. Thus, society rallied around this unfortunate figure.

And then, there was Mother Teresa. If there ever was a saint in the 20th century, it was probably her. Yet, she died a week after Princess Diana, and the media coverage and public outcry were much more subdued. Why? Perhaps because it was "right" for Mother Teresa to die at that time. After all, she had lived well beyond three score and ten years, so unlike Princess Diana. Mother Teresa's death seemed natural. And it was easy to accept her death because she had lived a life that many would not want, but that all would appreciate. She was not a "glamorous" person whom perhaps, many would like to have been like in life. Also, maybe it's easier to accept the death of such a saintly person, knowing that if there were a heaven, she would be there. Would Diana? Who knows?

How we understand these deaths and how we explain our reactions to these deaths is a major theme throughout this book. We will also look at how sociologists approach the study of death. The juxtaposition of these two deaths within a week of each other may lead some to examine death in the death-related behavior and institutions in the contrasting worlds these two women occupied. Death in the elite world of the rich and famous is an experience vastly different from that in Mother Teresa's world of poverty. Others may examine the rituals surrounding the deaths of these two women. Although the differences between the two women and their deaths are obvious, some may want to explore the common themes that are evident in death rituals in all societies.

A doctor may focus on the biological aspects of death, most concerned with the specific trauma that caused Diana's death and the exact point at which she died. An autopsy might be required to answer these questions. Mother Teresa's death raises few questions that would interest a doctor. We mentioned above that the intensity of the reaction to Diana's death may be due to her age. In fact, many approach the study of death from a developmental perspective, examining death at all stages of the life cycle. But in the end, what we all want to know is the meaning of death, and we feel that this examination of the meaning of death and death-related behaviors is one of the most important ways to approach the study of dying, death, and bereavement.

CURRENT INTEREST IN DEATH AND DYING

For the past 20 years or so, however, a fascination with dying and death phenomena has been evidenced within much of the U.S. population (Crase, 1994). The fascination has been fueled by print and electronic media, by directives or various legal bodies, and by legal gymnastics surrounding several prolonged and much-observed deaths of individuals. Americans are finally willing to acknowledge that death is part of life, and they want to talk about it (Foderaro, 1994). As Muriel Gillick notes, "Death happens 55 million times each year throughout the world and 2.3 million times annually in the United States. Of the tens of billions of people that have ever existed, everyone born before 1880 has died, and nearly everyone currently alive today will perish in this century." (Gillick, 2000)

Today's thanatology student is bombarded by pressing issues of the day that involve death and death-related matters: the acquired immunodeficiency syndrome (AIDS) crisis, the prolongation of dying from cancer, the growing incidence of chronic illnesses with uncertain courses, murder, ecological disasters, fetal transplants, and abortion. The topic of death is "alive and well" in today's contemporary society. Let's look at some of the reasons why dying and death as topics of discussion have come into their own in recent years.

Why the Increased Interest?

Though death has been around since the beginning of humankind, in recent years an almost fascination with death has evolved. Such an increased interest in **thanatology** (the study of dying, death, and bereavement) is due to several reasons: an aura of mystery surrounding death, an interest in ethical issues concerning death and dying, and increased media coverage of deaths, especially violent deaths.

The Mystery of Death

With the majority of Americans dying in hospitals and nursing homes—not the case a century ago—an aura of mystery developed. One is taken to the hospital or nursing home and is brought back dead. Thus, a child may begin to wonder what is going on with this "thing of death" that takes people away, not to return. As one little boy said in writing a letter to God, "Dear God, What is it like when you die? I don't want to do it. I just want to know." Often small children are not allowed to go into certain parts of a hospital; thus the taboo nature of a hospital setting makes one wonder what is going on in there—"I just want to know." Hospitals do not allow such entry in part because of contagious diseases to which small children would be more vulnerable than adults. Such rejection is not unlike Schulz's saying "no dogs allowed," as Snoopy tries to enter a forbidden area. "No children allowed" is indeed the strong message of rejection here. To say "no" to a child often increases his or her curiosity tremendously. "Why is this place such that I cannot enter?" the child may ask.

A desire to examine this "mysterious thing" called death has contributed to a growing interest in thanatology. Our society has done little to achieve formal **socialization** of its members to deal with death on personal and emotional levels. Even though the hospital may not allow children to enter certain areas or at particular times, parents have often tried to shield their "innocent" children from death scenes. Medical and theology schools, which train persons who will work with the dying, have not had significant curricular offerings to prepare their students for this death-related work. Overall, our socialization to dying and death situations has been unsystematic and ineffective.

LISTENING TO THE VOICES
Death on the Farm

As one brought up exposed to a farm environment, George Dickinson remembers on one occasion helping his father dispose of the body of a polled Hereford bull. He had been dead for about a day before he was found in the woods. The buzzards circling overhead were a good indication of where the lost bull would be. When he was actually found, about a dozen buzzards were already there. Just shooing those aggressive vultures away from the death scene was indeed a grim portrayal of death. Since the bull weighed several hundred pounds, they decided against earth burial and proceeded to cremate his remains. After soaking him in kerosene and putting numerous pieces of timber over the corpse (not completely unlike a pyre, as discussed in Chapter 10), they set him on fire. The corpse was still smoking several days later, though the buzzards had long since gone away.

Today's lack of familiarity with death may be in part due to fewer individuals being raised on farms than was the case in the early 20th century. Being brought up in a rural environment gave one direct exposure to birth and death as everyday events. Children were surrounded by the alpha and omega of the life cycle. Kittens, puppies, piglets, lambs, calves, chicks, and colts were born—and also died. Thus, it was commonplace to make observations of death and to deal with these situations accordingly.

Death can be very visible down on the farm. With less than 10 percent of the U.S. population engaged in farming today, birth and death scenes have largely been removed from the personal observations of most individuals. The mystery of these events at the beginning and at the end of life is at least somewhat addressed with farm dwellers—less likely the case with urbanites.

Ethical Issues

Individuals are living longer today because of life-support equipment, organ transplants, penicillin, and other antibiotics and "miracle drugs" (see chapter 8 for a detailed discussion). Such prolongation of life has raised ethical issues dating back to the 1970s (e.g., Karen Ann Quinlan and Baby Jane Doe) that involve the right to die, causing a furor in philosophy, law, and medicine. With the media highly publicizing these cases, the public was alerted to moral and legal questions on death not previously posed. Whether or not to "pull the plug" and disconnect life-supporting equipment presented questions for which ready answers were not found. An elderly woman summed up this dilemma when she spoke to Jinny Tesik of Compassion in Dying (Sturgill, 1995), an organization providing education regarding terminal care: "We used to be afraid to go to the hospital because that's where you went to die; now we're afraid to go because that's where they won't let us die."

Emigrants crossing the Plains in 1869 faced a challenging life ahead as they moved westward.

The whole issue of when death occurs evolves from these medical developments. The question of who determines when one is alive or dead has been addressed by physicians, lawyers, philosophers, and theologians. These questions, along with the controversy over abortion rights, were a few of the significant ethical issues of the 1970s that provided an open forum for discussion and debate concerning the topic of dying and death. Specific definitions of death were not as necessary, prior to the coming of these 20th century medical breakthroughs.

The Media

As a child growing up in Texas, George Dickinson has fond memories of spending the night with his grandparents on Fridays and going to the movie with his grandfather. The movie was always a "Western" with basically the same theme—the "good guys" (the cowboys) wore white hats, and the "bad guys" were the Indians. Though toward the end of each movie it appeared that the "good guys" were going to be wiped out, military reinforcements would always come to the rescue just at the last minute. The guys in the white hats would be rescued, and the "bad guys" would receive their "just reward" of death. Similarly, recent research (Yokota and Thompson, 2000) of G-rated animated feature films released in theaters between 1937 and 1999 revealed that characters portrayed as "bad" were much more likely to die of injury than other characters.

In the early 1970s, however, Hollywood began to additionally produce films revolving around the theme of death in which the "good guy," and the box office star, died. One of the first of these movies to deal with death was *Love Story*, in which one of the two main characters is dying throughout much of the film and eventually does die during the film. More recently, Tom Hanks dies of AIDS in *Philadelphia Story*, and Susan Sarandon dies of cancer in *Stepmother*. Other movies provide a realistic portrayal of historical events involving mass deaths, such as *Schindler's List*, *Saving Private Ryan*, and *Armageddon*. Almost an obsession, and certainly a fascination, with death seems to occur in today's society—especially on the screen in a somewhat imaginative world.

Likewise, many situation comedies on prime time television in the 1970s, 1980s, and 1990s and now the 21st century, have addressed the topic of death. Some have viewed death in a serious vein, whereas others have taken a humorous approach. Among the first situation comedies to talk about death was *The Cosby Show*, starring Bill Cosby. This popular late 1980s/early 1990s program showed an episode in which a goldfish had died. Bill Cosby gathered the family around the toilet bowl in the bathroom and insisted that they have a proper funeral for the deceased fish. One read scripture, one said a eulogy, and then another flushed the fish out into its "eternal resting place" in the sewer. The episode made a serious effort to explore some issues involving the death of pets, even though one might not experience the death of a goldfish in the same way as the death of a dog.

Likewise, *Sesame Street* devoted a 1-hour program to the death of Mr. Hooper, a few days after his death. However, rather than explore the feelings of loss, they focused on all the good qualities of Mr. Hooper. Only in the end did Big Bird, who seemingly was having a more difficult time with the death of his friend say, "We're going to miss you, Mr. Hooper." Indeed, well said. We do miss individuals when they die. Big Bird's comment was a most realistic way of expressing his feelings about the death of a friend on a popular children's television program. Although these two programs were both rather serious, we often respond to both death and sex as topics about which we joke. They are uneasy topics; thus to laugh about them is a way to cope.

Some of the earliest television programs to discuss death appeared in the early 1970s. *Living with Death* presented various death-related situations observed through the eyes of a CBS reporter. ABC's *The Right to Die* addressed moral questions of mercy killing and suicide. The National Endowment for the Humanities sponsored a program entitled *Dying*. For 2 hours this program very sensitively portrayed four cancer patients, ranging in age from the late 20s to early 70s. Each died during the course of the filming. A PBS documentary in 1979 showed the last three years of Joan Robinson's life. This film revealed the experience of a woman and her husband as they tried to live with her cancer of the breast and uterus. In the fall of 2000, *On Our Own Terms: Moyers on Dying*, a four-part, 6-hour series, explored issues related to death and dying, including candid conversations with people dying in their homes and in hospitals. This series was followed by a companion show, *With Eyes Open*, a four-part, half-hour PBS series of interviews examining grief, mortality, caregiving, and the afterlife.

The topic of dying and death maintained a high profile throughout the 1990s in part because of Dr. Jack Kevorkian. His "suicide machine" to terminate the life of a terminally ill patient with Alzheimer's disease in Michigan in 1990 drew headlines from the media. Subsequent suicide deaths assisted by Dr. Kevorkian through the 1990s (Kevorkian claimed over 100) have fanned the fire of this controversy. Kevorkian's actions peaked with the video showing on *60 Minutes* in 1998, after which he ended up in prison. Much was aired about the whole controversy of CBS's allowing this to be put on television. This "shot" was heard (and seen) around the world. The various trials of Dr. Kevorkian brought extensive coverage in the media. What more drama could the media request? Although some individuals might have viewed Kevorkian more-or-less as a "serial killer," he continued to be allowed to practice his physician-assisted suicide in defiance of the law. His goal was to change the law and make this behavior legal. In standing up for that which he felt to be right and at the risk of his own life, Kevorkian fought on and gained many supporters. After all, here is a man who will risk life and reputation (no matter how obnoxious he seemed to many) to prove his point. The media loved it and gave him the publicity he desired.

In addition to Dr. Kevorkian's actions, the issue of physician-assisted suicide made headlines when voters in Washington, California, and Oregon went to the polls to decide the legality of this issue. Although the final vote counts in Washington and California were close, such action was not approved. On the other hand, the voters of Oregon successfully approved physician-assisted suicide. The Oregon voting was followed by court action and ultimately action by the Supreme Court (see chapter 8). Talk shows on both radio and television debated the pros and cons of assisted deaths. Individuals tend to have strong opinions regarding topics such as abortion and assisted deaths. Therefore, such controversial topics solicit heavy media coverage.

Peter Nardi (1990) analyzed how the media handle AIDS and obituaries. He noted that one of the main issues raised by obituary reporting is personal privacy versus journalistic ethics. Unlike other diseases often unreported in earlier generations such as tuberculosis and cancer, AIDS raises questions of both medical and sexual ethics. The dilemmas faced by the media in AIDS-related deaths draw attention to the continuing stigma attached to AIDS. Sociologists and other social scientists have the task of identifying the circumstances and social psychological processes underlying these negative attitudes.

The media have been a positive force in bringing the topic of AIDS "out of the closet." For example, when a celebrity or sports hero, such as Rock Hudson or Earvin "Magic" Johnson, is infected with human immunodeficiency virus (HIV), the media have helped to make AIDS and HIV household words in many households where the words had never been spoken (Lemelle, Harrington, & Leblanc, 2000). In 1991, when Magic Johnson publicly announced that he is HIV-positive, this hero for many, a heterosexual professional basketball star, produced new ways to respond to AIDS from both the media and the public. Today, the public is more accepting of the identification of HIV infection and the testing procedures for detecting HIV antibodies. Yet, these discoveries were scientific breakthroughs and altered life for many, note Lemelle and colleagues.

Death Education

The topic of dying and death did not come into its own in recent times until the 1970s. Certainly death, like sex, was not a new event, but both had been rarely discussed openly. Sex, as a subject of discussion, came "out of the closet" in the 1960s, followed by death in the 1970s.

Thanatology Classes

Thanatology flourished in the academic world in the 1970s and continues to be popular today. Literally hundreds of courses on dying and death are offered in high schools and colleges. According to Dan Leviton (1977), the number of death education college courses increased from 20 in 1970 to 1,100 in 1974. However, thanatology offerings at the secondary school level declined, perhaps because of the pressure to return to the basics (Pine, 1986). V. A. Cummins (1978) in a survey of 1,251 colleges found that 75 percent were offering courses in dying and death and that these courses were offered in many departments. The majority of courses were housed in sociology and social work departments, followed by psychology, religious studies, philosophy, and health education departments.

Fergus Bordewich (1988) observes that a generation ago death was less likely than sex to be found as a subject in the curricula of American public schools. If death was acknowledged at all, it was usually discussed within the context of literary classics. In the past decade courses treating dying and death far more explicitly have appeared in schools across the country. Though the actual number is not known, thousands of schools are involved, according to groups that either oppose or support death education. Many schools have blended some of the philosophies and techniques of death education into health, social studies, literature, and home economics courses. Others have introduced suicide-prevention programs.

An argument for death education noted by Bordewich (1988, p. 31) from *The School Counselor* in 1977 said:

> An underlying, but seldom spoken, assumption of much of the death-education movement is that Americans handle death and dying poorly and that we ought to be doing better at it. As in the case of many other problems, many Americans believe that education can initiate change. Change is evident, and death education will play as important a part in changing attitudes toward death as sex education played in changing attitudes toward sex information and wider acceptance of various sexual practices.

Death education provides an opportunity to familiarize students and professionals with the needs and issues surrounding dying and death. This is important because of the pervasiveness of formal death institutions, media exposure to death, and haphazard experiences with death professionals, notes Vanderlyn Pine (1986). One of the goals of courses on dying and death for all age groups is to increase knowledge about death and about the professions involved with death—funeral directors, medical personnel, and governmental organizations, for example (Gordon

& Klass, 1979). Other goals are to help students learn to cope with the deaths of significant others, to deal with one's own mortality, and to be more sensitive to the needs of others. A more abstract goal is to understand the social and ethical issues concerning death as well as the value judgments involved with these issues.

What effects do courses on dying and death have on students? As noted in pre- and posttests of the attitudes of 184 students enrolled in a college course on dying and death (Dickinson, 1986), the majority of responses revealed less fear of death and death-related events at the end of the course. These students had higher death anxieties on entering the course than did subjects in four similar studies of other populations, including one of coal miners just prior to entering the mines (Dickinson, 1986). After the course, these students revealed death anxieties much lower than those of the other four populations. Perhaps students coming into an elective course in thanatology have a high fear of the topic anyway and enter the course to alleviate their fear. Nonetheless, this study did reveal less anxiety in the students after discussing, reading, seeing films, visiting a funeral home, and going to a cemetery. Results of a similar high school study of before and after comparisons of adolescents (Rosenthal, 1980) showed that their level of anxiety decreased as a result of an 18-week death education course. Thus, according to these two studies, exposure to dying and death seems to have made the participants less anxious about death-related matters.

The emotional conflicts related to dying and death are particularly acute for primary care health professionals. Although limited emphasis has historically been placed on death education in schools for health professionals in the United States, the current status of death education offerings in nursing, medicine, paramedicine, pharmacy, dentistry, and social work is somewhat encouraging (Dickinson, Sumner, & Frederick, 1992; Smith & Walz, 1995). Though few schools offer a full course in death education, the majority (with the exception of dentistry schools) have some emphasis on this topic in their curricula. Most began their offerings after 1975.

As more emphasis is placed on relating to terminally ill patients and their families in the health professions, we would hope that more positive attitudes toward treatment of the dying patient will emerge when these students become practitioners. Helping students deal with their own anxieties about death at the time of actually facing terminally ill patients would seem to be an appropriate time for intervention. In the end, the young professional, the patient, and the patient's family all should benefit from this emphasis on death education.

Thanatology Publications

In the 1970s Michel Vovelle (1976) published an article entitled "The Rediscovery of Death" in which he documented the sudden flurry of publications on the subject of death that appeared in the Western world from the 1950s on. Geoffrey Gorer's 1955 essay on "The Pornography of Death" seemed to open the door for publications on the subject of death. Gorer argued that death had replaced sex as contemporary society's major taboo topic. With death in the community becoming rarer and with individuals actually seeing fewer corpses, a relatively realistic view of death

---

LISTENING TO THE VOICES
Physicians Need Preparation for Relating to Dying Patients

I think the proper role for a physician with a dying patient is to enter into partnership where there is a sense that that partnership is not going to end with the end of medical therapy. But we're not trained for that. Most of us feel totally inadequate and have found our own defenses, including aban-

doning the patient. It's been confusing to me and a constant source of frustration.

Dawn McGuire, a fourth-year medical student at Columbia Presbyterian Medical Center in New York City (1988, p. 341)

---

had been replaced by a voyeuristic, adolescent preoccupation with it, observed Gorer. Vovelle argued that the focus of attention represented by these texts amounted to nothing less than the "displacement of a deeply seated taboo on a subject that had lain hidden in the shadows of the western psyche since some unspecified point in the 19th century" (Prior, 1989, p. 4).

One of the early "texts" for thanatology, published in 1959, was an anthology by psychologist Herman Feifel entitled *The Meaning of Death*. Feifel's work was an interdisciplinary attempt to restore death to cultural consciousness. In 1963 Jessica Mitford's *The American Way of Death* was very critical of the funeral industry. Elisabeth Kübler-Ross's *On Death and Dying*, published in 1969, advised Americans that they can play a significant role in the lives of the dying. Ernest Becker's *The Denial of Death*, published in 1973, argued that denying death is commonplace in our society. Both Kübler-Ross's and Becker's books became best sellers. It was Kübler-Ross, a physician, who was a real catalyst in making the medical profession realize that terminally ill patients are more than the cancer patient in Room 713 and are warm, living human beings who have personal needs. Both of these books alerted the public to the issue of dying in America. A more recent book discussing dying that was on the best-seller list from 1998 through 2001 is Mitch Albom's *Tuesdays with Morrie* (1997), in which the author writes about his former college professor who is dying of amyotropic lateral sclerosis (ALS). Two professional journals on thanatology emerged in the United States in the 1970s: *Death Studies* (formerly *Death Education*) and *Omega: The Journal of Death and Dying*. In the 1990s in England, *Mortality* evolved as a thanatology journal. Whereas the U.S. thanatology journals are psychology-oriented, *Mortality* has more of an orientation toward sociology. Other related journals that have evolved in recent years in the United States and United Kingdom are: *The Hospice Journal, Progress in Palliative Care, Illness, Crisis & Loss, Palliative Medicine*, and *Suicide and Life-Threatening Behavior*. The number of articles on death also expanded considerably in journals of education, family, medicine, health, nursing, psychology, social work, and sociology.

Death education should not only prove useful in coping wit¹
situations, but also should actually improve the quality of our
Kübler-Ross noted, relating to the dying does not depress he
her both appreciative of each day of life and thankful each n
with the potential of another day. Learning more about dyi
voke one to strive to make each day count in a positi
about death will tend to make one "look for the good in otn
the late author Alex Haley suggested on his letterhead stationery.

MORTALITY STATISTICS

Ask an individual how he or she wishes to die. With the exception of the comical reply, "When I am 92 and at the hands of a jealous lover," most people will respond,
"When I am very old, at home, unexpectedly, in my own bed, while sleeping—and with
my full mental and physical capabilities." Unfortunately for most of us, we will not die
as we would like. For some, this fact may be a source of apprehension and anxiety.

Death Etiology and Life Expectancy

Whereas in the 19th century the **mortality rate** was particularly high among children
and continued at a high level throughout adult life, the incidence of death is now
heavily concentrated among the elderly (Mulkay, 1993). Thus, average life expectancy is now much longer. For most individuals, death approaches slowly over
years of gradual decline. As noted in Table 1.1, the cause of death (etiology) for
most of us (65 percent) will be from one of two chronic diseases—heart disease or
cancer. With these **chronic diseases**, deaths are usually prolonged and are not sudden and unexpected, as most people might desire.

In the United States, cigarette consumption remains the single most preventable
cause of sickness and premature death (Anderson, 1996). American women did not
begin smoking in large numbers until after World War II, and their rates of lung cancer are now matching those of men (Cockerham, 2001). Indeed, this preventable
cause of death, cigarette smoking, has increased 75 percent worldwide in the last
two decades, according to Robert Anderson (1996) and is now responsible on a
global basis for about 5 percent (a conservative estimate) of all deaths (approximately 20 percent of all deaths in the United States).

Life expectancy has increased considerably in the past 50 years. Life expectancy
for males in the United States has increased from 54 years to more than 70 years over
the past half century. Females' life expectancy during this same period has increased
from 55 to nearly 80 years. Whites live nearly 7 years longer than blacks in the United
States. According to a study reported in the *New York Times* ("Is Life Expectancy Now
Stretched to Its Limit?" 1990), however, there may be a limit to how long our bodies
can hold out. The study concluded that science and medicine have pushed human life

Leading Causes of Death in the United States, 1900 and 2000 (in Death Rates per 100,000 Population)

Causes of Death	Death Rates Per 100,000 Population	% of all Deaths
*1900**		
1. Pneumonia	191.9	12.5
2. Consumption (tuberculosis)	190.5	12.5
3. Heart disease	134.0	8.3
4. Diarrheal diseases	85.1	5.6
5. Diseases of the kidneys	83.7	5.5
6. All accidents	72.3	4.7
7. Apoplexy (stroke)	66.6	4.3
8. Cancer	60.0	3.9
9. Old age	54.0	3.5
10. Bronchitis	48.3	3.2
2000		
1. Major cardiovascular diseases	265.45	31.4
2. Malignancies (cancer)	197.02	23.3
3. Stroke	58.35	6.9
4. Chronic obstructive pulmonary disease (lungs)	39.81	4.7
5. Accidents	34.92	4.1
6. Pneumonia/Influenza	31.57	3.7
7. Diabetes	22.87	2.7
8. Suicide	11.15	1.3
9. Nephritis, nephrotic syndrome and hephrosis (kidneys)	9.25	1.1
10. Chronic liver disease and cirrhosis	9.19	1.1

SOURCES: *Abstract of the Twelfth Census of the United States, 1900.* Table 93. Washington, DC: U.S. Government Printing Office, 1902, National Center for Health Statistics, Deaths/Mortality. Hyattsville, MD: Division of Data Services, February 14, 2000, and Monthly Estimates of the U.S. Population. Population Estimates Program, Population Division. Washington, DC: U.S. Census Bureau, November 1, 1999.

---

*These data are limited to the registration area that included 10 registration states and all cities having at least 8,000 inhabitants. In 1900 this composed 38 percent of the entire population of the continental United States. Since accidents were not reported in the 1900 census, this rate was taken from Lerner (1970).

expectancy to its natural limit of about 85 years. The researchers note that even if a cure were found for most fatal diseases such as heart disease and cancer, the natural degeneration of the body puts a cap of about 85 years on the average life span.

White Americans who live to age 80 and older, however, can expect to keep on living longer than octogenarians in other industrialized countries. The United States,

Though some early parish registers did not list the deaths of babies, this early 20th century portrait of an Appalachian family reveals the importance of every member of the family, even the deceased one. This was obviously a sad time for the family, yet the somber expressions on the members' faces were typical of photos from this period.

which ranks 15th among nations in life expectancy at birth, is having its average life expectancy lowered by a high **infant mortality rate** and higher death rates until middle age (Kolata, 1995). For example, infant mortality rates in Japan and Sweden are half those in the United States. Though the reason is not known for sure, demographers suggest that the reason why the oldest Americans are so long-lived is because their universal health insurance through Medicare may provide better care than old people receive elsewhere. Another factor might be the effect of education. Early in the 20th century, Americans were among the best educated in the world (Kolata, 1995). A more educated society means a higher standard of living, which ultimately contributes to longevity.

Though medical and scientific breakthroughs have obviously contributed to an increased life expectancy, the American public has also strived to change its lifestyle and improve its health in recent years. Priorities for health promotion are smoking cessation, physical exercise, nutrition and weight control, stress management, and appropriate use of alcohol and other drugs (Hyner & Melby, 1987). Certainly with the increasing number of persons with AIDS, safe sex is also being promoted. Although one cannot change his or her heredity, one can have an impact on life expectancy through exercise, nutrition, and patterns of living.

Gender Differences in Mortality Rates

Beatrice Gottlieb in *The Family in the Western World: From the Black Death to the Industrial Age* (1993) observes that in spite of childbirth always being risky for women—riskier in the past than now—men do not as a rule outlive women, nor did they in the past. The only exceptions worldwide are in southern Asia in countries such as Bangladesh, India, Nepal, and Pakistan (Cockerham, 1991). Nutritional deprivation and lessened access to medical care are among the possible reasons for this reversal of the usual female superiority in life expectancy.

In modern Western countries life expectancy is longer for women than for men, but this is not a new development. Although female infants initially are almost always outnumbered by male infants, a more equal balance is reached now, and also in the past, because male infants had a higher death rate. Indeed, the danger of childbirth is an important experience of families in the past; yet whatever is said about women dying in childbirth has to be put into this context of overall female survival (Gottlieb, 1993). The notion that women in the Western world ran a disproportionate risk because of childbearing and that they were more likely than men to die in early adulthood turns out to be mistaken, argues Gottlieb (1993).

In 19th-century United States, however, women typically died younger than men, according to Peter Freund and Meredith McGuire (1995). Reasons include women incurring risks in pregnancy and childbirth and women often getting what food was left after men and children ate their share. However, by the 1920s the gender patterns in mortality rates in the United States had changed, and women generally were living

DEATH ACROSS CULTURES
Abkhasians, the Long-Living People

While the debate may continue as to why some individuals live longer than others, it is well known that the Abkhasians on the coast of the Black Sea in Abkhasia live to ripe old ages. Some 125,000 in numbers, these people live in a country about half the size of New Jersey. They do not have a phrase for "old people." Those older than 100 years of age are called "long-living people"—most work regularly performing light household tasks, working in the orchards and gardens, and caring for the animals.

Not only do a large percent of the Abkhasians live a very long time, they also seem to be in fairly good health in their "old age." Close to 40 percent of aged men (older than 90) and 30 percent of aged women have good vision—do not need glasses for any sort of work. Nearly half have "reasonably good hearing." Most have their own teeth. Their posture is unusually erect.

Why do these people live so long? First, they have no retirement status but simply decrease their expected workload as they grow older. Second, they do not set deadlines for themselves; thus there is no sense of urgency except in emergencies. Third, they consider that overeating is dangerous; fat people are regarded as sick. Milk and vegetables make up 74 percent of their diet. The aged average an intake of 1,900 calories per day—500 less than the U.S. National Academy of Science recommends for those older than 55. They do not use refined sugars; they drink water and honey before retiring in the evening. They eat a lot of fruit, and meat is eaten only once or twice per week. They usually cook without salt or spices. The wine they drink is a dry red wine not fortified with sugar and has a low alcohol content. Few of them smoke, and they do not drink coffee or tea. Their main meal is consumed at lunch (between 2 and 3 p.m.), and their supper is light. Between meals they eat fruit or drink a glass of fermented milk (which is like buttermilk) with a high food value and useful for intestinal disorders. They have a relaxed mood at mealtimes and, when eating, take small bites and chew slowly, thus ensuring proper digestion.

Fourth, they avoid stress by reducing competition. Fifth, they exercise daily. Sixth, their behavior is fairly uniform and predictable. Seventh, they practice moderation in everything they do.

Much of the above is advice we have often heard but failed to put into practice. Perhaps we should adhere to some of the Abkhasians' practices. Who knows, we might even live longer and be more healthy—if that is a goal to be sought.

Adapted from Benet, Sula. (1974).
Abkhasians: The long-living people of the Caucasus. *New York: Holt, Rinehart, & Winston.*

longer than men, as had seemingly been the case in the rest of the Western world. Today, mortality rates for the two leading causes of death in the United States, heart disease and cancer, are higher for men than for women (Weiss & Lonnquist, 1994). Higher death rates from heart disease in men could be attributed to a higher rate of smoking and a harder-driving personality; also, sex hormones secreted by women's ovaries may provide some protection against heart disease. Higher cancer rates in men are likely attributed again to smoking cigarettes, drinking alcohol more excessively, and being exposed to cancer-causing agents in the workplace, according to G. L. Weiss and L. E. Lonnquist (1994). Indeed, men are more likely than women to die from almost all of the most common fatal diseases.

Biological advantages contributing to greater longevity for women may come from hormonal differences. Also, the monthly menstrual cycle of women reduces their iron count; thus women are less likely than men to build up a surplus of iron in their bodies. In addition, the conception ratio, projected to be higher than 120 males per 100 females, favors males in the United States, yet the **sex ratio** at birth is down to 105. Thus, male babies are the "weaker" of the sexes and die off more quickly. In the teens the sex ratio levels off, and after age 80, the ratio is less than 50 males per 100 females. There is evidence that female life expectancy is also higher among many other animal species (Sagan, 1987). Perhaps females simply have better-built bodies. Their bodies have the potential of carrying and supporting a fetus/embryo. Thus, females are the "Porsche" model, whereas males are the more "thrown-together" model.

Regarding cultural differences between the sexes, conceivably females watch their diet more carefully than do males due to their traditional knowledge about food and a special cultural emphasis on weight maintenance. With more current emphasis on diet and exercise for both males and females, however, this difference may diminish.

Anthropologist Ashley Montagu (1968) suggests that women have a superior use of emotions because they are more likely to cry than men. Because it is not "macho" to cry, men generally refrain from such behavior, resulting in more psychosomatic disorders such as peptic ulcers. Montagu asks, "Is this a superior use of emotions?" Perhaps being freer to express themselves through the release of emotional feelings contributes to a decrease of stress for women.

Males have historically been involved in more risk-taking activities, as mentioned earlier, through "masculine" behavior such as smoking (the rugged Marlboro man, for example) and drinking. Males have also been more inclined to drive fast cars, live the James Dean devil-may-care life, and participate in violent sports. Accidents cause

THE WIZARD OF ID   by Brant parker and Johnny hart

more deaths among males than among females (Cockerham, 2001). Males also have had more dangerous jobs, such as coal mining. Males have been in higher stress-producing positions, such as CEOs and other high-ranking administrative positions, than have females.

Despite the fact that men die earlier and have more life-threatening illnesses, women have higher **morbidity** (illness) rates than men (Freund & McGuire, 1995). Chronic illnesses are more prevalent among women than men, but they are less severe and life threatening. When sex differences are considered, an inverse relationship appears to exist between mortality and morbidity (Cockerham, 2001). Women may be sick more often but live longer. Men may be sick less often but die sooner. Women report more episodes of illness and more contact with physicians. Perhaps women are more willing than "macho" men to report that they are sick. Because illness indicates "weakness," men might be less likely to visit a physician or admit that they are anything less than "healthy."

As sex roles continue to change, as females are found in greater numbers and in a greater variety of nontraditional occupations, as males share more in domestic tasks, and as the sexes come together in more unisex behaviors (more women smoking and drinking, having more stressful jobs, and living the "fast" life), stresses and strains of life should be more equally distributed between men and women. The argument of biology versus culture as influencing life expectancy by sex can then be better addressed.

APPROACHES TO THE STUDY OF DYING AND DEATH

There are several ways to approach the study of dying and death. The two areas of study are primarily the natural sciences and the social sciences. Natural sciences include the biological approach. Social sciences include the sociological approach. The anthropological and psychological approaches mainly fall within the social sciences arena—certainly the case for studying death and dying—though they have subareas that lean toward the natural sciences.

In following the scientific method for research purposes, all the approaches use the same empirical methodology—the specific ways of gathering data simply vary. Nonetheless, the methodology is based on observation and reasoning, not on supernatural revelation, intuition, appeals to authority, or personal speculation.

The theoretical approach of this book is largely sociological, though somewhat slanted toward social psychology (the discipline that bridges sociology and psychology). We will take an anthropological approach at times, particularly when we discuss different cultures. The psychological orientation is webbed into the social psychology bent of the book. The biological approach is evident when medical issues pertaining to the human body (structure and function) are discussed. Medical issues regarding causes of disease, for example, are biological (genetic causes), yet social (environmental causes).

The Biological Approach

Biology is "the study of life." Though this is a book on dying and death, it is really about living and life. Thus, the biological approach indeed has merit in a book like this. The process of dying is primarily a biological process—something that the body does to the person. In this book, however, we are more concerned with what people *do* with that process. For example, biologists and medical personnel (or anyone else, including social scientists) respond to the *meaning* of the biology rather than to the biology per se. The physician's decisions are made on the basis of what the biological condition means to that physician. Making a medical diagnosis is the process by which the physician decides the meaning of the biological factors. This diagnosis, then, represents the process of transposing biological factors into meaning factors.

A recurring dramatic illustration of the fact that the behavior of the physician and others stems from the meaning rather than from the biology per se can be seen each time the news media report that a corpse in the morgue has come back to life after being wrongly pronounced dead by the experts. The physician's belief that a body is dead does not guarantee that it is. Behavior follows from the meaning, not from the biological factors per se—a *living* body was sent to the morgue.

A later discussion of dying in the American health care system (see chapter 7) will focus on the medical model. The medical model suggests that when sick, we go to a physician to be "made well." Yet, the dying patient in the later stages of the illness cannot usually be "made well." The dying patient then becomes the "deviant" within the medical system—the body is beyond healing. Chapters 5, 6, and 7 will discuss the dying process and will bring in biological issues (chapter 8 is about biomedical issues) but will also bring out relationships between physicians and patients and families (more of a social science orientation). Indeed, the biological approach is about "life," yet life includes dying and death.

The Psychological Approach

A psychological approach to dying examines, among other issues, the experiences of pain (see chapter 6), death anxiety (discussed in chapter 1), and emotional stages in dying (in chapter 5). Research on death anxiety has been almost exclusively conducted by psychologists. Death anxiety (fear) might involve the unknown, the nature of one's identity, fear of growing old (**gerontophobia**), immortality, and across the life span. The psychological approach looks at dying from a developmental perspective (sometimes called "life stages") and examines attitudes toward dying and death from the cradle to the grave ("the womb to the tomb"). Chapter 3 in this book takes such a developmental perspective; thus psychology plays a particularly large role in looking at death through the eyes of children, adolescents, and adults (younger, middle-aged, and older).

A psychological approach to dying and death looks at death denial in different cultures, examines the influence of mass media desensitizations, and studies death denial in both patients and physicians. As noted in chapter 5, a very common reaction to the announcement of a terminal prognosis or of the death of someone is denial (shock and disbelief). Such a reaction was experienced when Princess Diana's death was announced. This kind of reaction is sort of a defense mechanism: "I must have heard incorrectly." Psychology also analyzes emotions around dying and death and thoroughly studies the process of grief both before and after the death of a significant other, as discussed in chapters 13 and 14.

The psychoanalytic perspective comes out of psychology and bases much of its argument on the unconscious portion of the mind. This approach looks back into the patient's past to help explain current behavior. Death fear is often taken to be the result of repressed guilt, traced to unresolved earlier experiences. Such a perspective has an applied function and can be useful in helping one to cope with dying and death.

The Anthropological Approach

Anthropologists, particularly *cultural* anthropologists, study rituals with which people deal with death and hence celebrate life. As anthropologists R. Huntington and P. Metcalf (1979) observe in their excellent book entitled *Celebrations of Death; The Anthropology of Mortuary Ritual*, the study of death rituals is a positive endeavor. In all societies, the issue of death throws into relief the most important cultural values by which people live their lives and evaluate their experiences. Life becomes transparent against the background of death—fundamental social and cultural issues are revealed. It is evident in the Death Across Cultures box that death for this Native American is very much a part of life. As we shall see, this is a very different view than that held by most Americans.

Chapter 10 relies heavily on the research of cultural anthropologists in looking at death rituals in different cultures. Cultural anthropologists also look at emotional responses in different cultures, how different groups prepare the body for final disposition, and how they dispose of the body. Whereas in the United States we primarily bury a body in the ground, other cultures might place the dead person in a tree, and others are likely to cremate the body. Cultural anthropologists study these kinds of rituals.

Another subfield of anthropology, physical (biological) anthropology, however, leans in the direction of the biological sciences. Such a specialist, for example, might work with bone identifications (from humans and other animals) both from the past and present. This specialty often assists in identifying the remains of a human body and helps to determine age, sex, and other physical characteristics. Thus, anthropology (the study of humankind) has subfields in the social and the natural sciences.

DEATH ACROSS CULTURES
The Visit

by Phil George

Grandmother, I dreamed of you again—
I burned the sage you
   sent when I was in Vietnam.
Before you died—
   and I could not come home.
Beside my bed is the Spruce Bough
   To keep nightmares away.
Today I will visit you beside
   Our Stream—at Arrow.
I will sprinkle tobacco on waters:
   They will thank me.
Grandmother, let's visit—
   Talk and laugh when
   Sun is shining:
So I may sleep, rest at night

*A Nez-Perce-Tsimshian, Phil George's
Indian name is Two Swans Ascending
from Still Waters. "The Visit," from*
Voices of the Rainbow: Contemporary
Poetry by American Indians, *edited by
Kenneth Rosen, 1975, New York:
Viking Press. Reprinted by permission.*

The Sociological Approach

The sociological approach to the study of dying and death generally includes four theories: structural-functional theory, conflict theory, social exchange theory, and symbolic interaction theory. Social exchange and symbolic interactionism differ from structural functionalism and conflict theory on two crucial points. The first is that social exchange and symbolic interactionism examine the individual in society, whereas structural functionalism and conflict theory examine social facts, institutions, and forces. Second, interactionists and exchange theorists contend that the essential feature of society is its subjective character, in contrast to the focus on the objective approach taken by the other two.

In the view of interactionists and exchange theorists, social facts do not have any inherent meaning other than what humans attribute to them. W. I. Thomas argues that if people define situations as real, the situations will be real in their consequences. Robert Merton (1968) takes Thomas' idea a step further and argues that individuals *act* on their perceptions of the situation; thus the prediction of the situation comes true. This is known as the **self-fulfilling prophecy.** Max Weber (1966) defined social action as human behavior to which the acting individual attaches subjective meaning, and that takes into account the behavior of others.

Researchers using a social exchange or symbolic interactionist approach are often required to develop empathy for the subjects whom they study. At times this will require the researcher to enter the subjective world of the subject by participating in this person's life experience. A Native American proverb encourages us not to judge the behavior of others until we have "walked a mile in their moccasins." In this context, social behavior is explained from the perspective of the subjective meanings of the actors' intentions for their behavior.

Structural-functional theory assumes that society is in a state of equilibrium and that the various social institutions (e.g., family, religion, economy, and politics) function on behalf of each other. This approach might be used to explain how a particular death ritual maintains social structure. The conflict perspective assumes that society is in a state of disequilibrium (imbalance) and tends to focus on inequalities in society. This approach would examine, for example, the unequal access to medical care for the terminally ill. Let's look more carefully at these four sociological approaches to studying dying and death.

Structural-Functional Theory

French sociologist Emile Durkheim's work served as the primary foundation for theoretical frameworks relating to group actions and societal structures. Durkheim defined sociology as the study of social facts, which are external to the individual. Language, religion, and money exchange systems, for example, are all social facts that were in the world before we were born and will be here after we die. Social facts are constraining—they place limitations on what we can do (rules and regulations that we must follow).

Durkheim (1964) believed that "society is a social system which is composed of parts which, without losing their identity and individuality, constitute a whole which transcends its parts." From this point of view, social groups or collectivities (e.g., a particular nuclear family) cannot be reduced to merely a collection of individuals, and social phenomena have a reality of their own that transcends the constituting parts. This perspective, therefore, studies group-related phenomena (e.g., death rituals, structures that provide care for dying patients, and professional groups of funeral functionaries) rather than behaviors of particular individuals.

Group behavior provides strength to the individuals involved and shows "togetherness" to outsiders. After the death of a police officer in the line of duty, for example, hundreds of police officers from around the area will attend the funeral and ride in the funeral procession. As a large group, in essence, they are saying to the world, "Though one of our members is gone and a void occurs within our ranks, we are banding together to cover for the dismemberment within the police family. We remain strong and ready to function to protect society." The message is loud and clear. And in their behavior, they are consoling each other. With the death of Princess Diana, many individuals brought flowers to a central place, not only in England but in countries around the world. Being together and "sharing grief" is helpful. Being in the presence of other individuals is indeed consoling.

As stated earlier, the structural functionalists believe that society is in a state of balance, of equilibrium. Death itself is functional to the overall "balance" of a society. As Charlotte Perkins Gilman so wisely put it, "Death? Why this fuss about death? Use your imagination and try to visualize a world without death. Death is the essential condition of life, not an evil" (Gilman, 1935).

Death is a normal aspect of society. Deaths create jobs—the funeral industry and cemeteries, for example. Deaths open up jobs within the structure of organizations—someone "dies off" and must be replaced; thus death contributes to a smooth-running social system. The structure of society is such that deaths are expected to occur—they help to give equilibrium.

Structural functionalists view society as a social system of interacting parts in which death-related behavior is analyzed from two perspectives:

1. How do death-related meaning systems and death institutions contribute to the maintenance of the larger social system?

2. In what ways are death-related meaning systems and death institutions affected by their relationships to the larger social system?

Structural functionalists are interested in positive (**eufunctional**) and negative (**dysfunctional**) results of social interaction as well as the intended (**manifest**) and unintended (**latent**) consequences of death-related behavior. When family members commit themselves to caring for a dying family member, their behavior can be eufunctional for emotional ties within the family; however, the behavior can be dysfunctional (especially if family members must interrupt their employment) for the security of family financial resources. A **manifest function** of attending a funeral is

THE WIZARD OF ID   parker and hart

REPRINTED BY PERMISSION OF JOHNNY HART AND NAS, INC.

to support the bereaved as family members attempt to adapt to the loss of a loved one, but a **latent function** is to strengthen the relationships that exist within social groups (a funeral becomes a "family reunion").

If structural-functional theorists were interested in the funeral rites and rituals, they might investigate one or more of the following research questions:

1. How do funerals help to celebrate and maintain society's most salient social values?

2. How do funerals help to promote relationships within kinship groups (grandparents, parents, children, aunts and uncles, cousins, brothers, sisters)?

3. How do funerals contribute to and/or affect the relationships between bereaved families and the larger society?

4. How do funerals facilitate the grieving process as one mourns the death of a loved one?

5. How do death-related rituals help return bereaved persons to their normal social responsibilities?

6. How do differing methods of body disposition and funeral rituals socially differentiate families regarding social status?

Utilizing a structural-functional perspective, sociologist Kathy Charmaz (1975) describes the strategies used by coroner's deputies in maintaining the routine character of their work. The coroner's deputies attempt to get surviving family members to take over the responsibility for the care of the dead body and the financial obligations related to final disposition and to minimize personal involvement of the coroner's deputies. In Charmaz' description we can observe that if each party performs his or her socially prescribed role (or social function), the social system runs efficiently and social equilibrium is maintained.

Strategic control of the encounter between the coroner's deputies and the relatives of the deceased is enhanced by making the announcement in person. Part of the self-protection strategy is to remain polite and sincere, yet authoritative. The deputies must create the kind of ambiance wherein the announcement of the death is effective and believable. Unlike physicians in giving "bad news," the coroner's deputies lack the prestige of physicians and also do not have a relationship with the person. Deputies also lack the structural supports provided by the hospital situation; thus they must devise tactics to get their work done without incident. Typically, their objectives are to announce quickly and then to turn to the responsibility of the body and its subsequent disposal. Deputies feel that they get a better response when they successively lead the relative into questioning them.

Charmaz notes that the deputies usually avoid the word "dead," since it is such a harsh word, and use euphemisms instead. They try to manipulate the situation so that the relative actually uses the word "dead," thus making the death more "real." The deputies then reaffirm the statements and elaborate on them. When the strategies work, the transition from perceiving one's relative from alive to dead can be made rapidly. If the encounter is successful, the relative is likely to express appreciation for the deputies' "sensitivity." Relatives ask about the circumstances of dying. The deputies give the information they have and can release and then turn to the funeral and burial arrangements. Thus, the deputies have played the role of officials who cut through the survivor's grief and shock by pointing to the work that has to be done and have strategically structured the situation in ways that foster the relatives' acceptance of their directives.

Conflict Theory

Whereas structural-functional theory focuses on the issue of societal maintenance and social equilibrium, conflict theory focuses on issues related to social change and disequilibrium. Conflict theorists focus on competition, conflict, and dissension resulting from individuals and groups competing for limited societal resources.

In emphasizing social competition for limited resources, conflict theorists interested in death-related behavior would point out the inequality in the availability and quality of medical care and the differential death rates. For example, the poor are deprived of optimal care, in general, and of life-saving procedures, in particular (Charmaz, 1980). Relying on the work of Herbert Marcuse, Allen Kellehear argues that those who control life are the same as those who control death. According to Kellehear (1990a), the power elite in any society determine one's life and death chances. Society determines not only what types of death will occur, but when deaths occur and to whom and the chances, if any, of doing anything about it.

As will be discussed in chapters 6 and 8, organ transplantations are a good example of the conflict perspective. There simply are more recipients waiting for organ and tissue donations than there are donors. Thus, the issue becomes one of deciding which individuals receive the organs and which ones do not. There is competition for this limited supply of resources.

If conflict theorists were interested in funeral rites and rituals, they might investigate one or more of the following research questions:

1. What are the dysfunctional consequences of attending funerals?

2. What role conflicts and family disputes arise as a result of planning a funeral for family members?

3. How does not attending a funeral create conflicts between adults in neighborhood, friendship, and occupational groups?

4. What are the problems created by the presence of children at funerals?

5. How do particular family relationships contribute to increased competition for status among family members as they participate in the funeral of a family member?

6. How do methods of planning a funeral and the related expenditures contribute to increased family conflict and competition for scarce financial resources within the family?

Many sociologists trace the intellectual roots of the conflict theory to Karl Marx. Marx is buried at Highgate Cemetery in London, England. The epitaph on his tomb reads: "Workers of all lands unite. The philosophers have only interpreted the world in various ways; the point is to change it."

7. In what ways might members of the clergy and funeral directors engage in social conflict due to the fact that family members allocate to each functionary differing amounts of authority and social rewards (money) for conducting the funeral?

8. How is it possible for unethical funeral industry personnel to exploit bereaved survivors during a time of emotional distress?

9. How does the death of a parent create sibling rivalry among the children, and how does the death of a child create marital problems between the parents?

In distributing property at death through a will or inheritance, if all beneficiaries want fair treatment, conflict may occur because beneficiaries have different perceptions of what is fair. Thus, conflict theory applies to this "fairness" issue after death (Titus, Rosenblatt, & Anderson, 1979). Even if the will seems fair, to some individual recipients a particular item, such as a rocking chair or antique clock, might be personally more valuable. Individuals simply have different perceptions of what is fair. Fairness can mean that something is divided equally, but fairness also takes into account various principles of deservingness or right. Because there are so many possible interpretations of what is fair or what is equal and because people often seek fairness or equality, a dispute may not be resolved easily. Disputes over inheritance may be one of the major reasons for adult siblings to break off relationships with each other. In some cases the inheritance dispute may be the final battle between competitive siblings, and in that sense it resembles the "last straw" reported in breakups in other close relationships.

Social Exchange Theory

Two traditions are followed by social exchange theorists. The first tradition is consistent with principles of behavioral psychology and stresses psychological reductionism and behavioral reinforcement techniques. The second tradition has been influenced by the work of Peter Blau (1964) and is committed to many of the assumptions held by symbolic interaction theorists. Social exchange theories of this type contend that human behavior involves a subjective and interpretative interaction with others that attempts to exchange symbolic and nonsymbolic rewards. It is important that such social exchange involves reciprocity so that each interacting individual receives something perceived as equivalent to what is given.

From this perspective, individuals will continue to participate in social situations as long as they perceive that they derive equal benefits from their participation. For example, the social exchange theorist would contend that individuals will attend funerals (even though they tend to feel uncomfortable in such situations and find viewing the body as distasteful and anxiety-producing) because they perceive social benefit in being supportive of bereaved friends.

If social exchange theorists were interested in funeral rites and rituals, they might investigate one or more of the following research questions:

REPRINTED BY PERMISSION OF JOHNNY HART AND NAS, INC.

1. Why do some individuals attend funerals—what are the social rewards acquired by attending?

2. Why do some individuals not attend funerals—what are the social punishments or sanctions acquired by not attending?

3. What are the social and personal costs and benefits for families when they provide wakes, funerals, and other death-related rites of passage?

4. What are the social and personal costs for families when they do not provide wakes, funerals, and other death-related rites of passage?

5. Why would the average American family spend more than $5000 to bury its dead when it could accomplish the same purpose at a fraction of the cost?

Symbolic Interaction Theory

The foundation of symbolic interaction theory is that **symbols** (meaning) are a basic component of human behavior. People interact with each other based on their understanding of the meanings of social situations and their perceptions of what others expect of them within these situations. Stressing the symbolic nature of social interaction, Jonathan Turner (1985, p. 32) says: "Symbols are the medium of our adjustment to the environment, of our interaction with others, of our interpretation of experiences, and of our organizing ourselves into groups." For example, sociologist and physician Nicholas Christakis (1999) explains how self-fulfilling prophecy applies to a physician–patient relationship in the use of symbols. Physicians, in making predictions about a terminally ill patient's prognosis, can change the patient's perception in a way that promotes the predicted outcome. If the doctor were to give a favorable prognosis to a cancer patient, it would probably lead to a favorable outcome. If the patient's morale is good, she or he will probably have better nutrition habits and be stronger to fight off potentially fatal infections better than might otherwise be true. The patient will not be cured, but his or her life may be prolonged.

Another example of self-fulfilling prophecy applying to doctors working with patients relates to compliance with therapeutic regimens (Christakis, 1999). If positive feelings are inculcated in an adolescent patient with newly diagnosed diabetes, research data suggest that this is one of the best predictors of a good outcome. But if the person is not upbeat, then the physician's job becomes changing the patient's negative view of the illness. Likewise, if a physician assumes (perhaps a false assumption) that a particular patient does not wish to be told "the truth" regarding prognosis, and the physician then does not "tell the truth," the definition of the situation by the physician then becomes reality, and the patient is *not* told and remains officially unaware. Thus, the social construction of reality: We define situations as real, and they become real in their consequences.

The symbolic interactionist perspective can be summarized in what some have called the **ISAS** statement—*individual*-level behavior is in response to *symbols*, relative to the *audience*, and relative to the *situation*. ISAS stands for the initial letter of each of the four basic components (Vernon & Cardwell, 1981). Death-related behavior of the dying person and of those who care about that person is in response

---

LISTENING TO THE VOICES
Self-Fulfilling Prophecy

When George Dickinson was a graduate student, he went to the hospital to donate blood to a fellow graduate student. Prior to taking his blood, two nurses each checked his temperature (this was back in the "old days" when glass thermometers were used and then washed, and reused). On both occasions his temperature was 104 degrees, a rather high temperature for an adult. Before going to donate blood, he was feeling very good, other than being a little anxious about the "blood thing." Now that two separate nurses, two authoritarian figures with medical credentials, had determined that he had a rather high temperature and therefore would not take his blood, he began to *feel*

"badly." Thus, the self-fulfilling prophecy was at work—he was "defined" as being sickly, began to *feel* sickly, and indeed was getting sick! Prior to going home and "going to bed" with his "illness," he decided to purchase a thermometer, because he really did not feel "feverish," though otherwise questioning his health. After holding the thermometer in his mouth for the recommended 3 minutes, his temperature read 98.6°F! (Obviously, the two nurses did not shake down the thermometers and had placed them in water with a temperature of 104°F to sterilize them.). He began to *feel* better immediately. What others say and how we perceive what they say indeed affects our behavior.

---

to meaning, relative to the audience and to the situation. Death-related behavior is shared, symboled (given meaning), and situated. It is socially created and not biologically predetermined.

Symbols.   A symbol is anything to which socially created meaning is given. Indeed, language itself is a system of oral symbols. For example, a group of Southwest Native Americans did not have a future and past tense in their language, only the present tense. Indeed this would have an impact on meaning and perception! The Yanomamo Indians in Venezuela and Brazil have a numerical system in their language, which is 1, 2, and "more than 2." Thus, if you were asking adults their ages, they would all be "more than 2," making it difficult from our perspective in the United States to perceive adult age differentials. Another group had only three colors in their color spectrum, thus "limiting" somewhat their perception of colors. If, for example, you go to buy a new car or a new pair of shoes, they probably have many colors and numerous variations on each color—a brown shoe is not just "brown, " it is "something, something brown." Thus, meaning assigned to a situation or object will vary significantly according to the perception.

Meaning is socially created and socially perpetuated; it is preserved in symbols or words. However, preserved words have to be rediscovered and reinterpreted if they are to be continually used in human interaction. Interaction is a dynamic, flexible, and socially created phenomenon. What death means to each of us results from our socially shared ideas. Generation after generation repeat the process with a somewhat

For the patient entering *Saint Christopher's Hospice in London, the average stay is 3 weeks. Knowing the meaning of entering a hospice program confirms that death is forthcoming.*

different content—no book means the same thing to every reader. Death-related behavior and meanings are dynamic phenomena. For example, death to baby-boomers is perceived somewhat differently than by earlier generations in the United States, as discussed in chapter 3.

As human beings, we are more than stimulus-response creatures. Indeed, we have the ability to use symbols—we have the ability to *interpret* the stimulus, assign meaning to it, and then *act* upon our interpretation. What is the meaning of this stimulus? The telephone ringing (stimulus) typically invokes a response of answering. However, one may not be in a frame of mind to talk on the phone, and thus responds by "not answering." The interpretation of the stimulus, given the moment or situation, was to *not* respond. We do not always respond in the same way to the same stimulus. The physician may give the patient "bad news" (stimulus), yet the patient may respond by interpreting what was said (the stimulus) with denial—the doctor made a mistake. She or he did not really mean that which was said. In other words, the patient believes what she or he wants to believe, and most of us in fact do not want to hear the doctor say that we have a terminal condition. Thus, through the interpretation (denial), the individual concludes that everything is okay.

Audience-Related Behavior. As a teenager, you probably did not behave in the same way in the presence of your parents or grandparents as you did when among

The meaning of "exit" from Saint Christopher's Hospice for patients typically means death. Thus, the meaning of "exit" varies significantly for patients and visitors.

peers. Among our peers, we were likely be "cool," while the presence of "older" adults might have brought out more sophisticated behavior. Our audience may even affect certain meanings we attach to interactions and words. In the same way, our audience affects our death-related meanings. The audience involved may be family, physicians, clergy, nurses, peers, or even strangers walking down the hall of the hospital. A dying person observes the behavior of people around them and listens to what they do or don't say. Decisions about the meaning of the dying person's own death and behavior while dying is often heavily influenced by the words and behavior of this audience.

The Situation.   Like other meanings, the definition of the situation is an attempt by the individual to bring meaning to the world. Because the situational definition always involves selective perception, the terminally ill patient will assign meaning to the environment and will respond to this symbolic reality. The thing to which the patient responds does not have existence independent of his or her definition. Therefore, each terminally ill patient will interpret the dying environment differently. This accounts for the different experiences of dying patients (Leming, Vernon, & Gray, 1977). This point will be elaborated upon in chapter 6.

If symbolic interactionists were interested in funeral rites and rituals, they might investigate one or more of the following research questions:

1. What are the social meanings that give rise to the attendance of funerals?

WORDS OF WISDOM
Finding the Meaning in Words

Blood,
  and Pus . . .
  Entrails,
  Vomit,
  Dung, Spit and Afterbirth.
  Disgusting words.

Love,
  And Soft.
  Mother.
  Kiss, Mood, and Friendship.
  Tender words.

Spring,
  And Smile.
  Dance,
  Play,
  Fun, Sing, and Beachball.
  Happy Words.

Death,
  And Fire.
  Pain,
  War,
  Divorce, Poverty, Hospital.
  Sad Words.

No. That's not right at all. You cannot
  String words together
  And say They're bad
  Or good.

Where are the verbs?
  Who are we talking about?
  What are the circumstances?

Vomit is beautiful to a mother whose child
  Had just swallowed a pin.
  Love is pain if you are a third party,
  Outside, looking in.
  Death is very nice for someone very old,
  Very ill, and ready.

And surely you've danced with a clod.
  Or had a sad spring.

No, Words aren't sad
  Or glad.

You are.
  Or I am.
  Or he is.

From A Bag of Noodles *(p. 8),*
by W. A. Armbruster, 1972, St. Louis:
Concordia, p. 8. Copyright 1973
by Concordia Publishing House.
Reprinted by permission.

---

2. How are death-related behaviors influenced by the social audiences of the American funeral?

3. In what ways do families use the funeral service to project an image of family value commitments and group cohesiveness?

4. How do families use the funeral to demonstrate status differentiation?

5. What is the influence of location of the funeral upon the meaning of the funeralization process?

6. How does the size of the funeral audience convey meaning to the bereaved and attribute social importance to the life of the deceased?

CONCLUSION

Why study death from a sociological perspective? The answers are many.

First, although it is not the primary goal of this book to prepare you for your own death or the deaths of loved ones, you may still learn many things that will be personally beneficial. Given the fact that you will have death-related experiences, it might be useful for you to employ sociological and social psychological perspectives to acquire valuable insights into your own life and the lives of people about whom you care.

Second, as you read about the history of death-related customs in the United States and elsewhere and as you comprehend social class and ethnic variations in bereavement customs, the sociological perspective may provide you with a new appreciation for your own life experiences and family traditions as they relate to issues of dying, death, and bereavement.

Third, as you consider the ways by which the family socializes its members to develop death conceptualizations, attitudes, and feelings, the sociological and social psychological perspectives may help you to understand more fully how your own abilities to cope are a function of your upbringing. Furthermore, as you study the material in this book, you may attain a more objective perspective on some of the decisions that may confront you and your family as, one day, you attempt to deal with death-related situations. The sociological and social psychological perspectives, then, can make you more aware of the many options available to you regarding death-related decisions.

Finally, the study of dying, death, and bereavement is an important and interesting area of study to which social and behavioral scientists have much to contribute. A good reason to study dying and death from a sociological perspective is to learn more about the discipline of sociology. Social thanatology provides the student with an opportunity to investigate the meaning and application of many concepts developed in sociology and the other behavioral sciences.

SUMMARY

1. The American way of dying is typically confined to institutional settings and removed from usual patterns of social interaction.

2. American society has done little to formally socialize its members to deal with dying and death on the personal and emotional levels.

3. The "thanatology movement" was a concerted effort in the 1970s to bring about an open discussion and awareness of behaviors and emotions related to dying, death, and bereavement. The movement continues into the 21st century.

4. An increased emphasis on dying and death is reflected in today's media.

5. The issue of when death occurs is a difficult one to resolve because consensus does not exist in America regarding the meaning of life and death.

6. The American way of dying has changed considerably since 1900. Relative to earlier times, Americans are less likely to die of acute diseases. Currently 65 percent of the deaths taking place in the United States can be attributed to two chronic diseases: heart disease and cancer.

7. In the past 50 years, life expectancy has increased approximately 16 years for males and 25 years for females. The increase in life expectancy has had many effects upon Americans' understanding and ability to cope with dying and death.

8. Death and dying can be approached from the perspectives of biology, psychology, anthropology, and sociology.

9. Meaning consists of symbols that are socially created and socially used. Some symbols refer to something else or have an empirical referent. Some do not. The major function of both types of symbols is to permit humans to relate to each other and thus create shared behavior and meaning.

10. From a sociological approach, four theories are used in the study of death and dying: structural-functional, conflict, social exchange, and symbolic interaction.

11. Structural functionalists view society as a social system of interacting parts. Structural-functional theory focuses upon the issue of societal maintenance and social equilibrium.

12. Conflict theory is primarily concerned with issues related to social change and disequilibrium. Conflict theorists focus upon competition, conflict, and dissension that result when individuals and groups compete for limited societal resources.

13. Social exchange theory contends that human behavior involves a subjective and interpretative interaction with others that attempts to exchange symbolic and nonsymbolic rewards. Social exchange will always involve reciprocity so that each individual involved in the interaction receives something perceived as equivalent to what is given.

14. From the symbolic interactionist perspective, human beings are autonomous agents whose actions are based on their subjective understanding of society as socially constructed reality.

Discussion Questions

1. Why did death "come out of the closet" in the 1970s? What events related to the "thanatology movement" helped change the American awareness of dying and death?

2. What effect has the media had on present interest in death?

3. Why is death education from a sociological perspective important?

4. Compare the leading causes of death in 1900 with those today.

5. What affects life expectancy?

6. Discuss why women outlive men in the United States and most countries of the world.

7. How do the biological, psychological, anthropological, and sociological approaches to death differ?

8. Compare and contrast the structural-functional and conflict orientations in death-related research.

9. Compare and contrast the symbolic interactionist and social exchange orientations in death-related research.

10. Explain what is meant by the ISAS statement and describe the four components.

GLOSSARY

Chronic Disease: A noncommunicable self-limiting disease from which the individual rarely recovers, even though the symptoms of the disease can often be alleviated. Chronic illnesses usually result in deterioration of organs and tissues that make the individual vulnerable to other diseases, often leading to serious impairment and even death. Examples of chronic illnesses include cancer, heart disease, arthritis, emphysema, and asthma.

Dysfunction: An event that lessens the adjustment of a social system. Dysfunctional features of a society imply strain or stress or tension. A dysfunction of modern medicine might be that it prolongs life, contributing to an overcrowded population, and also contributes to a frail and disabling existence for many in their old age.

Eufunction: Consequences of behavior that are positive. For example, a eufunction of attending a funeral is that it shows concern and care for the survivors.

Gerontophobia: The fear of growing old.

Infant Mortality Rate: The number of deaths of children under 1 year of age per 1,000 live births.

ISAS: A shorthand presentation of the symbolic interactionist's statement: "Behavior of the individual is in response to symbols, relative to the audience, and relative to the situation."

Latent Function: Latent functions are consequences that contribute to adjustment but were not intended. For example, a latent function of attending a funeral is that it becomes a family reunion. Distant relatives from various geographical locations often show up for a funeral. Latent functions are the sociological explanations of a given action.

Life Expectancy: The number of years that the average newborn in a particular population can expect to live.

Manifest Function: Manifest functions are objective consequences that contribute to adjustment and were so intended. They are the official explanations of a given action.

Morbidity: The rate of occurrence of a disease.

Mortality Rate (Death Rate): The number of deaths per 1,000 population.

Self-Fulfilling Prophecy: A situation defined as real that becomes real in its consequences when individuals act to make it so.

Sex Ratio: The number of males per 100 females.

Socialization: The social process by which individuals are integrated into a social group by learning its values, goals, norms, and roles. This is a lifelong process that is never completed.

Symbol: Anything to which socially created meaning is given.

Thanatology: The study of death-related behavior including actions and emotions concerned with dying, death, and bereavement.

SUGGESTED READINGS

Charmaz, K. (1980). *The social reality of death*. Reading, MA: Addison-Wesley. An important book that provides phenomenological, interactionist, conflict, and Marxist interpretations of death-related behavior.

Charmaz, K., Howarth, G., & Kellehear, A. (Eds.). (1997). *The unknown country: Death experiences in Australia, Britain and the USA*. London: Macmillan. A book providing novel portrayals of death experiences in three Western countries that seemingly share cultural roots. However, sociological scrutiny reveals some sharp differences among these countries in cultural beliefs and practices; this book focuses on death within the borders of the heretofore unknown country.

Clark, D. (Ed.). (1993). *The sociology of death: Theory, culture, and practice*. Oxford: Blackwell Publishers. An anthology of social thanatology articles written from a sociological perspective offering a range of theoretical and empirical topics, together with a discussion of some of the key methodological questions. The issues covered include the concept of social death; the relationship between death and high modernity; death across the life course; cultural and historical perspectives on death and dying; the social management of death; the meaning of ritual; the role of the caring services; and problems of research, method, and analysis.

Fulton, G. B., & Metress, E. K. (1995). *Perspectives on death and dying*. Boston: Jones and Bartlett Publishers. Outstanding discussion of death and dying that takes a unique, multidisciplinary approach to the topic. Discussed are not only the traditional subjects of philosophy, psychology, sociology, anthropology, law, and medicine, but also other disciplines that are affected by the rapid changes taking place in the health-care field, such as economics, marketing, and management.

Kellehear, A. (1990). *Dying of cancer: The final year of life*. Chur, Switzerland: Harwood Academic Publishers. Provides a sociological analysis of the process of dying with cancer.

Rogers, R. G., Hummer, R. A., & Nam, C. B. (2000). *Living and dying in the USA*. San Diego: Academic Press. A discussion of behavioral, health, and social differentials of adult mortality.

Seale, C. (1998). *Constructing death: The sociology of dying and bereavement*. Cambridge, UK: Cambridge University Press. A broad-ranging text that provides a comprehensive overview of most of the major social studies of dying and death.

Smith, J. M. (1996). *AIDS and society*. Upper Saddle River, NJ: Prentice-Hall. Well-documented book written by an attorney with AIDS that is a study of the history of science and medicine, health and illness, and sex and death.

Stine, G. J. (1993). *Acquired immune deficiency syndrome: Biological, medical, social, and legal issues*. Englewood Cliffs, NJ: Prentice-Hall. AIDS from numerous academic perspectives.

Subedi, J., & Gallagher, E. B. (1996). *Society, health, and disease: Transcultural perspectives*. Upper Saddle River, NJ: Prentice-Hall. Book presenting the interrelationship between health, disease, and different social settings in countries around the world.

C H A 2 P T E R

The American Experience of Death

I mourn not those who lose their vital breath
But those who, living, live in fear of death.

—The Ancient Greek Anthology

Death is not the enemy; living in constant fear of it is.

—Norman Cousins, *Head First*

How do you react when you first learn of a major tragedy such as the shooting at Columbine High School, the crash of TWA Flight 800, the burning of the Branch Davidian headquarters in Waco, the explosion of the space shuttle Challenger, the Oklahoma City bombing, tornadoes ripping through Midwestern and Southern communities, hurricanes on the East Coast, or the death of someone you did not know personally but felt that you did (e.g., Charles Schulz of *Peanuts* fame or Princess Diana of England)? Our "gut reaction" is often one of disbelief, mixed with feelings of sorrow for those affected. We often get caught up in the event, especially when the media continue their focus. We may initially wish that we could in some way "help them." But with the passing of time, alas, sometimes sooner than later, we are "over" the event and go on with life.

Journalist Harry Schatz (1976), writing a quarter of a century ago, recalls his mother weeping bitterly around 1940 over a news report on radio about a tornado in Texas that killed several children. She kept saying over and over, "The poor children." Schatz wondered what his mother, an immigrant from Russia who came to Brooklyn in 1900, had in common with these children in Texas, since she was so far removed from them in culture, language, and thought. Yet her compassion reached out over thousands of miles to share personally in the heartfelt grief and anguish of a distant tragedy to unknown children. Schatz was moved to this recollection of 35 years before while reading a newspaper item about a particularly brutal murder. He found himself casually turning the page and doing a double take as he realized that the event had not triggered any emotional response in him. He was startled and overcome by a sense of guilt at his reaction. He asked himself if he were slowly and insidiously being programmed to accept as commonplace and normal such events and if his human qualities were disintegrating into moral decay. Certainly, he did not react as his mother had years before.

Imagine what Harry Schatz's mother's reaction would have been to that same event today with the visual images of television! She was grieving over the radio account of the children's deaths with only a picture in her head as to what was happening. Perhaps we would all like to think that we would react like Harry Schatz's mother and show concern and compassion in such a situation. After all, a tragedy has occurred. But in our fast-paced world of today, do we really "have time" for such reflection? It worried Harry Schatz that maybe his reaction is the direction toward which our society is headed. A distant death, be it of a single individual or a large number of individuals, can indeed be "brought home" via visual images such as television. The hype of the media, particularly television, to sensationalize events such as death to improve their "ratings" and to be No. 1, obviously plays into their coverage. Television reporters are notorious for holding the camera on a grief-stricken person when a much more humane approach might be to give the person privacy at that point. For the moment, we may be "caught up" in the event, as was Schatz's mother, but nowadays we soon move on.

Indeed, "moving on" in life is what might be necessary in our postindustrial society. We cannot take the time to "get involved" with the many distant deaths, some very tragic, that are reported to us by the media. Yes, we may occasionally "send flowers" or "leave a note" by the fence in Oklahoma City or in London. Probably we mean well with such an action, and the behavior helps us in a therapeutic way to move on through this event. Perhaps moving on does not mean that we do not care or are callous, but responding and getting involved with these happenings simply is not fitting for our secularized, heterogeneous, specialized, fairly densely populated society of the 21st century.

Anthropologist Colin Turnbull (1972) notes that in our society the individual is paramount. The family has lost much of its value as a social unit, and religious beliefs and practice no longer bind us into communities, says Turnbull. Kathy Charmaz (1980) argues that this individualistic perspective on death has led American society to abdicate most of its social responsibilities to the dying and their family members. She asserts that in America we are assured of an inequitable distribution of health care and a lack of social concern for the dying precisely because Americans value individualism and privacy.

> ## WORDS OF WISDOM
> ### I Heard a Fly Buzz
>
> I heard a Fly buzz—when I died—
> The Stillness in the Room
> Was like the Stillness in the Air—
> Between the Heaves of Storm—
>
> The Eyes around—had wrung them dry—
> And Breaths were gathering firm
> For that last Onset—when the King
> Be witnessed—in the room—
>
> I willed my Keepsakes—Signed away
> What portion of me be
> Assignable—and then it was
> There interposed a Fly—
>
> With Blue—uncertain stumbling Buzz—
> Between the light—and me—
> And then the Windows failed—and then
> I could not see to see—
>
> *Emily Dickinson*

Charmaz (1980) notes that beliefs in individualism, self-reliance, privatism, and stoicism are ideological and justify the ways in which dying is handled. The ideological view of dying as a private affair, something that should be the responsibility of the family, relieves other social institutions, notably health and welfare organizations, from the necessity of providing comprehensive services.

Despite the fact that the individual is responsible for taking care of dying, 80 percent of deaths in the United States today occur in institutional settings—hospitals and nursing homes (Stryker, 1996). The figure has gradually risen since 1949, when it was 50 percent, to 61 percent in 1958, and to 70 percent in 1977 (Nuland, 1994). Grandfather seldom dies at home today, where he has spent most of his life, but instead in a rather impersonal institutional setting. Yet, studies (Hays et al., 1999) in the United States and various other countries reveal that the majority of individuals consistently express a preference to die at home. The removal of death from the usual setting prompted Richard Dumont and Dennis Foss (1972, p. 2) to raise the question: "How is the modern American able to cope with his own death when the deaths he experiences are infrequent, highly impersonal, and viewed as virtually abnormal?" In some cases it makes the dying feel insignificant, like the person in Emily Dickinson's poem, "I Heard A Fly Buzz."

DEFINING DEATH

"The Wizard of Id" defines *death* as a "once in a lifetime experience." Although one cannot refute this definition, it would not suffice in the medical or legal professions. Until the 1960s there was little real controversy about what it means to be dead in the public policy sense (Veatch, 1995). Now, for the first time, however, it is a matter of real public policy significance to decide precisely what is meant

REPRINTED BY PERMISSION OF JOHNNY HART AND NAS, INC.

when someone is labeled "dead." Today, matters have changed because technologies have greatly extended the capacity to prolong the dying process and to use human organs and tissues for transplantation or research.

A headline in a newspaper ("Doctors 'kill' patient," 1988) reads "Doctors 'Kill' Patient to Save Her Life." The article went on to say that surgeons had put the patient into a coma, stopped her heart, chilled her by 40 degrees, and drained her body of blood for 40 minutes to save her from the aneurysm pressing on her brain. She was placed into a sort of suspended animation that allowed surgeons to cure a hard-to-reach, high-risk aneurysm once considered inoperable. As one surgeon noted, "It may be the surgery of the future in cases where bleeding poses the greatest risk to the operation." Thus, the headline said the physicians "killed" to "save" the patient. Defining death is indeed a confusing issue!

The diagnosis of death has vacillated throughout history between the centralist theory and the decentralist theory (Powner, Ackerman, & Grevnik, 1996). The centralist theory proposed that a single organ contains the vital life force and, if it fails, the person would die, a prominent view before the 18th century. Decentralism proposed that the entire body and every organ and cell possess the life force. This theory gained precedence when physicians discovered that a failed organ could be resuscitated. The modern theory of brain death, however, has resurrected the centralist theory.

The definition of death has changed significantly from the 1960s from strictly physical criteria to a debate about the value of life and the essence of human qualities (Prior, 1989). That debate relocated the origin of death in the brain rather than in the respiratory system, and the first test of death became one that questioned the receptivity and awareness of the dying patient (Gervais, 1987). Death, once so clearly defined as a physical fact, became entangled in a discussion about the value of life.

International Definitions

Defining death is a difficult assignment. In the 1950s the United Nations and the World Health Organization proposed the following definition of death: "Death is the permanent disappearance of all evidence of life at any time after birth has taken place" (United Nations, 1953). Thus, as death can take place only after a birth has occurred, any deaths before a (live) birth cannot be included in this definition. The latter is called a *fetal death* and is defined in the following way (Stockwell, 1976):

> Death (disappearance of life) prior to the complete expulsion or extraction from its mother or a product of conception irrespective of the duration of pregnancy; the death is indicated by the fact that after such separation the fetus does not breathe or show any other evidence of life, such as beating of the heart, pulsating of the umbilical cord, or definite movement of voluntary muscles.

Not all countries observe the definition of death recommended by the United Nations. In some countries, infants dying within 24 hours after birth are classified as stillbirths rather than as deaths or are disregarded altogether. In some other countries infants born alive who die before the end of the registration period (which may last several months) are considered stillbirths or are excluded from all tabulations. As Robert Hertz (1960) observes, they are considered still a part of the spirit world from which they came, therefore their death is often not accorded ritual recognition, and no funeral is held. In some areas, the baby is not given a name until he or she lives 1 year. The child does not "exist" until given a name. Thus, if the child dies before achieving 1 year of age and therefore was not named, then the child did not exist. If the child did not "live," he or she could not "die!" Thus, one is not "alive" until officially registered or named, and one cannot be legally "dead" if never alive!

In the early days of record keeping in the United States (Gottlieb, 1993) some parish registers did not list the deaths of babies; babies were remembered as abstractions rather than as persons. In the latter part of the 19th century and early 20th century, however, some families showed that the dead babies were not forgotten and had them appear in the family picture with parents and living children (see photo on p. 13).

For the Kaliai of Papua, New Guinea (Counts & Counts, 1991, p. 193), dying is almost complete "if the breath smells of death, if the person stares without blinking or shame at another person's face, is restless and must be moved frequently, or loses bladder or bowel control." Dying is complete "when breathing stops, when the heart ceases to beat, and when the eyes and mouth hang open." The Kaliai also believe that persons dying or dead may return to life at any time after the dying process begins. The Kaliai have no generic term for *life*. Thus, the whole question of life and death is more complicated than it might initially appear.

American Definitions

By traditional definition, death is when heartbeat and breathing stop. But medical advances have made this definition of death obsolete. Under this definition many people living today would have been considered dead because their hearts stopped beating and their breathing stopped as a result of heart attacks. With respirators to artificially breathe for a patient, along with other life-support equipment, many lives are saved. A diagnosis of **brain death** allows respirators to be turned off when the brain is "totally and irreversibly" dead (Cope, 1978).

The gravestone marking of a crossbone and skull dates to the 17th and early 18th centuries and depicts the stark reality of death—later becoming the symbol for "poison" on containers. After death, muscle and skin deteriorate, and the skeleton remains.

Leon Kass (1971, p. 699) defines *death* simply as "the transition from the state of being alive to the state of being dead." Simple enough, yet the issue is clouded by many factors. Traditionally, death has clinically meant "the irreparable cessation of spontaneous cardiac activity and spontaneous respiratory activity" (Ramsey, 1970, p. 59). The functions of heart and lungs must cease before one is pronounced dead by this definition.

In 1968 a committee of the Harvard Medical School (Beecher, 1968) claimed that the ultimate criterion for death is brain activity rather than the functioning of heart and lungs. The Harvard report notes that death should be understood in terms of "a permanently nonfunctioning brain," for which there are many tests. With brain death the body would not be able to breathe on its own because breathing is controlled in the brain. If the brain is dead, any artificially induced heartbeat is merely pumping blood through a dead body.

In 1981 President Ronald Reagan created the President's Commission for the Study of Ethical Problems in Medicine and Biomedical and Behavioral Research to study the ethical and legal implications of the matter of defining death (Crandall, 1991). The commission examined three possible ways of determining death: death of part of the brain, death of the whole brain, or nonbrain ways to determine death, such as absence of heartbeat. The commission decided on a definition of death that includes the traditional nonbrain signs of death and a "whole-brain" definition. Its proposal to determine death, which was accepted by both the American Bar Association and the American Medical Association, is as follows (Crandall, 1991, p. 565):

> An individual who has sustained either (1) irreversible cessation of circulatory and respiratory functions, or (2) irreversible cessation of all functions of the entire brain, including the brain stem, is dead. A determination of death must be made in accordance with accepted medical standards.

The majority of states have adopted the proposal or one similar in content. Had the commission decided on death of part of the brain, then neocortical death (the irreversible destruction of neural tissue in the cerebral cortex, critical for intellectual functioning) could have been operative. Likewise, brain stem (which controls breathing and heartbeat) death would not qualify as "death," as happened in the case below.

A case in Florida ("Baby born," 1992) involved a baby girl, Theresa Ann Campo Pearson, born without a fully formed brain. She had only a partially formed brain stem and no cortex, the largest part of the brain. Her parents asked the courts to declare her brain dead, but to no avail. The Florida Supreme Court said it did not have the constitutional authority to hear the case. Theresa Ann did not have an "entire brain" and thus did not fit the legal definition for declaring one dead: "irreversible cessation of all functions of the entire brain, including the brain stem." The parents had wanted to donate their daughter's organs, but by the time the 10-day-old girl died, her organs had deteriorated so much that they could not be transplanted.

Various tests can be given to determine any signs of brain activity (Cope, 1978). The determination can be made reliably by competent physicians. They search for any hint of normal brain-controlled reflexes. They turn the head and even put cold water into the ear to look for any sign of movement. They search for any sign of the pupils of the eye responding to light. One test is to touch the cornea of the eye and see whether this triggers a blink. The respirator is also stopped briefly to look for any sign of spontaneous breathing. Absence of brain activity as signified by a flat electroencephalogram (EEG) is another criterion for brain death. The tests can be repeated 24 hours later to make certain that the absence of these life signs is not temporary. The brain-death prognosis is totally predictable and uniform: Brain dead people never regain consciousness, much less any recovery, and suffer cardiovascular failure within a short time (Jasper, Lee, & Miller, 1991).

According to one neurologist (Cope, 1978), when death is declared on the basis of cessation of breathing and heartbeat, the brain is deprived of blood supply and dies within a matter of minutes. Death of the brain is what is final and absolute. On the other hand, one may lose all brain function, but it is not irreversible. For example, a person having taken an overdose of barbiturates may have a temporary absence of any signs of brain activity, but this is not brain death because it is reversible.

Perhaps as a result of President Reagan's commission, there has come to be increasing acceptance of a brain-oriented criterion for determining death as the "irreversible loss of all brain function" (Youngner et al., 1989, p. 2205). Nonetheless, Raanan Gillon (1990) argues that the criteria for brain death are a compromise between extreme versions of two concerns. On the one hand, the compromise is with a concept of death as complete disintegration of the human organism's biological functioning, with all living components dead or at least totally disintegrated and dissociated from each other. On the other hand, brain death criteria are also a compromise with the concept of death as cessation of personal existence because brain death criteria will classify as alive some humans who are dead as persons. A clinical example is a vegetative state that involves permanent loss of consciousness and of the capacity and potential for consciousness. Because a capacity for consciousness is a necessary condition for being a person, an individual in a permanent vegetative state cannot be a person, yet according to brain death criteria (whether brain-stem death criteria or whole-brain criteria) that same human being is unequivocally alive.

THE MEANING OF DYING AND DEATH

The meaning of death varies from culture to culture. In some societies, an individual who dies is not recognized as "dead" until a year has passed. This "wait" gives the family time to prepare for the funeral. In the meantime the body is "kept" in the house (hut) and "given" food, drink, and cigarettes. Socially, the person is not dead, because he or she has yet to be defined as such. On the other hand, in a few cultures an individual may be alive and talking (biologically alive), but is socially dead. Indeed, in our own culture, an individual in a vegetative state is technically alive, yet socially dead.

The philosophical meaning of death suggests that the objective reality of death, the biological process that occurs when someone dies, is very different from the individual, subjective experience of death. For the one who is dying, death is momentous. It is the end of the world, the demise of one's very existence. Yet, objectively, the death of an individual is just a part of the cycle of living and dying. It is neither momentous nor particularly noteworthy. The philosopher Fingarette observes that "death itself is nothing, but the thought of it is like a mirror" (Fingarette, 1996, p. 5). Let's take a closer look at social and philosophical meanings of death.

The Social Meaning

Social meaning is the most important component of every aspect of dying and death considered in this textbook. Recall from chapter 1 the discussion of symbols and meaning in our presentation of the symbolic interactionist perspective. Social meaning consists of symbols that are socially created and socially used. Some symbols refer to something else or have an empirical referent. Some do not. The major function of both types of symbols is to permit humans to relate to each other and thus create shared behavior and meaning.

If dying is perceived to be primarily a biological process, then defining dying and death simply as biological or physical processes would seem to be appropriate. Most people would probably agree with the "biology-is-primary/social meaning-is-secondary" interpretation. We do not.

Dying, in fact, is much more than a biological process. It is truly one of the most individual things that can happen to the body, and what happens takes place exclusively within the skin of the one person. However, with reference to the meaning of dying, the dying process is one of the most social experiences that one can have—all human bodies exist within a social and cultural context. When a person dies, many things other than internal biological changes take place. For nearly all human beings, every act of dying influences others. At the same time, the dying are influenced by their environment and those around them. Consequently, the act of dying is a social or shared event.

Deriving Meaning from the Audience

In chapter 1 we discussed the importance of the audience in establishing meaning. It is essential for those who interact with the dying to understand the acute awareness of the dying person of those around him or her. This is especially true given the American tendency not to talk directly about death or dying. The dying person is therefore always trying to interpret the behavior and non–death-related words spoken in his or her presence. This awareness of the audience includes, but is not limited to, the following:

1. What people are willing to talk about with me—and what they avoid

2. Whether they are willing to touch me, and how they touch me when they do

3. Where I am, or maybe where others have located me—hospital, nursing home, intensive care unit, isolation unit, or my room at home

4. Tangible and verbal gifts that others give me

5. What people will let me do, or expect me to do, or will not let me do

6. The tone of voice that people use when they talk to me

7. The frequency and length of visits from others

8. Excuses that people make for not visiting

9. The reactions of others to my prognosis

The audience to which the dying person relates may also be supernatural. Symbol users are not restricted to the natural-empirical world, nor are they restricted to the world of the living. If believers realize that those who have died have an existence in another realm or that there is a life after death, this belief is real to them and has consequences for their behavior.

People who are approaching death may involve themselves in a gradual replacement of a living audience with a supernatural or other-world audience. As we have demonstrated, the dying relate to many audiences.

Deriving Meaning from the Situation

Where a person dies is also given meaning. As patients come to grips with their terminal condition, the manner in which they define the situation (and respond to it) will have a tremendous impact on their dying. Dying in a nursing home or hospital is different from dying in the home, in bed, surrounded by a loving family and feelings of belonging (see chapter 7). If patients view the institutional death setting as a supportive environment, their coping behavior may be helped. On the other hand, if they feel all alone and if they have defined the place of dying as a foreign environment, adjustment will not be facilitated (Leming, Vernon, & Gray, 1977).

Death as a Lost Relationship

The sociological perspective emphasizes the social-symbolic nature of human interaction. The key factor that unites biological entities into a social group entity is shared meaning: Many of the goals that one person wants to achieve and many of the experiences that one person wants to have require shared and coordinated cultural meanings. Symbols are the means by which socially created meaning is shared in the process of human interaction. Symbols are words or gestures that stand for something else by reason of association.

A specific death has distinct meanings wherever the deceased had meaningful relationships, and the death of one person has extensive social consequences. With a death in a husband-wife **dyad,** one half of that entity dies. If the couple has two children, one fourth of the family dies. One thirty thousandth of the community dies, and one two-hundred and eighty millionth of the nation dies. When one person dies, his or her spouse loses a husband or wife, children lose a parent, and friends lose a friend. Yet, the person who dies loses all human relationships. Thus, one could argue in numerical terms, that the dying person is losing more than anyone else.

If a person is identified with many social roles or positions and that person's death creates a situation where many roles or positions are vacated, then many persons or role occupants die in a single death. Furthermore, we may agree that, although one biological body dies, ownership of that body is difficult to determine. "Who owns my body?" is a question often posed by people who are dying. Related to the basic ownership question are "Who is qualified to make decisions about my body?" and "If I am the one who is dying, what right do any others have to tell me what to do with my body?" Because humans are social animals, ownership is a creation of symbol-using people. Joint ownership patterns are created and exist for most people. Therefore, body ownership might be considered shared, and the decisions related to it would be joint decisions.

The elimination or departure of a person from the ranks of "the living" leaves a hole in the midst of the living. Certain meaning is lost, while new meaning is added—behavior that previously involved the dead person is literally no longer possible because it also has ceased to exist. Established interaction or behavior patterns are disrupted, an event that demands attention. The funeral process involves activities, rites, and rituals associated with the final disposition of the dead person. This process usually reduces the social disruption caused by death insofar as the rituals are acceptable to those involved and are performed according to societal norms. The living are concerned not only with the death of a person, but also with what happens to them as a result of that death. Although the biological person may be gone, the meaning remains just so long as the living grant the "symboled immortality" or "meaning immortality" to the deceased (Lifton & Olson, 1974).

In the social commentary immediately following the death of Diana, Princess of Wales, a British broadcaster was asked by an American interviewer if the funeral ceremony would be a "royal funeral." He replied,

> Diana is much more than a royal—royals are two a penny. Princess Di is a star the likes of which we have only seen in James Dean, Marilyn Monroe, Elvis Presley, and John Kennedy. Her immortality will transcend the House of Windsor, not because she was born to it, but because it has been bestowed upon her by the British people and the people of the world.

Granting or creating such immortality is one of the things symbol-using beings can do. A person may even be granted considerably greater significance than he or she had while alive, as in the case of Princess Diana. For the bereaved, changing the meaning of a lost relationship is essential to the process—and is accomplished with varying degrees of ease—depending on the nature of the relationship.

Creating and Changing Death-Related Meaning

Biological bodies are created; they live, and they die. Bodies of meaning are also created, live, and die. Biological continuity occurs through a process of biological transmission or transference. Meaning (culture) continuity occurs through a process of social-symboled transmission. The socialization process occurs as biological bodies are transformed into social beings and as we teach our children how to behave in what our society considers to be a human way.

The flowers given in tribute to Diana, Princess of Wales, at Kensington Palace. The image exemplifies the words of the song, "Candle in the Wind," sung by Elton John at her funeral. Rewritten from a song originally composed to commemorate the life of Marilyn Monroe, another star who died tragically at the age of 36, the lyrics include this line, "Goodbye, England's Rose. Your candle burned out long before your legend ever will." More than 2.5 billion people (one third of the world's population) observed the funeral of Princess Diana via television.

It is important for caregivers and loved ones of dying persons to realize that their own death-related meanings will shape their behavior toward the dying. And, as we discussed in chapter 1, these behaviors may affect the meanings that the dying ascribe to their own death. To avoid thoughtless actions that may affect an individual's ability to die with dignity and integrity, those administering to dying persons should always question the following commonly held assumptions:

1. Those with whom you work necessarily share your meaning of death.

2. Meanings that were helpful to earlier generations are equally functional today.

3. Meaning remains constant and does not change.

4. Dying biologically is all that is happening.

5. Knowing about the biological aspects of dying will in and of itself provide knowledge about how humans expect to behave in death-related situations.

6. The terminally ill person is the only one having death adjustment problems.

> ## WORDS OF WISDOM
> ### Learning from the Dead
>
> Most people preserve the dead and learn from them. We like to distinguish ourselves from the simpler beasts by declaring that we are the only animals to bury our dead and so to rescue the past. Elephants may put branches on a dead elephant, or sprinkle the animal with dust; bears will bury another animal to ripen it for eating; dolphins hold formal mourning rites, as even cows do. Still, human speculation about death has been historically self-centered and self-flattering, rejoicing in unique anguish and burden. What other animals may do has seemed unimportant, because to us they have neither history nor myth, which our dead have given us.
>
> *Taken from* Aspects of Death in Early Greek Art and Poetry, *by Emily Vermeule, 1979, Berkeley: University of California Press. Reprinted by permission.*

7. The dying person has somehow stopped meaningful living during the terminal period.

8. Talking is the only way for the caregiver to communicate that she or he cares.

Creation of new meaning is always possible. Death-related meaning is no different than any other type of meaning. It is important to remember that this meaning is created by humans, not discovered in the world. All meanings are subject to change. However, well-established meaning is difficult to change. Meaning, for example, is frequently defined as sacred and is, therefore, more likely to be protected and perpetuated than changed. Crises or traumatic conditions may be necessary for a change in death-related meaning. As noted in chapter 1, many of the contemporary death-related meanings, including the rituals involved in adjustment, were created by ancestors who experienced dying in quite different social circumstances than those found today. Furthermore, considering the dramatic changes in health, longevity, and health care, our ancestors experienced death in somewhat different biological bodies. Therefore, it is not surprising that discontinuities have developed in American death-related meanings and experiences.

Any aspect of death-related behavior can be changed if there is enough societal (or subsocietal) support. One person can change death-related meaning for himself or herself, but it is difficult to maintain and sustain the new meaning if a **significant other** does not support, legitimatize, or validate it.

Like the ripples caused by dropping a stone into a lake, changes in death-related meaning will inevitably have consequences that penetrate other areas of living. Change in the sacred components of dying and death may come in through the back door, so to speak. Cremation may gain increased acceptance, not as a direct result of changes in religion, but rather as a result of the unavailability of space for earth burials. Likewise, changes in life-prolonging, or death-prolonging, procedures may result more from the availability of technological devices than from changes in religion or **mores**—the "must" behaviors that a society believes are for the good of that society.

The Philosophical Meaning

Socrates said that the true philosopher welcomes death. Death is not an end, but a transition. We human beings are unique among the inhabitants of the earth in being creatures both of emotion and of reason (Rosenberg, 1983). As feeling beings, it is fitting for us to be touched and moved by the death of someone close to us, and through death they are lost to us irredeemably and forever. Thus, we should seek some comfort and solace in the face of such a loss. It is also fitting for us, as thinking beings, not to find our comfort and our solace only in myth and in muddle and in self-delusion. Rather, we should search out our consolations from the standpoint of a clear and reasoned understanding of the truth, which only rational beings such as us humans uniquely could ever achieve.

An **existentialist philosophical** approach to death is a rather practical one and suggests that we must all face death. We can face death. Death becomes an extraordinary event to each of us. No matter how surrounded by others we may be at the time, we will face death alone. Not unlike the little engine that could, "I think I can, I think I can, I think I can," and "I did." We can "do it." In anthropologist Colin Turnbull's account in *The Mountain People* (1972) of his friend Lomeja's dying he observes that in the end Lomeja turned away from Turnbull and "in true Ik style, he died alone." We are not unlike the Ik in Uganda—we too "die alone"— from an existentialist philosophical perspective.

An offshoot of the existentialist philosophical perspective is **phenomenology.** This aspect of philosophy studies "the thing itself"—the phenomenon. The phenomenological study of dying would, for example, try to discover what dying is to those experiencing it. It studies the phenomenon directly and how it is constituted. If one were studying near-death experiences, who better could discuss these than persons having had near-death experiences? Or if one wished to study suicide and wanted to know what it is like, who better to ask than individuals who have attempted to take their own lives? What was the person thinking when trying to commit suicide? Study the phenomenon—the suicide attempt, the person who had the near-death experience. Obtain a first-hand account straight from the "horse's mouth."

The reflections on death of philosophers Georg Hegel and Martin Heidegger focus on its relation to consciousness and to the Being of the individual human being (Secomb, 1999). Using metaphor, literary figures, and abstract concepts, these theorists draw on empirical everyday life and speculate about the underlying significance, effect, and meaning of mortality for human existence. Hegel suggests that the encounter with the absolute negation of death enables human consciousness to attain a consciousness-for-itself (or self-consciousness), and Heidegger proposes that the human being is the being which is defined by its projection-towards-death, notes Secomb.

THE AMERICAN EXPERIENCE OF DEATH

The experience of death varies from individual to individual, from culture to culture, and through history. This diversity is due to both psychological and sociological factors. In this section, we will explore the sociological factors that have influenced the dramatic changes in attitudes from the Puritans to contemporary society. These changes reflect, in part, a changing way of life in America. We can identify three major shifts in perspective in the American experience of death: "the living death" (1600–1830), "the dying of death" (1830–1945), and the "resurrection of death" (1945–the present).

Living Death (1600–1830)

Historically, changes in the American way of life contributed to how we viewed death and responded to dying and death. Between 1600 and 1830, for example, death was a living part of the American experience. Rural Americans during these earlier years were well acquainted with death. People died in their homes and would be reminded of death by the tolling of the funeral bell in the village churchyard or simply passing by the church, surrounded by the cemetery, as they walked to and from their shops and homes. Symbols on gravestones in the early 18th century cemeteries included the crossbone and skull (stark reality of death), which was replaced by the soul face and later the soul face with wings attached (symbolic of an afterlife). The New England Puritans believed that a sovereign God ruled over the earth and displayed his sovereignty by intervening in the natural or social world—one such providence was death. They knew that they deserved death and that the last enemy on earth was death: Death was painful, a punishment for sin, and brought separation from their loved ones. Unlike modern thanatologists, the Puritans encouraged each other to fear death. They increasingly used that strong human emotion to rouse people from their psychological and spiritual security. Puritans knew that they would die, but not *when* they would die, or if they were among the elect. Therefore they admonished themselves and each other to be constantly prepared for death. "It will do you no hurt. You will Dy not One Minute the sooner for it," argued Cotton Mather, "and being fit to Dy, you will be the more Fit to Live" (Geddes, 1981, pp. 64–65). Over and over again they prayed, "Lord, help me to redeem the time!" Like modern thanatologists, they felt that an awareness of death could improve the quality of their lives and that they had a role to play in the work of God's redemption. Indeed, for them, dying, death, and bereavement were opportunities to glorify God by demonstrating human dependence on divine providence.

The deathbed was the final place of preparation. Dying Puritans received visitors who wanted to help provide "a lift toward heaven." Unlike Catholics, who relied on the sacramental rite of Extreme Unction, the Puritans focused on the state of the dying person's soul and mind. They prayed together, read the Bible, and urged active acceptance of the will of God. In the same way, "the funeral was another opportunity for the bereaved to turn affliction into spiritual growth through resignation to, and joyful acceptance of, the will of God" (Geddes, 1981, p. 104).

Most Puritan deaths occurred at home. The family sent out for midwife-nurses to care for the corpse, ordered a coffin, and notified friends and relatives not already present. Puritans considered the corpse a mere shell of the soul. They simply washed it, wrapped it in a shroud, and placed it into the coffin. Friends and relatives visited the home to console the bereaved. They brought gifts of food and reaffirmed the Gemeinschaft (community) spirit of the village. Prayers were offered, not for the soul of the deceased, but for the comfort and instruction of the living and to glorify God. The Puritan funeral acknowledged the absence of the deceased and drew the community members together for mutual comfort.

After the funeral, the mourners were to return to their calling and resume their life's work. In 1802 Nathaniel Emmons delivered a magnificent funeral sermon entitled "Death without Order." In it, he reviewed the Puritan orthodoxy of death, observing that "in relation to God, death is perfectly regular; but this regularity he has seen proper to conceal from the view of men." Emmons saw the uncertainty of death as a demonstration of God's sovereignty and human dependence and as a way of teaching people "the importance and propriety of being constantly prepared for it." However, he also saw that, despite the fact of death's disorder, multitudes of Americans had resolved "to observe order in preparing to meet it" (Emmons, 1842, pp. 3, 29). Between the 1730s and the 1830s, such orderly Americans were influenced by the Enlightenment, the American Revolution, Unitarianism, and Evangelicalism—and all of these were influenced by an underlying market revolution. These movements slowly but surely reformed the Reformed Tradition and gave Americans an eclectic tradition from which to fashion new beliefs and behavior about dying, death, and bereavement.

The Dying of Death (1830–1945)

Between 1830 and 1945, American thinking toward dying and death, rather than a "living death" as previously, was more of a "dying of death," as noted by an English author of 1899. This process brought "the practical disappearance of the thought of death as an influence bearing upon practical life" ("The Dying of Death," 1899) and the tactical appearance of funeral institutions designed to keep death out of sight and out of mind. This change was in part due to the rise of a new American middle class.

Both ideas and institutions were the product of a new American middle class—a group of people trying to distinguish themselves from the European aristocracy and from the American common people, somewhere in the "middle." Alexis de Tocqueville saw the middle class as "an innumerable multitude of men almost alike, who, without being exactly rich or poor, possess sufficient property to desire the maintenance of order" (de Tocqueville, 1835/1945, pp. 2, 145); historians today see them as substantial property owners, professionals, businessmen and merchants, shopkeepers and skilled artisans, commercial farmers, and their families. They possessed property, but their property (and the hope of increasing it) also possessed them, intensifying their fear of death. For the acquisitive member of the middle class, "The recollection of death is a constant spur. . . . Besides the good things that he possesses, he every instant fancies a thousand others that death will prevent him from trying if he does not try them soon." Therefore, the middle class wanted death with order.

One strategy for achieving death with order was the ideology of separate spheres. In the course of the 19th century, middle-class people separated management from labor, men's work from the home, and women's work from men's. They also tried to separate death from life, both intellectually and institutionally. Increasingly, specialists segregated the funeral from the home and the cemetery from the city. The rural, landscaped cemetery movement began, for example, with Mount Auburn Cemetery in Cambridge, Massachusetts, in 1831.

Both separation and specialization were strategies of control, an increasingly important idea in Victorian society. Nineteenth-century Americans worked for self-control, social control, and control over nature. They saw self-control as the key to character and sexual control as the key to marriage. In separating their homes from their shops, they tried to control both spheres of their lives. The home would be a controlled environment for reproduction and socialization, while the workplace would be a controlled environment for increased production and time discipline. Schools served as a transition from one controlled environment to another, while asylums provided controlled environments for societal deviants. Science and technology attempted to make the whole continent a controlled environment. Therefore, it should not surprise us that the same class that practiced birth control should also devise forms of death control (Hale, 1971; Howe, 1970; Rosenberg, 1973).

Belief in the beauty of death and the funeral was new to the 19th century as the middle class used its aesthetic awareness to beautify corpses, door badges, caskets, casket backdrops, hearses, horses, funeral music, cemeteries, monuments, mourning customs, and death itself. Death was made so artistic that it almost became artificial, thus less fearful. By the end of the 19th century the **obituary** began to serve as a funeral notice, and later the telephone allowed people to notify others about funerals without leaving their house, a more impersonal means than face-to-face, as was the case in the mid-1800s. Consolation literature, which showed mourners that they were not alone in their sorrow and showed the emotional and moral benefits of their sorrow, allowed people to share their grief without sharing it with people they knew. Such literature included obituary poems and memoirs, prayer guidebooks, hymns, and books about heaven. These writings inflated the importance of dying and the dead by every possible means.

Because of scientific breakthroughs, an insistence that death should be painless developed and that all people can eventually attain an "easy," natural death at an advanced age. With the discovery of ether in the 1840s and the introduction of the word "painkiller" in the 1850s, Americans applied physical and mental anesthetics to kill the pain of death. A major reason for people to think seriously about death was in part eliminated. Death began to be treated as an occurrence of old age, an idea that encouraged people to postpone or preempt preparation for death. Some medical researchers even asserted that old age is a curable congenital disease and posed the question, "Why not live forever?" Americans were advised not to take death personally, but rather to accept it as a part of human progress. Thus, the emphasis was turned from death and the deceased to survivors and posterity. The whole idea of life insurance developed out of these ideas (see chapter 11). By defining death as part of evolutionary progress propelled by a merciful God, Americans were allowed to approach death optimistically. Thus, we have the "dying of death" in America.

The Resurrection of Death (1945 to the Present)

On August 6, 1945, the United States dropped a single atomic bomb on the Japanese city of Hiroshima. The bomb destroyed everything within 8000 feet, blowing bodies at 500 to 1000 miles per hour through the air, and killing 70,000 people. The heat of the explosion ignited clothing within a half-mile radius, and trees up to a mile and a half away. The subsequent shock wave ruptured internal organs. The effects of the radiation disfigured or killed thousands more and the whole world bears the scars of the blast. This event ushered in the era of atomic weapons and our mode of thinking about death began to change, wrenched violently back to the basic fact of death. And so America experienced the "resurrection of death."

In a prescient 1947 article on the social effects of the bomb, Lewis Mumford had this to say about the Atomic Age:

> Life is now reduced to purely existentialist terms: existence towards death. The classic other worldly religious undergo a revival; but even more quack religions and astrology, with pretensions to scientific certainty flourish; so do new cults. . . .

> ## WORDS OF WISDOM
> ### Fear of an Atomic War
>
> I ask myself why am I so frightened of an atomic war wiping out the bulk of mankind? It is true that I am frightened, and deeply upset by the thought—so deeply that I have to ration myself how long I think about it, no more than a few seconds at a time. And I guess that most people are the same. Yet, after all, whereas an atomic war would leave some humans alive, my death, so far as I am concerned, would leave none whatever. . . . It can only be, I think, that I would feel even more intensely about my own coming death, were it not that I am prevented from picturing it.
>
> *From "A Note on Death,"*
> *by P. N. Furbank (born 1920),*
> *February, 1978, Protus, 2.*

> The belief in continuity, the sense of a future that holds promises, disappears; the certainty of sudden obliteration cuts across every long-term plan, and every activity is more or less reduced to the life-span of a single day, on the assumption that it may be the last day. . . . Not a single life will yet have been lost in atomic warfare; nevertheless death has spread everywhere in the cold violence of anticipation. (Mumford, 1947, pp. 9–20, 29–30)

"For the first time in six centuries," wrote Edwin Shneidman (1973, p. 189), "a generation has been born and raised in a thanatological context, concerned with the imminent possibility of the death of the person, the death of humanity, the death of the universe, and, by necessary extension, the death of God." "The bomb," said philosopher William Barrett (1958, p.65), "reveals the dreadful and total contingency of human existence. Existentialism is the philosophy of the atomic age." Indeed, the bomb did lead many postwar adults to the philosophy of existentialism, which began with the reality of death, worked through anxiety and alienation, and culminated in the "God is dead" theology of the 1960s. The threat of "megadeath" was a constant reminder of the fragility of life and uncertainty of existence—ideas that the Puritans could surely appreciate. The sheer fact that atomic weapons are available to numerous countries places massive destruction of entire societies in the limelight.

The threat of the end of the world took away some of the traditional consolations of dying, including three conceptions of immortality: immortality through reproduction, immortality through the continuity of nature, and immortality through creative endeavors. As Albert Einstein, whose theories helped give birth to the bomb, said, "The atomic bomb has changed everything except the nature of man." By the 1960s and 1970s, a resigned fatalism had set in. Television ads showed children no longer asking what they wanted to do *when* they grew up, but *if* they grew up.

Thus began the effort to avoid a nuclear war, which, throughout the 1950s and 1960s, seemed almost inevitable. This fear of nuclear war culminated in the Cuban Missile Crisis in 1962, where John F. Kennedy, the president of the United States, faced down Nikita Khrushchev, the premier of the Soviet Union, in a battle of nerves that could easily have led to the annihilation of life as we know it.

Today, despite a slowdown in the arms race, the threat of a nuclear war is always present. Although the immediacy of this threat has waned, there are other, smaller wars that are a reality in many countries, claiming lives on a daily basis. And there are many other reminders of our mortality, and the painful and brutal form that death can take. Beginning with the Vietnam War in the 1960s, we have seen the devastation of AIDS, ethnocide in Africa and the Balkans (a grisly reminder of the Holocaust), the interminable round of deaths and retaliations in the Middle East, and the deaths from the drug wars in Central America. With the media reporting daily on the rising death toll from disease and warfare all over the world, we are only too aware that death surrounds and shapes our very existence. To die a peaceful death seems a rare gift in this world of violence.

This photograph is much more than a man digging a hold in the ground. Consider your emotional response to this photograph.

Contemporary Attitudes toward Death

Although most people in our society do not need to fear the violent death that is so much a part of life in many parts of the world, Americans are afraid of death, and seek to deny it. This overwhelming fear is palpable in the words of Ken Walsh (1974), in *Sometimes Sleep*, "Lord if I have to die, Let me die;/But please, Take away this fear"(p. 18).

But as we will see, many of the attitudes and practices developed during the "dying of death" period in our history are still very much a part of the American experience of death and dying today.

Denying Death

Many may look at the prevalence of life insurance policies and wills as evidence that Americans are quite accepting of death and are taking measures to deal with that eventuality. However, as discussed earlier, this may, in fact, be an attempt to deny death by avoiding the one aspect of death that can be controlled, social death. Through life insurance, we continue the accumulation of wealth, and through wills we control the distribution of property. Despite the "resurrection of death" in the 1940s, there are many who still see much evidence that we are a death-denying society (Dumont & Foss, 1972).

First, we prefer to obscure the dying process. Instead of talking about death directly, we employ **euphemisms** for death as buffers rather than the stark words *dying, dead,* and *death* (Harvey, 1996). Examples of these euphemisms are: *succumbed, passed away, was taken, went to heaven, departed this life, bit the dust, kicked the bucket, croaked, was laid to rest, cashed in, expired, ended his or her days, checked out, ran out of time, brought down the curtain, went on to glory, is pushing up daisies, was taken by the Grim Reaper, heard the trumpet call,* and *is six feet under.*

The following joke captures this American approach to death:

> John came home from a trip and said, "How's the cat?" His brother Bob replied, "The cat's dead." John then said, "Don't be so blunt. You should ease into such information." "Okay," said Bob," I opened the window, the cat got on the ledge, and the next thing I knew, it had fallen to its death." Then John said, "By the way, how is Mom?" Bob answered, "I opened the window. . . ."

Second, we have a taboo on death conversation. A friend of George Dickinson recently wrote a two-page letter describing her family's summer activities. She noted that she and her husband were on vacation and that he became ill. He was taken to the hospital for surgery. She said, "He *had* the best of care. I *loved* him so." She never said that her husband had died, but it was obvious. It is difficult to say those words—*dead, dying,* and *death.*

Third, **cryonics** (body freezing, derived from the Greek word meaning "cold") suggests a denial of death. In this procedure, the person's body is frozen in dry ice and liquid nitrogen. A very expensive process initially with additional high annual costs, cryonics has not caught on; fewer than 50 individuals are currently in a cryonic state. Cryonics is based on the idea that someday a "cure" will be found for the "deceased" and that he or she can then be treated with the "cure" and thawed out. Thus, one does not die, but rather he or she is put into the cooler and brought back at a later date.

The first body frozen was that of James Bedford (Adams, 1988), a 73-year-old psychologist from Glendale, California, who was frozen in 1967. Soon some 13 cryonics centers were in operation; however, only three remained open by 1990 (Selvin, 1990). Current price for whole body is $120,000 or preservation of just a head for $35,000.

Fourth, we do not die in America; rather, we simply take a long nap. Caskets have built-in mattresses, some strawlike and others innerspring, and the head of the deceased person rests on a pillow in the casket. The room in the funeral home where the body is "laid out" is referred to by some as the "slumber room." Certainly, no one ever heard of having pets *killed* by the veterinarian; we have them "put to sleep." As is discussed in chapter 11, the word cemetery derives from the Greek word *koimeterion*, which means "sleeping place" (Crissman, 1994). Cemetery markers sometimes read "rest in peace" or make other references to sleeping. Some markers for burials from the 19th and early 20th centuries have a crib or bed effect—headboard, footboard, and sideboards, sometimes with a "double bed" for a married couple. Dying does not occur in America; one just goes to sleep.

Fifth, unlike nonliterate societies, we call in a professional when someone dies. The funeral director takes the body away, and we do not see it again until it is ready for viewing. When we see the body again, the cosmetic job is such that the person looks as "alive" as possible under the circumstances. When viewing a body at a funeral home, George Dickinson once heard a woman make the comment that the dead person "looks better than I have seen her look for years!" This is the "dying of death" idea discussed earlier.

Sixth, in many places in the United States, the casket is not lowered into the ground until after the family and friends have left the cemetery. It is not easy to watch a casket lowered into the ground because this reminds the viewers of the finality of death. There are several "bad trips" when a significant other dies—initially being told of the death, seeing the body for the first time, seeing the casket closed for the last time, and seeing the casket lowered into the ground. One can avoid the last "trip" by leaving the cemetery.

Thus, examples of denial of death can be found in the United States. We are not suggesting that this is all bad. It is a way of coping and simply seems to be evident in the American way of dying.

Fearing Death

In this book the point is made often that death per se has no meaning other than that which people give it. And some Americans give it a positive meaning. For

---

WORDS OF WISDOM
Nothing is Lacking in Death

Chuang Tzu was dying, and his disciples wanted to give him a grand burial. But Chuang said that all the ceremonial needs were already there: "Heaven and earth will be my outer and inner coffin, the sun and moon my pair of jade discs, the stars in their constellations will be my pearl jewels, the myriad will be my mourners. What could you add?"

Taken from Death: Philoς Soundings *(p. 123), by H. Fingareι 1996, Chicago: Open Court.*

---

example, B. F. Skinner, the psychologist who popularized behavior modification, approached his death from leukemia with no apparent regrets. The 86-year-old Skinner, just a few weeks before he died, said with a laugh (1990), "I will be dead in a few months, but it hasn't given me the slightest anxiety or worry or anything. I always knew I was going to die." Likewise, 75-year-old psychiatrist Timothy Leary noted, a few months before his death from prostate cancer in 1996, that he was "waiting for his graduation." Leary (Mansnerus, 1995) stated that he was "looking forward to the most fascinating experience in life, which is dying." He said that dying must be approached the way that life is lived—with curiosity, hope, fascination, courage, and the help of your friends.

Anthropologists will tell you that there are many cultures that share the attitude of these prominent Americans. The Chinese look upon death with pleasure rather than fear (Rzhevsky, 1976). This view is evident in the quotation from a fourth-century Chinese philosopher shown in the Words of Wisdom box.

Although there are certainly many cultures and many Americans who approach death without fear, a surprising number of cultures, including our own, attach fearful meanings to death and death-related situations. Why is it that so many people have less-than-positive views of death? Philip Slater (1974) suggests that in societies where individuality is a high cultural priority, the fear of death logically follows. People in industrial and postindustrial societies lose the connectedness based on community provided in other societies. Thus, the fear of death may be the price paid for living in a society whose ideology rests on the type of individualism experienced in the United States.

Death in the United States is viewed as fearful because Americans have been systematically taught to fear it. Horror movies portray death, ghosts, skeletons, goblins, bogeymen, and ghoulish morticians as things to be feared. Sesame Street tries to create a more positive view of monsters by showing children being befriended by Grover, Oscar, and Cookie Monster. With the possible exception of Casper "the Friendly Ghost," death-related fantasy figures have not received much in the way of positive press. Instead of providing positive images, our culture has chosen to reinforce fearful meanings of death. Cemeteries are portrayed as eerie, funeral homes are to be avoided, and morgues are scary places where you "wouldn't be caught dead."

The fear of death may be incorporated in the minds of some individuals with the fear of being buried alive. To be taken into the blackness when life is extinct is a dreadful enough prospect, notes Robert Wilkins (1996), but for the Grim Reaper to come calling before the "appointed hour" is to condemn the yet-living to a seeming eternity of suffocative horror. Though scattered accounts of individuals being "buried alive" are noted over the past several centuries (Wilkins, 1996), most medical personnel remain skeptical. When the Paris cemetery Les Innocents was moved from the center of the city to the suburbs in the latter part of the 19th century, the number of skeletons found face down in their coffins convinced the lay public that premature burial was common. Skeptics explain such "movement" as being the result of externally imposed movements such as the coffin bumping into walls as it is being negotiated down narrow stairwells or of sudden movements produced by bolting funeral horses. Nonetheless, the fear of being buried alive has persisted over the centuries. Such fear indeed might contribute to the overall fear of death.

One of the reasons why classes in dying and death often incorporate field trips to funeral homes and cemeteries is to confront negative death meanings and fantasies with firsthand, objective observations. The preparation room at the mortuary is a good example. If you have never been to one, think about the mental image that comes to your mind. For many, the preparation room is a place that one approaches with fear and caution—something like Dr. Frankenstein's laboratory, complete with bats, strange lighting, body parts, and naked dead bodies. The great disappointment for most students as they walk through the door is that they find a room that looks somewhat like a physician's examination room. "Is that all there is?" is a comment often heard after visiting the preparation room.

Sociologist Erving Goffman (1959) observed that first impressions are unlikely to change and tend to dominate the meanings related to subsequent social interaction patterns and experiences. Some of us want to retain untrue fearful meanings of death, even when confronted by positive images. On a recent field trip to a funeral home, one student refused to enter the preparation room. While other students were hearing a description of embalming procedures, she discovered a bottle labeled "skin texturizer" and became nauseated. Other students had a difficult time understanding her problem because they had had positive experiences.

Fearful meanings can also be ascribed to death because of a traumatic death-related experience. Being a witness to a fatal auto accident, discovering someone who has committed suicide, or attending a funeral where emotional outbursts create an uncomfortable environment for mourners can increase death anxiety for individuals. However, such occurrences are rather uncommon for most people and do not account for the prevalence of America's preoccupation with death fears.

Content of Death Fears

When one speaks of death fear or death anxiety, it is assumed that the concept is unidimensional and that there is a consensus about its meaning. Such is not the case, however, because two persons may say that they fear death, and yet the content of their fears may not be shared. Death anxiety is a multidimensional concept and is based upon the following four concerns: (a) the death of self, (b) the deaths of significant others, (c) the process of dying, and (d) the state of being dead. This model's more elaborate form (Leming, 1979–1980) shows eight types of death fears that can be applied to the death of self and the death of others:

1. Dependency
2. The pain in the dying process
3. The indignity in the dying process
4. The isolation, separation, and rejection that can be part of the dying process
5. The leaving of loved ones
6. Concerns with the afterlife
7. The finality of death
8. The fate of the body

From the Eight Dimensions of Death Anxiety (Table 2.1) we can see that the content of fear will be influenced by whose death the individual is considering. From a personal death perspective, one may have anxiety over the effect that one's dying (or being dead) will have on others. There might also be private worries about how one might be treated by others. From the perspective of the survivor, the individual may be concerned about the financial, emotional, and social problems related to the death of a significant other.

Because many factors related to the experience of death and death-related situations can engender fear, we would expect to find individual differences in the type and intensity of death fear, including social circumstance and past experiences. However, with all of the potential sources for differences, using a death fear scale developed by Michael Leming, consistently high scores are found for the fears of dependency and pain related to the process of dying, and relatively low anxiety scores are found for the fears related to the afterlife and the fate of the body (see Practical Matters box).

Approximately 65 percent of the more than 1,000 individuals surveyed had high anxiety relative to dependency and pain, and only 15 percent had the same high level of anxiety relative to concerns about the afterlife and the fate of the body (Leming, 1979–1980). Thus, it is the process of dying—not the event of death—that causes the most concern.

Death Fears, Gender, and Age

Is there a difference in death fear between males and females? Gender differences in death fear are one of the most frequently noted and poorly understood findings in thanatology (Dattel & Neimeyer, 1990). In a survey of research studies on death fear by gender, J. M. Pollak (1979) reported that the majority of findings concluded that women reveal more death fear than men, yet the other studies find virtually no differences between the sexes. Judith Stillion (1985) suggests that the discrepancy in these findings simply reflects the greater tendency of women to admit troubling feelings.

TABLE 2.1   **The Eight Dimensions of Death Anxiety as They Relate to the Deaths of Self and Others**

Self	Others
Process of Dying	
1. Fear of dependency	Fear of financial burdens
2. Fear of pain in dying process	Fear of going through the painful experience of others
3. Fear of indignity in dying process	Fear of not being able to cope with physical problems of others
4. Fear of loneliness, rejection, and isolation	Fear of being unable to cope emotionally with problems of others
5. Fear of leaving loved ones	Fear of losing loved ones
State of Being Dead	
6. Afterlife concerns	Afterlife concerns
Fear of unknown	Fear of the judgment of others—"What are they thinking?"
Fear of divine judgment	Fear of ghosts, spirits, devils, etc.
Fear of the spirit world	Fear of never seeing the person again
Fear of nothingness	
7. Fear of the finality of death	Fear of the end of a relationship
Fear of not being able to achieve one's goals	Guilt related to not having done enough for the deceased
Fear of the possible end of physical and symbolic identity	Fear of not seeing the person again
Fear of the end of all social relationships	Fear of losing the social relationship
8. Fear of the fate of the body	Fear of death objects
Fear of body decomposition	Fear of dead bodies
Fear of being buried	Fear of being in cemeteries
Fear of not being treated with respect	Fear of not knowing how to act in death-related situations

How is age related to death fear? Literature (Fortner & Neimeyer, 1999) on the relationship between death anxiety and age demonstrates an inverse relationship (i.e., the older the person, the less death anxiety). This is especially true when middle-aged persons are compared with the elderly—death anxiety declines from middle age to older age. Within the elderly **cohort,** however, this trend with age does not hold as death anxiety stabilizes during the final decades of life, according to B. V. Fortner and R. A. Neimeyer. Other research (Rasmussen & Brems, 1996) confirmed that age and psychosocial maturity are significantly and inversely related to death anxiety. Psychosocial maturity was found to be a stronger predictor than age, perhaps suggesting why previous studies have revealed only moderate correlations between age and

PRACTICAL MATTERS
Leming Fear of Death Scale

Read the following 26 statements. Decide whether you strongly agree (SA), agree (A), tend to agree (TA), tend to disagree (TD), disagree (D), or strongly disagree (SD) with each statement. Give your first impression. There are no right or wrong answers. In any given subscale, a score of 3.5 or higher means slightly fearful of death.

I. Fear of Dependency

1. I expect other people to care for me while I die.

SA	A	TA	TD	D	SD
1	(2)	3	4	5	6

2. I am fearful of becoming dependent on others for my physical needs.

SA	A	TA	TD	D	SD
6	5	4	3	(2)	1

3. While dying, I dread the possibility of being a financial burden.

SA	A	TA	TD	D	SD
6	5	4	(3)	2	1

4. Losing my independence due to a fatal illness makes me apprehensive.

SA	A	TA	TD	D	SD
6	5	(4)	3	2	1

Total of 4 scores ___11___ divided by 4 = __2.75__

II. Fear of Pain

5. I fear dying a painful death.

SA	A	TA	TD	D	SD
6	5	4	(3)	2	1

6. I am afraid of a long, slow death.

SA	A	TA	TD	D	SD
6	5	(4)	3	2	1

Total of 2 scores ___7___ divided by 2 = __3.5__

III. Fear of Indignity

7. The loss of physical attractiveness that accompanies dying is distressing to me.

SA	A	TA	TD	D	SD
6	5	4	3	(2)	1

8. I dread the helplessness of dying.

SA	A	TA	TD	D	SD
6	5	(4)	3	2	1

Total of 2 scores ___6___ divided by 2 = __3__

(continued)

(continued from previous page)

IV. Fear of Isolation/Separation/Loneliness

9. The isolation of death does not concern me.

SA	A	TA	TD	D	SD
1	2	3	(4)	5	6

10. I do not have any qualms about being alone after I die.

SA	A	TA	TD	D	SD
1	2	3	(4)	5	6

11. Being separated from my loved ones at death makes me anxious.

SA	A	TA	TD	D	SD
6	5	(4)	3	2	1

Total of 3 scores _12_ divided by 3 = _4_

V. Fear of Afterlife Concerns

12. Not knowing what it feels like to be dead makes me uneasy.

SA	A	TA	TD	D	SD
6	5	(4)	3	2	1

13. The subject of life after death troubles me.

SA	A	TA	TD	D	SD
6	5	4	(3)	2	1

14. Thoughts of punishment after death are a source of apprehension for me.

SA	A	TA	TD	D	SD
6	5	(4)	3	2	1

Total of 3 scores _11_ divided by 3 = _3.66_

VI. Fear of the Finality of Death

15. The idea of never thinking after I die frightens me.

SA	A	TA	TD	D	SD
6	5	4	(3)	2	1

16. I have misgivings about the fact that I might die before achieving my goals.

SA	A	TA	TD	D	SD
6	5	(4)	3	2	1

17. I am often distressed by the way time flies so rapidly.

SA	A	TA	TD	D	SD
6	5	(4)	3	2	1

18. The idea that I may die young does not bother me.

SA	A	TA	TD	D	SD
1	2	3	(4)	5	6

19. The loss of my identity at death alarms me.

SA	A	TA	TD	D	SD
6	5	4	(3)	2	1

Total of 5 scores _18_ divided by 5 = _3.6_

(continued)

(continued from previous page)

VII. Fear of Leaving Loved Ones
20. The effect of my death on others does not trouble me.

SA	A	TA	TD	D	SD
1	2	3	4	(5)	6

21. I am afraid that my loved ones are emotionally unprepared to accept my death.

SA	A	TA	TD	D	SD
6	5	4	(3)	2	1

22. It worries me to think of the financial situation of my survivors.

SA	A	TA	TD	D	SD
6	5	4	(3)	2	1

Total of 3 scores __11__ divided by 3 = __3.66__

VIII. Fear of the Fate of the Body
23. The thought of my own body decomposing does not bother me.

SA	A	TA	TD	D	SD
1	2	(3)	4	5	6

24. The sight of a dead body makes me uneasy.

SA	A	TA	TD	D	SD
6	5	4	3	(2)	1

25. I am not bothered by the idea that I may be placed in a casket when I die.

SA	A	TA	TD	D	SD
1	2	(3)	4	5	6

26. The idea of being buried frightens me.

SA	A	TA	TD	D	SD
6	5	(4)	3	2	1

Total of 4 scores __12__ divided by 4 = __3__

Developed by Michael Leming
(1979–1980). "Religion and Death:
A Test of Homans' Thesis."
Omega, 10(4), 347–364.

death anxiety. Possibly age alone cannot account for the decrease in death anxiety among the elderly; the combination of aging and the achievement of greater psychosocial maturity may serve to decrease death anxiety.

In summary, death fears are not instinctive; they exist because our culture has created and perpetuated fearful meanings and ascribed them to death. They are also a function of the fact that death is not an ordinary experience challenging the order of everyday life in society. They are a function of occasional firsthand encounters with deaths so unusual that they become traumatic.

Horror movies gross large profits for the movie industry. Not unlike "rubber-neckers" on the highway, the sight of a bloody scene in a movie appears to be fascinating to many people. To be "frightened" by death and horror seemingly is "thrilling" to some individuals.

Relieving Death Anxiety Through Religion

George Homans (1965) suggests that when individuals encounter death, the **death anxiety** they experience is basically socially ascribed (learned). Death fears can be likened to the fears of other things—snakes or electricity, for example. If we believe that we are in a dangerous setting, we react accordingly. Some religions, with an emphasis on immortality of the soul and a belief in a coming judgment, increase the level of death anxiety for individuals. However, after individuals have fulfilled the requisite religious or magical ceremonies, they experience only a moderate amount of anxiety. Homans brings the perspectives of anthropologists Bronislaw Malinowski and A. R. Radcliffe-Brown together by drawing four conclusions:

1. Religion functions to relieve anxiety associated with death-related situations.

2. Death anxiety calls forth religious activities and rituals.

3. To stabilize the group of individuals who perform these rituals, group activities and beliefs provide a potential threat of anxiety to unite group members through a "common concern."

WORDS OF WISDOM
Of Death

Men fear Death as children fear to go in the dark; and as that natural fear in children is increased with tales, so is the other.

A man would die, though he were neither valiant nor miserable, only upon a weariness to do the same thing so oft over and over.

From Of Death *by Francis Bacon*
(1561–1626)

---

4. This secondary anxiety may be effectively removed through the group rituals of purification and expiation.

Summarizing the relationship between **religiosity** and death anxiety, we can arrive at the following theoretical assumptions:

1. The meanings of death are socially ascribed—death per se is neither fearful nor nonfearful.
2. The meanings that are ascribed to death in a given culture are transmitted to individuals in the society through the socialization process.
3. Anxiety reduction may be accomplished through social cooperation and institutional participation.
4. Institutional cohesiveness in religious institutions is fostered by giving participants a sense of anxiety concerning death and uniting them through a common concern.
5. If the religious institutions are to remain viable, they must provide a means for anxiety reduction.
6. Through its promise of a reward in the afterlife, and its redefinition of the negative effects of death upon the temporal life of the individual, religion diminishes the fear that it has ascribed to death and reduces anxieties that are ascribed to death by secular society.

To test the empirical validity of these assumptions, Michael Leming (1979–1980) surveyed 372 randomly selected residents in Northfield, Minnesota, concerning death anxiety and religious activities, beliefs, and experiences. Subjects were divided into four categories based on a religious commitment scale developed by C. Glock and R. Stark (1966) and J. Faulkner and G. F. DeJong (1966). Approximately 25 percent of the respondents were placed into each category—the first consisted of those who were the least religious, and the fourth consisted of those persons who were the most religious.

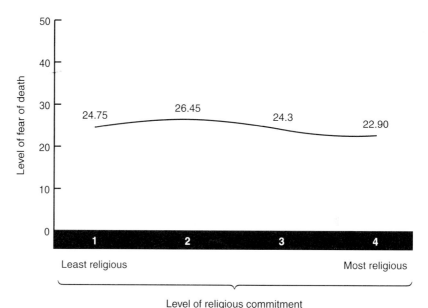

Mean Fear of Death Scale by Level of Religious Commitment

The graph shown here gives the mean fear of death scores for each of the levels of religious commitment. The relationship between the variables of religiosity and death anxiety is **curvilinear**—persons with a moderate commitment to religion have added to the general anxiety that has been socially ascribed to death from secular sources. The persons with a moderate commitment receive only the negative consequences of religion—what Radcliffe-Brown calls a "common concern." These persons acquire only the anxiety that religion is capable of producing, and none of the consolation. On the other hand, the highly committed individual has the least anxiety concerning death. Religion, as Malinowski predicts, provides individuals with a solace when they attempt to cope with death attitudes.

In conclusion, religiosity seems to serve the dual function of "afflicting the comforted" and "comforting the afflicted." We have discovered that religion, when accompanied with a high degree of commitment, not only relieves the dread that it engenders, but also dispels much of the anxiety caused by the social effects of death.

Upon further investigation, Leming found that factors of education, age, and even religious preference did not affect this relationship. With the exception of the fear of isolation, persons who had the strongest religious commitment were the least fearful with regard to the various types of death concern. Furthermore, in each of the eight death fear types the strength of commitment was the most significant variable in explaining the relationship between religion and the fear of death. Robert Kavanaugh (1972, p. 14) seems to have empirical support for his statement, "The believer, not the belief, brings peace."

*According to anthropologist Bronislaw Malinowski, death anxiety
(fear) is reduced for the individual through religion. For many individuals confronting death, religion "gives them strength" when medical
science has no answers.*

Contemplating One's Own Death

Psychiatrist Robert E. Neale in his book *The Art of Dying* (1973) suggests that the
reader look in the mirror and imagine herself or himself as a corpse. Such a suggestion
should certainly catch the attention of the reader! Neale asks the reader to reflect on
her or his own intimacy with death by answering a series of questions.

> ## PRACTICAL MATTERS
> ### Questions to Reflect on One's Own Intimacy with Death
>
> 1. Have you ever seen an embalmed body? Yes ✔ No ____
> 2. Have you ever seen an unembalmed body? Yes ✔ No ____
> 3. Have you ever seen a person die? Yes ✔ No ____
> 4. Have you ever been in a situation in which you thought you would die?
>   Yes ____ No ✔
> 5. Have you had a member of your family or close friend die? Yes ✔ No ____
> 6. Have you ever attended a funeral? Yes ✔ No ____
> 7. When you were a child, was death talked about in your family? Yes ____ No ✔

A lack of intimacy with death is not irrevocably bad, notes Neale, but it can lead to making too much or too little of death. There are no right answers to whether one thinks too often of death or thinks of death too infrequently. Perhaps some individuals make too little of death while others make too much of it. We are unable to distinguish between them in any concrete case, says Neale.

To contemplate the meaning of one's own death, philosopher Herbert Fingarette (1996) likes Tolstoy's story "The Death of Ivan Ilyich." Initially, Ivan Ilyich reacts to the possibility of impending death in a way that perhaps most of us do: with denial. He thinks that maybe this is not death after all, but the pain is merely the effect of some malfunctioning organ which the doctors can probably correct. Soon Ilyich realizes that these hopes are feeble straws, yet he is impelled to grasp at them to continue hoping. At the end of the story Ilyich is forced to confront his terror, and a voice within him asks what he wants. He answers that he wants to live, pleasantly, as before. This dialogue with himself leads Ivan Ilyich to a reexamination of his life, a review of the "pleasures" for which he has lived. He realizes that he has deceived himself and now sees that his life has been a commitment to a selfish inhumanity. He had enjoyed exercising power without compassion. Facing death, he comes to understand the truth of his life. Fingarette notes that Ivan's "confrontation with death" turns out to be a retrospective exploration and revelation to him of the meaning of his life. Seeing the truth of his life, he is able at last to face death and see its truth. At the end, the dying Ivan Ilyich asks himself, "Where is it? What death?" Tolstoy answers, "There was no fear because there was no death."

Philosopher Martin Heidegger (1927/1962) acknowledges that we know we are going to die, yet we accept this inauthentically. Our social existence represents an evasive attitude to death. If we are to live authentically, therefore, we must realize that death is not the end of life, but rather the inner possibility of being. Accordingly, death, despite its universality, has an individualizing effect: We each have to die (and therefore live) in our own way.

Malcolm Boyd (1997) states that the worst way to live is to fear living, and the worst way to die is to fear dying. He suggests confronting fears and staying open to others, opting for service instead of selfishness, learning to forgive, relinquishing hatred, and practicing loving.

At this point in your reading, take time out to write your own obituary. This will literally force you to consider that someday you indeed will die. In your obituary consider how you would like to be remembered by others. Were your goals in life to control others, to be rich, to give orders rather than take them, to be a status-seeker, etc.? Or would you prefer to be remembered in other ways, such as a compassionate individual who "went about doing good," trying to help those in life who had less than you did, devaluing material aspects of life and seeking something more "meaningful" than tangible products? After writing your obituary, stop and reflect on what you have said about yourself. Do you feel good about how you have said you wish to be remembered? If yes, then perhaps you are content with your life and can die feeling good about self, having reached Erik Erikson's last stage of the life cycle of senescence, as discussed in chapter 3. If no, then maybe you would wish to reevaluate where you are going with your life. An exercise such as writing one's own obituary is not an easy assignment and one that you are not often asked to do, but it can be most helpful in deciding if you are heading in the direction in life in which you wish to go. As with Ivan Ilyich, you may see the "truth" of your life and thus be able to face death and indeed have "no fear."

CONCLUSION

Although the American way of dying is being discussed and researched more today than in previous decades, this discussion often poses as many questions as answers. Aristotle (trans. 1941, p. 975) observed long ago that "death is the most terrible of all things"; yet it will not go away and, thus, we must learn to cope with it. The following questions do not have simple and straightforward answers: When does death take place? Who should determine the timing of a particular death? Who in society should be responsible for defining the meaning of life and death? When is a death an accident, and when is it a suicide?

Americans have developed a paradoxical relationship with death: We know more about the causes and conditions surrounding death, but we have not equipped ourselves emotionally to cope with dying and death. The American way of dying is such that avoiding direct confrontation with dying and death is a real possibility for many persons. What we need is the ability to both understand and cope with dying and death. The purpose of this book is to provide an understanding of dying, death, and bereavement that will assist individuals to better cope with their own deaths and with the deaths of others. As Rabbi Earl Grollman notes (Foderaro, 1994): "The important thing about death is the importance of life. Do what you have to do now. Live today meaningfully."

SUMMARY

1. The definition of death varies from culture to culture and throughout history. It has vacillated between the centralist theory that places the life force in a single organ and the decentralist theory that places the life force in every cell of the body. The most recent definition uses the centralist theory of brain death. However, establishing the criteria for brain death has not been a simple process. The definition of death is further complicated by the fact that there is not a consensus on the meaning of life and death.

2. Death-related meaning permits sharing death-related behavior. The death-related behavior of the dying individual and of those who are significant to that individual is in response to meaning, relative to the audience and to the situation. It is a phenomenon that is socially created, not biologically predetermined.

3. Because most people participate in various social groups, they are involved in many different interaction patterns. Consequently, even though it is but one biological body that dies, many "role holes" or vacancies are left by the death of a single individual. Bodies die and so do social relationships and social networks.

4. Death meaning includes evaluations of whatever those involved decide to evaluate. Evaluators may include values believed to be absolute, abstract, and situational. Defining values such as "living is always preferable to dying" as abstract rather than as absolute helps explain the relativity of situational values and is likely to lead to fewer adjustment problems when an individual confronts dying.

5. Dying is a social process. The person who is dying is living and is involved in living experiences with others.

6. Evolving death meanings are part of the general cultural changes taking place in contemporary society. Much of this change is centered around a discounting of biological influences upon social behavior.

7. The American experience of death has taken us from "living death" (1600–1830), characterized by a constant preparation for and fear of death, to the "dying of death"(1830–1945), an era in which death was denied, removed from daily activities, and relegated to the funeral homes, and the emphasis was on ways to postpone or preempt death. The next stage, the "resurrection of death" was prompted by the dropping of the atomic bomb, when death was once again on everyone's mind, and nuclear disaster was imminent. Today, we see evidence of the denial of death from the Victorian era, as well as the fear of death from the Puritans.

8. Evidence supporting the view that the United States is basically a death-denying society includes the widespread use of euphemisms for death, taboos on death conversations, the fascination with cryonics, caskets built for comfort, cosmetically enhanced corpses, and internment after guests leave the gravesite.

9. Death in the United States is feared because we have been taught to fear it.

10. Death anxiety is a multidimensional concept. There are eight types of death fears: dependency; pain of dying; indignity of dying; isolation; separation; and rejection; leaving loved ones; concerns with the afterlife; finality of death; and fate of the body.

11. Religion can relieve death anxiety if an individual is highly committed to it. A moderate commitment will increase anxiety about death.

12. One way to prepare for death and reduce anxiety is by contemplating one's own death.

DISCUSSION QUESTIONS

1. How has the definition of death changed over the years? What complications has this created for the American way of dying?

2. Discuss the differences between biological and symbolic death.

3. What arguments are offered in the rejection of this premise: "In death, biology is primary; meaning is peripheral"? Evaluate and discuss.

4. Each act of dying has three interconnected characteristics: shared, symboled, and situated. How does this relate to the statement that more than a biological body dies?

5. Discuss the implications of the following quote: "Even though it is but one biological body that dies, many 'role holes' or vacancies are left by the death of that one person."

6. In making decisions about the death meaning, how does the treatment of the dying patient affect that patient's understanding of death and his or her role in the dying process?

7. Discuss the evolution of death experiences in American life from 1600 to 2000.

8. What effect has the dropping of the atom bomb had on American death conceptions?

9. What factors have contributed to the American avoidance of dying and death?

10. Discuss whether you think the United States is basically a death-denying or a death-accepting society.

11. What is meant by this statement: "Death anxiety is a multidimensional concept"?

12. As one faces imminent death, one becomes increasingly aware of the social nature of life. This increase in awareness can lead to a life review during which one realizes how extensively one lives with, through, and for others. Speculate as to why this change in perspective takes place.

GLOSSARY

Brain Death: The brain is totally and irreversibly dead. This is sometimes referred to as the Harvard definition of death.

Cryonics: A method of subjecting a corpse to extremely low temperatures through the use of dry ice and liquid nitrogen.

Cohort: Persons of a similar age group.

Curvilinear: Referring to a type of nonlinear relationship between two variables where at a certain point, associated with the increasing values in the independent variable, the relationship with the dependent variable changes. A scattergram graph of this relationship will look like either a *U* or an inverted *U*.

Death Anxiety: A learned emotional response to death-related phenomena characterized by extreme apprehension; used synonymously with *death fear*.

Dyad: Two units regarded as a pair (e.g., a husband and a wife).

Euphemism: A word or phrase that is considered less distasteful than other words or phrases.

Existentialist Philosophy: Based on the doctrine that existence takes precedence over essence and holds that human beings are totally free and responsible for their acts, and that this responsibility is the source of the dread and anguish that encompass them.

Mores: Ways of society that are felt to be for the good of society. These are "must" behaviors that have stronger sanctions than a folkway (e.g., eating three meals per day) but that are not as severe as laws.

Obituary: Notice of a death, usually with a brief biography.

Phenomenology: The philosophical study of phenomena without any attempt at the nature of being; studying "the thing" itself.

Religiosity: The extent of interest, commitment, or participation in religious values, beliefs, and activities.

Significant Other: A person to whom special significance is given in the process of reaching decisions.

Temporal: Referring to a "here and now" or "this worldly" orientation that does not take into account the afterlife or a supernatural existence.

SUGGESTED READINGS

Anderson, P. (1996). *All of us: Americans talk about the meaning of death*. New York: Dell. Exploration of American perspectives on death through 60 interviews with Americans ranging from gang members to movie makers.

Buckman, R. (1992). *I don't know what to say: How to help and support someone who is dying*. New York: Viking Press. Discussion from the author's own clinical experiences as an oncologist of many issues that evolve in the process of watching someone die, including the topic of how to communicate with a dying person.

Dickinson, G. E. & Leming, M. R. (Eds.). (2002). *Annual editions: Dying, death, and bereavement 02/03* (6th ed.). Guilford, CT: Dushkin/McGraw Hill. An anthology covering topics on the American way of dying and death, developmental aspects of dying and death, the dying process, ethical issues of dying, death, and suicide, and funeral and burial rites.

Fingarette, H. (1996). *Death: Philosophical soundings*. Chicago: Open Court. Includes an original essay by the author and a collection of 12 classic statements on death written by individuals such as Tolstoy, Schopenhauer, Chuang Tzu, Camus, and Freud.

Freund, P. S., & McGuire, M. B. (1995). *Health, illness, and the social body: A critical so ology* (2nd ed.). Englewood Cliffs, NJ: Prentice-Hall. Sociological work discussi power issues in health and healing.

Gillick, M. R. (2000). Lifelines: *Living longer, growing frail, taking heart.* New York: W.\ Norton & Company. Takes the reader on a journey into the end-of-life stories of fou composite people. Public policy discussions of frailty and disability in an aging world ar also included.

Kaizer, B. (1997). *Dancing with Mister D: Notes on life and death.* New York: Doubleday. Probing by a Dutch doctor, with training in philosophy as well as medicine, of the intensity of the American preoccupation with death and dying as he shares his extraordinary experiences among the terminally ill.

Huntington, R., & Metcalf, P. (1979). *Celebrations of death: The anthropology of mortuary ritual.* Cambridge,UK: Cambridge University Press. Shows that American deathways are less irrational than they have appeared to some social critics and presents a new anthropological synthesis on death.

Oaks, J., & Ezell, G. (1993). *Dying and death: Coping, caring, understanding* (2nd ed.). Scottsdale, AZ: Gorsuch Scarisbrick, Publishers. A helpful book for teachers, counselors, and health professionals in better understanding dying and death.

Sloan, D. C. (1991). *The last great necessity: Cemeteries in American history.* Baltimore: Johns Hopkins University Press. Excellent historical account of changing attitudes and changing landscapes from churchyards to urban cemeteries to suburban parks in the United States.

Tolstoy, L. (1960). *The death of Ivan Ilyich and other stories.* New York: New American Library (Original work published 1886). Provides an insightful view of the difficulties and confusions that may have an impact on a survivor and/or the dying person.

Weiss, G. L., & Lonnquist, L. E. (2000). *The sociology of health, healing, and illness.* Englewood Cliffs, NJ: Prentice-Hall. Discusses the influence of the social environment on health and illness, health and illness behaviors, practitioners and their relationship with patients, and the health-care system in the United States and other countries.

Wilkins, R. (1996). *Death: A history of man's obsessions and fears.* New York: Barnes and Noble Books. Examines the lengths to which people throughout history have gone to cope with the five principal fears regarding death: the fear of being buried alive, of the body being defiled in the grave, of disintegrating, of being forgotten, and of suffering an ignominious death.

CHAPTER 3

GROWING UP WITH DEATH

Of death I try to think like this—
The Well in which they lay us
Is but the Likeness of the Brook

That menaced not to slay us,
But to invite by that Dismay
Which is the Zest of sweetness
To the same Flower Hesperian,
Decoying but to greet us—

I do remember when a Child
With bolder Playmates straying
To where a Brook that seemed a Sea
Withheld us by its roaring
From just a Purple Flower beyond
Until constrained to clutch it
If Doom itself were the result,
The boldest leaped, and clutched it—

—Emily Dickinson (No. 1558)

In the Emily Dickinson poem above, death is seen not as the "menace," but rather as that which "invites" us (Cooney, 1998). Invites us where and to do what?: To the "Flower Hesperian" (the evening flower); to the "Zest of sweetness" (is this not life itself?). The "boldest" among us leap to clutch at the flower of life. Death disguised ("decoying") as a "roaring" "Sea" is there "to greet us" and to provide that obstacle to overcome. In this chapter we will talk about death with the hope that we can help each other better understand death conceptualizations at different stages of the life cycle—from childhood through old age.

Sigmund Freud traced conceptions of death to our earliest feelings concerning sexuality and to our fears of being punished for them. Alfred Adler had several brushes with death himself and, as a child, suffered from a debilitating disease. When Adler formulated his theories concerning the human psyche and its development, he attributed our need to strive and overcome to our early sensitivity to weakness and death.

The **ego psychologists** later departed somewhat from Freud and credited the individual with a greater amount of ability to manage the stresses and problems of life. Yet, they recognized that humans raise a whole set of defenses against the idea of death. Both children and adults have the power to distort their perceptions according to their needs. For example, individuals have ways of denying harsh or painful thoughts, and people often believe and see what they want to believe and see, thus transforming images of their imaginations into reality.

In his Pulitzer Prize–winning book entitled *The Denial of Death*, Ernest Becker (1973) argued that fear and denial of death are basic dynamics for everyone. He asserted that we struggle to find meaning in life through heroic efforts. If we discover that heroic efforts are not possible, the dilemma is avoided by building elaborate systems to explain the dilemma away. Some even flee into neurosis or a psychotic break. Becker felt that the fear of death is a basic problem of meaning with which we all struggle.

Though the subject of death was not a major concern in his writings, Swiss developmental psychologist Jean Piaget was probably instrumental in nudging psychologists to employ better methods of research in the developmental approach toward understanding concepts of death (Ginsberg & Opper, 1979). Through keen observations of his children and others, Piaget postulated that it is not until the early teen years that one is capable of genuinely abstract thought processes.

Since the publication of Herman Feifel's *The Meaning of Death* in 1959, interest and research on the conceptualization of death have grown. Representing **psychoanalysis, behaviorism, humanism,** and other points of view, much has now been written on a **developmental approach** to death attitudes and awareness.

It is not the intent of this chapter to suggest that age should be seen as the sole determiner of one's death concept. Many other factors influence cognitive development, such as level of intelligence, physical and mental well-being, previous emotional reactions to various life experiences, religious background, other social and cultural forces, personal identity and self-worth appraisals, and exposure to death or threats of death. Though an age-based approach is followed in this chapter, other important factors should not be ignored.

FOR THEIR OWN GOOD

It isn't that difficult to shield children from death.

> Mommy? Where's Grandpa?

With a little ingenuity, you can keep them in the dark for 15, 16 years - or even more.

> Huh?

We'll start with an easy one: you're watching Bambi, and Bambi's mom gets shot.

Simply look her in the eye and say:

> She's fine! I saw her LAST WEEK!! In the woods behind GRAND UNION!!!

Let's say he or she gets curious about cemeteries:

> It's a STONE STORE.

Pet death? Watch and learn:

> What's wrong with Fifi?

> Quick! Look over there.

Do the old switcheroo -

> HURL

> What? Where? I don't see anything.

- and it's like nothing ever happened.

> Fifi had brown stripes

> No, she didn't.

> MEOW

Why not keep it on a "need to know" basis? After all - who needs to know?

> Grandpa is in the Belgian Congo.

LISTENING TO THE VOICES
Not Dead, Just "Hibernating"

A 6-year-old gives this account of her pet's death and burial and suggests that she is not aware of the permanency of death.

I had a cute little hamster. I think we had been talking about hibernation in school, because when I came home and found the hamster lying there, I thought he was hibernating! My next door neighbor, Sharon, came over and said that if we warmed him up then he would come back to life. So I put him in my electric blanket to see if he would wake up. He never did and by that time my mom realized what had happened, so we had to bury him. My brother and I put him in the shoe box, put a blanket in there to keep him warm and a picture of us to keep him company. Then we buried him in the backyard.

From "First Childhood Death Experiences (pp. 176–177)," by G. Dickinson, 1992, Omega, 25.

CHILDHOOD

Despite a new openness to death in the United States, children are generally not encouraged to express themselves on the topic of dying and death. As Jane Sahler (1978, p. XV) in writing about children and death notes, "One of the fascinating things about children is the naive, simplistic way in which they approach the unknown. Yet it is this very naiveté that makes their questions the most difficult to answer." Children usually have no prior experience on which to base their reactions; thus, their earliest experiences are unique to each of them.

To even be talking about children and death within the same sentence seems inappropriate, because death among children is much less frequent than death among older adults. One learns to crawl before walking; typically one is supposed to "grow up" (and indeed grow old) before dying. Nonetheless, children are confronted with death, probably not their own but that of others, and they need to understand and cope. Dying and death have an impact on our lives throughout the life cycle. Death is on the news, in the movies, in popular songs, and on television. The subject of death at all ages is difficult to avoid. It is ubiquitous.

How Do Children Learn About Death?

Children learn about death in a variety of ways. The first death experience is often that of a dead bird or any "road kill" or a dead insect (ant, roach, or beetle) or spider or a pet. In a study of college students' recollections of their first death experiences, George Dickinson (1992) found that 28 percent of his subjects' first death experiences involved a pet.

WORDS OF WISDOM
The Woman Who "Had" Three Sons

There was an old woman who had three sons,
Jerry and James and John,

Jerry was hanged, James was drowned,
John was lost and never was found;

And there was an end of her three sons,
Jerry and James and John!

Mother Goose Nursery Rhyme

The death of an animal is a good opportunity for parents to help children learn their first lessons of death by answering questions. The children may have had the pleasure of watching the family pet grow from a small puppy, kitten, or guinea pig into a large dog, cat, or guinea pig. Such an experience is typical of the maturation process: After birth, one grows, finally reaches "full size" and then ages and dies. With pets having a shorter live span than humans, children often observe a pet going through the life cycle. Thus, a pet's death is a realistic orientation to death within the animal kingdom.

Of course, the death of a family member, such as a grandparent (the greatest number of earliest death memories involve grandparents, according to Dickinson, 1992), can bring death "too close" to home. The child then learns quickly by observing others and experiences the various feelings she or he has. Or that first death experience can occasionally be that of a classmate or a teacher, thus validating the importance of school systems socializing teachers to death issues and encouraging them to serve as good role models for the students by being communicative and open to their questions.

More impersonal death experiences would include those in the mass media: movies, television, and the news. Though exposure to death by these means may occur often and sometimes continuously, the experience is usually less personal. Nonetheless, deaths publicized by the media can be bothersome to children, especially if open communication is not available to vent feelings.

Children are also introduced to death through religion. No matter what religious orientation one has, death plays a role. For Jews or Christians, for example, the Bible is filled with death accounts and talk of death.

Children's storybooks sometimes contain accounts of death, some of which are rather gruesome, for example, Mother Goose nursery rhymes. Other more "innocent" stories may include death themes and may be written specifically for children to teach them about death.

In a thorough review of the literature on children's death concepts, C. Randy Cotton and Lillian M. Range (1990) conclude that several factors have been shown to influence the development of death concepts. They note that children's death concepts change with age—older children are considered to have more accurate death concepts, are affected by cognitive level (those more cognitively mature have relatively more accurate death concepts), and are affected by experience (those having more experience with death are more likely to believe in personal mortality).

Personal Experiences

Children most often confront death at an early age. Their vivid recollections of these first death experiences testify to the impact that such experiences had on these individuals (Hall, 1922; Tobin, 1972; Kastenbaum, 1991). George Dickinson (1992) found, in analyzing 440 essays of college students (average age of 24) about their first childhood death experiences, that the average age at the first experience was 8 years and that their recollections were quite clear. The children's reactions upon learning of the death of a significant other encompassed a wide range of feelings and emotions: relief, confusion, disbelief, fear, happiness, anger, sadness, emptiness, shock, and guilt. One 10-year-old "had extreme fear" until someone explained to her about what would happen to the body. Some children were upset because they were unable to go to the hospital and say goodbye to the deceased.

Guilt experienced at an early age can stay with a person into adulthood. For example, the responsibility of a 4-year-old girl (Dickinson, 1992, p. 172) was to feed the dog, Charlie, every night. One night she forgot to feed Charlie, and the next day Charlie did not come home. She said:

> I just knew he had run away because I had not fed him. I felt terribly guilty, but I kept my mouth shut and did not tell my parents of my grave mistake. A few days passed, and Charlie still did not come back, even though every night I tried to feed him. I can literally remember calling his name and crying and yelling that I was sorry. My parents finally told me that Charlie would never be coming back because he had been run over. Well, this information devastated me. I put all the blame on myself. I just knew that Charlie had been run over because I had failed to feed him that night. I kept all this to myself until recently when my family was discussing the history of our family pets. Even though now I know that I am not responsible for Charlie's death, I still wish that I would have fed him that night.

Within the family setting one's primary socialization and perhaps the most intense experiences of a lifetime occur. Death is one of these experiences. When a family member dies, the family rallies around, and extended family members appear who have not been seen since the last death in the family. Indeed, a **latent function** of a death in the family is that of a family reunion.

Within the family we often first experience a death, most typically that of a grandparent. With the death of a grandfather, for example, a small granddaughter or grandson soon becomes aware that Grandpa is immobile, noncommunicative, cold, and lying in a box called a casket. The child observes the reactions of others, then usually responds accordingly. Sometimes adults may be laughing in the kitchen during the "mourning period," something that often is repulsive to the small child.

THE FAMILY CIRCUS
by Bil Keane

"Well, yes — we'll see Granddad someday when we go to heaven."

"Could I just wait in the car?"

"How could they be so irreverent?" the child may ask. "Why aren't they sad all the time, over Grandpa's death?"

It is very helpful when parents pay special attention to small children at the time of a death in the family, as indicated by this account of a 9-year-old boy on the death of his grandfather (Dickinson, 1992, pp. 175–176):

> The day of my grandfather's death, my dad came over to my aunt and uncle's house where my brother and I were staying. He took us into one of the bedrooms and sat us down. He told us Granddaddy Doc had died. He explained to us that it was okay if we needed to cry. He told us he had cried, and that if we did cry we wouldn't be babies, but would just be men showing our emotions.

Thus this parent made the little boys feel comfortable about crying, if they so felt the need (a healthy orientation from a parent). Without parents being especially attentive to small children, a child may needlessly suffer. For example, one child was afraid that her dead uncle would come back and bother her, so she slept with the covers over her head at night.

A 7-year-old girl (Dickinson, 1992, p. 173) felt that her parents had prepared her for her grandfather's funeral when they told her that he had gone to heaven and that she would meet him there someday. However, when they went to the funeral home and saw his body, she said, "You told me that Grandaddy was in heaven. No he is not, he is right here." She was upset that her grandfather was lying there not talking to anyone. Thus, parental "explanations" do not always suffice in the concrete thinking of small children.

Personal experiences with death today are for some children "too close to home," if they happen to live in communities besieged by chronic violence (Barrett, 1996). Daily exposure to episodes of violence and death increases children's risk of trauma and potentially undermines their psychological and emotional well-being. The cumulative effect of the exposure of young children to episodes of chronic violence and trauma cannot have anything but negative consequences for their learning potential and interpersonal conduct at school, at home, or in the community, notes Barrett (1996). Children in urban environments are increasingly at risk of becoming victims of homicidal violence as they are likely to be both victims and perpetrators of lethal violence. Indeed, such an atmosphere of death is more realistic than that portrayed on television and in other media sources.

Mass Media

Television and movies expose children to countless deaths each week, whether of the cartoon characters Cow and Chicken or Wile E. Coyote, or in feature films or other TV episodes. The orientation here, however, is often one of death being temporary. How many times does Wile E. Coyote die within a 30-minute period? Road Runner continues to escape death, but barely (the "bird" with "nine lives"). One does not "die off" on soap operas but simply reappears in a few weeks on another network and playing another role. Or someone killed in a movie appears again in a few months in another movie. Death is not permanent, but only a temporary situation, as a child might interpret it from television and the movies.

Television also highlights sensational national and world events such as school shootings, major natural disasters, and war. These realistic, though often gruesome, scenes bring death "live" into the home; this death is not "reversible." However, in some fictional television programs, death has a fairy-tale quality (Silverman, 2000). The protagonists are not really hurt by the blows inflicted on them and frequently come back alive, like Wile E. Coyote. Media deaths do not allow children to learn about the true consequences of someone dying or to learn that death is part of the real world. They also do not learn how to cope with death when such a reality does occur.

Newspapers, like television, display death across the front pages, in addition to other parts of the paper. Pictures of death often accompany the write-ups in the newspaper. Certainly the image of violent death—the picture of the fireman carrying the body of the small child out of the federal building in the Oklahoma City bomb disaster—is ingrained in the memory of U.S. residents. Indeed, the numerous incidents of elementary and high school children being killed by classmates in school are all too common stories in the media today.

Books often have "death scenes" in them. Death is presented in both the book and movie about Bambi, for example. Books can be sources of comfort and can provide valuable information to children. Books are written both for parents to read to their children and for the children to read themselves. Books can serve as good sources for broaching the subject of death.

Nursery rhymes and fairy tales expose children to the topic of dying and death. At the time when these children's classics came into existence, death of family members, often siblings, was a much more immediate reality than it is today and they reflected this grim reality.

Mother Goose nursery rhymes not only describe peaceful deaths, but include deaths by drowning, hanging, boiling, decapitation, devouring, choking, squeezing, and starving. Such gruesome deaths should be a cause for concern among parents, yet we might read one of these "stories" to our children after putting them to bed, only to find later that they have difficulty going to sleep. No wonder! For example, the following prayer, with a suggestion of fatalism, has kept many a child awake at night:

> Now I lay me down to sleep,
> I pray the Lord my soul to keep;
> And if I die before I wake,
> I pray the Lord my soul to take.

Other popular stories include the characters Hansel and Gretel who roast the witch in her own oven and Dorothy who is trapped in Oz until the wicked witch of the west could be liquefied. Nature's destruction of the body is pointed out in this nursery rhyme:

> She saw a dead man on the ground;
> And from his nose unto his chin,
> The worms crawled out, the worms crawled in.

The tale of Solomon Grundy points out the brevity of life.

> Solomon Grundy,
> Born on a Monday,
> Christened on Tuesday,
> Married on Wednesday,
> Took ill on Thursday,
> Worse on Friday,
> Died on Saturday,
> Buried on Sunday.
> This is the end
> of Solomon Grundy.

Mother Goose nursery rhymes are a source of death education for many children, although concerns about the themes have led to nursery rhyme reform. In 1937 Allen Abbott urged nursery rhyme reform, as did Geoffrey Hall in 1949. In 1952 Geoffrey Handley-Taylor (Baring-Gould, 1967, p. 20) published a brief biography of the literature of nursery rhyme reform and noted that the average collection of 200 traditional nursery rhymes contains approximately 100 rhymes which "personify all that is glorious and ideal for the child." He noted, however, that the remaining 100 rhymes "harbor unsavory elements."

Thus, the media are filled with orientations toward death and dying, a primary source of death information for children.

LISTENING TO THE VOICES
God Wants a Dead Kitten?

The father of one 4-year-old girl told her that her kitten "went to heaven to be with God." She responded, "Why does God want a dead kitten?"

A 3-year-old became angry when told that her grandmother had gone to heaven, and responded, "I don't want her in heaven, I want her here!"

From "First Childhood Death Experiences" (p. 173) by G. Dickinson, 1992, Omega, 25.

Religion

It appears that children's religious beliefs are determined by their cognitive-developmental status (Wass, 1995). What children mean by concepts such as "heaven" and "God" reflects their thinking and reasoning modes at the particular developmental level at which they are functioning. Hannelore Wass notes that some young children view God very concretely as a father figure, except that he is much taller, stronger, and older, and has larger hands, than Dad. God gives out food and lets dead people be angels and fly around. Other children view God as imposing restrictions. Some children do not believe in an afterlife.

Children, depending on their particular religious upbringing, if any, may be exposed to ghosts, the devil, hell, and concepts that may be scary for them. One 6-year-old girl (Dickinson, 1992) was afraid of her great-grandmother who had "big brown moles all over her face and sat in a wheelchair." When told that she had died, the little girl became "infatuated with her ghost." When the television shows went off the air at night and the "black and white fuzzy stuff" came on, she used to imagine that she would hear her great-grandmother's voice or see her face in the television screen.

To tell a child of an ill-defined place where everyone goes after death could be frightening. However, the majority of children in a study (Dickinson, 1992) of 440 college students seemed content with the explanation that the deceased had gone to heaven, which was described as a "happy place." Given a child's world of fantasy—the magic world of Disney mentality—the response "gone to heaven" seemed to suffice for the majority of these respondents. Thus, religion can play a part in children's understanding of death.

Children's Understanding of Death

As noted earlier, children growing up in an inner city experience violence and death all too often. Other children in the United States, however, have less frequent exposure to death as a natural part of the life cycle; nonetheless, death is an aspect of the life cycle from early childhood to old age.

Largely gone are the days when most children lived with or near their grandparents, experiencing their death in the home. Urbanization and mobility have contributed to a separation from grandparents and, consequently, from the process of their dying. Death has been removed from the home to more institutionalized settings—the hospital or nursing home. This is summed up by the statement of the little boy who said, "I don't want to go to the hospital because that is where you go to die." That is where Grandfather was taken the last time that the little boy saw him alive. After "going to the hospital," Grandfather turned into a cold, stiff corpse. The process of the grandfather's dying was not part of the daily routine in the family's home.

As was pointed out in chapter 1, gone are the days when the majority of children were reared on farms and experienced the life cycle daily—animals were born, and animals died. Birth and death were commonplace in the socialization of children reared on farms. Gone are the days when siblings and relatives were born at home and died at home.

Not being exposed to dying and death makes today's children view this process and event as very removed and "foreign." This underscores the natural tendency to deny the facts of death. Just how do children conceptualize death? In the following discussion children are divided into age groups, but one must remember that ages and stages are in no way absolute. Each child's understanding of death is affected by his or her experiences and the family's cultural heritage.

Birth to Age 3

By age 6 months an infant perceives differences in caregivers and the degree to which physical and emotional needs are being met. One cannot have a concept of death until the beginning of thought, as evidenced by the emergence of symbolic function between 18 and 24 months. It has been suggested that the origin of death anxiety is in the traumatic separation from mother at the time of birth (Hostler, 1978). The delight of a child's playing peek-a-boo lies in the relief of the intermittent terror of separation. Under age 5 death is perceived as separation, but separation from one's caregiver is a terrifying thought.

Erik Erikson (1963) observed that an infant "decides" early in life whether the universe is a warm and loving place to be. Upon this primitive, yet momentous, subconscious decision is based to a large degree the ability to deal with threats and difficulties of later life. We must not assume that the small child has no concept or grasp at all of death, and we must be concerned about the effects of a given death upon his or her life. Children's capacity to grieve is related to their level of **cognitive development** (Ginsburg & Opper, 1979). Until a child has developed a sense of what Piaget called "object constancy," she or he cannot feel loss. Object constancy refers to the notion that objects continue to exist, even when they are no longer visible. Very young children have no concern for objects, or people, that are not visible to them. In their minds, these objects or people do not exist, so they cannot experience any sense of loss. It is only at about 18 months that children have reached a stage where they do understand the significance of objects and can grieve over something that has been "lost"—whether it is a toy or a loved one.

A toddler recognizes that a pet is alive and a table is not. The toddler's vocabulary normally includes "to die" by age 2½ and "to live" by age 3 (Hostler, 1978). Although the 2-year-old child dying in a hospital has no real concept of his or her death, the child has a real appreciation of the altered patterns of care and the separation from usual caregivers.

A critical analysis of the literature concerning the development of the concept of death in early childhood by Michael Stambrook and Kevin Parker (1987) suggests that some children, under certain conditions (such as experiencing a loss through death), can understand death to be final much earlier than is typically suggested. It has even been hypothesized (Yalom, 1980) that children under 3 years of age "know about but deny death." Perhaps the lack of research consensus found here simply confirms that we cannot depend solely on age categories regarding children's conceptions of death but must also consider individual experiences of the child.

Ages 3 Through 5

The clearest and most consistent observation from the literature on death concepts of children is that children as young as 3 have very definite ideas of what death means, though there is considerable disagreement as to the nature of those ideas, according to Stambrook and Parker (1987). Most children 3 to 5 years old lack an appreciation that death is a universal phenomenon and is a final and complete cessation of bodily functions. Though death is seen as temporary and/or partial, Stambrook and Parker (1987) note that the child is very curious to know and understand the practical and concrete aspects of all that surrounds death. For some children, death is believed to be life under changed circumstances. For example, the dead in caskets can still sense what is happening in the environment but cannot move.

The permanency of death is not clear in early childhood as evidenced by this story told to George Dickinson about a 3-year-old child whose father had been killed six months earlier in an auto accident. One day the little boy's mother came in and said, "I have a surprise for you." The little boy anxiously replied, "Is Daddy coming home?" On another occasion this same little boy was playing with his LEGOs and built a house. He placed a person inside the house and was asked to identify the individual. He replied, "Oh, that is Daddy. He is asleep for 100 years." Daddy is not dead; he is simply away on a trip or engaged in a deep sleep.

When George Dickinson's daughter was 5 years old, she attended the funeral of an elderly friend of the family. After the service and the final viewing, the casket was closed and sealed. At this point she said, "But, Daddy, how is Mrs. Kirby going to breathe in there?" Again, it is difficult for a 5-year-old to grasp a concept of death. Death is not permanent, since one still needs to breathe!

The practicality of young children is apparent in this story of a 5-year-old whose mother had died. While waiting in the airport to fly from Kentucky to Texas to bury her mother, she looked up at her father and said, "Daddy, can you cook?" This was a traditional family in which the father "brought home the bacon," and the mother cooked it. The child's concern was whether or not the father could fill the vacant role.

Calvin and Hobbes
by Bill Watterson

Ages 6 to 12

Israel Orbach and associates' study (1985) of 6- to 11-year-old children's concepts of death in humans and animals showed that death in humans is easier to comprehend than death in animals. They note that seemingly the comprehension of animal death is acquired chronologically later than that of death in humans. Apparently animal death is a more difficult concept than human death. Because animals are known to be a favorite identification object of young children, the death of a pet may be more meaningful and more personal to a child than that of a generalized other human being. Thus, children may erect more defenses against death of animals. Another explanation noted by Orbach et al. is that human and animal deaths can be regarded as two specific examples of the complicated concept of life and death. While a child may comprehend the meaning of death to a specific example of life, such as among humans, he or she may not comprehend it in another sphere of life, such as plants and animals. This reflects a lack of understanding of the broader concepts of life, birth, and death and animate and inanimate objects.

In this age range the child has such motor mastery that he or she can ride a bicycle without holding the handles and can sing a song at the same time. The child begins to view the world from an external point of view, and language skills are becoming communicative and less egocentric. The first major separation from home occurs, and the child enters the world of school where teachers and adults, other than parents, become the models for identification. Although magical thinking persists, the child's ability to test reality increases (Hostler, 1978).

A sense of moral judgment continues to develop in this age group. The child attains an understanding that rules are of human origin and that he or she can participate in their origin and modification. The child is mastering school and social skills and has an interpretation of that experience based on an external point of view—schoolmates, teachers, other adults, readings, and the media.

Between the ages of 6 and 12, the evolution of the concept of death as a permanent cessation of life begins. However, cognitive obstacles to abstract thought—egocentrism, **animism,** and magical thinking—persist and prevent its completion. Psychosocial experiences at home and in one's community influence the development of the death concept (Hostler, 1978). Death is linked to forces in the outside world and becomes (Lonetto, 1980):

> Scary, frightening, disturbing, dangerous, unfeeling, unhearing, or silent. Death can be invisible as a ghost, or ugly like a monster, or it can be a skeleton. Death can be a person, a companion of the devil, a giver of illness, or even an angel.

Maria Nagy concluded from her research (1948) over a half century ago in Hungary that children aged 5 to 9 personified death, representing death as a live person or some variation such as an angel, a skeleton, or a circus clown. Death is viewed as a taker, something violent that comes and gets you like a burglar or a ghost or the grim reaper. These children felt that they could outmaneuver death so that its universality is supposedly not yet acceptable to them. Children older than age 9, however, saw death not only as final but inevitable and universal.

In analyzing death drawings of 431 children between 9 and 18 years of age in Sweden, Maare Tamm and Anna Granqvist (1995) found that personification of death is present to some extend through all ages, but is more pronounced in boys.

For young children, death is often viewed as the taker of life, like the Grim Reaper, as depicted in this drawing by a 10-year-old child.

There is a fear that death is contagious, something that can be caught like a cold (Schaefer and Lyons, 1986). George Dickinson remembers in the early 1950s, when he was a child of 11 or 12 years of age, that polio was rampant. When polio was diagnosed in two of the children living next door to him, he waited for the "thing" to visit his house next. Though polio usually did not kill the victim, nonetheless, this experience was frightening for a child, as "it" weaved throughout the neighborhood.

Swedish researchers Tamm and Granqvist (1995) concluded that children between the ages of 9 and 12 represented death predominantly in biological terms. They were concerned with the physical features of death and the dead and the causes and conditions of death. They were especially concerned with violent death and the rituals of the dead. These children dealt with death in a realistic and worldly manner, depicting what they knew and perceived daily, largely via television. Besides television, their interest in violent death could be related to killings of worms and insects by young children or by their playing cops and robbers or other gun-related games involving killing.

Explaining Death and Dying to Children

Robert Kavanaugh (1972) refers to children as "little people." He sees them as "compact cars instead of Cadillacs," traveling the same roads of life and going the same places as big cars. Although they are more vulnerable and fragile, they have all the parts and purposes of big people. They are ready and capable to talk about anything within the framework of their own experience. Little people can handle any situation adults can handle comfortably and should do anything big people should do, as long as they are physically able, notes Kavanaugh (1972).

The fact is, children want to know about death. In speaking to a group of fourth graders a few years ago on the topic of dying and death, George Dickinson's frustration was trying to decide whom to call on when a dozen or more hands at a time were up from these 40 children wanting to ask questions. Each had his or her own concerns, and they were sincere, legitimate interests. He remembers one little boy asking if it were true that one would die from getting embalming fluid on the skin. Dickinson tried to assure him that this would probably not be a cause of death, but it was a genuine concern for this child.

When a child experiences death directly, whether through a pet or a human, it is an emotional time and must be dealt with immediately (Seibert and Drolet, 1993). Children are "excellent observers" but "poor interpreters." Adults need to support, comfort, and help the child to express grief. An attempt by adults to protect children from the reality of death reinforces the perception that death is either not real or too frightening to examine or that the ending of life is not worth noting with reverence and respect, observe Dinah Seibert and Judy Drolet. Children also have vivid imaginations, which may lead to more fear and confusion if death is not discussed.

When we "big people" fail to give "little people" credit and try to shield them from information because they are "unable to take it," could it not be that we "big people" feel uncomfortable talking to children about dying and death, and thus avoid the topic altogether? Avoidance is not the best solution, as noted in the example of the needless worry and frustration of a 4-year-old girl who was not told about her puppy's traumatic death by a mowing machine until 2 or 3 weeks afterwards (Dickinson, 1992). She was allowed to search for him "frantically" every day. She even put out food and would worry at night that he was cold or hungry.

John Bowlby (1980) argues that young children can mourn in a way similar to that of healthy adults. For this to happen the child should participate fully in the awareness of what is taking place in the time during and after the death. Rather than withhold information, questions should be asked and answered. Kavanaugh (1972) suggests that we allow the children to talk freely and ask their own questions, without any adult speeches or philosophic nonanswers, let them ramble, talk crudely if they wish, change the subject, or present unanswerable questions without being squelched. The child should always be supported in a "comforting way" with an assurance of a continued relationship. Joy Ufema (1996), a nurse who works with dying patients, observes that children who are afraid of dying can be comforted with a talk on their religion's thoughts on death or on the near-death experiences of others. Their sensitivity on the issue of death comes with growing up, as they realize that death is an inevitable part of life.

Be Honest and Open

Being honest and straightforward with children in talking about dying and death is a good rule of thumb to follow. Respond to the child's questions immediately with concrete, rather than abstract, answers. If your answer is unsatisfactory, the child will probably ask a follow-up question in a few seconds or a few hours. If the child asks something you cannot answer, be honest and say you do not know. If you cannot explain something, find someone who can.

Answer directly, but do not be too detailed with your responses. When a small child asks where he or she came from, you do not go into minute detail about the sperm and the ovum forming a zygote. You state that you came from your mother. If the child is not satisfied, the next question may then be, "How did I get in my Mommy?" At this point, George Dickinson refers the question to the mother!

The same is true if the child asks, "Daddy, what makes the car run?" Dickinson responds that the car has an engine. If the follow-up question is, "What makes the engine run?" he refers the child to the local mechanic, since his technical skills are severely limited!

If parents are open to discussing death, opportunities will present themselves. As noted earlier, dead flies, mosquitoes, birds, and animals beside the road can lead into a general conversation about death, which will not be fraught with emotional turmoil. When the explanation is postponed until the death of someone or something deeply loved by the child, the acceptance of the reality of death will be more difficult.

The burial of a deceased pet by parents and children together can be a positive learning exercise in relating to death. The animal is cold, still, and not alive—it is dead. That is reality. When George Dickinson's children were young, they had guinea pigs. These animals have a short life span; thus, many guinea pig funerals were held in the backyard with the entire family participating. For other family members to be present at a time of grief helps an individual to better bear this burden.

Avoid Euphemisms

Try to avoid **euphemisms** when talking to a child about death. Use words like *dead, stopped working,* and *wore out*—simple words to establish the fact that the body is biologically dead (Schaefer and Lyons, 1986). One child was told that Grandfather's heart was bad and stopped working. This seemed to satisfy the child. Couched in a different way, however, another child was told, "Grandfather can breathe easier now," implying that Grandfather is still breathing. This child wanted to join his grandfather because he had asthma and would welcome the chance to "breathe easier." He was told, however, that he was too young to join Grandfather (Dickinson, 1992)!

Dead is a difficult word to say, but to use a euphemism such as *went to sleep* may make it difficult for the child to go to sleep at night. If Grandad "went to sleep" and is then buried in a box underground or destroyed through cremation, that situation is probably not what the small child's goals in life are! Thus, the objective is to stay awake and *not go to sleep*! (And parents wonder why the child is still awake at 10 p.m., after having been put to bed at 8!)

To use euphemisms such as *went away* or *departed this life* or *passed away* may cause the child to expect the deceased to return. After all, he or she is simply on a trip! Even to suggest that "he's gone to heaven and will live forever" is confusing to a child when this is said through heavy tears and with upset, emotional feelings.

A 4-year-old girl was told that her younger sister was "too sick to live with us so she went to visit Grandmother in heaven and would never come home again." She was frightened for months afterward because "whenever anyone did not feel well, I thought they would go away forever too" (Dickinson, 1992, p. 173). Rabbi Harold Kushner (1981) cautions, however, that to try to make a child feel better by stating how beautiful heaven is and how happy the deceased is to be with God may deprive the child of a chance to grieve. By doing such, we ask a child to deny and mistrust his or her own feelings, to be happy when sadness is desired. Kushner notes that the child's right to feel upset and angry should be recognized. One should also be cautious that the child not feel that the deceased *chose* to leave and "go to heaven." It should be made clear that the person did not wish to leave or "desert," as the child may feel.

WORDS OF WISDOM
Lament

Edna St. Vincent Millay

Listen children:
Your father is dead.
From his old coats
I'll make you little jackets;
I'll make you little trousers
From his old pants.
There'll be in his pockets
Things he used to put there,
Keys and pennies
Covered with tobacco;
Dan shall have the pennies

To save in his bank;
Anne shall have the keys
To make a pretty noise with.
Life must go on,
And the dead be forgotten;
Life must go on,
Though good men die;
Anne, eat your breakfast;
Dan, take your medicine;
Life must go on;
I forget just why.

Anthropologist Colin Turnbull (1983) describes death as being like it was before birth—a state of nothingness. Ask someone to describe what it was like before birth, and you will probably not receive much of an answer. To suggest that after death there is "no place" to describe—a nothing—would deny children a defined place to imagine. But then with the vast imagination of children, perhaps this might be more creative for them than a vague description of a "big house in the sky."

After you have made an effort to avoid euphemisms and gotten across the fact that the person is dead, the next step is to explain what is going to happen next. Tell the child about the body being moved from the hospital to the funeral home. Alert the child that funeral arrangements will be made (if that is the case), and a funeral will follow. Outline the format of a funeral itself, then talk about going to the cemetery (if that is the case) and burying the body in the ground.

Show Emotion

It is important that the child knows it is okay to show emotion when someone dies. Because it is a very sad time, the child should be told that everyone is upset and that many individuals may be crying. If the child feels like crying, he or she should be assured that crying is okay. Crying is normal. It should also be explained that because some people do not show emotion does not mean that they did not love the dead person (Schaefer and Lyons, 1986).

You probably cannot "overhug" a child during these times. It is important to always reassure the child that you care for him or her. Hugs and tears are very compatible expressions. Do not apologize for crying. Your crying in front of the child gives assurance that crying is okay.

Rabbi Earl Grollman (1967) defines crying as the sound of anguish at losing a part of oneself in the death of one whom he or she loves. Because children often cry when they get hurt, would it not follow that it would be natural to cry when "losing a part of oneself?" Grollman observes that tears and sorrowful words help the child feel relieved. Displaying emotions makes the dead person or pet seem more worthy. Tears are a natural tribute paid to the deceased. The child misses the one who is gone and wishes the person were still around.

Unfortunately, some parents do not allow their children to cry when death occurs (Dickinson, 1992). A 15-year-old recalls being spanked with a hairbrush for crying over the death of her puppy. A 10-year-old was "smacked" by her uncle to "make her stop crying at the funeral." Some children were told that it was not "grown-up" to cry. For many, their first childhood memories of death included recalling that it was the first time they had seen their fathers cry. No one commented that it was the first time they remembered seeing their mothers cry.

Several college students recalled (Dickinson, 1992) that they watched others' reactions to a death, then responded accordingly. Because children are watching adults, a parental role model of expressing one's self in front of children might be very beneficial to the their socialization. Children then would not feel a need to hide in the closet to cry or cry in their pillows at night, as some reported. One child noted the positive experience of his parents and sibling sitting down together to have a good cry at the death of his pet—this pointed out to him the warmth of the family in sharing this event. It was not a burden he carried alone; his family grieved with him. It is comforting to know that others care.

According to former professional football player Rosie Greer, it is okay to cry. He sings a song in which he notes it is all right for women, men, girls, *and* boys to cry. For little boys to cry is *not* "sissy." If Rosie Greer says it is all right, it must be okay! Anthropologist Ashley Montagu (1968) agrees with Rosie Greer when he states that women have a superior use of their emotions because of their ability to express themselves through crying.

Some men feel that "real men" do not cry because it is not macho to cry. Such men will wear dark shades on the cloudiest of days, if they fear they may cry at a funeral. Several years ago, George Dickinson's teenage daughter attended a funeral of a friend with him. After leaving the cemetery on a *very cloudy day*, she said, "Dad, you were right. Nearly all the men at the graveside services were wearing dark sunglasses to hide the fact that they may be crying." Males hold back the tears and develop psychosomatic disorders such as peptic ulcers. "Is this a superior use of emotions?" Montagu would ask.

What favors are we doing our children by teaching them that it is not okay to cry? It is okay to laugh, why should it not also be okay to express feelings through crying?

DEATH ACROSS CULTURES
Death Without Weeping

Nancy Scheper-Hughes

In Northeastern Brazil extremely high infant and child mortality rates exist in shantytowns. The average woman experiences 9.5 pregnancies, 3.5 child deaths, and 1.5 stillbirths. Infant and child deaths become routine. Part of learning how to mother is learning when to let go of a child who shows that it "wants" to die or that it has no "knack" or no "taste" for life. The church contributes to the routinization of, and indifference toward, these deaths. When anthropologist Scheper-Hughes asked the local Catholic priest about the lack of church ceremony surrounding infant and childhood death today, he replied: ". . . The new church is a church of hope and joy. We no longer celebrate the death of child angels. We try to tell mothers that Jesus doesn't want all the dead babies they send him." Similarly, the new church has changed its baptismal customs, now often refusing to baptize dying babies brought to the back door of a church or rectory. The mothers are scolded by the church attendants and told to go home and take care of their sick babies. Thus, current Catholicism is denying traditional comfort to these mothers.

From "Death without Weeping" (pp.106–110) by Nancy Scheper-Hughes, 2000. In E. Angeloni (Ed.), Anthropology 00/01 Annual Editions, Guilford, CT: Dushkin/McGraw-Hill. Reprinted by permission.

ADOLESCENCE

Adolescence is the "training period" between childhood and adulthood in the life experience of humans. Adolescence—from age 12 to age 19—is composed of two significant periods (Gordon, 1986). The period from 12 to 15 (with 11 as the threshold) encompasses the acquiring of formal logical thought, the onset of biological sexuality, the growth of physical structure, and development of myriad psychosocial tasks. The second period, from 16 to 19, with transition around 15, is characterized by the completion of physical maturation, increasing intimacy with the opposite sex, continued acquisition of adult social skills, clarification of ethics and values, and ability to make long-term commitments to persons and goals. Adolescence is often a time fraught with anxiety, rebellion, and indecision.

Adolescents generally do not think in terms of their distant future. Their struggle with present life experiences, especially the concern with their own identity and anxiety about successes during the immediate future, evidently occupies so much mental energy that thinking about what life will be like at age 45 or 70 is nearly

Adolescence is a time of transition from childhood to adulthood. Adolescents do not often think about death, since the aging process and death seem so far away.

impossible for them. As psychologist Hannelore Wass (1995) notes, adolescents often say, "Death is a long way off, why worry now?" Standing on the threshold to adulthood, adolescents are more focused on the years of living that lie ahead.

Sherry Schachter (1991) states that a high percentage of adolescents must adjust to the death of a peer and that is very difficult. The death of a peer emphasizes the adolescent's own vulnerability and mortality. The adolescent experiences the death of a peer as a grave injustice. How could he or she die before reaching the fullest potential and opportunity to experience life? Because society does not acknowledge the bereaved friend in the same manner that it acknowledges the bereaved family, Schachter observes. This experience is compounded by isolation, as each friend grieves alone and they become the forgotten mourners. Because adolescents' identities shift away from parents and family as they tend to see themselves as members of the peer group, the loss of a peer can upset identity formation that may already be unstable (O'Brien, Goodenow, & Espin, 1991).

Schools today are aware of the importance of having an opportunity to express one's self when a classmate dies and are doing a commendable job of dealing with the death of a student. They bring in counselors and have them available for anyone desiring to talk. They provide numerous opportunities for classmates to mourn and deal with the death. George Dickinson, though not a counselor, has been invited into high schools following the death of a student simply to serve as a catalyst in the classroom to encourage the students to "talk about" the deceased and express their feelings.

According to Jean Piaget (1958) and his followers, by 11 or 12 years of age people are able to move from the use of language and ideation that is concretely oriented to an abstract level of thought. No longer wedded to the concrete, the adolescent can now use conditional statements such as "if-then." Ideas can be taken apart and put back together in new ways. Adolescents often describe death in abstract concepts such as darkness, eternal light, transition, or nothingness, notes Wass (1995). They formulate their own theologies about life after death, which include beliefs in reincarnation, spiritual survival on earth, and spiritual existence at another level in a state of peace and beauty, in addition to beliefs about heaven and hell or total annihilation at death.

Our society's discomfort with the process of aging, illness, and death does not contribute positively toward the adolescent's image of a future (Gordon, 1986). Because being young is envied and growing old is feared in America, to "grow old along with me" may not mean that "the best is yet to be" in the minds of adolescents. Consequently, a callousness toward physical deterioration and death often develops in adolescence. When the death of a significant other occurs, previous experience has probably not prepared the adolescent for the feelings of rage, loneliness, guilt, and disbelief that accompany a personal loss. If peers have not experienced a similar loss, they may have difficulty being supportive of others who have lost a loved one.

Identity Crisis and Death Anxiety

Being gripped by questions such as "Who am I?" and "How do I fit into the scheme of things?" the adolescent struggles with good and evil, love and hate, belonging and loneliness, and thoughts of life and death. These can be disturbing issues to the adolescent who is not sure if he or she is a child or an adult because identity signals come from all directions, adding to the confusion of one's self-concept.

The adolescent has a vivid awareness of the dialectic between being and not being, according to psychologist Robert Kastenbaum (1986). Daily experience provides occasions for assertions of a new self, of "dyings" of the old self, and of apparent abortions of the new self, notes Kastenbaum. Thus, it is precisely when the adolescent is most endangered by the polarities of being and not being that the concept of self emerges with singular force. The perception of a self under reconstruction is granted most keenly as one experiences the possibility of failure, loss, catastrophe, and death.

One research study (Koocher et al., 1976) demonstrated that high school students experience much more anxiety, depression, and death fear than do either junior high students or adults. They propose that the "identity crisis" of the adolescent years could be largely responsible for this. Psychiatrist Robert Lifton (1976) also notes that adolescence is the time when a sense of great potential for disintegration, separation, and instability occupies the mind and brings about greater death anxiety.

LISTENING TO THE VOICES
Did I Really Love My Brother?

Elizabeth Richter

"Everybody told me to be strong for my parents and to be quiet because I might make them more upset. Everyone told me to put my feelings aside, like my feelings weren't as important as my parents'. I never cried until about a year later, when it hit me that my sister Sandy wasn't around." That's how 18-year-old Lisa describes her own grieving for an older sister who was murdered.

"When I was told my brother had died, I just left the room. I couldn't even cry. My father cried, although he had never cried in his life. Because I couldn't cry, I began to wonder if there was something wrong with me. Did I really love my brother?" That is the way 18-year-old Sharon reacted to her little brother's dying of a brain tumor.

Adapted from E. Richter, 1986, U.S. News & World Report, August 4.

Like masturbation, adolescent **death anxieties and fears** are universally experienced and discussed with equally uninformed peers, notes Audrey Gordon (1986). One of the tasks of adolescence is to begin to grapple with the meaning of life and death and to emerge with a philosophical stance that promotes optimism for the future. This is not an easy assignment, especially because one's attitude toward the future may involve more pessimism than optimism.

The Experience of Death in Violent Neighborhoods

Concepts of death are powerfully influenced by experience with death or threats of death. Mental health, anxiety management, and meaningfulness in life are the most powerful factors in developing concepts of death. Broad cultural-religious influences also enter significantly into each of these factors.

As discussed earlier in this chapter, for children and adolescents living in certain inner cities in the United States, violent deaths are fairly routine in their neighborhoods. Especially in neighborhoods with heavy drug traffic, shootings and knifings may be all too common. For example, in a study (APACVY, 1993) of eighth-graders in Chicago, 75 percent reported they had seen someone robbed, stabbed, shot, or killed. Many inner-city children and adolescents show symptoms of posttraumatic stress. After they witness a violent death, such an image will probably be imprinted on their brains. Indeed, if this happened once, it could happen again, and perhaps the next time, the child or adolescent may be the victim—being in the wrong place at the wrong time. Images on the evening news and pictures in the newspapers of small children or adolescents being caught in cross-fire or being shot in their beds at night as retaliation by a juvenile gang or an older sibling are all too vivid for television news watchers or newspaper readers. Though they may not have been the object of the murder, nonetheless, if they are "in the vicinity," they may become a victim. Having to bury children and adolescents is typically not normal in the life cycle, but not uncommon in a subculture of violence.

Socialization in a milieu of violence may make it "okay" for violence to be considered an appropriate response in many circumstances. Such violence may not be strongly punished if it occurs under provocation and if the people toward whom it is directed are stigmatized or viewed as disreputable. Thus, violent deaths may become "normal" in a subculture of violence. At any given time in history, there are neighborhoods (even entire countries) warring with other groups, often for "religious reasons." In Northern Ireland, for example, for many years now there has been an off-and-on state of warfare between Protestants and Catholics. To live in circumstances in which terrorist bombings occasionally occur or sniper fire kills individuals must make for "long nights and days" for children and adolescents growing up in such an environment. They are all victims because they live in constant fear for their lives. Carrying weapons may be a way many young people attempt to deal with this fear (Wass, 1995). Certainly, their perceptions of death would probably be different from those of individuals brought up in a more peaceful atmosphere.

Media Influences

Adolescents often see dying and death depicted in the media as violent, "cool," distant, or unnaturally beautiful. By the time adolescence is reached, one will have witnessed thousands of violent deaths in movies and on television (Wells, 2000). The "lesson" often is that aggression is used by the "good guys" to gain rewards that are unavailable to others. Death abounds in horror and action films, with body counts sometimes exceeding a death per minute of running time. The *Scream* and *Last Summer* series of movies, popular among adolescents in the late 1990s and into the 21st century, are examples of movies with violent deaths. With videotapes, adolescents' opportunities to view violence and uncensored death increase significantly in the privacy of their own homes.

Adolescents are fascinated by these movies about violent death. One perhaps builds up a sense of invulnerability by seeing other people dead or dying, notes psychologist Fred Hinker (1985). Hinker says that teenagers watching these movies are like teenagers riding motorcycles at extremely high speeds. They are saying, "It can't happen to me. I'm proving it by flirting with death."

Yet, other films in which death plays a major role are less violent. A favorite movie industry theme now is a person dying with some part of his or her life unfinished, thus posing a problem (Wells, 2000). *Flatliners*, a film about medical students experimenting with near-death experiences, is really concerned with their own deeply repressed and unresolved problems as they "die." Lingering and romantic deaths reminiscent of 19th century literature, though now caused by cancer rather than tuberculosis, have been popular in films such as *Dying Young* and *Beaches*. A block-buster 1999 movie entitled *The Sixth Sense* is about a little boy who "sees" and "relates to" ghosts. Our society seems to have a renewed fascination with ghosts.

Destructive themes in rock music are not unlike those of many movies viewed by adolescents. According to Hannelore Wass, D. Miller, and C. A. Redditt (1991), however, death themes in rock music might be therapeutic. The themes should be taken metaphorically rather than literally. Rock lyrics may sometimes provide the means for dealing with issues of death and for managing the anxieties created by these themes. Wilson Frank and the Cavaliers' "Last Kiss," for example, is about a teenager's

Movies with death and violence seem to fascinate adolescents. Movies may represent a fantasy world, not the real world, and one can get "lost" in the violence that "will not happen to me."

girlfriend dying in an auto accident in which he was driving: "Oh where oh where can my baby be? The Lord took her away from me. She's gone to heaven so I've got to be good, so I can see my baby when I leave this world." He talks about giving her a last kiss, as he held her close. He closes with, "I lost my love, my life that night."

Pornographic rock, where fantasy has taken over and death is dehumanized and distorted is evident in some of today's lyrics (Nordhcim, 1993). Groups that produce hateful and destructive lyrics have become popular. For example, Guns N' Roses sold 15 million copies of their album *Appetite for Destruction.* Everlast's "What It's Like" is about violence and death via drugs and guns. Mentioned in the lyrics is a "kid named Max" who pulled out his gun and wound up dead, leaving his wife and kids caught in the midst of pain. Everlast's "What It's Like" is about loss and emptiness being the result of death. "Ends" (both of these songs are on their *Whitey Ford Sings the Blues* album) basically says that the end justifies the means—to take a life to achieve one's goal is okay. It is the end (result) that counts, no matter how one gets there ("Sometimes kids get murdered for the ends").

In Korn's "Dead Bodies Everywhere" a son talks about his parent who seems to view him as a nothing, a "cipher." "You make me feel like no one," he says, "Dead bodies everywhere." The teenager seems to be saying that life is depressing and oppressive. Verve Pipe's "The Freshmen" is about death via an accident, "We fell through the ice." Marcy Playground's "One More Suicide" depicts Christopher O'Malley's jumping off the bridge into the river. "One more suicide" the lyrics say over and over. Other song titles popular today with teenagers include: Onyx's "Betta Off Dead," Paula Cole's "Road to Dead," Scarface's "Hand of the Dead Body," Stone Temple Pilots' "Dead & Bloated," Street Military's "Dead in a Year" (from their *Don't Give a Damn* album), House of Pain's "Back from the Dead" ("I'm the resurrected, skip the autopsy 'cause I never O.D.'d, I'm back from the dead"), and Insane Clown Posse's "Dead Body Man."

B. Plopper and E. Ness (1993) identified 90 death-related songs that appeared in the Top 40 during the previous 37 years. They concluded that death songs constitute a disproportionately popular subset of Top 40 music, that males dominate as the person who dies in the songs, and that grief responses in the songs are confused and restricted.

Learning Adult Rituals

Society is created and held together by **rituals**—morning wake-up rituals, going to school/work rituals, eating rituals, religious rituals, political rituals, and others (Gordon, 1986). Ritualization provides sanctioned boundaries within which the self can be safely expressed. Dying, death, and bereavement all are ritualized by society as a way of containing and giving meaning to feelings of loss. The granting of adult responsibility and privilege to adolescents varies from culture to culture and group to group. Because older people all too often do not teach children the adult rituals for handling dying, death, and bereavement, it becomes one of the personal tasks of adolescence to acquire this knowledge. Being unfamiliar with the proper rituals to follow at the time of a death, an adolescent may feel especially anxious. What do I do? What do I say? George Dickinson recalls his own son, as a high school student, getting ready to go to the funeral home for visitation after the death of a classmate. He asked what he should say to the parents. Having not previously experienced the death of a peer, he was not sure what the "correct" behavior was. Though ritualization helps an individual successfully complete a task, somewhere along the way, we have to be socialized as to the sanctioned rituals in our society.

Just as for younger children, adolescents learn the "proper" way to respond in dying and death situations by watching others. For example, Dickinson has had numerous students tell him that they did not know how to react to a death, but if others were crying, then they would usually mimic such behavior. On the other hand, if others were crying and the individual could not, she or he often felt guilty. "Why am I not crying, since that seems to be the thing to do?" they may ask themselves.

According to Audrey Gordon (1986), the African-American community is far more likely to keep deceased family members at home "to lie in state" for the wake than the white community, perhaps a reflection of socioeconomic variables as well as a strong tradition of family and religious networking. Because of this practice, black adolescents are typically more socialized at an earlier age to the rituals surrounding death and dying than are most white adolescents. Young black children are routinely taken to funerals and encouraged to interact with family members who are dying in the home. A strong religious background in most African-American families helps to develop a belief about survival after death that makes death less threatening than it is for the more secularized white middle class, notes Gordon. Although the black culture certainly has its fears and superstitions about death and dying, these are not taboo subjects.

Communicating About Death

Whenever an individual is going through an unstable or stressful period—especially when the stress has to do with basic feelings about self-worth, identity, and capability—the thought of death is particularly difficult to manage. Crucial to positive outcomes is the manner in which parents, friends, peers, teachers, and others enable the adolescent to deal with the anxiety caused by the thought of death and the actual death of friends, acquaintances, and loved ones. Particularly in the adolescent years, parent-child relationships are often fraught with conflict and alienation, and adolescents often prefer to discuss important ideas and emotional issues with their peers, not their parents (Wass, 1995).

Though parents may feel uncomfortable talking about death with their children, it is important that they try, notes psychologist Hannelore Wass (1995). Open and honest communication is far more helpful than silence or evasion and is best achieved in an atmosphere of mutual trust and respect. Audrey Gordon (1986) says that adolescents need not stumble around in a darkness that adults have helped to

---

PRACTICAL MATTERS
Improving Family Communication About Death

1. Be aware of the adolescent's concerns about death and be open to discuss anything that he or she feels like exploring.

2. Listen actively and perceptively, keeping your attention on the adolescent and the apparent feelings underlying his or her words.

3. Accept the adolescent's feelings as real, important, and "normal."

4. Use supportive responses that reflect your acceptance and understanding of what the adolescent is trying to say.

5. Project a belief in the adolescent's worth by indicating that you are not attempt-

ing to solve his or her problems, but are instead trying to help the adolescent find his or her own solutions.

6. Be willing to take time to enjoy each other's company and to provide frequent opportunities for talking together.

Adapted from "In Talking about Death: Adolescents, Parents, and Peers." (pp. 197–198) by J. N. Mc Neil, 1986. In C. A. Corr & J. N. McNeil (Eds.), Adolescence and Death, *New York: Springer Publishing Company.*

---

create. With proper preparation and support, younger adolescents can be helped to become aware of death in manageable ways, and older adolescents can be helped to impart a meaning to death and life that transcends everyday events and infuses the future with hope.

ADULTHOOD

We have already observed that one of the main problems with following a developmental scheme in the explanation of how people think, feel, and integrate life's experiences is that there are so many possible combinations of factors in any given life. This is particularly true in studying the adult stage of life. The longer one lives, the more complex the picture becomes as the probability of additional factors influencing a particular person increases. It is important to keep in mind that research findings describing a particular population, age group, or cross section of people should not be seen as more than a description of that particular sample of people. Individuals may vary widely within a single sample.

Also, different studies may portray different profiles of a single population. This reflects the theoretical approach of the researcher and limitations of present knowledge, methodologies, and conclusions.

Another limitation to studying adults' concepts of death includes the lack of a satisfactory definition of "maturity" in relation to the concepts of death. Yet

> ### WORDS OF WISDOM
> #### A Season for Everything
>
> For everything there is a season, and a time for every matter under the heaven,
> a time to be born, and a time to die;
> a time to plant, and a time to pluck up what I planted;
> a time to kill, and a time to heal;
> a time to break down, and a time to build up;
> a time to weep, and a time to laugh;
> a time to mourn, and a time to dance;
> a time to cast away stones, and a time to gather stones together;
> a time to embrace, and a time to refrain from embracing;
> a time to seek, and a time to lose;
> a time to keep, and a time to cast away;
> a time to rend, and a time to sew;
> a time to keep silence, and a time to speak;
> a time to love, and a time to hate;
> a time for war, and a time for peace.
>
> *"Ecclesiastes 3:1," The Holy Bible (Revised Standard Version), 1962, New York: Oxford University Press.*

another limitation of the developmental perspective is that researchers have not focused on adult conceptualizations of death. There has been a lack of a satisfactory definition of "maturity" in death concepts. While Erik Erikson and others have studied development over the life span, few have studied the relationship between the stages depicted in these studies and adult ways of thinking about death. The major exceptions to this trend are found in studies of mentally and terminally ill populations.

Finally, a limitation is faced in determining what "adult" actually means. Biologically, though we may stop growing taller at a certain age, great changes continue in the human body. Psychologically, it is even more difficult to arrive at any definition of adulthood that will be generalizable to a significant percentage of the population. One must conclude that the word *developmental* takes on a more individualistic character. Thus, one should expect that age will be less influential in explaining death conceptualizations for adult populations than for other age groups.

Young Adulthood

From the discussion of the development of the adolescent's intellectual understanding of death, one could expect the young adult, described generously by B. Strauss and R. Howe (1991) as encompassing the ages of 22 to 43, to have a good grasp of the universality, inevitability, and finality of death. The young adult should know, at least intellectually, that death is an entirely possible event for anyone at any moment, yet death probably seems far away to most young adults, especially during the early years of young adulthood. Certainly, the untimely death in 1997 of Princess Diana at the age of 36 forced many young adults to think about death, if only for

---

WORDS OF WISDOM
Death Can Come at Any Moment

In his book entitled *Man's Concern With Death*, historian Arnold Toynbee (1968) wrote:

> From the moment of birth there is the constant possibility that a human being may die at any moment, and inevitably this possibility is going to become an accomplished fact

sooner or later. Ideally, every human being ought to live each passing moment of his life as if the next moment were going to be his last. He ought to be able to live in the constant expectation of immediate death and to live like this, not morbidly, but serenely. Perhaps this may be too much to ask of any being.

---

a few days. Unless they are forced to do so, we should not expect that all young adults would normally think of death constantly nor take it into consideration with each decision of importance.

Members of certain monastic Christian orders have practiced greeting each other daily with the words, "Remember death!" The fact that one tends to recoil from such a practice, however, is evidence that one would rather not take this advice.

The young adult especially would appear to reject the advice to remember death. At this stage of life, one is just entering the arena of a somewhat independent life where capabilities and skills can be tested and pride can be taken in positive results. Hopes, aspirations, challenges, and preparation for success in life are the focus at this age. This means that dealing with dying or death would mean to face rage, disappointment, frustration, and despair (Pattison, 1977).

The four types of death most salient for young adults are abortion, the death of a child, death from AIDS, and death by violence, according to Judith Stillion (1995). The death of a grandparent is also a common event for young adults, but the deaths of elderly persons do not violate young adults' sense of a just world because such deaths are generally viewed as being in the natural order of things. Abortion, however, is a special, very personal, emotional issue. The death of a child after birth seems unjust. AIDS is threatening because it "strikes at the very heart of their ability to work and procreate," primary tasks for this age group (Stillion, 1995, p. 310).

Males in the age category of late adolescence and young adulthood are dying by violence, a situation that seems so unnecessary and wasteful. Violent crimes in this age group have increased dramatically, especially beginning in the 1980s. The rate of death by violence has increased most steeply in the 15- to 24-year-old age group with homicide being the second highest cause of death after accidents in this age group (Ewing, 1990). Homicide has been the primary cause of death for African-American males for more than a decade (Hammond & Yung, 1993). A black male is 11 times more likely than a white male to be killed (Wass, 1995).

Young adults struck by serious disability or life-threatening disease have demonstrated that an almost universal sense of injustice and resultant anger exist when a young person is forced to "remember death." Along with all of the international political issues involved, the protests against the Vietnam War in the 1960s and 1970s probably contained a good deal of repugnance for the idea of risking one's personal future at this stage of life. A walk through the wards of any major veterans' hospital or pediatric oncology floor would soon convince the most stubborn observer that significant physical and mental losses, and death itself, seem most abhorrent when occurring in adolescents or young adults.

Daniel Levinson and colleagues (1978) labeled the young adult stage the "novice phase" because there is a strong sense of the need to learn, practice, and train oneself in the art of reaching one's fullest potential as a person and contributor to self-fulfillment, family, and society. Achieving something worthwhile would be devastatingly contradicted by any thought of serious limitation, sickness, or death.

M. F. Lowenthal, M. Thurnher, and D. Chiriboga (1975) evaluated 216 people grouped in four stages of the life cycle: high school seniors, young newlyweds, middle-aged parents, and an older group about ready to retire. Each person was asked specific questions concerning concepts and thoughts about death. Older people thought of death mainly in connection with specific and personal circumstances, such as the death of a friend, whereas younger people were likely to have death thoughts in response to general events such as accidents, earthquakes, or war.

Again, it is evident that though age may have some influence over the way that one thinks, individual circumstances and external forces have a more powerful influence upon one's thoughts about death. The older that one grows, the more apparent it becomes that one needs to reflect on life as much as one needs to engage in it. Such a practice can provide the individual with greater life satisfaction. As one moves into middle adulthood, however, it can have both positive and negative results.

Middle-Aged Adulthood

Although there appears to be no agreement among social scientists as to the exact beginning of middle age, most seem to suggest that this period of life starts between ages 40 and 45. The U.S. Census Bureau defines middle age as being ages 45 to 64, whereas Vera in the Broadway play *Mame* describes middle age as being "somewhere between 40 and death." Middle-aged individuals are sometimes referred to as the "sandwich generation" because they may be supporting and/or caring for their own children as well as for their parents. They are "caught between" these two groups.

Carl Jung (1923) suggested that the primary goal of the second half of life is to confront death. Jung (1971) believed that middle adulthood is the time of greatest growth for most people—a time of integration of undeveloped dimensions of personality. He described the major task of the middle years to be reassessing and giving up the fantasy of immortality and omnipotence that carries us through earlier years when our own death is incomprehensible. Through the increasing awareness of

---

WORDS OF WISDOM

The idea is to die young as late as possible.

—Ashley Montagu

---

body changes, loss of significant others, and children moving toward independence, we shift from a future-oriented perspective to an inescapable confrontation—a conscious awareness of death and mortality (Douglas, 1991). Erik Erikson's theory of ego development (1963) in these years suggests a concern for the next generation and an acknowledgment of mortality.

For the person who has lived 40 or 50 years, life brings with it the advantages of experience. Promotion to supervisor, foreman, or analogous status rankings in one's work or social milieu demonstrates that one gains greater political and social power during middle age. Not everyone is promoted, however, and even those who are, as well as their less fortunate colleagues, become gradually aware that physical vitality has now begun to wane.

Robert Fulton and Greg Owen (1988) describe this group born after World War II (**Baby Boomers**) as primarily experiencing death at a distance. Unlike earlier generations, they were probably born in hospitals and are no longer likely to die from infectious diseases. Unlike their parents and grandparents, this Baby Boom generation experienced the maximum benefits of an urbanized and technologically advanced society. Fulton and Owen further note that as the commercial meat processing industry removed the slaughtering of animals from the home, so did modern health-care institutions shield this group from general exposure to illness and death. Death became invisible and abstract. This generation was the first in which a person could reach adulthood with only a 5 percent chance that an immediate family member would die.

With an estimated 77 million Baby Boomers reaching the age of 50 between 1996 and 2015, a funeral home in Toronto, Ontario, recently held an open house as a marketing ploy aimed at the middle aged (O'Hara, 1996). It is likely that death will become a major topic, perhaps an obsession, with many people of this generation. Hali Weiss (1995) predicts that the Baby Boomers will demand more meaningful deaths and burials in the 2000s, just as they took control of the childbirth experience in the 1970s. With cremation rates rising, a gradual rejection of traditional burial practices is on the horizon, notes Weiss.

Thanks to improved diets, healthier lifestyles, and unprecedented advances in medical care, the Baby Boomers are living longer than their parents did. But even nonsmokers who fasten their seat belts, exercise regularly, and eat plenty of vegetables must die eventually, notes Hank Cox (1996a). The images of "being struck down" in one's prime and of the "untimely accident or death" are pervasive for this age group as stated by Ruth Purtilo (1990). Denial, hostility, and depression are factors that often accompany an illness of the middle aged, notes Purtilo.

Panic and Denial

The "panic" begins when one realizes that the idealized self, fulfilled and successful, may never be attained. Gail Sheehy (1976) observes that earlier pangs of the panic begin for most men between the ages of 35 and 45 (the "deadline decade") when they confront the certainly of their own death for the first time. Those whose job or self-concept depends upon youthful physical vigor especially suffer from this recognition of possible unfulfilled dreams. No amount of jogging reverses the effects of time. No patent medicine can undo the damage from wear and tear. The only hope is to make the best of what energies and experience remain and to focus on what one does best.

According to developmental theorists, death is a salient issue for midlife adults. Indeed, one of the major midlife tasks to be completed is to accept death as a reality (Waskel, 1995). Failing health, deaths of parents, loss of close friends, and changes in physical appearance contribute to a heightened awareness of death. Midlifers watch their peers die quick deaths from cardiovascular disease and prolonged deaths from cancer; thus, fitness becomes a necessity to help them cope with the increasingly familiar threat of death in their own lives (Stillion, 1995). Individuals in the middle years may begin an exercise program in the hope of contributing to their longevity. For example, ethicist Ruth Purtilo (1990, p. 251) overheard a middle-aged man, who for years had enjoyed running just for the sport of it, tell his friend, "Yeah, my running will probably guarantee that I live five years longer, but I will have spent that five years running!" With heredity being a given, an individual can at least try to increase longevity by behavior and eating habits. Exercise, though it takes time, may contribute to the prolongation of life and may indeed help to make the "last of life the best to be."

Another developmental influence on death concerns among adults is the longevity of their parents. As long as one's parents are living, there is a buffer between the person and death—one's parents are "supposed" to die first. After one's parents die, however, this buffer is gone, and one's own generation becomes the genealogical line of descent to die. What effect does that have? Does it affect the level of death anxiety?

Socioeconomic issues also affect the degree of anxiety experienced by adults. Those whose strengths lie in intellectual and social skills will be less threatened because there is a cultural-social clock that one is able to impose over the biological clock (Neugarten, 1968). This seems to make more actual difference in the way that one lives and how much satisfaction one is able to derive from life. Jack Riley (1968) found a stronger correlation between education and more positive views concerning death than between age and positive death views. Stated negatively, people with less education appear more often to have more negative views concerning death. One must remember, however, that a correlation does not prove a causal connection.

In analyzing the relationships between death anxiety, age, developmental concerns, and socioeconomic status in a sample of 74 middle-aged women, V. Richardson and R. Sands (1987) found that developmental factors were the salient issues with death concern, death as an interpersonal loss, and death as a dimension of time. Age was the sole predictor of death anticipation, and death denial and income were significant with death as physical. No variable predicted death as depressing.

LISTENING TO THE VOICES
In the Autumn of Our Lives

A recent personal correspondence (Dickinson, 1996) in the late autumn from a relative in her 50s reflects on the changing season of the year and, accordingly, on changes in the life cycle with the passing of time:

When the leaves began falling we covered the pond with latticework and screens on top. . . . The fish seem lively still but soon the cold weather will plunge them into dormancy. I feel fairly dormant myself. I have such trouble getting up early in the morning. I think hibernation is slipping up on me. As I look out the window I realize it has only been a short while since the bright leaves were infant buds and I marveled at spring. Probably the mirror would tell me the same thing.

These results revealed the multidimensionality of death attitudes and the significance of considering both developmental and socioeconomic influences in predicting death attitudes.

Reflection and Acceptance

Elliott Jacques (1965) observed that awareness of death changes people's lives in middle age, causing them to become more philosophical about their lives and to reevaluate values and priorities. In the process of confronting their own mortality, people deepen their capacities for love and enjoyment and ultimately acquire more meaning in their lives. Carl Jung (1933) also believed that adults become more introspective and concerned with meaning during midlife and concluded that they experience an inner transformation after recognizing previously suppressed aspects of their personality. He maintained that those who successfully come to grips with life become more individuated.

Daniel Levinson and colleagues, in *The Seasons of a Man's Life* (1978, p. 215), learned from data gathered from men in their study that for middle-aged men death is often contemplated:

> At mid-life, the growing recognition of mortality collides with the powerful wish for immortality and the many illusions that help to maintain it. A man's fear that he is not immortal is expressed in his preoccupation with bodily decline and his fantasies of imminent death. At the most elemental level, he feels that he is fighting for survival. He is terrified at the thought of being dead, of no longer existing as this particular person.

Age differences in types of death concerns were observed by M. Stricherz and L. Cunnington (1982). Middle-aged adults emphasized the pain of dying and dying before being ready. The young adults focused on losing persons they cared for, and elderly persons worried about being helpless, taking a long time to die, and having to depend on others. This study found that all age groups consider death, but they do so in different ways and contexts.

Communing with nature is another way to enjoy the world around us.

Middle-aged adults are aware of getting closer to the "day of reckoning" and will need to evaluate values, meanings, and sense of self-worth in the face of finitude. The individual becomes increasingly conscious of thoughts of painful death, of the dying process, and of ceasing to be as a person. An awareness develops of the meaning of absence to significant others—spouse, children, other relatives, and friends.

Thus, death at this point in life often carries with it some of the same sense of injustice and anger that the younger adult experiences with thoughts of death. Now, however, these thoughts are tempered with the recognition that many more forces exist that could bring death "home" to the individual.

Personal Growth

Although the mature adult needs to give up fantasies of immortality, omnipotence, and grandiosity, there still needs to be a sense of accomplishment—a fulfilling of oneself and one's plans for family and personal enterprises. Consciousness of time and death, therefore, makes little difference for the middle-aged adult. One strives to put all of one's skills and experiences to the best possible use.

After one is into the decade of the 50s, a turning point is reached at which one's finitude becomes even more evident. One develops an increasing consciousness that one no longer measures time from birth as much as one measures time until death or until the end of one's most productive years. Focusing upon what one wants most to do before retirement or death causes one to avoid those things considered extraneous and/or uninteresting. All of this is not to say that increasing age means cessation of personal growth. On the contrary, as long as there is life, growth can occur for the individual who has the will to live and the will to give of self to others.

However, all is not lost in middle age. Bernice Neugarten (1968) points to evidence that people in their 40s see the world in a more positive way than do those in their 60s. Possibilities for the 60-year-old are more likely to be faced in passive modes of coping than in active modes. She suggests that women tend to cope increasingly in affective and expressive terms, whereas men at this stage increasingly employ abstract and cognitive modes of coping.

Older Adulthood

Death is expected for the elderly person, notes Robert Kastenbaum (1992b). Indeed, in contemporary American society, death is primarily something that happens in old age. Almost two thirds of the two million persons who will die in the United States this year will be 65 years of age or older (Morgan, 1995). H. N. Sweeting and M. L. M. Gilhooly (1992) state that death is seen as appropriate for very elderly persons who have lived their allotted span of life. They also note that elderly individuals are more likely than other members of society to be living alone and are unlikely to be working.

Although some have observed that "growing old is hell," others look forward to the autumn of their lives. As Goethe stated, "To grow old is in itself to enter upon a new venture." Aging professional athlete Satchel Paige noted many years ago that growing old is really mind over matter—as long as you do not mind, it does not matter. In the movie *Citizen Kane*, Mr. Bernstein said to Mr. Thompson, a reporter, "Old age. It's the only disease, Mr. Thompson, that you don't look forward to being cured of." Thus, individuals tend to have different reactions to growing old.

Another division of old age has appeared in recent years—a division between the young-old and the old-old (Neugarten, 1974). The young-old are in the age group of 55 to 75; the old-old are 75 and over. Neugarten pictures the young-old as possessing relatively good health, education, purchasing power, and free time and being politically involved.

The young-old are distinguished from the middle aged primarily by the fact of retirement. Whereas 65 has been the marker of old age since the beginning of the Social Security system, age 55 is becoming a meaningful lower age limit for the young-old because of the lowering age of retirement. Obviously, employment and health status have a tremendous effect upon placement in either category.

Life can be compared with a train traveling through a tunnel. There is a point at which the train is leaving the tunnel rather than entering it. As older relatives and friends die, one cannot help but become aware that he or she is not immune to death. As noted earlier, when no older generation exists and the members of one's own generation die with increasing frequency, the scarcity of time becomes a reality. Therefore, for the person older than age 60, the end of the tunnel is in sight.

The letter from Lucie Reid (Listening to The Voices box), sent to George Dickinson and his wife, suggests that the "end of the tunnel is in sight" for her. She seems accepting of the fact and ready to die.

LISTENING TO THE VOICES
Accepting Death

My Dear Children,

This seems a strange gift for this time of the year. This lovely ball was given me by the friend who made it, and I want it to hang for many years on your happy tree. If I live until another Christmas, I would be 97 which is too long to stay in this devastated world which my generation has made. I am ready to depart any time. God has been wonderfully good to me. I have had all any one could ask for—love and care and now every comfort in this shadowing time. I say with Cardinal Newman, "So long thy hand hath led me, sure it will lead me on."

I know that you are leading lives full of meaning, and my blessings go with this bright ball. I believe a circle has no beginning and no end.
Sincerely your friend,

Lucie Reid

After living next door to Lucie Reid for two years in western Pennsylvania, George Dickinson's family moved. For several years after that, they received a beautiful hand-decorated Christmas ball from Mrs. Reid. The ball usually arrived in mid-December. One year, however, the ball arrived in mid-July. This letter was enclosed in the package. Lucie Reid did indeed die before "another Christmas."

The majority of individuals in the age group born in the earlier part of the 20th century were reared on farms or in nonurban environments. Those who lived through the Great Depression often had cause to wonder where tomorrow's meal would come from. Robert Fulton and Greg Owen (1988) noted that illness, dying, and death took place at home and were observed by children and adults alike. Animals were slaughtered for food. Death was visible, immediate, and real. Fulton and Owen observed that to an extent these individuals had lived in terms of the simple round of life known historically to humankind: birth, copulation, and death.

Achieving Integrity
Erik Erikson's theory of human development (1959) divides the life cycle into eight stages from infancy to old age. Each stage of development presents a crisis in understanding of oneself, of one's purposes, and of relationships with others. The developmental task at each stage is to resolve the crisis successfully; then the person can progress to the next stage of maturity. The eight stages are infancy, early childhood, play age, school age, adolescence, young adulthood, adulthood, and **senescence.** Erikson argues that a person must have been successful to some degree in resolving the seven crises that come before in order to resolve the eighth stage—senescence.

The task of the final stage of life is to achieve integrity—a conviction that one's life has meaning and purpose and that having lived has made a difference. Having reached senescence, one can look back on her or his life and indeed have a feeling, not unlike that of the Apostle Paul in the Bible, that he or she has "fought a good fight," completed the tasks, and feels a sense of completeness or wholeness. Having the task of life now behind one (completed) helps to reduce anxiety that there is yet more to accomplish and therefore puts one in a "ready" position to die.

C. D. Ryff's findings (1991) in studying the elderly fit well with Erikson's idea of integrity: As people achieve integrity, they view their past less critically and become content with how they have lived their lives. Ryff found that older adults seem to see themselves as closer to really being the persons they wanted to become than does any other age group. With senescence, one should have reached the acceptance stage of the dying process, as will be noted by Kübler-Ross in chapter 5, and feel pretty secure about self. A good self-concept, which completion of Erikson's eight stages should bring, allows the dying individual to have self-confidence, whether facing death or some other challenge in life. If the argument is correct that one dies as one lives, then the dying process for the individual who has reached senescence should be rather positive, as one's demeanor carries over to the dying process. Such an attitude and behavior would make the dying patient much easier to relate to for caregivers.

A death notice in the Philadelphia Gazette in 1752 (Archives) of Mr. William Bradford suggests that he probably reached senescence, as he neared the end of his

In the "autumn of their lives" for some elderly people "the best is yet to be," if they have good health and adequate financial means and housing.

WORDS OF WISDOM
Socrates on the Fear of Death

To fear death is nothing other than to think oneself wise when one is not; for it is to think one knows what one does not know. No man knows whether death may not even turn out to be the greatest of blessings for a human being; and yet people fear it as if they knew for certain that it is the greatest of evil.

"The Apology (The Defense of Socrates)" (p. 435), by Socrates, 1971. In W. H. D. Rouse (Trans.), Great Dialogues of Plato, *Bergenfield, NJ: Mentor Books.*

---

94 years of life. He came to the United States around 1680 and settled where Philadelphia now stands. He was "a Man of great Sobriety and Industry, a real Friend to the Poor and needy, and kind and affable to all. . . ." "Being quite worn out with old Age and Labour, his Lamp of Life went out for want of Oil" (a beautiful euphemism for death!). It is very likely that Mr. Bradford felt good about his self and the life that he had lived.

Research (Kail and Cavanaugh, 1996) shows a connection between engaging in a **life review,** a concept advanced by Robert Butler, and achieving integrity, Erikson's eighth stage of the life cycle. The life review is triggered by the realization that one has reached the end of life and that death is near. The life review serves to prepare a person for dying—preparation that may decrease the fear of death. In one study (Haight, 1992), homebound older adults who were part of a program that assisted people in remembering and reviewing their lives showed significant improvements in life satisfaction and positive feelings and a decrease in depressive symptoms, compared to homebound older adults who did not participate. These changes were still evident two months after the program ended.

A life review sometimes presents an opportunity to return to places one had frequented earlier in life and of which the individual has fond recollections. In the public television production from the 1970s entitled *Dying,* mentioned in chapter 1, the Reverend Bryant, dying of liver cancer, takes his two adult sons and his wife back to the places he lived as a child: the old homeplace, the cemetery where his parents were buried, and the place where he worked as a young boy. He was re-living his earlier life with his children. "This is where your dad spent his formative years of life, and I want to introduce them to you, as I am about to exit this life," the Reverend Bryant seemed to be saying. He was having a life review ("This is your life, Reverend Bryant," just like the old television show entitled *This Is Your Life*). The life review gives an individual an opportunity to achieve "closure" on her or his life (a way of "putting the ribbon" on the package of life, drawing down the final curtain of the play). Life is over, and, with integrity, the actor feels that she or he has performed well.

Diminishing Death Fears

Like their younger contemporaries, older adults have some anxieties and concerns as they approach death. They fear a long, painful, and disfiguring death or death in a vegetative state, hooked to sophisticated machines while hospital bills devour their insurance and savings. They fear that their families will be overburdened by their prolonged care and the expense that it involves. And they dread losing control of their lives by consignment to nursing homes (McCarthy, 1991). Empirical evidence suggests that older persons think of death more often than do younger adults, but that older persons appear to have less fear and anxiety concerning death (Bee, 1992; Kalish & Reynolds, 1981; Leming, 1980).

According to Robert Neimeyer and David Brunt (1995), several factors account for elderly adults' more positive attitudes toward dying and death: These attitudes could reflect the diminished quality of elderly persons' health and lives; their greater religiosity; their more extensive experience in having worked through the death of parents, peers, and partners; or the fact that their expectation to live a certain number of years has been met.

Richard Dumont and Dennis Foss (1972) suggest that, because older people are more likely to have fulfilled their goals in life, they are less fearful of death. For others, death might threaten personal achievement. Those older people who have not fulfilled their goals are more likely to have either made their goals more modest or somehow to have rationalized their lack of achievement. It is also possible that older people come from an age cohort whose members were better socialized as children to deal with death. They are more likely to come from rural backgrounds and to have had earlier encounters with deaths of siblings, other family members, friends, and animals.

Although elderly adults tend to have less fear of death than younger groups, they have had more experience with death (Moss & Moss, 1989). They have probably known more people who have died, have been to more funerals, and have visited more cemeteries than have younger persons. Thus, they may have become somewhat more able to imagine a world without them in it as they experience the death of others and consider what they will leave behind them. Many older persons have, therefore, in some ways come to terms with their own finitude.

Differences in Prevalence of Death Fears.   Hannelore Wass (1979) reports that few studies have examined sex differences in the fear of death among the elderly. Most studies generally show no sex differences in noninstitutionalized, in institutionalized, and in **geriatric** patients with acute conditions. A study by Hannelore Wass and S. Sisler (1978) of noninstitutionalized older persons, however, showed significant sex differences indicating that women have more fear of death than do men. Or perhaps women, as suggested earlier (Stillion, 1985), have a greater tendency to admit troubling feelings. In a more recent study by R. W. Duff and L. K. Hong (1995) of 674 residents from West Coast retirement communities, however, women showed less anxiety than men. A possible explanation is that as the proportion of widows increases in the retirement community, a supportive subculture for widows may emerge that helps buffer them from anxiety over death, note Duff and Hong. In addition, the retirement community may be a very different environment and population, and thus not all observations derived from the outside society are applicable to the retirement community. When only institutionalized versus noninstitutionalized elderly persons have been studied to determine if death fears vary with these two populations, Wass (1979, p. 193) cites studies that show that elderly persons institutionalized in nursing homes tend to be more fearful than those living out in the community in an apartment or house.

Wass (1979) further reports that some evidence shows that widowed individuals tend to be more fearful of death than are the elderly persons who are married or remarried. Persons living alone have a greater fear of death than those living with a family. As noted earlier in this chapter, education seems to have an impact on views of death: Elderly persons with only a grade school education have a greater fear of death than do those with a college education. Elderly persons with low incomes exhibit a greater fear of death than do those with higher incomes. High levels of death anxiety for the elderly seem to be associated with poor physical and mental health.

In a study of patients with terminal cancer, D. A. Lund and Michael R. Leming (1975) found that older patients had less anxiety concerning their diseases and terminal conditions than did younger patients. Older patients tended to experience greater depression, however.

There is evidence (Leming, 1980; McKenzie, 1980; Norman & Scaramelli, 1980) that differences in death anxiety and fear appear to be more a function of religiosity than of age. Because older persons are more likely to be religious, a sense of comfort should be provided as they approach death. The older person is more likely to believe in an afterlife and to rely on a faith in God as a coping strategy in dealing with death. Duff and Hong's study (1995) of residents in West Coast retirement communities concluded that the ceremonial act of regular attendance at religious services is associated with lower death anxiety, whereas private religious practices such as prayer or scripture reading do not have a significant impact on death anxiety. A longitudinal study of elderly people in New Haven, Connecticut, by E. L. Idler and S. V. Kasl (1992) also reported that "public religious involvement" has a significant protective effect against functional disability, whereas "private religious involvement" does not. Years ago French sociologist Emile Durkheim in *The Elementary Forms of Religious Life* (1915/1947) argued that it is the ceremonial acts that are mainly responsible for the life-enriching effects of religion.

Accepting Death.   Despite numerous concerns of the elderly, many elderly Americans die peaceful and relatively painless deaths. A National Institute of Aging study (McCarthy, 1991) of 1,000 persons older than age 65 who had died concluded that their health did not deteriorate until fairly near the end in most cases. Over half were in good or excellent health a year before they died. Ten percent were in good health the day before. About one third knew that death was approaching. On the other hand, a study at five teaching hospitals in the United States of nearly 1,200 seriously ill elderly patients by Joanne Lynn and colleagues (1997) found that these patients often were in severe pain, had shortness of breath, dysphoria, and other

WORDS OF WISDOM
Warning

Jenny Joseph

When I am an old woman I shall wear purple
With a red hat which doesn't go, and doesn't suit me.
And I shall spend my pension on brandy and summer gloves
And satin sandals, and say we've no money for butter.
I shall sit down on the pavement when I'm tired
And gobble up samples in shops and press alarm bells
And run my stick along the public railings
And make up for the sobriety of my youth.
I shall go out in my slippers in the rain
And pick the flowers in other people's gardens
And learn to spit.

You can wear terrible shirts and grow more fat
And eat three pounds of sausages at a go
Or only bread and pickle for a week
And hoard pens and pencils and beermats and things in boxes.

But now we must have clothes that keep us dry
And pay our rent and not swear in the street
And set a good example for the children.
We must have friends to dinner and read the papers.

But maybe I ought to practice a little now?
So people who know me are not too shocked and surprised
When suddenly I am old, and start to wear purple.

"Warning," by J. Joseph, 1987.
In S. Martz (Ed.), When I Am an Old
Woman I Shall Wear Purple:
An Anthology of Short Stories and Poetry,
Watsonville, CA: Papier-Mache Press.

symptoms that were difficult to tolerate. The patients preferred treatments that focused on comfort, even if the use of these treatments shortened life.

Though Robert Kastenbaum (1992a) highlights the diversity of attitudes toward dying and death found in elderly persons, he notes certain themes and characteristics found in three studies: J. A. A. Munnichs (1966) in the Netherlands, A. D. Weisman and Robert Kastenbaum (1970) in the United States, and Allen Kellehear (1990a) in Australia. Munnichs's interviews with 100 men and women older than age 70 revealed that fear did not dominate, relatively few were apprehensive about dying, and fewer were obsessed by the idea of death and finitude. Munnichs found that most had worked out some kind of accepting orientation toward death. Death was no longer a threatening stranger or mysterious external force. Kastenbaum (1992a) notes that these individuals in Munnichs's study had become "philosophical" about death.

Likewise, in a study of hospitalized geriatric patients in the United States (Weisman & Kastenbaum, 1970), anxiety was not the predominant response to the prospect of dying and death. The ego integration hypothesis of Erikson and the life review theory of Butler were consistent with the pattern of findings. In the Australian study of 100 dying persons, most of them were making "final preparations" and seemed concerned about the needs of survivors.

Choosing a Place to Die

Death for the older person becomes a normal and acceptable event. The crisis for the elderly is not so much death itself, but how and where the death will take place. The prospect of dying in a foreign place in a dependent and undignified state is a very distressing thought for older adults. They do not wish to be a financial or physical burden on anyone, yet the options of care may be limited.

The question of physically relocating the elderly to another place is a sensitive issue among **gerontologists** (Crandall, 1991). Those favoring relocation believe that relocating older persons is often less detrimental than leaving them where they currently live. They believe that moving older persons from substandard facilities, for example, will be beneficial. Those opposed to relocation of elderly persons argue that relocating is traumatic and will generally increase their risk of death.

Although most elderly individuals would prefer to die in their homes in familiar surroundings, the majority continue to die in hospitals and nursing homes (Cox, 1996b). Marty Zusman and Paul Tschetter (1984) point out that to die at home requires substantial resources, including money, space, and time; thus, the well-to-do are sometimes allowed to die at home because of their private control. Many people simply cannot afford the luxury of dying at home. Richard Kalish (1965) notes, however, that institutionalized settings are better equipped to handle dying individuals and that the family is removed from the considerable strain of caring for a dying family member in the home. Thus, although elderly persons might prefer to die at home, most are not likely to be allowed to do so.

The elderly are more likely to be separated from family and friends as they die. For them, the dying process may involve a fear of isolation and loneliness.

CONCLUSION

Rather than trying to "protect the young from death talk," we need more open communication channels with children on this topic. Children from an early age have a concept of death. We need to stop pretending that children cannot handle this topic. It is okay for children (and adults!) to show emotions such as crying when they feel so inclined. It is okay to feel the way one feels—moral judgments should not be attached to feelings. We should try not to be judgmental of others' feelings.

Whoever or whatever (in the case of a pet) has died, children, especially, need support at this time. One college student recalled to George Dickinson the occasion of her mother's unexpected death in an accident. She was 5 years old at the time and said she felt so insecure when she heard the news. She said, "It was like my whole security blanket crumbled." Indeed, her world had collapsed.

Psychosocial studies of developmental concepts of death are themselves in a stage of infancy. Significant progress is being made, however, in understanding how children and adolescents experience loss at various stages of their development. Adolescence is a particularly vulnerable period with regard to facing death.

People can and do manage much of what happens within their minds. One gathers information and insights, helps himself or herself to cope, and gives aid to others in need. At whatever stage of the life cycle, one can be helped to face both life and death more positively.

More research is needed to explore the relationships between death conceptualizations, gender differentiation, and place within the adult life cycle. Some stereotypes concerning older men and women are being discredited, but our conclusions are tentative, and more empirical research is needed.

Growing older pushes one to depend more upon educational, intellectual, and social skills than upon physical prowess. Feeling that one is useful and contributes to the well-being of others, as well as having a healthy understanding of death, contributes significantly to meaningful living and dying. As psychologist and thanatologist Robert Kavanaugh (1972, p. 226) wrote:

> I am ashamed how little I know about death and dying, but never have I enjoyed life more, dreamed more beautiful dreams each night, than when I began having courage to begin facing death.

Some pessimism can be found in older people, but disengagement from life is not necessarily a universal trait. The fear of death lessens with age, but the thought of death increases. The way that time is used changes, as does the meaning that one finds in life's experiences. Yet, living a happy and meaningful life is one of the ways to develop a positive and healthy view of death. Perhaps Alex Keaton of the television show *Family Ties* best summed up the developmental approach to death when he said, "Children die with opportunities and dreams; old people die with achievements and memories."

SUMMARY

1. Due in part to an urbanized and industrialized society, children are removed from death, other than through the rather artificial means of the media.

2. The permanency of death is unclear to young children who tend to see death as reversible.

3. Children can take about anything adults can dish out to them, including the topics of dying and death.

4. Honesty and openness in relating to children about death is very important.

5. Avoid euphemisms when talking to children about death.

6. It is okay to express emotions through crying. Adults can sanction this behavior by not hiding their own tears when they are sorrowful. By seeing big people cry, children can know that crying is normal behavior.

7. During adolescence the sense of personal identity is most vulnerable, and concepts and feelings of death are powerfully influenced by that vulnerability.

8. The young adult stage has been labeled the "novice phase" because a strong sense of need to train oneself in the art of reaching one's fullest potential is contradicted by any thought of death.

9. The "panic phase" begins during the middle years when one realizes that the idealized self that one longed to develop may not actually happen.

10. Research concerning the conceptualization of death in middle adulthood is difficult to find.

11. According to Erik Erikson, the task of the final stage of life, senescence, is to achieve integrity.

12. The crisis for elderly persons is not so much death, but rather how and where the death will take place.

DISCUSSION QUESTIONS

1. Discuss your first childhood memory of death. How old were you? Who or what died? What do you remember about this event?

2. As a child growing up, how was death talked about in your family?

3. Discuss the various perceptions of death as one goes from birth through age 12. What shortcomings do you find with a life cycle approach?

4. List as many euphemisms as you can to identify dying and death. Why do you think euphemisms are used with death?

5. Discuss why adults tend to avoid talking about death with children.

6. What are some of the factors, other than age, that influence death conceptualizations? Why are these factors important in understanding the ways that people conceptualize death?

7. Why do high school and college students have a higher level of death anxiety than junior high school students?

8. In your opinion, have death conceptualizations changed much in the past few decades? Do you see death conceptualizations changing much in the next few decades?

9. What are some death themes in contemporary adolescent music? How do you explain death themes in music?

10. How do you explain the popularity of some of the movies with brutal death scenes?

11. Why would education tend to reduce one's anxiety about death?

GLOSSARY

Adolescence: Stage of life commonly defined as the onset of puberty when sexual maturity or the ability to reproduce is attained.

Animism: The tendency of young children to attribute qualities such as motives and feeling to inanimate objects.

Baby Boomers: Individuals born between 1946 and 1964.

Behaviorism: A school of psychology that focuses chiefly on overt behavior rather than on inner psychological dynamics that cannot be clearly identified or measured.

Cognitive Development: Development of processes of knowing, including imagining, perceiving, reasoning, and problem solving.

Death Anxieties/Death Fears: Learned emotional responses to death-related phenomena that are characterized by extreme apprehension.

Developmental Approach: Branch of social sciences concerned with interaction between physical, psychological, and social processes and with stages of growth from birth to old age.

Ego Psychologists: Theorists and therapists who moved away from Freud toward putting more emphasis in their therapy upon the coping strategies and strengths of the person than upon the more elusive dynamics of the libido and the unconscious.

Euphemism: The use of a word or phrase that is less expressive or direct but considered less distasteful or less offensive than another word or phrase.

Geriatrics: The study of the medical aspects of old age.

Gerontology: The study of the biological, psychological, and social aspects of aging.

Humanism: Psychological model that emphasizes an individual's phenomenal world and inherent capacity for making rational choices and developing to maximum potential.

Latent Function: Consequences of behavior that were not intended (for example, a funeral brings the family together, usually in an amiable way).

Life Review: Robert Butler's term suggesting a reverence for what once was and for a time for judgment. It includes looking back over one's life and perhaps tracing back one's steps in earlier years, a review of one's life, as death draws near, and is a therapeutic technique to help elderly persons.

Psychoanalysis: A school of theory and therapy that concentrates upon the unconscious forces behind overt behavior, dealing principally with instinctual drives and their dynamics in the individual's inner psyche. Its principal focus is the interplay of transference and resistance between psychoanalyst and client.

Ritual: The symbolic affirmation of values by means of culturally standardized utterances and actions.

Senescence: Erik Erikson's last stage of the life cycle. The task is to achieve integrity, a conviction that one's life has meaning and purpose and that having lived has made a difference.

SUGGESTED READINGS

Abel, E. K. (1991). *Who cares for the elderly? Public policy and the experiences of daughters*. Philadelphia: Temple University Press. Discusses family care for disable erly people, with emphasis on the changing role of women's caregiving responsibiliti the future.

Avery, G., and Reynolds, K. (Eds.). (2000). Representations of Childhood Death. New Y St. Martin's. An anthology describing how the death of children has been managed ᵢ __ depicted over the centuries and how attitudes toward death are culturally constructed.

Barry, R. L., & Bradley, G. V. (Eds.). (1991). *Set no limits: A rebuttal to Daniel Callahan's proposal to limit health care for the elderly*. Urbana, IL: University of Illinois Press. A response to Callahan's Setting Limits in which he suggests possible age-based rationing schemes to deny medical care and treatment to the elderly when they reach the end of their "natural life span."

Chesler, M. A. and Chesney, B. K. (1995). *Cancer and self-help: Bridging the troubled waters of childhood illness*. Madison, WI: University of Wisconsin Press. Provides good information for parents and professionals who wish to organize and provide leadership or support to local self-help groups. This is also an excellent book for students of social science.

Christ, G. H. (2000). *Healing children's grief*. New York: Oxford University Press. Presents the stories of 88 families who participated in a parent-guidance intervention through the terminal illness and death of one of the parents from cancer; discusses the various needs of the children for help and support.

Cox, H. G. (2001). *Later life: The realities of aging*. Upper Saddle River, NJ: Prentice-Hall. An interdisciplinary approach to aging, with discussions of dying and death among the elderly.

Dickinson, G. E., and Leming, M. R. (2002). *Annual editions: Dying, death and bereavement 02/03* (6th ed.). Guilford, CT: Dushkin/McGraw-Hill. An anthology with several articles relating to children, the elderly, and death.

Doka, K. J. (1995). *Children mourning, mourning children*. Washington, DC: Hospice Foundation of America. An anthology covering these topics: the child's perspective of death, the child's response to life-threatening illness, children mourning and mourning children, and innovative research on children and death.

Solomon, D. H., Salend, E., Rahman, A. N., Liston, M. B., & Reuben, D. B. (1992). *A consumer's guide to aging*. Baltimore: Johns Hopkins University Press. A practical guide to assist the elderly with staying healthy, maintaining an emotional balance, planning a financial future, changing family roles, and living in the leisure lane.

Stevenson, R. G., & Stevenson, E. P. (1996). *Teaching students about death: A comprehensive resource for educators and parents*. Philadelphia: Charles Press. A collection of experiences and ideas from professionals who work with children on the topic of death.

Szinovacz, M., Ekerdt, D. J., & Vinick, B. H. (Eds.). (1992). *Families and retirement*. Newbury Park, CA: Sage Publications. This book covers issues such as marital relationships in retirement, extended kin relationships in retirement, and the timing of retirement.

Wass, H., & Neimeyer, R. A. (Eds.). (1995). *Dying: Facing the facts* (3rd ed.). Washington, DC: Taylor and Francis. An anthology with 20 contributors that has especially good discussions of death in different stages of the life cycle.

Young, M., & Cullen, L. (1996). *A good death: Conversations with East Londoners*. London: Routledge. The authors' extensive interviews with cancer patients, diagnosed with only 3 to 4 months to live, are a sensitive, sociological account of what it is like to die of cancer in London today.

C H A P T E R 4

PERSPECTIVES ON DEATH AND LIFE AFTER DEATH

Religion provided me with answers to problems I didn't even know I had.
—St. Olaf College student, 1977

*Death radically challenges all socially objectivated definitions of reality—
of the world, of others, and of self. . . . Death radically puts in question
the taken-for-granted, "business-as-usual" attitude with which one exists
in everyday life. . . . Religion maintains the socially defined reality by
legitimating marginal situations in terms of an all-encompassing sacred
reality.*

—Peter Berger, *Sacred Canopy*

All living creatures are part of a cycle of life and death. This is a natural process that occurs with regularity and is fairly easy to explain in biological terms. However, humans have never been satisfied with just a biological explanation for this cycle of life and death. Across cultures and throughout history, humans have constructed belief systems that explain death in cosmological, spiritual, and/or religious terms. These explanatory systems, even though widely diverse, usually deal with death in some way because, like no other social event or situation, death inherently challenges the "taken-for-granted" meanings of all societies.

Some belief systems view death as a continuation of life, with spirits of the dead inhabiting the world alongside the living. For example, the traditional African attitude toward death is "positive and accepting and comprehensively integrated into the totality of life" (Barrett, 1992). Other belief systems regard death as a transition

125

to another level of existence—an afterlife—which has very little to do with the living. The Christian concepts of heaven and hell represent one example of this type of belief. For the Navajos, heaven is earth, where all goodness resides. Death leads to an underworld that is similar to earth and considered the place of all evil. Precautions must be taken to ensure that the dead are buried properly so they do not come back to haunt the living.

There are also those who believe that, after death, there is nothing, and death is meaningless. While this view is held by many people, there is no culture or dominant belief system in which this nihilistic perspective on death is normative. One of the most feared, distressing, and anxiety-provoking deaths is a death that is perceived as meaningless. The search to find meaning in life and death is a task that all people share.

THE NEED TO LOOK BEYOND DEATH

From a symbolic interactionist perspective (see chapter 1), meanings are created and reproduced by humans. These meanings supply a base for activities and actions (behavior is in response to meanings) and provide order for the people who share a given culture. Peter Berger (1969) suggests that the human world has no order other than that created by humans. To live in a world without the order contributed by one's culture would force an individual to experience a meaningless existence. Sociologists refer to this condition as **anomie**—"without order."

Many situations in life challenge the order on which social life is based. Most of these situations are related to what Thomas O'Dea (1966) refers to as the three fundamental characteristics of human existence: uncertainty, powerlessness, and scarcity.

Uncertainty refers to the fact that human activity does not always lead to predictable outcomes. Even after careful planning, most people recognize that they will not be able to achieve all of their goals. Less optimistically, the 21st century's Murphy's Law states "Anything that can go wrong will go wrong." The human condition is also characterized by *powerlessness*. We recognize that there are many situations in life and events in the universe over which humans have no control—among these situations are death, suffering, coercion, and natural disasters. Finally, in *scarcity*, humans experience inequality with regard to the distribution of wealth, power, prestige, and other things that make a satisfying life. This inequality is the basis for the human experience of relative deprivation and frustration. The three experiences of uncertainty, powerlessness, and scarcity challenge the order of everyday life and are, therefore, marginal to ordinary experiences. According to O'Dea (1966), such experiences "raise questions which can find an answer only in some kind of 'beyond' itself." Therefore, **marginal situations,** which are characteristic of the human condition, force individuals to the realm of the transcendent in their search for meaningful answers.

Berger (1969, pp. 23, 43–44) claims that death is the marginal situation par excellence:

> Witnessing the death of others and anticipating his own death, the individual is strongly propelled to question the ad hoc cognitive and normative operating procedures of "normal" life in society. Death presents society with a formidable problem not only because of its obvious threat to the continuity of human relationships, but because it threatens the basic assumptions of order on which society rests. Death radically puts in question the taken-for-granted, "business-as-usual" attitude in which one exists in everyday life. Insofar as knowledge of death cannot be avoided in any society, legitimations of the reality of the social world in the face of death are decisive requirements in any society. The importance of religion in such legitimations is obvious.

It is **religion,** or a transcendent reference, which helps individuals remain reality-oriented when the order of everyday life is challenged. Contemplating death, we are faced with the fact that we will not be able to accomplish all of our goals in life. We also realize that we are unable to extend the length of our lives and/or control the circumstances surrounding the experience and cause of our deaths. We are troubled by the fact that some must endure painful, degrading, and meaningless deaths, whereas others find more meaning and purpose in the last days of their lives than they experienced in the years preceding "the terminal period." Finally, the relative deprivation created by differential life spans raises questions that are unanswerable from a "this world" perspective.

Religious-meaning systems provide answers to these problems of uncertainty, powerlessness, and scarcity created by death. O'Dea (1966, pp. 6–7) illustrates this function of religion:

> Religion, by its reference to a beyond and its beliefs concerning man's relationship to that beyond, provides a supraempirical view of a larger total reality. In the context of this reality, the disappointments and frustrations inflicted on mankind by uncertainty and impossibility, and by the institutionalized order of human society, may be seen as meaningful in some ultimate sense, and this makes acceptance of and adjustment to them possible. Moreover, by showing the norms and rules of society to be part of a larger supraempirical ethical order, ordained and sanctified by religious belief and practice, religion contributes to their enforcement when adherence to them contradicts the wishes or interests of those affected. Religion answers the problem of meaning. It sanctifies the norms of the established social order at what we have called the "breaking points," by providing a grounding for the beliefs and orientations of men in a view of reality that transcends the empirical here-and-now of daily experience. Thus, not only is cognitive frustration overcome, which is involved in the problem of meaning, but also the emotional adjustments to frustrations and deprivations inherent in human life and human society are facilitated.

Religious systems provide a means to reestablish the social order challenged by death. Our society has institutionalized the continued importance of religion by creating funeral **rituals** that have a religious quality about them. To demonstrate the widespread tendency toward religious perspectives, the 1990 General Social Survey, conducted by the National Opinion Research Corporation, discovered that less than 10 percent of all Americans surveyed claimed to have "no religion." It appears that in the United States most people rely upon religious and spiritual paths in coping with life and its challenges.

Although many people do not follow a formal religion, there are very few who do not have some kind of spiritual concern with transcendent meanings that may be shared by a number of religious groups. These nonaffiliated individuals, or what Glenn Vernon called the "religious 'nones'" (people who responded "none" when asked the question "To which religion do you belong?"), often borrow from many religious traditions and spiritual perspectives as they formulate supernatural and transcendent interpretations for death and after death possibilities. There are even some religious groups that have institutionalized this broader, more inclusive spiritual orientation to religion and death. Included among these groups are Unitarian-Universalists, Baha'is, Unity Church members, and even some Quakers. From a functional point of view, people with a spiritual point of view are similar to the religious, but they are more individualistic and private or subjective in the way in which they apply their transcendent views.

DIVERSITY IN PERSPECTIVES

William Cowper once said, "Variety's the very spice of life, that gives it all its flavour." If Cowper is correct, a look at various cultures and how they view dying and death indeed will produce a lot of "flavours." If we did not have cultural diversity, however, would not life be rather dull?

Cross-Cultural Views

Cross-cultural studies of death reveal that most societies seem to have a concept of soul and immortality. For some, a belief in a soul concept explains what happens in sleep and after death. For others, a belief in souls explains how the supernatural world becomes populated (Tylor, 1873). Much of the death ritual performed is related to the soul and is often an appeasement of the beings in the spiritual world. This spiritual dimension of death plays a significant role in the social structure of societies that hold such beliefs.

In studying death cross-culturally we are reminded of other similarities, such as beliefs in ghosts. For many groups, souls of the dead become ghosts, who may take up residence indefinitely, or after a short period of wandering may cease to exist. Some people converse with ghosts and make offerings to them. This is sometimes referred to as the "cult of the dead" (Taylor, 1988). These ghosts are not worshiped but simply maintain a relationship with a person. They act as guides and protectors and may also confer power. In some societies, to have a guardian spirit is a positive experience. For others, like Navajo and Apache Native Americans, ghosts are responsible for sickness and death and are to be avoided (Cox & Fundis, 1992).

This Japanese funeral emphasizes the social status of the man who has died. The living honor the dead and thereby create a symbolic community consisting of their ancestors and living members of their communities.

Ghosts are not unique to nonliterate cultures. Many sane, sober individuals in the United States have had "encounters" with ghosts.

The continuity between the living and the dead is elaborated in ideas of reincarnation and in other ways in non-Western as well as Western societies. For example, many Japanese today think that a person's spirit belongs to the same family and the same local community before and after death (Nagamine, 1988). A person's spirit gradually fades away from its family as time goes on. The Japanese do not clearly distinguish the dead from the living and seem to recognize the continuity between life and death. Especially in rural Japan, the dead remain an integral part of life and offer constant solace to the living (Kristof, 1996).

The Japanese are not alone in believing that the dead remain a vital part of their lives. For instance, in some parts of southern India, where many Hindus spend 13 days of mourning, each person serves part of a meal on a banana leaf to provide for the needs of the dead person's soul (Tully, 1994). The meal is laid out in the open, and if the crows eat it, that means that the offering has been accepted.

The Hopi Indians viewed life and death as phases of a cycle (Oswalt, 1986). Death was an important change in individual status because it represented an altered state for the person involved. The Hopi believed that a duality of being existed in each person—a soul and a body expressed as a "breath-body." A person dying on earth literally was reborn in the afterworld. Corpses were washed and given new names before burial. The breath-body's pattern of existence would supposedly be the same in the domain of the dead as it had been on earth—when it died in the afterworld, it would be reborn on earth.

The Igbo in Nigeria (Uchendu, 1965) believe that death is important for joining their ancestors. Without death, there would be no population increase in the ancestral households and thus no change in social status for the living Igbo. The lineage system is continued among the dead. Thus, the world of the "dead" is a world full of activities.

The Ulithi of Micronesia are not morbid or defeated by death, according to William A. Lessa (1966). Their rituals afford them some victories, and their mythology provides a hope for a happy life in another realm. Though their gods are somewhat distant, these gods assure that the world has an enduring structure, and their ancestral ghosts stand by to give more immediate aid when merited. Thus, after the Ulithi express their bereavement, rather than retreat, they spring back into their normal work and enjoyment of life.

For the Dunsun of northern Borneo, few events focus on the beliefs and acts concerned with the nonnatural world more than the death of a family member (Williams, 1965). Death is considered a difficult topic that everyone fears talking about, yet it must be prepared for because it causes great changes.

The Abkhasians on the coast of the Black Sea (Benet, 1974) view death as irrational and unjust. The one occasion when outbursts of feeling are permitted is at a funeral—wailing and scratching at one's flesh are permitted.

Case Study: The Sacred World of Native Americans

While religion is a word that has no equivalent in the languages spoken by Native Americans, they are intensely spiritual people whose spirituality is exhibited in their art and everyday living (Mander, 1991). In the daily life of the Native American, spirituality is an inevitable part of every facet of life. Spirituality and prayer are inseparable to the Native American. To separate spirituality and prayer would mean cultural death for the Native American. Every day is a holy day because all that Native Americans undertake begin and end with the influence of a spiritual kinship with nature.

The Native American looks to this spirituality to explain the successes and failures of daily life. The guiding spirit or force, which rules nature, can help him or her through daily life and its struggles. As a consequence, Native Americans do not view death as something to be ignored, but rather as a natural part of life. For the Native American, death is the natural end of life—it needs to be natural and the dying person should be allowed to die without restrictions. Therefore, the confrontation with one's destiny is the confrontation with death and with one's spirituality.

Attitudes toward life and death are learned and spiritual teachings are one source of learning. The concepts of mortality, soul, and afterlife are important for Native Americans and, as a consequence, they place a great emphasis upon prayer and spiritual rituals. While the individual is essentially spiritual, the Native American will turn to a priest or a shaman or other qualified person to deal with the sacred world because they have the knowledge of how to deal with the sacred. The ritualistic prayers and chants of the Navajo have their parallels in other religions, as does the centrality of belief and faith for religious practice. However, a major difference for Native American spirituality is that the sacred words of the Navajo are passed down in an oral tradition, even though in recent years, these sacred words have been collected in a written form (Wyman, 1970; Gill, 1981).

While generalizations are dangerous, the variety of cultural expressions of dying and death do have some commonalties for the various Native American tribes. Most tribes express a willingness to surrender to death at any time without fear. The Lakota Chief Crazy Horse was noted for his chant before going into battle, "Today is a good day to die." Every day is to be lived as if it were one's last day. One must enjoy life and live fully. Just as one cannot buy land, one cannot buy life. Death is waiting. One cannot escape. One does not seek death before its time, nor does one avoid death or try to delay its occurrence. No one is ever truly alone. The dead are not altogether powerless. There is no death, but rather simply a change of worlds (Steiger, 1974).

Once one dies, the ghost becomes of great concern. One does not use the name of the dead. Ghosts may be too busy, and to interrupt them may cause anger (Stoeckel, 1993). Not to honor the dead also invites their wrath. For the Navajo, all sickness and disease, whether mental or physical, is believed to be due to supernatural causes. To appease their ancestors, the Apache set aside part of their fields as the fields of the dead. They would not cultivate these fields for 3 years, even if it meant starvation for the family (Mails, 1974).

To the Lakota everything was filled with spirits and powers that controlled or otherwise affected the lives of the people. Humans are no better or worse than other living things. All things have a spirit, and illness and death are a part of life, which must be accepted. If one is to be in balance or harmony with nature, one must die when it is one's time. A person should not have to sacrifice dignity by crying out for interventions that will only prolong the natural process of dying because it is better to die today than to be shamed by living tomorrow.

The lessons of the Native American are clear. A natural death is better than an artificial one involving futile interventions. For the Native American, medicine heals the spirit as well as the body. When the Native American dies, it was his or her time to die. It was not the fault of the shaman or the family. In the past, white physicians did not have the techniques and medicines that they now have to prolong the dying process. Then, they too relied on supernatural interventions. Today, when a person dies, the physician loses a patient, and death is being portrayed as the enemy of modern medicine. For the Native American, death is natural and not an enemy, just as suffering is a part of life that cannot be avoided. Modern medicine's emphasis upon winning has led to guilt, experienced by the patient, family, and physician. From a Native American perspective, neither the physician, the family, nor the patient has lost. Rather, the patient died, as will we all.

Thus, attitudes toward death vary significantly among societies around the world—from seeing death as a continuation of life to the end of everything. We will now turn to the role of death in creating religious beliefs and rituals.

(Adapted from an article written by Gerry Cox and used with his permission)

PRACTICAL MATTERS
Using Religious Symbols in A Funeral Home

Thirty years ago while on a class field trip to a funeral home, Michael Leming complimented the funeral director on the beautifully painted pastoral scene hanging on a wall. The funeral director said it was a very unusual wall hanging and took it down to show the framed, velvet, reverse side, which could also be displayed. He then took out a box containing a cross, a crucifix, and a star of David that could be hung on the velvet backing. The funeral director told the class that he changed the hanging whenever the religious affiliation of the deceased varied. Since the time of that field trip, Leming has become very conscious of the way that funeral homes extensively use religion to provide comfort for the grieving. Consider the following:

1. Within the funeral home, "chapel" is the name given to the room where the funeral is held.

2. Most memorial cards have the 23rd Psalm on them.

3. The music that one hears on the sound systems within most funeral homes is religious in nature.

4. Wall hangings found in most funeral homes usually have religious content.

5. Funeral homes often provide Christmas calendars, complete with Bible verses and religious scenes, for certain religious groups and other interested members of the community.

RELIGIOUS INTERPRETATIONS OF DEATH

Until this point, we have provided a functional perspective on religion—we have defined religion in terms of what religion does for the individual and society. This perspective focuses on the consequences of religion rather than on the content of religious belief and practice.

We will now attempt to provide substantive perspectives on religion and death. The substantive perspective to religion attempts to establish what religion is. Substantive definitions of religion endeavor to distinguish religious behaviors from nonreligious behaviors by providing necessary criteria for inclusion as religious phenomena. The substantive definition of religion most frequently used by sociologists is the one formulated by Emile Durkheim in his *The Elementary Forms of Religious Life,* first published in 1915:

> A religion is a unified system of beliefs and practices relative to sacred things; that is to say, things set apart and forbidden—beliefs and practices that unite into one single moral community called a church, all of those people who adhere to them.

In this definition, Durkheim designates four essential ingredients of religion—a system of beliefs, a set of religious practices or rituals, the sacred or supernatural as the object of worship, and a community or social base. Most substantive definitions of religion used by contemporary sociologists will incorporate these four essential ingredients. However, some sociologists have argued that in order to be inclusive of phenomena that most people consider religious, it might be proper to exclude the necessity of having a sacred point of reference. Buddhism, for example, does not have a supernatural being to which beliefs and rituals are oriented. We will now consider five religious traditions and explain how each tradition attempts to interpret the meaning of death-related experiences and provides funeral rites and rituals for the bereaved. The five religious traditions considered are Judaism, Christianity, Islam, Hinduism, and Buddhism.

Judaism

Death in the Jewish tradition came into being as a result of Adam's and Eve's sin, which caused them to be expelled from the Garden of Eden. When Adam and Eve ate the fruit from the "Tree of Conscience," they received the curse of pain in childbirth, the burden of work, and the loss of physical immortality. According to the biblical account in Genesis (2:4–3:24), although death was a punishment, it also brought the ability to distinguish between good and evil as well as the power and responsibility to make decisions that have a future consequence. According to J. Carse (1981):

> Adam and Eve lost their immortality, but acquired consciousness instead. God drove them out of Paradise into death, but also into history. God's design for the people of Israel is to save them from their enemies in order that their history might continue.

In this description we gain an understanding of the importance of God's covenant with Abraham—that Abraham would become the father of many nations and that God would have a special relationship with his descendants forever (Gen. 17). Consequently, immortality was to be found in one's identity within the group.

Among contemporary Jews, including those who consider themselves religious, opinion differs regarding personal immortality. Some contend that there is no after-*life*, only an after*death*—the dead go to Sheol, where nothing happens, and the soul eventually slides into oblivion. Other Jews believe in a resurrection of the soul, when

THE WIZARD OF ID   Brant parker and Johnny hart

REPRINTED BY PERMISSION OF JOHNNY HART AND NAS, INC.

individuals are brought to a final judgment (Carse, 1981). Still, for others there is a real ambivalence regarding the immortality of the soul. Carse cites the writings of Rabbi Leona Modena (1571–1648), who states:

> It is frightening that we fail to find in all the words of Moses a single indication pointing to man's spiritual immortality after his physical death. Nonetheless, reason compels us to believe that the soul continues.

Regardless of the content of Jewish beliefs regarding the immortality of the soul, Jewish funeral customs and rituals emphasize that God does not save us, as individuals, from death, but saves Israel for history, regardless of death (Carse, 1981).

Christianity

Although Christianity shares much of the historical and mythical foundations of Judaism, there are many distinct differences in the Christian approach to death, afterlife, and funeral rituals. For the Christian, death is viewed as the entrance to eternal life and, therefore, is preferable to physical life. There is a strong belief in the immortality of the soul, the resurrection of the body, and a divine judgment of one's earthly life after death, resulting in the eternal rewards of heaven or the punishments of hell. For the Roman Catholic there are four potential dispositions of the soul after death—heaven, hell, limbo, and purgatory. According to McBrien (1987, p. 443), "some will join God forever in heaven; some may be separated eternally from God in hell; others may find themselves in a state of merely natural happiness in limbo; and others will suffer in purgatory some temporary 'punishment' still required of sins that have already been forgiven."

For the Christian, the teachings of Jesus and the Apostle Paul are the most important sources in arriving at a theology of life after death. Jesus declares to his followers:

> I am the Resurrection and the Life, he who believes in Me, though he die, yet shall he live, and whoever lives and believes in Me shall never die. (John 11:25, RSV)

In the 15th chapter of the first letter to the Corinthians, the Apostle Paul discusses the significance of Jesus' resurrection for the Christian believer. In this chapter Jesus is portrayed as the "first born from the dead" and as "the one who has destroyed death" (1 Cor. 15:26, RSV). Paul declares (1 Cor. 15:52-58, RSV) that at the end of history:

> The dead will be raised imperishable, and we shall be changed. For this perishable nature must put on the imperishable, and this mortal nature must put on immortality. When the perishable puts on the imperishable, and the mortal puts on immortality, then shall come to pass the saying that is written:
> Death is swallowed up in victory.
> O death, where is thy victory?
> O death, where is thy sting?
> The sting of death is sin, and the power of sin is the law. But thanks be to God, who gives us the victory through our Lord Jesus Christ.

Christians employ two basic, and somewhat paradoxical, perspectives when facing death. The first has just been described—that through faith in Jesus Christ, the Christian has victory over death and gains eternal life with God. The following passage (Rom. 8:31–39, RSV) provides us with an example of this orientation:

> What then shall we say to this? If God is for us, who is against us? He who did not spare his own Son but gave him up for us all, will he not also give us all things with him? Who shall bring any charge against God's elect? It is God who justifies; who is to condemn? Is it Christ Jesus, who died, yes, who was raised from the dead, who is at the right hand of God, who indeed intercedes for us? Who shall separate us from the love of Christ? Shall tribulation, or distress, or persecution, or famine, or nakedness, or peril, or sword? As it is written,
>
> "For thy sake we are being killed all the day long; we are regarded as sheep to be slaughtered."
>
> No, in all these things we are more than conquerors through him who loved us. For I am sure that neither death, nor life, nor angels, nor principalities, nor things present, nor things to come, nor powers, nor height, nor depth, nor anything else in all creation, will be able to separate us from the love of God in Christ Jesus our Lord.

The second perspective on death employed by Christians emphasizes the experience of true human loss. This approach is exemplified by Jesus (John 11:32–36, RSV) as he responds to the death of his friend Lazarus:

> Then Mary, when she came where Jesus was and saw Him, fell at his feet, saying to Him, "Lord, if you had been here, my brother would not have died." When Jesus saw her weeping, and the Jews who came with her also weeping, He was deeply moved in spirit and troubled; and He said, "Where have you laid him?" They said to Him, "Lord, come and see." Jesus wept. So the Jews said, "See how He loved him!" (John 11:32–36, RSV)

C. S. Lewis, in his book *A Grief Observed* (1961, p. 24), illustrates how Christians use this paradoxical double perspective:

> What St. Paul says can comfort only those who love God better than the dead, and the dead better than themselves. If a mother is mourning not for what she has lost but for what her dead child has lost, it is a comfort to believe that the child has not lost the end for which it was created. And it is a comfort to believe that she herself, in losing her chief or only natural happiness, has not lost a greater thing, that she may still hope to "glorify God and enjoy Him forever." A comfort to the God-aimed, eternal spirit within her. But not to her motherhood. The specifically maternal happiness must be written off. Never, in any place or time, will she have her son on her knees, or bathe him, or tell him a story, or plan for his future, or see her grandchild.

"At that time, the disciples came to Jesus saying, 'Who is the greatest in the kingdom of heaven?' And calling to Him as a child, He put him in the midst of them, and said, 'Truly, I say to you, unless you turn and become like children, you will never enter the kingdom of heaven. Whoever humbles himself like this child, he is the greatest in the kingdom of heaven.'" (Matthew 18:1–4)

Islam

As in Christianity, life after death is also an important focus within the Islamic tradition. Earthly life and the realm of the dead are separated by a bridge that souls must cross on the Day of Judgment. After death, all people face a divine judgment. Then they are assigned eternal dwelling places where they will receive either eternal rewards or punishments, determined by the strengths of their faith in God and the moral quality of their earthly lives. According to the Qur'an, there are seven layers of heaven and seven layers of *alnar* ("Fire of Hell"), and each layer is separated from the layer above by receiving fewer rewards or greater punishments. The fundamental reason why individuals might be condemned to a life of torment in the Fire of Hell is a lack of belief in God and in the message of his prophet Muhammad. Other reasons include lying, being corrupt, committing blasphemy, denying the advent of the Judgment Day and the reality of the Fire of Hell, lacking charity, and leading a life of luxury (Long, 1987).

Like Jews and Christians, followers of Islam believe that God is fundamentally compassionate and place a similar emphasis on God as just. Therefore, individuals are held accountable for moral integrity at the time of their death. The primary expression of the Islamic concern for justice and accountability is found in the belief in assignment to paradise or damnation. Accordingly, the Qur'an provides very vivid sketches of both paradise and hell. However, many Islamic theologians also stress that God's judgment is tempered with mercy, that the angel Gabriel will intercede on behalf of those condemned to punishment, and that they will eventually be pardoned.

The following prayer (the opening of the Surah of the Qur'an) illustrates the Islamic perspective on divine justice and mercy:

> In the name of God, the Merciful, the Compassionate. Praise belongs to God, the Lord of all Being, the All-merciful, the All-compassionate, the Master of the Day of Doom.
>
> Thee only we serve; to Thee alone we pray for succour, Guide us in the straight path, the path of those whom Thou has blessed, not of those against whom Thou art wrathful, nor of those who are astray.

Hinduism

Unlike the religious traditions that we have previously discussed, Hinduism does not have a single religious founder nor a single sacred text. Although the *Vedas* are recognized by almost all Hindus as an authoritative source of spiritual knowledge, Hinduism is not dogmatic. Many theologies and religious approaches exist within the Hindu tradition. Many gods also exist, each viewed as an aspect or manifestation of a single ultimate reality, but it is *not* essential to believe in the existence of God to be a Hindu (Srinivas & Shah, 1968). Essentially, Hinduism is a system of social customs imbued with religious significance.

Three concepts are central to an understanding of Hinduism—*karma, dharma,* and *moksha. Karma* refers to a moral law of causation; it suggests that an individual's actions produce results for which the individual is responsible. *Karma* also refers to the balance of good and bad deeds performed in previous existences. *Dharma* are the religious duties, requirements, and/or prescriptions. The extent to which one fulfills one's *dharma* determines one's karma. In turn, *moksha* is the reward for living a saintly life. The main ways of achieving *moksha* are overcoming spiritual ignorance by acquiring true knowledge, performing good deeds, and living a life of love and devotion to God (Srinivas & Shah, 1968).

DEATH ACROSS CULTURES
Teaching Reincarnation in the Bhagavadgita

Death and birth are seen by us Hindus as gateways of exit and entry to the stage of this world. This process is described in one of the most striking and famous analogies in the entire Bhagavadgita. It is a verse which we movingly recite when the body is being cremated, and it has given comfort and solace to the grieving Hindu heart for centuries.

> Just as a person casts off worn-out garments and puts on others that are new, even so does the embodied-soul cast off worn-out bodies and take on others that are new.

This analogy is full of suggestions, but I can only briefly draw your attention to a few of these. First, a suit of clothing is not identical with the wearer. Similarly, the body, which is likened here to worn-out garments, is not the true being or identity of the human person. Second, there is the similarity of a continuity of being. When a worn-out suit of clothing is cast off, the wearer continues to be. Similarly, with the disintegration of the physical body, the in-dweller, that is, the Self, continues to be. Finally, there is the parallel with the timing of the change. A suit of clothing is changed when it no longer serves the purpose for which it was intended. Similarly, the physical body is cast off when it no longer serves the purpose for which it came into existence.

This does not mean that the Hindu is not saddened by death or does not lament the loss it brings. There is a mysterious element in death which will always sadden and hurt. While we ache and long for the beauty, tenderness, and companionship of a dear one, we are deeply comforted in the knowledge that all that is good and true and real in him or her continues to be. What has passed away is what has always been limited by time.

Annant Rambachan, Department of Religion, St. Olaf College. By permission of the author.

The central doctrine affecting death-related attitudes and behavior in the Hindu religion is reincarnation and the transmigration of souls *(samsara)*. For the Hindu, one's present life is determined by one's actions in a previous life. Furthermore, one's present behavior will shape the future. According to R. W. Habenstein and W. M. Lamers (1974, p. 116), "The ultimate goal of the soul is liberation from the wheel of rebirth, through reabsorption into or identity with the Oversoul *(Brahma)*—the essence of the universe, immaterial, uncreated, limitless, and timeless." The way in which one would experience reabsorption is to overcome spiritual ignorance *(avadya)* by ultimately coming to know the truth of oneself and the truth of God. It is *moksha* that brings reabsorption and ends *samsara,* the cycle of reincarnation and transmigration of souls.

By way of contrast, whereas Jews, Christians, and Muslims believe in the immortality of the soul and hope for an afterlife, Hindus hope that their soul will be absorbed at death. For the Hindu, the goal is not to experience life after death, but rather to have one's soul united with the Oversoul. Punishment for the devout Hindu might be to *have* "everlasting spiritual rebirth."

Within the Hindu tradition, death brings two possibilities—liberation or transmigration of the soul. Neither is inherently fearful, even though separation from one's friends and loved ones may cause sadness and personal loss. Somewhat analogous to the dual perspectives of Christianity, death for the Hindu individual brings hope of something better but also the human loss of being separated from a dead loved one.

Buddhism

There are many types of Buddhism, but the most popular are *Theravada, Mahayana,* and *Tantrayana.* According to Peter Pardue (1968), *Theravada* Buddhism is the dominant religion in Southeast Asia (Myanmar, Thailand, Laos, Vietnam, Cambodia, and Sri Lanka). *Mahayana* Buddhism is primarily practiced in Korea, China, Japan, and Nepal. *Tantrayana* has traditionally been the dominant form of Buddhism in Tibet, Mongolia, and parts of Siberia. Although there are differences between each of these sects and many others found throughout the world, Buddhists in general find a common heritage in the life of Siddhartha Gautama ("The Enlightened One" or "Buddha").

Buddha was born in 563 B.C. as a prince in northern India. When he was a child, his thoughts were preoccupied with the finitude of human existence. Unsuccessfully, his family tried to shelter him from human suffering and death. At the age of 29 he left his life of privilege and began to search for personal salvation. After rejecting physical asceticism and abstract philosophy, he attained a state of enlightenment through the process of intense meditation. For the remaining 50 years of his life, he served as a missionary and preached his message of salvation to all people regardless of social position and gender (Pardue, 1968).

The Buddhist message of salvation, taught by Buddha in his first sermon, is the "Four Noble Truths." The first of these truths is that all human existence is characterized by pain and suffering in an endless cycle of death and rebirth. The second truth is that the cause of the agony of the human condition is desire for personal satisfaction, which is impossible to obtain. The third truth is that salvation comes by destroying these desires. By completely destroying ignorance, one experiences enlightenment, and the cycle of transmigration of the soul is broken. Finally, in the fourth truth, one can experience perfect peace through the eightfold path to enlightenment. In the words of Pardue (1968, p. 168):

> For this purpose the proper [meditation] is the "eightfold path," an integral combination of ethics *(sila)* and meditation *(samadhi),* which jointly purify the motivations and mind. This leads to the attainment of wisdom *(prajna),* to enlightenment *(bodhi),* and to the ineffable *Nirvana* ("blowing out"), the final release from the incarnational cycle and mystical transcendence beyond all conceptualization.

Buddhism assists its practitioners by encouraging them to detach from striving for living, holding onto relationships, and attempting to avoid pain and sadness. From a Buddhist perspective it is by accepting the impermanence of life, relationships, and desirable things and then letting go of them that one finds enlightenment.

In the Buddhist meditation entitled "The Way of the Mystic," Sister Medhanandi (1998) claims,

> There are two kinds of death for the living, one that leads to death and one that leads to peace, to enlightenment. When we carry around a lot of wreckage in the mind, we are not putting down the burden. We are identifying with and caught in self-view, "I am an abused person" or "a grief-stricken person," or "Five of my friends have died from AIDS and I just can't face life." That's a death that leads to death.
>
> But if we can meet the present moment with mindfulness and wise reflection, we can begin to put down that burden, surrender it, and allow ourselves to receive the next moment with purity of mind, letting the conditions that arise and our attachment to them die. That kind of death leads to enlightenment.

She goes on to say in an article entitled "The Joy Hidden in Sorrow" (Medhanandi, 1996):

> In pain we burn but, with mindfulness, we use that pain to burn through to the ending of pain. It's not something negative. It is sublime. It is complete freedom from every kind of suffering that arises; because of a realization—because of wisdom—not because we have rid ourselves of unpleasant experience, only holding on to the pleasant, the joyful. We still feel pain, we still get sick and we die, but we are no longer afraid, we no longer get shaken.
>
> When we are able to come face to face with our own worst fears and vulnerability, when we can step into the unknown with courage and openness, we touch near to the mysteries of this traverse through the human realm to an authentic self-fulfillment. We touch what we fear the most, we transform it, we see the emptiness of it. In that emptiness, all things can abide, all things come to fruition. In this very moment, we can free ourselves. We are not diminished by tragedy, by our suffering. If we surrender, if we can be with it, transparent and unwavering—making peace with the fiercest emotion, the most unspeakable loss, with death—we can free ourselves. And in that release, there is radiance. Jelaluddin Rumi wrote: "The most secure place to hide a treasure of gold is some desolate, unnoticed place. Why would anyone hide treasure in plain sight? And so it is said: Joy is hidden in sorrow."

Like the Hindu, for the Buddhist the goal is not to experience life after death but rather to experience *nirvana*—which has the property of neither existence nor nonexistence. According to Margaret Ayer (1964, p. 52), nirvana is the "state of peace and freedom from the miseries of the constantly changing illusion which is existence."

The location of nirvana is to be found in the image of the flame when a candle is "blown out"—it is in a place beyond human understanding. According to Ayer (1964, p. 53), whenever people achieve nirvana "they 'will be seen no more.' It

DEATH ACROSS CULTURES
The Story of Kisagotami

Sister Medhanandi

In the story of Kisagotami, this teaching is beautifully brought to life. She had married into a wealthy family, in spite of her poverty and unattractive appearance, and finally won the acceptance of her in-laws when she bore a little son. Suddenly, the child died. Nothing could be more tragic for Kisagotami. She refused to accept that her little son was dead. In her desperation, she came to see the Buddha, cradling the infant in her arms, believing that the Blessed One could revive him.

The Buddha asked her to procure a small quantity of mustard seeds from a house where no one had ever died. When she could find no household that had been spared death's unremitting hand, the insight into the impermanence of all conditioned phenomena arose in her mind. And so, Kisagotami was able to go beyond "the death of sons," beyond sorrow. (S.N. [Samyutta Nikaya] 5)

From "The Way of the Mystic"
by Sister Medhanandi, 1998
[On-line] . . . Available:
http://www.buddhanet.net/mystic.htm.

---

is through loss of desire, selfishness, evil, and illusion that this state of wisdom, holiness, and peace is reached." Buddhism contends that whereas physical death causes one to experience life again in a transmigrated form, "death to this world" (via nirvana) provides the gateway for ultimate happiness, peace, and fulfillment. Unlike Hinduism and the other religious traditions that we have considered, the ultimate goal of Buddhism is a state of consciousness and not a symbolic location for the disembodied soul.

TEMPORAL INTERPRETATIONS OF DEATH

Even though the funeral industry and most people in the United States tend to merge religious and death meanings, temporal and existential interpretations of death also provide a means for protecting social order in the face of death. Such interpretations tend to emphasize the empirical, natural, and "this world" view of death.

According to Glenn Vernon (1970, p. 33), "when death is given a temporal interpretation and is seen as the loss of consciousness, self-control, and identity, the individual may conclude that he or she can avoid social isolation in eternity by identifying him or herself with specific values, including religious ones." If we define religion as a system of beliefs and practices related to high-intensity value meanings and/or meanings of the supernatural (Vernon, 1970), then it is possible for individuals with temporal orientations to be "religious" in their outlook without affirming an afterlife. Furthermore, because any death has many consequences for the persons on whom it impinges, we would expect that even religious persons would assign some temporal meanings to death.

Vernon has pointed out that individuals whose interpretations are primarily temporal share the following beliefs and attitudes:

1. They tend to reject or de-emphasize a belief in the afterlife.

2. They tend to believe that death is the end of the individual.

3. They tend to focus upon the needs and concerns of the survivors.

4. They tend to be present-oriented for themselves, but present-and-future-oriented for those who will continue after them.

5. Any belief in immortality is related to the activities and accomplishments of the individual during his or her lifetime—including biological offspring and social relationships that the individual has created.

There is a strong temptation to view the person who has a temporal orientation as being very different from the person who finds comfort in a religious interpretation of death. In fact, both will attempt to restore the order in their personal lives and that found in society, by placing death into the context of a "higher" order.

For the individual with religious commitments, protection from anomie and comfort for anxiety are to be found by being in a relationship with the supernatural. For the person with a secular or temporal orientation, these same benefits are found in becoming involved with other people, projects, and causes. These involvements, although not pertaining to the supernatural, still provide a frame of reference that transcends the finite individual—a person may die, but his or her concerns will continue after death.

An existentialist perspective on death, as discussed in chapter 2, would suggest that one creates meaning for one's own life and death. Because one lives in a hostile world where one will eventually die, the primary human task it to find meaning in an otherwise meaningless world. This perspective is well illustrated by the article "On Death as a Constant Companion," which appeared in *Time Magazine* in 1965:

> Alone with his elemental fear of death, modern man is especially troubled by the prospect of a meaningless death and a meaningless life—the bleak offering of existentialism. "There is but one truly serious philosophical problem," wrote Albert Camus, "and that is suicide." In other words, why stay alive in a meaningless universe? The existentialist replies that man must live for the sake of living, for the things he is free to accomplish. But despite volumes of argumentation, existentialism never seems quite able to justify this conviction on the brink of a death that is only a trap door to nothingness.

WORDS OF WISDOM

You can keep your things of bronze and stone, just give me one person who will re-member me once a year.

Damon Runyon

Symbolic Immortality

From an existentialist perspective, the creative efforts that comprise one's life can provide individuals with personal meaning for both life and death. One way this is accomplished is through the creation of a symbolic immortality, which actually includes surrogate forms of immortality—such as the continuity of history, the permanence of art, or even the biological force of sex. **Symbolic immortality** (Lifton & Olson, 1974) refers to the belief that the meaning of the person can continue after he or she has died. For the religious, symbolic immortality is often related to the concept of soul, which either returns to its preexistent state, goes to an afterlife, is reincarnated in another body, or is united with the cosmos. For the person whose primary orientation is temporal, symbolic immortality is achieved by being remembered by others, by creating something that remains useful or interesting to others, or by being part of a cause or social movement that continues after the individual's death.

One of the reasons that many parents give for deciding to have children is the need for an heir—someone to carry on the family name. Research has demonstrated that in the United States, families with only female children are more likely to continue having children (in hopes of producing a male offspring) than are families with only male children. For the ancient Hebrews, the cultural institution of the **levirate marriage** required that a relative of the deceased husband have sexual intercourse with his dead relative's widow in order to provide a male heir. If it is possible to pass on something of oneself to one's children, then children are one method of providing symbolic immortality.

Within a given society there are high-status and low-status types of death. Giving one's life in defense of family or country is generally conceived as being of the high-status type. In previous times, dying in childbirth was considered to be a high-status death for females. As the dying of martyrs dramatically illustrates, a death may be willingly entered into if it is meaningful. Given the right configuration of meaning components (see chapter 2), not to die would be more difficult. For example, dying may be preferable to defining oneself and being defined by others as a coward or a traitor. Dying is acceptable if it furthers "the cause." People may, in fact, literally work themselves to death to obtain a promotion, an artistic achievement, or public recognition. Whatever the specific content, the key factor of concern is the meaning involved. If the meaning is right, dying may be evaluated as a worthwhile thing.

The death of an individual brings about a change in influence over the members of his or her group or society. Heroes are more likely to come from the ranks of the dead than of the living. Hero meaning is a symboled-meaning component. One cannot attain the status of hero by oneself. Society bestows this rank only upon certain of its members, one type of symboled immortality. Death is often a part of the process by which heroes are made. When something dies, something else is born or created. The death of Jesus Christ has expanded in significance over time. It is likely that Christ's death has had a greater impact upon humanity than did His life. The Roman Catholic Church grants sainthood only to persons who have been dead for many years. For that reason Mother Teresa, undoubtedly destined for sainthood, could only be given a funeral "fit for the best of mortals." Conversely, governments may grant pardons to convicted persons years after their deaths. Meaning is a flexible, yet powerful, thing.

Investing oneself in relationships with others also ensures that one will be remembered after death. Some will argue that if we have influenced the lives of others, something of us will continue in their lives after we die. Organ donations supply a tangible method for providing this type of symbolic immortality. In this way, one can even ensure that a part of his or her physical self can continue in another person. Currently there is an increasing tendency for individuals to donate their organs and tissues upon death to the living. In many urban areas, kidney foundations, eye banks, and transplant centers will supply donor cards and, when death occurs, will arrange for transplants.

The body of Mother Teresa, "the once and future saint," lying in state before her funeral.

WORDS OF WISDOM
Ozymandias

Percy Bysshe Shelley

I met a traveler from an antique land
Who said: Two vast and trunkless legs
of stone
Stand in the desert . . . . Near them, on
the sand,
Half sunk, a shattered visage lies,
whose frown,
And wrinkled lip, and sneer of cold
command,
Tell that its sculptor well those passions read
Which yet survive, stamped on these
lifeless things,
The hand that mocked them, and the
heart that fed;
And on the pedestal these words
appear:

"My name is Ozymandias, king of kings:
Look on my works, ye Mighty, and
despair!"

Nothing beside remains. Round the decay
Of that colossal wreck, boundless and
bare
The lone and level sands stretch far
away.

One of the reasons why people write books is to promote their own symbolic immortality. As long as their books can be read, their influence will outlive their biological body. The same is true for television and motion picture stars. Each year the youthful Judy Garland is resurrected from the dead as *The Wizard of Oz* is shown on television.

Great inventors, political leaders, and athletic "hall of famers" are also given immortality when we use their products, remember their accomplishments, and celebrate their achievements. In the case of medical practitioners and bionic inventors, not only do the living remember their accomplishments, but also these accomplishments extend the lives of those who provide the dead with immortality.

The person's finite identity is protected by the group's permanence. Just as Standard Oil is the legacy of John D. Rockefeller, John H. Leming and Son's Insurance Agency will remain even after John H. Leming and his sons are dead (providing that the new owners feel that it is in their business interest not to change the name of the company).

The same can be said for people who give themselves to political movements and causes. Marx, Lenin, Stalin, and Mao—as leaders of communism—will be remembered by future communists, despite the efforts by present officials to accomplish the contrary. We even provide infamous immortality to villains, murderers, and traitors. It seems ironic that most of us spend a lifetime working for immortality, when the guns of John Wilkes Booth, Lee Harvey Oswald, Sirhan Sirhan, and James Earl Ray supplied their owners with our everlasting remembrance.

For the existentialist, a lifetime is spent searching for and creating meaning. The search for meaning is a task that all people share. Furthermore, significant others are involved, as individuals try to create meaning for themselves. In many respects, the meaning created turns out to be meaning for the group or society. Symbolic immortality is something that only the living can give the dead. Yet people live with the faith that their survivors will remember them and perpetuate the meaning of their lives after they die. Like religious interpretations of death, temporal meanings enable individuals to protect themselves and their social order from death.

For the existentialist, the construction of symbolic immortality can give life a purpose and a sense of fulfillment, but for the critics of existentialism, even symbolic immortality cannot outwit death. From their point of view, there is also a more looming problem for those whose immortality depends upon others: What would happen if a nuclear holocaust were to occur, and there were no survivors?

NEAR-DEATH EXPERIENCES

In the past, one of the assumptions that most people made about the field of thanatology was that the real "experts" were not among us—they were dead. With the publication of Raymond Moody's *Life After Life* (1975), many people have stepped forward to challenge this assumption. Having been near death or having been declared clinically dead by medical authorities, a number of survivors of these experiences have "returned from the dead" to tell us that they now know what it is like to be dead and that they possess empirical evidence to support a rational belief in the afterlife.

Defining A Near-Death Experience

The near-death experience (NDE) is defined as an experience of a person near death (e.g., a person in a coma or a person who has experienced a heart attack), in an unconscious state, reporting that "something happened." In his now famous book, *Life After Life,* Raymond Moody (1975) brought national attention to NDEs. Throughout the book, Moody creates a composite of the many accounts by individuals who have survived the experience of being near death or being declared clinically dead. This ideal model was constructed by interviewing more than 150 people who had had NDEs. Moody reports, and other scientists support (Ring, 1980; Johnson, 1990), the similarity of most of these "life after death" accounts. Although no two NDEs are identical, there appears to be a pattern of experiences that include many, if not all, of the following nine traits (Perry, 1988):

1. *A sense of being dead.* At first many people don't realize that the experience that they are having has anything to do with being near death. They find

themselves floating above their body and feeling confused. They wonder, "How can I be up here, looking at myself down there?"

2. *Peace and painlessness.* An illness or accident is frequently accompanied by intense pain, but suddenly during a near-death experience the pain vanishes. According to research by psychologist Kenneth Ring (1980), 60 percent of people who have had a near-death experience report peace and painlessness.

3. *Out-of-body experience.* Frequently people feel themselves rising up and viewing their own bodies below. Most say that they are not simply at a point of consciousness but seem to be in some kind of body. Ring says that 37 percent have out-of-body experiences.

4. *The tunnel experience.* This generally occurs after an out-of-body experience. For many, a portal or tunnel opens, and they are propelled into darkness. Some hear a "whoosh" as they go into the tunnel, or they hear an electric vibration or humming sound. The descriptions are many, but the sense of heading toward an intense light is common to almost all tunnel experiences. Twenty-three percent of Ring's subjects reported entering darkness, which some described as entering a tunnel.

5. *People of light.* After people pass through the tunnel, they usually meet beings of intense light that permeate everything, as these beings fill the people with feelings of love. As one person said, "I could describe this as 'light' or 'love,' and it would mean the same thing." They frequently meet with friends and relatives who have died, though the glowing beings can't always be identified. In Ring's research, 16 percent saw the light.

6. *Being of Light.* After meeting several beings of light, there is usually a meeting with a Supreme Being of Light. To some this is God or Allah, to others simply a holy presence. Most want to stay with him forever.

7. *The life review.* The Being of Light frequently takes the person on a life review, during which his or her life is viewed from a third-person perspective, almost as though watching a movie. Unlike watching a movie, however, the person not only sees every action, but also its effect on people in his or her life. The Being of Light helps put the events of life into perspective.

8. *Rising rapidly into the heavens.* Some people report a "floating experience," in which they rise rapidly into the heavens, seeing the universe from a perspective normally reserved for satellites and astronauts.

9. *Reluctance to return.* Many find their unearthly surroundings so pleasant that they don't want to return. Some even express anger at their doctors for bringing them back.

The phenomenon is so widespread that in 1978 the International Association for Near-Death Studies (IANDS) was founded by researchers and those who have

had NDEs. Today, the organization is found on every continent but the Antarctic and has many local chapters in 38 states in the United States. IANDS publishes a quarterly newsletter *Vital Signs,* maintains a Speaker's Bureau, sponsors an annual North American conference, and produces a wide variety of educational materials. IANDS also encourages interest in research and professional applications through its publication of the quarterly *Journal of Near-Death Studies* and a program of small grants to encourage scholarly research on NDEs.

Explaining Near-Death Experiences

The following three questions become relevant as we contemplate the relationship between near-death experiences and other material presented in this chapter.

1. Are NDEs real, and what do they tell us about the dying process and/or being dead?

2. Can afterlife beliefs be empirically supported by these NDE accounts?

3. How do religious beliefs affect the content of near-death or clinical death experiences?

To all of our questions, we must temper our responses by saying that if not real, these experiences are very real to those who have had them. Not unlike being in love, NDE phenomena are extremely subjective and do not lend themselves to verification by others. Research by I. Stevenson, E. W. Cook, and N. McClean-Rice (1989) concluded that, of those claiming the near-death experience, only 45 percent were judged by medical experts to have had serious, life-threatening illnesses or injuries; the rest were judged to have had no life-threatening condition. It is difficult to doubt that the individual has experienced something; however, we are unable to prove, or disprove, that the individual has died. All that we can say scientifically is that the individual claims to have had the experience of "being dead."

Returning to our analogy of being in love, all that can be known is that a person says that he or she is in love. Whether or not the person really is cannot be determined empirically. From the perspective of the individual, it does not really matter because a situation that has been defined as real will have very real behavioral consequences.

Barbara Walker (1989) claims that many having a near-death experience have post-NDE depression due to the fact that health-care providers and family members often deny the validity of the NDE experience. It is not uncommon for persons having a near-death experience to tell a nurse or psychiatrist about their near-death event, only to be told that they were hallucinating (Walker, 1989). Kenneth Ring (1980) observed that a considerable number of divorces resulted from a spouse's inability to relate to the NDE.

If we try to determine the validity of a belief in an afterlife, we are confronted with problems similar to those encountered in the near-death experience. What can science say about beliefs in the afterlife? Scientifically, afterlife beliefs cannot be proved or disproved with NDE evidence. **Science** is based upon the principle of **intersubjectivity**. This means that independent observers, with different subjective orientations, must agree that something is "true" based upon their separate investigations. Unfortunately, the opportunity to experience the afterlife (and to return) is not uniformly available to all observers. Research by Tillman Rodabough (1985) gives many metaphysical, physiological, and social-psychological alternative explanations to account for the near-death experience. Therefore, although those who have had these experiences may feel rationally justified in their beliefs, the evidence that they use is not scientifically based. Science, at this point, can neither verify nor refute afterlife beliefs.

In responding to our last question concerning the effects of religious beliefs on the content of near-death or clinical death experiences, there is some scientific evidence that can provide limited answers. According to Raymond Moody (1975) and R. R. Canning (1965), the content of afterlife experiences is largely a function of the religious background, training, and beliefs of the individuals involved. Only Roman Catholics see the Virgin Mary and the host of Catholic saints, whereas encounters with Joseph Smith are reserved for Mormon believers. Like dreams, continuity exists between experiences in "this world" and experiences in the "afterlife." For example, personages in the afterlife are dressed in attire that would conform to the individual's cultural customs and beliefs. Also, relatives appear to be at the same age that they were when they were last seen. In fact, there is so much continuity between this world and the "other world" that persons with afterlife experiences report few, if any surprises. In response to this evidence, Kathy Charmaz (1980) raises the following question: Is the consciousness reported in these near-death experiences a reflection of a shared myth or evidence for it? In a more recent Australian study, Allen Kellehear (1996) reported that NDEs are confined largely to societies in which historic religions are dominant. His study argued that social and historical accounts are the best explanations for similarities of NDEs.

Even with the many disagreements regarding the near-death experience and its etiology, there is consensus that NDEs have profound effects upon the lives of most of those having the experience. According to P. M. Atwater (Johnson, 1990), 65 percent of individuals who have had a near-death experience make significant changes in their lives, and 10 percent make radical changes. According to Skip Johnson (1990), people with a near-death experience have the following aftereffects as a result of the phenomenon:

1. *Loss of fear of death.* People report that they no longer fear the obliteration of consciousness of self.

2. *Sense of the importance of love.* A near-death experience can radically change people's value structure. They see the importance of brotherly love.

3. *Sense of cosmic connection.* People feel that everything in the universe is connected. Many have a newfound respect for nature and the world around them.

4. *An appreciation of learning.* People gain a newfound respect for knowledge, but not self-gain. Many often will embark on new careers or take up serious courses of study.

5. *A new feeling of control.* People feel that they have more responsibility for the course of their lives.

6. *A sense of urgency.* Some people realize the shortness and fragility of their lives.

7. *A better-developed spiritual side.* This leads to spiritual curiosity and abandoning of religious doctrine purely for the sake of doctrine.

8. *Reduction in worries.* Some people feel more in control of life's stresses and are able to be more forgiving and patient.

9. *Reentry syndrome.* Some people have difficulty adjusting to normal life, especially those who undergo intensive changes in values that disrupt their former lifestyle. Some people report developing psychic abilities that can be scary to them and/or family and friends.

Finally, there is evidence that individuals (even if they have attempted suicide) who have had near-death experiences do not try to bring about an end to their lives to return to the "life beyond." In fact, most individuals find new reasons for living as a result of these experiences (Flynn, 1986; Greyson, 1983).

CONCLUSION

Religion is a system of beliefs and practices related to the sacred—to what is considered to be of ultimate significance. The cultural practice of religion continues because it meets basic social needs of individuals within a given society. A major function of religion, in this regard, is to explain the unexplainable.

For most "primitive," or less-complex societies, events are explained by a supernatural rather than by rational or empirical means. Thus, an eclipse of the sun or moon was said to be a sign that the gods had a message for humankind. These supernatural explanations were needed, in part, because there were no competing rational or scientific explanations. Technologically and scientifically advanced societies tend to be less dependent upon supernatural explanations. Yet such explanations are still important, especially when scientific explanations are incomplete. It is not uncommon to hear a physician say, "Medical science is unable to cure this patient; it's now in God's hands." Or, a physician might say, "It's a miracle that the patient survived this illness; I can't explain the recovery." Thus, one might suggest that religion takes over where science and rationality leave off. We depend less on religious or supernatural explanations than do nonliterate societies, but nonetheless we rely on them when knowledge is incomplete.

Yet even in secular societies, religion and spirituality often play significant roles by helping individuals cope with extraordinary events—especially illness, dying, and death. Not only does religion help restore the normative order challenged by death, but also strong religious and spiritual orientations can enable individuals to cope better with their own dying and the deaths of their loved ones. For others, strong commitments to a temporal orientation may fulfill many of the functions provided by a religious or spiritual worldview.

SUMMARY

1. Religion helps individuals, when the order of everyday life is challenged, by providing answers to problems of uncertainty, powerlessness, and scarcity created by death.

2. Religious systems provide a means to reestablish the social order challenged by death.

3. When one encounters death, the anxiety experienced is basically socially ascribed.

4. Religion provides individuals with solace when they attempt to cope with death.

5. Temporal interpretations of death provide a means for protecting the social order by emphasizing the empirical, natural, and "this worldly" view of death.

6. Symbolic immortality is evidenced by offspring carrying on the family name, by donating body organs, and by having accomplishments or achievements (positive and negative) remembered by others.

7. Near-death or afterlife experiences are influenced by the individual's religious background, cultural beliefs, and prior social experiences.

DISCUSSION QUESTIONS

1. How does religion function to provide a restoration of the order challenged by the event of death?

2. Explain the following statement: "Religion afflicts the comforted and comforts the afflicted."

3. How can symbolic immortality and temporal interpretations of death provide a source of anxiety reduction for those who face death?

4. How can organ donations provide symbolic immortality for donors and their loved ones?

5. Do accounts of near-death experiences provide empirical evidence for afterlife beliefs? Why or why not?

6. What are the similarities and differences in Jewish, Christian, Islamic, Hindu, and Buddhist beliefs about death and funeral practices?

GLOSSARY

Anomie: A condition characterized by the relative absence or confusion of values within a group or society.

Intersubjectivity: A property of science whereby two or more scientists, studying the same phenomenon, can reach the same conclusion.

Levirate Marriage: An institution typified by the Hebrew requirement that a relative of the deceased husband must have sexual intercourse with the deceased's widow in order to provide a male heir.

Marginal Situations: Unusual events or social circumstances that do not occur in normal patterns of social interaction.

Religion: A system of beliefs and practices related to the sacred, the supernatural, and/or a set of values to which the individual is very committed.

Rituals: A set of culturally prescribed actions or behaviors.

Science: A body of knowledge based upon sensory evidence or empirical observations.

Symbolic Immortality: The ascription of immortality to the individual by perpetuating the meaning of the person (the self).

Suggested Readings

Berger, P. L. (1969). *Sacred canopy: Elements of a sociological theory of religion.* New York: Doubleday. Excellent treatment of the role of religious worldviews as they relate to life crises. Death is discussed as the ultimate marginal situation to normal social functioning that calls forth religious meaning systems.

Eliade, M. (Ed.). (1987). *The encyclopedia of religion.* New York: Macmillan. Academic resource on religious rites, beliefs, and traditions. This reference tool is written from an interdisciplinary perspective by international scholars of religion.

Flynn, C. P. (1986). *After the beyond: Human transformation and the near-death experience.* Englewood Cliffs, NJ: Prentice-Hall. Provides a background for dealing with persons who claim to have had the near-death experience.

Moody, R. A., Jr. (1975). *Life after life: The investigation of a phenomenon—Survival of bodily death.* Boston: G. K. Hall. Patient accounts of near-death experiences shared with Moody, a physician and philosopher. Moody objectively presents the cases.

Parry, J. K., & Ryan, A. S. (Eds.). (1995). *A cross-cultural look at death, dying, and religion.* Chicago: Nelson-Hall Publishers. Religion and death are discussed in this anthology. Groups included are African Americans, Buddhists, Roman Catholics and other Christians, Chinese, Dominicans, Filipinos, Christian fundamentalists, gays, Muslims, Jews, Koreans, lesbians, Mexican Americans, and women.

Ring, K. (1980). *Life at death: A scientific investigation of the near-death experience.* New York: Coward, McCann and Geoghegan. Provides an understanding of the near-death experience.

Spiro, H. M., Curnen, M. G. M., & Wandel, L. P. (Eds.). (1996). *Facing death: Where culture, religion, and medicine meet.* New Haven: Yale University Press. Christian, Judaic, Islamic, Hindu, and Chinese perspectives on death and rituals of mourning are discussed.

Young, M., & Cullen, L. (1996). *A good death: Conversations with East Londoners.* New York: Routledge. This book takes four approaches to death: the rational, the spiritual, the humorous, and the poetic.

CHAPTER 5

THE DYING PROCESS

After the death of his wife, Jenny, Forrest Gump stated, "Mamma always said that dying is a part of life. I sure wish it wasn't."

—Forrest Gump

I'm not afraid of dying. I just don't want to be there when it happens.

—Woody Allen

We tend to fear the dying process more than the event of death: the pain, the isolation, the loss of physical strength and neutral activity, the loss of favorite activities, the possibility of institutional care, the loss of independence and regression to a state of extreme dependency. As Morrie Schwartz, who is dying of amyotrophic lateral sclerosis (ALS), or Lou Gehrig's disease, says in the best-selling book entitled *Tuesdays with Morrie* (Albom, 1997), "The ultimate sign of dependency is having someone wipe your bottom." The loss of one's physical capabilities can be demeaning.

Yet, being immobile and limited to one's bed leaves a lot of time for thinking. On a positive note, such time presents the opportunity for the patient to get her or his house in order and complete any unfinished business, such as saying goodbyes, obtaining a will (if he or she did not already have one), and making amends for any wrongdoings committed earlier in life. The time of idleness allows for reflection and recalling memories of days gone by. The patient may have some idea of the doctor's prognosis as to the amount of time left to live and thus may plan accordingly.

LISTENING TO THE VOICES
Teaching While Dying

Mitch Albom visited his former sociology professor, Morrie Schwartz, on Tuesdays for the last 14 weeks of Professor Schwartz' life and shares the account in *Tuesdays with Morrie*. In dying a lingering death from ALS, Morrie noted that you should "be grateful that you have been given the time to learn how to die." He showed a complete lack of self-pity. Professor Schwartz said that the way you get meaning into your life is by loving others, devoting yourself to your community around you, and devoting yourself to creating something that gives you purpose and meaning (Money is *not* the most meaningful thing!). Without love, "we are birds with broken wings."

Professor Schwartz reminded us of the Buddhist saying that each of us has a little bird on our shoulder every day that asks, "Is today the day? Am I ready? Am I doing all I need to do? Am I being the person I want to be?"

Morrie said, "Aging is not just decay. . . , it's growth. It's more than the negative that you're going to die, it's also the positive that you *understand* you're going to die, and that you live a better life because of it."

One achieves respect by offering something that one has, observed Schwartz. You do not have to have a big talent. There are lonely people in hospitals and nursing homes who only want companionship. Play cards with a shut-in and find new respect for yourself, *because you are needed.*

"Death is as natural as life," said Morrie. "It's part of the deal we made." Everything that gets born dies. Death ends a life, not a relationship.

Adapted from Tuesdays with Morrie *by Mitch Albom, 1997, New York: Doubleday.*

Because the process of dying is typically not something that we experience often in life, the dying person may have few clues as to how she or he should act. We want to behave as expected and follow the norms of our society, but not knowing the "rules for dying" may be frustrating for the individual. Indeed, we are typically not socialized to the process of dying. So, the dying patient may simply learn while in the process.

On the other hand, medical personnel may be limited in their ability to relate to the dying person, because they have been socialized to "prolong life and relieve suffering." Thus, we may find that both the patient and those around her or him may be uncomfortable with their roles as caregivers to the dying. This awkwardness is often compounded by the patient's lack of awareness of the prognosis. The situation is made even more uncomfortable by the fact that perhaps the medical personnel involved are not aware of what the patient and family know, and thus mum is the word: "I'm afraid to say anything for fear I might disclose information that was not previously known to others."

Friends and family may also be uncomfortable because, as discussed in chapter 1, they have had little exposure to death. In fact, they may be so uncomfortable that they do not assist the dying person at all.

Richard Kalish's study of college students' attitudes toward the dying (1966) perhaps reflects society's reactions to dying persons in general. He found that one third of the respondents would not willingly allow a dying person to "live within a few doors" of them. Only half of the sample would willingly become close friends with a dying person. Kalish notes that if these college students' reactions are in any way typical, then most people would avoid dying persons, as they do those who are grieving, often noting that they do not know what to say.

DEATH MEANINGS

The meaning of our dying will depend to a great extent upon the social context in which the dying occurs. Meanings are the basic component of human behavior because individuals respond to the meanings of phenomena rather than to the phenomena themselves. Meanings are both socially created and socially perpetuated (Leming, Vernon, & Gray, 1977).

Sociologist Robert Merton, in a classic article written in 1949 entitled "The Bearing of Sociological Theory on Empirical Research," illustrates the idea of responding to the conceptual situation or meaning, rather than to the physical object or phenomenon itself, with "gasoline drums." Objects conceptually described as "gasoline drums" will produce behavior of great care in people. But, if people are confronted with "empty gasoline drums," behavior is different. A smoker, for example, probably will not be very cautious around an "empty" drum. Yet, the "empty" drum is more hazardous because it contains explosive vapor. We respond not to the physical object but to the conceptualized situation or meaning of the phenomenon or object. "Empty" here is a synonym for "null and void" or "inert," without regard to vapor in the drum. Thus, empty gasoline drums become the cause of fires. Clarification of the meaning of "empty" would have a profound effect upon our behavior.

The major types of meanings to which individuals respond in death-related situations are the following: time meanings, space meanings, norm and role meanings, value meanings, object and self meanings, and social situation meanings (Vernon, 1972). We will now look at each of these types of meanings and their effects upon the dying process.

Time Meanings: Dealing With Prognosis

As anthropologist Colin Turnbull once told George Dickinson, "Americans want to live forever. If you ask individuals how long they wish to live, they will probably say somewhere in the upper 90s because they do not suspect they will reach 100. But if they reach 100, then they will aim for 101, 102, etc." Turnbull said that African groups know when it is time to die, and they accept this. Perhaps Richard Pryor

---

LISTENING TO THE VOICES
About Living

I invited to class a woman whose diagnosis was terminal cancer. Deeply impressed by the visitor's positive outlook on the time left to her, one student noted on the following day, "When I came into the session I expected to meet someone who was dying. Instead, I realized that woman is more alive than I am. The discussion wasn't about dying at all, it was about living."

"Dialectic on Dying" (pp. 182–189), by J. M. Boyle, 1981. In M. M. Newell et al. (Eds.), The Role of the Volunteer in the Care of the Terminal Patient and the Family, *New York: The Foundation of Thanatology/Arno Press.*

---

summed it up well when he said, "Even if I live for 100 years, I'll be dead a lot longer." Thus, one is dead longer than alive.

In thinking about the dying process, the first thing that comes to our awareness is the concept of time. We are confronted with the fact that time, for the terminally ill patient, is running out. Yet, when does the dying process begin? Are we all not dying, with some reaching the state of being dead before others? From the moment of our births, we are approaching the end of our lives. We assume that terminally ill patients will experience death before other individuals, but this is not always the case.

Because it is possible to diagnose diseases from which most people die, we can assume that patients with these diseases are "more terminal" than individuals without them. The terminally ill patient is very much concerned with the time dimension of his or her physical existence. "How long have I got to live?" is more than a question found on soap operas; it is a query that general practitioners hear, on average, six times a year and oncologists face as often as 100 times a year (Sharlet, 2000). In general, patients want to know if their illness is terminal. However, physicians traditionally may have felt unprepared to tell their patients of a terminal diagnosis. According to studies by Nicholas Christakis (1999), both a physician and a sociologist, doctors usually ignore prognosticating, and when they do, they overestimate their patient's remaining days by a factor of two to five. Christakis notes that prognosis is akin to prophecy. Prophets seek to shape the future they foresee by foretelling it, revealing good fortune or warning of doom. In so doing, prophets can alter the actions of those to whom they prophesy. Christakis found that only 14 percent of doctors in his study included substantive discussion of the future in their meetings with patients. Most physicians stick to an unspoken code: Don't foresee. If you do, do not foretell.

With more emphasis on consumer rights today, physicians are more likely to tell patients of their diagnosis than was the case in the early 1960s. In a 1961 survey of

PRACTICAL MATTERS
Breaking Bad News

Suggestions for breaking bad news to patients by Sean Morrison and Jane Morris (1995) follow a six-step protocol:

1. Arrange to meet in a private setting, where you will not be interrupted.

2. Establish what the patient (and/or family) already knows.

3. Identify how much the patient (and/or family) wants to know.

4. Share the diagnosis and prognosis with the patient (and/or family). Present the various treatment options available and provide a realistic appraisal of the benefits and burdens of each.

5. Respond to the patient's (and/or family's) feelings, and identify and acknowledge their reactions.

6. Formulate a plan of care and establish a contract for the future.

Adapted from I Don't Know What to Say: How to Help and Support Someone Who is Dying, *by R. Buckman, 1992, New York: Viking Press. Morrison, R. S. and Morris, J. (1995, July) When there is no cure: Palliative care for the dying patient. Geriatrics, 50, 45–50.*

physicians (Oken, 1961), only 10 percent favored telling a patient with cancer his or her diagnosis, whereas in 1994, 95 percent of physicians surveyed self-reported that they always disclose the diagnosis (Holland, 1994). This change could reflect the American Medical Association's shift in policy in 1980 (Dickinson & Tournier, 1994) of encouraging physicians to tell the patient his or her prognosis:

> The physician must properly inform the patient of the diagnosis and of the nature and purpose of the treatment undertaken or prescribed. The physician must not refuse to so inform the patient. Previously, the AMA had left the decision to the discretion of the physician. Additionally, under the Federal Patient Self-Determination Act of 1991, hospitals must inform patients of their rights to make their own decisions about their medical care (Blackhall et al., 1995); physicians today commonly tell their patients their diagnoses, no matter how dire.

A study by George Dickinson and Robert Tournier (1994) of physicians' attitudes toward terminally ill patients revealed more of an openness toward informing the patient of his or her prognosis after 10 years of practicing medicine than soon after graduation from medical school. A follow-up study (Dickinson, Tournier, & Still, 1999) 10 years later showed a continued openness to communicating with terminally ill patients and their families. A study by Seale (1991) in England suggests a general preference for more openness between physicians, dying patients, and their families about illness and death. This openness should be tempered by the consideration that bad news needs to be broken slowly in a context of support while recognizing that not everyone wishes to know all.

Kerry Gasperson (1996), in her study of first-year medical students' attitudes toward delivering bad news in a clinical context, observes that a prominent concern of these students is how to deliver news about the terminal nature of a condition. The professionals seem to differ regarding the use of euphemisms when delivering bad news. P. Maguire and A. Faulkner (1992) endorse the use of euphemisms to ease the patient into the truth, whereas Timothy Quill (1991) strongly advises against using euphemisms when communicating a diagnosis to the patient. When the message is not clear, some clinicians think, communication difficulties may result. For example, stories are abundant about patients, who when told that they have a malignancy, are relieved to know that it is not cancer (Sell et al., 1993).

Some ethnic groups in the United States prefer that the patient not be told the prognosis. A study of 800 elderly patients in California (Blackhall et al., 1995) found that immigrant families from South Korea and Mexico were far less willing to let their terminally ill family members make decisions about medical care than were either black or white Americans. Black and white respondents were about twice as likely as Korean Americans to be honest with the patient about his or her terminal status, and about one and a half times as likely as Mexican Americans. For immigrants from South Korea and Mexico, the role of the family was larger and that of the individual was smaller than is generally the case in the United States. In Japan a similar situation exists as doctors rarely tell dying patients that they are terminally ill (Kristof, 1996). According to J. Dillard (1983), in the giving of "bad news" to terminally ill patients, as a sort of "buffering" mechanism, many Asian Americans have been taught to mask behavior and avoid eye contact when self-disclosure is occurring.

Though many physicians in the past have not favored telling their patients that they are dying, Elisabeth Kübler-Ross (1969) suggests that the patient should be told, but that the prognosis should not include the amount of time the patient has left. As Janice Rosenberg (2000) observes in her article entitled "Art of Prognosis Becoming an Increasingly Valued Skill," experts agree that currently there is no accurate method for predicting when a patient will die. Prognosticating is especially difficult for noncancer diagnoses, though prognostication improves for any illness when physicians know their patients as people. Specifying an amount of time dispels hope and serves no function other than telling the patient that the condition is very serious and life threatening, advises Kübler-Ross (1969). If the patient presses for time specificity, a range of time should be given, such as: "Sixty percent of the patients with your disease live as long as 3 to 5 years." This statement provides hope without deluding the patient. Also, it is honest. Medical science cannot say with any certainty that a person will live 1 year—some die sooner, and others live longer (some even outlive the physician!). Physicians should tell patients they have limited time left when they still have enough energy to think about projects and goals, notes Rosenberg (2000).

Thus, after the physician has given the news to a patient, there is no wonder that when the physician asks the patient if there are any questions, the patient often has none. Many times the patient does not clearly understand what the doctor said.

Some of the most beautiful human interactions I have witnessed have occurred between dying patients and supportive families. Sometimes the quality of human interactions in the terminal phase far exceeds anything the patient or family experienced prior to diagnosis.

I strongly feel the dying patient should be told as much as he or she wants to know. The family should also be encouraged to share feelings with the patient in an open manner. Nothing is worse than dying alone. The terminal patient whose family won't broach the subject, or who is afraid to upset his or her family or doctors with fears and feelings, does die alone.

A physician's comment from a survey of 1,093 physicians, "Death Education and Physicians' Attitudes Toward Dying Patients" (pp. 167–174) by George Dickinson and A. A. Pearson, 1980–1981, Omega, 11.

Indeed, as patients we need to be more assertive and to inquire because it is our lives that are at stake and our right to know. However, it is not easy for a physician to give a patient a "bad prognosis." At a palliative care conference in Sheffield, England, on June 6, 1999, Dame Cicely Saunders said that a patient said to her, after she told him he was going to die: "It's hard to be told. It's hard to tell, too, isn't it?"

Perceptions about the course that dying will take are referred to as **dying trajectories**. Staff interaction with patients is closely related to the expectations formed about the patients' dying. Trajectories may range from lingering to sudden. Quick trajectories usually involve acute crises (Fulton and Metress, 1995). In real life, patients perceived to be in a lingering death trajectory may suffer a loss in social worth, and medical personnel may give up on them, contributing to a downward slide to the patient's health. On the other hand, even when nurses know that a patient is terminally ill, their behavior is determined by the physicians' "official message." The patient may then be treated as if there were little chance for recovery. This alteration in treatment is predicated upon the physician's decision but is not carried out until he or she communicates it. If the patient does not die "on schedule," the family and staff may experience stress (Shneidman, 1980).

Space Meanings: Isolation and Confinement

Even when the patient has not been told of a terminal condition, he or she will eventually become aware of it. Many times factors related to social space will give the patient clues that the condition is terminal. As E. P. Seravalli (1988, p. 1729) notes,

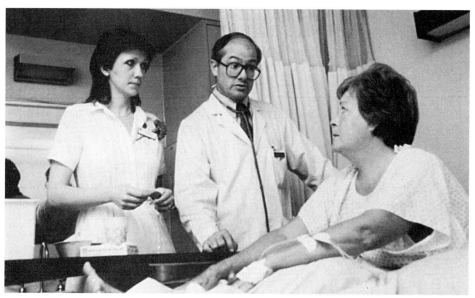

When a patient is told that the disease is terminal, the information must be communicated in a manner that is honest but does not rob the patient of a sense of hopefulness.

"The dying usually come to know when the end is near by observing themselves and the people around them." Within the hospital, there are areas where the very ill are treated. When one is moved into the intensive care unit or onto an oncology ward, it becomes obvious that all is not well and that death is a real possibility. When being moved from a double room to a single-patient room, 17-year-old Tom Nelson, who was dying of cancer (as seen on CBS's *Living With Death*), said, "Isn't this what they do to you when you are about to die?"

Confinement to a health-care institution conveys a tremendous amount of meaning to the patient. For the most part, the patient is alone. Through spatial meanings the individual is "informed" that he or she is removed from those things that give life meaning and purpose—family, friends, and job. For the terminally ill patient this is the first stage of **societal disengagement**—the process by which society withdraws from, or no longer seeks, the individual's efforts.

The following statement (Hale, 1971) demonstrates the significance of space when applied to the dying patient:

> As for the patient himself, he faces his life's ultimate crisis in a foreign place—the hospital. Usually, he dies alone, not in the presence of anyone who cares. Often the family is allowed to visit even the dying only during the hours set by the hospital. And since hospital visits usually lack privacy, they are of little comfort either to the patient or to the relatives who are entering the period of "anticipatory grief." There is little room or opportunity for sharing that most poignant moment of a lifetime—death! There is little opportunity for dignity in those settings.

Though many institutional settings have liberalized their visitation policies some-what since the early 1970s, unfortunately the preceding statement by Hale still ba-sically holds.

Within any health-care setting, the patient's confinement serves to diminish his or her social and personal power. According to Rodney Coe (1970), three processes occur within the institution to accomplish this—"stripping," controlling resources, and restricting mobility. The process of *stripping* takes place when the patient is is-sued a hospital gown and stripped of any valuables for safe keeping. When this is done, the patient's identity is also stripped. Most factors that differentiate patients with regard to status in the larger society are taken from the patients and create the primary status of patients—all patients look alike. According to Coe (1970, p. 300), "Every distinctly personalizing symbol, material or otherwise, is taken away, thus reducing the patient to the status of just one of many." Perhaps this helps to explain why physicians are known to be such bad patients—they are forced to relinquish their physician status when they are admitted to the hospital.

The second process is *controlling resources*. When the patient is denied access to medical records and important information about the events of the hospital, per-sonal power is greatly diminished. Personal power is the ability to make decisions that determine the direction of one's own life. Without all of the information con-cerning one's situation and the place of one's confinement, it may not be possible to make important decisions. One method used by hospitals in controlling resources is to deny all patients and their families access to medical charts and records—an "ig-norance is bliss" mentality.

The third process is *restricting mobility*. Not being able to leave one's room or bed further reduces the patient's personal power. The patient is put into a position of dependency upon others. Confinement of this type greatly affects the patient's au-tonomy. It also makes it possible for others to withdraw from the patient.

Social space is very important in the process of patient disengagement. This dis-engagement can be accomplished by two methods—the patient can withdraw from others, and others can withdraw from the patient. If the patient is debilitated by ill-ness, energy may not be abundant enough to continue normal patterns of social in-teraction. The loss of physical attractiveness can also cause the patient to withdraw. Some patients, knowing that their condition is terminal, may disengage themselves as a coping strategy to avoid having to see all that their death will take from them. They may disengage as a sign of their acceptance of social death—"I'm as good as dead" (anticipatory death).

When significant others withdraw from the patient, patient disengagement will take place. In this situation the process of disengagement is something beyond the patient's control. Family members and friends can refrain from visiting the patient as a sign of their acceptance of social death. The "terminal" label can stigmatize the patient, and others may treat the individual differently.

Orville Kelly, founder of Make Each Day Count (a support group for persons with cancer), tells the story of being invited by a friend to dinner. The table was set with the finest china and silverware—with one exception. Orville's place setting consisted of a paper plate and plastic fork, spoon, and knife. Mr. Kelly was the guest, but he had cancer, so he was supposed to use disposables so no one would be contaminated!

Randall Wagner, an active volunteer with the American Cancer Society, recalls a situation at a high school football game when his leukemia was in a state of remission. A friend had a Thermos of hot chocolate but had only two cups. The first cup was given to Randall. After he had finished drinking, he returned the cup to be filled for someone else. Nobody would drink out of Randall's cup, however. He was told to "just keep it."

Many individuals are afraid of "catching" cancer. One does not "catch" cancer. Likewise, many individuals are afraid to be physically near someone with AIDS for fear of acquiring the disease. Because AIDS is primarily passed through sexual intercourse, intravenous drug needles, blood transfusions, or from a mother to her fetus/embryo, one is not going to acquire AIDS by sitting next to or by physically being in the room with someone with AIDS. Because many do not know how to relate to persons with cancer or AIDS or to the terminally ill in general, they withdraw as a method of coping with their inadequacy.

Norm and Role Meanings: Expected Dying Behavior

Norms are plans of action or expected behavior patterns felt to be appropriate for a particular situation. **Roles** are plans of action or expected behavior patterns specifying what should be done by persons who occupy particular social positions. Applied to the death-related behavior of the dying patient, norm definitions would involve the general expectation that the dying patient should be brave and accept the fact that life will soon end. The patient is not supposed to cry or become verbal in regard to feelings about his or her death. Nurses often sanction such behavior by giving less attention to patients who deviate from this norm. Elisabeth Kübler-Ross (1969, pp. 56–57) gives the following example of one such deviant:

> The patient would stand in front of the nurses' desk and demand attention for herself and other terminally ill patients, which the nurses resented as interference and inappropriate behavior. Since she was quite sick, they did not confront her with her unacceptable behavior, but expressed their resentment by making shorter visits to her room, by avoiding contact, and by the briefness of their encounters.

Role meanings differ from norm meanings in that they specify, in a detailed fashion, what behavior is expected of persons who occupy specific social positions. For example, if a wife/mother who is a principal provider in the family is dying, it is expected that she do all that she can before death to provide for the financial needs of her family. It would probably be expected that she make arrangements for her funeral, ensure that her bills are paid, finalize her will, and establish a trust fund for her children.

As noted earlier, one of the important aspects of norm and role meanings is their relationship to the process of societal disengagement by which society withdraws from the individual (Atchley, 1994). The individual can also withdraw from societal participation and choose not to perform the roles that he or she performed before the terminal diagnosis. This type of disengagement goes beyond withdrawing from interaction patterns with others and refers to a withdrawal from the social structure (such as quitting one's job and taking a trip around the world).

Role disengagement has many consequences for the patient and his or her family members. N. J. Gaspard (1970, p. 78) notes the following about a family with traditional gender roles:

> If the father is ill, the mother must generally become the breadwinner, and children who are able often must take over household tasks sooner than they otherwise might. Each person in such a situation may feel both guilt and resentment at such a change whereby they can no longer adequately fulfill the expectations they have had of themselves. Conversely, if the mother is ill, household help may be hired, and problems may arise with regard to the mother's maintaining, in so far as is possible, her self-image in relation to caring for her family.

In addition to the disengagement process, patients are expected to acquire the **sick role** (Parsons, 1951). They are expected to want to get better and to want to seek more treatment even though everyone realizes that such treatment only prolongs death and not life. This role disengagement can create conflict within the family, especially when the patient has accepted his or her death (and even longs for it), whereas family members are unwilling to let the patient go. As Kübler-Ross (1969) documents, families many times cannot comprehend that a patient reaches a point when death comes as a great relief and that patients die easier if they are allowed and helped to detach themselves slowly from all of the responsibilities and meaningful relationships in their lives. John Harvey (1996, p. 173) states:

> There is a kind of freedom that comes with a recognition that your time is limited. Long-term consequences seem less important. A certain kind of fearless authenticity often comes from a confrontation with one's mortality.

Often, we, the survivors, are the stumbling blocks for dying persons. We do not want the dying person to die; thus, we will not "let go" and allow her or him to say goodbye and break away.

Value Meanings: Reassessing the Value of Life and Death

Values, like all other meaning systems, are socially created. They are not inherent in the phenomena, but rather they are applied by humans to the phenomena. Death per se is neither good nor evil. Humans do ascribe value, however, to the different types of deaths. Each of these value meanings has important behavioral consequences.

LISTENING TO THE VOICES
Don't Abandon the Patient

When I say I feel as comfortable with a dying patient as with any other, and that I do not find treating a dying patient unpleasant, I do not mean that I am anaesthetized to the fact that they are dying, and do not have feelings about the patient which are different from my feelings about a patient whom I know will get well. Anesthetizing of feeling is the method which we physicians employ initially in dealing with the pain—ours and theirs—involved in treating a dying patient. But this passes, and when one accepts the patient as part of life, and not someone who is no longer a real part of the world (or a fright-ening part of the world), then caring for the dying patient becomes (though often sad) neither unpleasant nor something one wishes to avoid. To abandon the dying patient is the worst thing that can be done—both for the patient and the doctor.

A physician's comment from a survey of 1,093 physicians, "Death Education and Physicians Attitudes Toward Dying Patients" (pp. 167–174) by George Dickinson and A. A. Pearson, 1980–1981, Omega, 11.

Most people in our society view death as being intrinsically evil and, therefore, something to fear. We tend to see death as an intruder—the spoiler of our best plans. Thus, in the past the medical profession has attempted to delay death in favor of life. People are kept on machines to prolong life, even if death is inevitable. We assume that people wishing to die are mentally ill or irrational because we see death as something to be avoided. Yet, the terminally ill eventually come to view death as a great blessing, when they have finally accepted the fact that they are going to die. This is possible because humans are able to create hierarchies of values in which value meanings take on relative meanings. To the terminally ill patient, dignity is more highly valued than is life with pain, indignity, and suffering. Consequently, death may be ascribed positive value for the dying patient who has accepted the inevitability of his or her death.

Daniel Goleman (1989) cites research showing that confronting individuals with the fact that they will die makes them cling tenaciously to their deepest moral values. They tend to become more moralistic and judgmental. They are harsher toward those who violate their moral standards and kinder toward those who uphold them. Open-minded persons, however, become even more tolerant of those whose values differ from theirs. Researchers (Goleman, 1989) note that these findings give the fear of death a central role in psychological life and that a culture's model of "the good life" and its moral codes all are intended to protect people from the terror of death. Cultures prescribe what one must do to lead a "good life," and if one leads a "good life," he or she will be protected from a tragic fate at death.

Object and Self Meanings: Accepting the Self as Terminal

The previous discussion focused on meanings that have been applied to an object—the dying patient. From a biological perspective the person is a living organism—a physical object. From a social-psychological perspective, the patient is a social object—a self.

An important part of accepting oneself as terminal is exploring the meaning of that condition with others. With death close, dying patients may still want to express what they experience and to convey their fantasies about death (Seravalli, 1988). The presence of the physician in this last phase of life may be crucial for a peaceful death. Such a presence may allow the patient not only to die with self-respect, but also to feel less lonely. Sociologist Arthur Frank (1991), who himself had two serious illnesses by the age of 40, believes that too many ill individuals are deprived of conversation. He notes that too many believe that they cannot talk about their illness: They simply repeat what they know from the medical staff. When ill persons try to talk in medical language, they deny themselves the drama of their personal experience. Frank notes that sick persons need to talk about their hopes and fears and the prospect of death. But because such talk embarrasses us, we do not have practice with it and, lacking practice, we find such talk difficult.

As death draws near, individuals may speak of preparation for travel, often recognizing one who has already died coming for them (Callanan & Kelley, 1992). Occasionally, angels or religious figures are mentioned as beckoning to them. A friend, who works with chronically and terminally ill children in a hospital setting, told George Dickinson about a terminally ill child who said that the "pretty woman with flowers standing in the corner of the room" was calling the child to come with her to play with the children. Though no one was "standing in the corner of the room," the friend acknowledged the child's "observation." "I am not ready to go yet," the little girl replied. A couple of days later, the child again said that the woman "in the corner of the room" was signaling for her to come and play with the children, who were laughing and having a good time. The child said, "I am now ready to go with them." She died an hour later. Thus, there is a connection between the living and the dead as one prepares for "the journey."

Detachment From the Living

Many patients who are defined as having a terminal condition begin to view themselves as being "as good as dead." Ideally, the dying person has reached Erik Erikson's final stage of the life cycle—senescence—believing that his or her life has had meaning and purpose, and the person is ready to break away. Life is over, and the dying patient is accepting of that. The patient has accepted the terminal label, has applied it to a personal understanding of who she or he is, and has experienced anticipatory death. Families tend to see their loved ones as being in bereavement. This symbolic definition of the patient is reinforced by the role disengagement process, the spatial isolation of the dying patient, and the terminal label placed upon the patient by the physician and other medical personnel. The patient seems to take on a status somewhere between the living and the dead.

Elisabeth Kübler-Ross (1969, p. 116) states that in the end terminally ill persons have a need to detach themselves from living persons to make dying easier. The following example illustrates this point:

> She asked to be allowed to die in peace, wished to be left alone—even asked for less involvement on the part of her husband. She said that the only reason that kept her still alive was her husband's inability to accept the fact that she had to die. She was angry at him for not facing it and for so desperately clinging on to something that she was willing and ready to give up. I translated to her that she wished to detach herself from this world, and she nodded gratefully as I left her alone.

When an individual's condition has been defined by self and others as terminal, all other self-meanings take on less importance. Although a given patient may be an attorney, a Democrat, a mother, a wife, or a Presbyterian, she tends to think of herself primarily as a terminally ill patient. The "terminal" label becomes her **master status** because it dominates all other status indicators. Consequently, most of the symbolic meanings previously discussed become incorporated into the individual's self-meaning.

Kübler-Ross's Five Stages of Dying

Acquiring the terminal label as part of the self definition is not an easy task for the individual. In her best-selling book *On Death and Dying* (1969), Elisabeth Kübler-Ross delineated the following five stages that patients go through in accepting their terminal self-meaning: denial, anger, bargaining, depression, and acceptance.

In the first stage, *denial,* the patient attempts to deny that the condition is fatal. This is a period of shock and disbelief (for example, the patient wants to believe that the physician must have read the wrong x-ray). The patient wants to believe that a mistake has been made. One may seek additional medical advice, hoping that the terminal diagnosis will be proven false. When the diagnosis is verified, the patient may often retreat into self-imposed isolation.

The second stage, *anger,* is a natural reaction for most patients. The patient may vent anger at a number of individuals—at the physician for not doing enough, at relatives for outliving the patient, at other patients for not having a terminal condition, and at God for allowing the patient to die. In other words, a **scapegoat** is sought—the patient seeks someone or something on which the blame can be placed.

When one has incorporated the terminal label into self-meaning, an attempt to bargain for a little more time may result. This third stage, *bargaining,* may include promises to God in exchange for an extension of life, followed by the wish for a few days without pain or physical discomfort. For example, in the movie *The End,* Burt Reynolds decides to commit suicide by drowning. He swims far out into the ocean but begins to change his mind about wanting to die. He bargains with God and prays to God to make him the best swimmer in the world so that he can swim back to shore. In return for God's granting his wish, Reynolds promises to give most of his income for the rest of his life to the church. As he draws nearer to shore and realizes that he is going to survive, his "promises" to God are reduced to 50 percent of his income, then 40, and downward, until he reaches the shore and reduces his promise to almost nothing. Bargaining generally includes an implicit promise that

Dying with dignity or self-respect does not always happen. Positive self-meanings can be created and sustained with the help of others. These others, who become the social audience for dying-related behaviors of the patient, include family, physicians, clergy, nurses, peers, or even strangers walking down the hall of the hospital.

the patient will not ask for more if the request to postpone death is granted. The promise is rarely kept, however, notes Kübler-Ross (1969).

In the fourth stage, *depression*, the patient begins to realize that a mistake was not made, the x-rays were correctly read, and the prognosis is not good. The patient realizes that meaningful things of life—family, physical appearance, personal accomplishments, and often a sense of dignity—will be lost as death approaches. This is a time to "get one's house in order" and to begin breaking away.

In the fifth stage, *acceptance*, the patient accepts death as a sure outcome. Although not happy, this acceptance is not terribly sad, either. The patient is able to say, "I have said all the words I have to say and am ready to go."

Kübler-Ross's stage theory of dying is not without its critics (Copp, 1998; Kellehear, 1990a; Charmaz, 1980; Garfield, 1978; Pattison, 1977; Kaufmann,

1976). Some reject the developmental nature of the sequential stage approach because it lacks universality—not all patients manifest all five stages of behaviors. Others have noted that the stages are not mutually exclusive—some patients may bargain, be depressed, and be angry at the same time. Finally, others have observed that the order of the stages is more arbitrary than Kübler-Ross would have us believe—dying patients may go from denial to acceptance, followed by depression and anger.

Kathy Charmaz (1980) argues that the stages emanate from preconceived psychiatric categories imposed upon the experiences rather than from the data. She notes that what originated as a description of a reality often becomes a prescription for reality.

The stages also do not adequately take into consideration the perspective of the patient. Anger, for example, may be vented at others because they have withdrawn from the patient in an attempt to cope with the loss of someone for whom they care. The bargaining behavior of the patient may be motivated by a need for moral and social support from caregivers, rather than by a hope for an extension of time (Gustafson, 1972). Depression may be a function of the severity of the physical condition of the patient rather than an emotional response to the terminal condition (Charmaz, 1980). As the disease progresses, the strength of the patient will diminish and be evaluated by others as psychological depression. In actuality, the patient may be depressed not by dying, but rather by the physical effects of the illness.

In a chapter entitled "On Death and Lying," W. Kaufmann (1976) identifies a contradiction between Kübler-Ross's perspective and her actual findings. He notes that in the introduction to her work she suggests that the denial of death is psychologically inevitable and yet she goes on to record interviews with literally hundreds of dying people who were not apparently "inevitably" and uniformly denying death.

With regard to the stages of dying, we must conclude that dying behavior is more complex than five universal, mutually exclusive, and linear stages. Kübler-Ross has helped us, however, in understanding that each of the five behaviors is a "normal" coping strategy employed by dying patients. It may be that it is the social situation that accounts for the similar coping strategies of dying patients and that in other cultural settings different patterns of behavior will be found.

Social Situation Meanings: Definition of the Environment

As the dying person comes to grips with a terminal condition, the way in which the social situation is defined will have a tremendous impact upon the process of dying. If the hospital is viewed as a supportive environment, patient coping may be facilitated. However, if the patient feels all alone in the place of confinement and if the hospital is defined as a foreign place, personal adjustment to dying and death will be hindered. Indeed, the hospital is not home by any sense of the imagination. Compared to a hospital, there is nothing like "home sweet home," when confined to a bed. In the environment of the home, the patient has familiar smells, sights, and perhaps sounds. In the privacy of the home the patient can likely be on her/his own schedule, rather than that of a hospital. In a hospital, for example, the patient may be awakened simply

---

WORDS OF WISDOM
The Care of the Dying Patient

No matter how we measure his worth, a dying human being deserves more than efficient care from strangers, more than machines and septic hands, more than a mouth full of pills, arms full of tubes and a rump full of needles. His simple dignity as man should merit more than furtive eyes, reluctant hugs, medical jargon, ritual sacra- ments or tired Bible quotes, more than all the phony promises for a tomorrow that will never come. Man has become lost in the jungle of ritual surrounding death.

From Facing Death *(p. 6)*
by R. E. Kavanaugh, 1972, Baltimore:
Penguin Books.

---

because it is the time that the hospital serves breakfast (with no regard for the possibility that the patient perhaps only an hour or two before finally fell asleep). At home, the schedule can be somewhat more flexible. Also at home, unlike in a hospital, the patient can have pets in the room—even in the bed, if she or he wishes.

The bureaucracy, with all its rules and regulations, may seem so inappropriate and inconvenient to a patient, yet the patient must abide by the rules—dying or not. In the hospital, the patient is basically told what to do. For example, George Dickinson was recently in the hospital for surgery. Feeling an urge to visit the toilet for the purpose of urinating, he got up out of the bed and attempted to walk to the bathroom (he was in for surgery but feeling very healthy). A nurse walked by his room about the time his feet hit the floor and literally screamed at him, "*Get back in that bed! What do you think you are doing?*" He immediately dived back into the bed to avoid further wrath from this "sergeant" nurse. A couple of hours later, the same nurse, in a calmer voice this time, said, "Mr. Dickinson, you can now 'empty your bladder.'" Little did she know that her previous verbal attack had canceled his desire to "empty the bladder." However, her commanding voice had quickly socialized him to the hospital rule of "bladder emptying"; thus, since the "rule enforcer" now said that he could go, he went (right on command!).

Like all other meaning systems, the definition of the social situation is an attempt by the individual to bring order to his or her world. Order contributes to efficiency. Thus, order is important to the bureaucratic setting of a hospital. Because situational meaning always involves selective perception, the terminally ill patient will create the meaning for the social environment and will respond to this meaning and not to the environment itself. Hospital rooms are generally bland and uniform. The "bland" environment of the room may carry over to the patient's perception of the entire hospital setting—including the staff's demeanor. They may be perceived as cold and unfeeling. Each terminally ill patient will not only experience death in a different environment, but also will have a unique interpretation of the social situation. This accounts for the different experiences of dying patients. The hospice movement (see chapter 6) is an attempt to create a more positive and supportive social situation in which dying can take place.

RELATING TO THE DYING PERSON

Relating to a dying individual is really not easy for any of us. For medical personnel there is the "failure" of not being able to make the person "well" again. For family members and friends there is the sorrow in anticipating a world without that special person. We turn now to a discussion of being in the presence of someone who is dying.

Medical Personnel

Physicians' unwillingness to deal with death and terminally ill patients is cited frequently in the literature (Kane & Hogan, 1986). Physicians as a group tend to express more conscious death anxiety than do groups of individuals who are physically ill. Anne Kane and John Hogan (1986) note that death anxiety is significantly linked

"MEDICAL ETHICS DO NOT ALLOW ME TO ASSIST IN YOUR DEATH. I AM, HOWEVER, PERMITTED TO KEEP YOU MISERABLE AS LONG AS POSSIBLE."

with both age and experience for physicians as a whole. The older a physician becomes, the lower the death anxiety. Likewise, the longer a physician has practiced medicine, the less death anxiety he or she has. As noted earlier in this chapter, similar findings were noted by George Dickinson and Robert Tournier (1994) in a 10-year follow-up, and again in a 20-year follow-up (Dickinson, Tournier, & Still, 1999), of a study of physicians soon after they graduated from medical school. The younger, less-experienced physicians reported higher levels of death anxiety and felt less comfortable with dying patients.

Physicians may have a greater need than others to reject death. Some physicians may compensate for their unconscious personal fear of illness and death by distancing themselves from dying patients (Black, Hardoff, & Nelki, 1989). American medicine is often indicted because physicians are educated to treat diseases rather than people and to deal with patients impersonally rather than holistically. However, in a study of 1,012 physicians, George Dickinson and Algene Pearson (1979a) concluded that those physicians with a high probability of dealing with dying patients (such as oncologists) were more open with their patients than were physicians who practiced medicine in areas in which there was a lower probability of dealing with death (for example, obstetricians and gynecologists). Because this study suggests that differences exist among medical specialties in relating to dying patients, it is possible that one factor influencing the selection of a medical specialty is the medical student's personal understanding and feelings concerning dying and death issues. Yet, Michael Bell (1996), in the latter stages of a fight against terminal cancer, observed that this priority of technical skills over bedside mannerisms may be exactly what we prefer that our doctors develop because one's very life may depend on such skills and knowledge. The death of a patient may be viewed by the physician as a professional failure and thus generate guilt. Therefore, it is important for medical students to develop technical skills necessary to deal efficiently with terminally ill patients and their families.

The Socialization of Physicians

Surgeon Sherwin Nuland, in his book *How We Die* (1994), shares the story (early in his practice) of one of his patients, a 92-year-old woman with heart disease who was being treated for an acute digestive tract disorder. Dr. Nuland persuaded the patient to have an operation for the disorder, though she argued that she had lived long enough and did not want any further intrusions into her body. The young doctor was following the medical code to "prolong life and relieve suffering." Yet, he was not listening to the patient's system of logic to use this sudden illness as a gracious way to die. The patient survived the operation, yet died 2 weeks later of a stroke. She told Dr. Nuland that he had let her down by not allowing her to die in due course without the pain and complications experienced from the operation. Dr. Nuland said he learned from this experience that for dying patients "the hope of cure will always be shown to be ultimately false." He learned that he needs to listen to the logic of his patients and not rely solely on his own medical code of logic.

Americans are poorly socialized with regard to issues of dying and death, as noted in chapter 1. One might expect that the early socialization experiences of physicians would be similar to those of other members of society. This is exemplified by the following story told to George Dickinson by the late Dr. Charles B. Huggins, Nobel recipient for cancer research and professor at the University of Chicago School of Medicine, about his first day as a student in medical school:

> After the gross anatomy professor finished the initial lecture, the class went to the laboratory to begin work on the cadavers. My **cadaver** was a female. After taking one look at the body (having never seen either a dead woman or a naked woman), I said to myself, "I should have gone to law school after all."

Even though this Nobel recipient entered Harvard Medical School in 1920 (at the age of 18), many first-year medical students today have had similar experiences and feelings and find that their first exposure to death is an impersonal experience in an anatomy laboratory. Human dissection in a gross anatomy laboratory, typically occurring during the first semester, is a rite of passage for all future medical doctors (Dickinson et al., 1997). Students must deal with the psychological and social aspects of cutting and touching a dead body and must struggle with questions of her or his own and others' mortality (Evans & Fitzgibbon, 1992). In addition to their own anxieties about death, students in gross anatomy classes are subjected to horror stories from upperclassmen and classmates. Cadaver stories portray, and thus help create, a world of outsiders and insiders and emotionally weak or strong medical students (Hafferty, 1991). A norm within medicine is that one must have emotional strength to survive the mental stress brought on by practicing medicine; strength begins and must be displayed in the gross anatomy laboratory.

Entry into medical school involves movement from a largely lay to a largely medical culture, notes Frederic Hafferty (1991) in his study of the socialization of first-year medical students. Experiences such as the dissection of human cadavers in the anatomy laboratory provide students with opportunities to internalize a variety of attitudes, values, motives, and rationales with respect to both their current role as student and their future role as physician. These experiences give students an opportunity to assess their progress in identity transformation and skill building (Hafferty, 1991).

Situational Adjustment.   Students entering the medical profession come in with certain attitudes and feelings toward patients that will be shaped and continually processed until their attitudes comply with those of the medical profession itself. Howard Becker (1964) refers to this "molding" as **situational adjustment.** As one moves in and out of social situations, learning the requirements for success in that situation, with a strong desire to deliver the required performance, the individual becomes the kind of person that the situation demands. Thus, the medical student is "molded" into the medical profession by learning what is expected and then "doing it." Becker notes that much of the change in an individual is a function of the interpretive response made by the entire group—the consensus that the group reaches with respect to its problems.

The process of situational adjustment accounts for changes that people undergo, but people also exhibit some consistency as they move from situation to situation. Becker (1964) refers to this consistent line of activity in a sequence of varied situations as "commitment." A variety of commitments constrain one to follow a consistent pattern of behavior in many areas of life.

Whether one's medical socialization took place many decades ago, as for Dr. Huggins, or whether it takes place in the twenty-first century, medical training transforms laypersons into something else—physicians (Hafferty, 1991). Robert Merton, G. G. Reader, and P. L. Kendall (1957, p. 287) noted that socialization is the collection of "processes by which people acquire the values and attitudes, the interests, skills, and knowledge—in short, the culture—current in the groups of which they are, or seek to become, a member." When we speak of socialization (Hafferty, 1991), we are talking about the structure, the method, and the route by which initiates move from one status to another and acquire the technical skills, knowledge, values, and attitudes associated with the new position or group. Thus, one must attain a new cultural base but must also facilitate movement away from the old status.

A study by Thomas Campbell, V. Abernethy, and G. J. Waterhouse (1984) comparing the attitudes of physicians and nurses toward death found differences that were tied to professional roles, not to gender. Nurses tended to see a more positive meaning in death than did physicians—death as rebirth rather than as abandonment and as tranquil rather than as frightening. Because physicians make the crucial decisions, whereas nurses carry out the orders, the implication of patient death as professional failure is reduced for nurses.

Becker's (1964) emphasis on situational adjustment, followed by a commitment to consistency, seems to suggest that socialization to occupational role overcomes earlier gender role socialization. This is the conclusion of George Dickinson and Sylvia Ashley-Cameron (1986) in explaining the different attitudes of female nurses and physicians regarding dying patients. Even if women have traditionally been socialized to be the more nurturant gender and more sensitive and responsive to the needs of others, female physicians tend to be less sensitive and responsive than female nurses. Thus, the differences between physicians and nurses regarding death and terminally ill patients tend to be more a function of the role expectations of the particular medical occupation than of gender.

Jack Kamerman (1988) cautions, however, that as nurses are given greater responsibility in diagnosis and treatment, their attitudes toward death may move closer to those of physicians. In addition, as nurses become more susceptible to the strains that physicians experience at a patient's death, it is possible that they will retreat further behind the shield of professional detachment, particularly if nursing follows the medical model of professional status.

The Effects of Gender Socialization.   By contrast, a study by S. C. Martin, R. M. Arnold, and R. M. Parker (1988) shows that gender differences do play a role in how male and female doctors communicate with patients. Perhaps this is because medical schools do not place heavy emphasis on patient interviewing, and female doctors therefore default to the less-directed communication style that is used predominately by females in American society. Women's styles of speech are typically less obtrusive, men are more likely to interrupt women than vice versa, and women are more likely to allow such interruptions. The conflicting paradigm to this less-directed communication style of women is that physicians are socialized to dominate the physician-patient interaction. The physician is to control the flow and topics of conversation. Female physicians tend to integrate these two models of communication by being more egalitarian in their relationships with patients, more respectful, and more responsive to patients' psychosocial issues (Martin, Arnold, & Parker, 1988).

The successful communication skills of female physicians are supported by George Dickinson and Algene Pearson (1979b). In their 1976 study of more than 1,000 physicians they found that female physicians related better to dying patients and their families than did male physicians. A 1986 follow-up to this study (Dickinson & Tournier, 1993) showed even more striking differences between males and females after a decade of practicing medicine. Perhaps the traditional "feminine characteristics" of gentleness, expressiveness, responsiveness, and kindness help to explain these differences.

Martin, Arnold, and Parker (1988), in studying gender and medical socialization, reported that medical students regard female faculty physicians with clinical responsibility for patients to be "more sensitive, more altruistic, and less egoistic" than males; nurses believe that female physicians are more "humanistic" and have greater "technical" skill in communicating with patients; and patients also perceive female physicians as more humanistic, more empathic, and better listeners. If more physicians and other medical personnel were to display these "human qualities" with a greater frequency, patients might receive more help and support in their dying.

Awareness Contexts

Whether medical personnel are male or female or nurse or physician, communication with terminally ill patients is of utmost importance. Barney Glaser and Anselm

Strauss (1965) present four awareness contexts in interacting with a dying patient. These researchers define the **awareness context** as what an interacting person knows of the patient's defined status, and his or her recognition of the patient's awareness of a personal definition. The awareness contexts are closed, suspicion, mutual pretense, and open.

Closed awareness is usually the first context. In order to maintain the patient's trust and yet to keep him or her unaware of the terminal condition, the staff may construct a fictional future biography. If the patient has not been told "the truth," or has been "told" but did not want to hear, this often places an additional burden on the hospital nursing staff in working with the patient for an 8-hour shift. Because the physician often directs the medical team, nurses are sometimes forced to work in a closed awareness context regardless of their own views. Because most patients are able to recognize death-related situational and spatial clues, this context tends to be unstable, and the patient usually moves to the suspicion or full awareness contexts.

Suspicion awareness is a context for control between the patient and the medical staff. The patient suspects that he or she is dying but receives no verification from the staff. Nurses must use teamwork to refute this challenge.

Mutual pretense often follows and requires subtle interaction with both patient and staff "acting correctly" to maintain the pretense that the patient is not approaching death. A game continues to be played whereby all concerned "act" as if they know nothing about the terminal condition of the patient.

If mutual pretense is not sustained, *open awareness* follows in which the patient and everyone else knows that the patient is dying. This is the context found in hospice programs (see chapter 6). Ambiguities may develop in this context, however. The patient is obligated not to commit suicide and to die "properly." Dying "properly" is difficult, however. Nurses and other medical staff expect "proper" dying, but the patient usually knows no model to follow, notes Robert Blauner (1966). If the patient is dying in an unacceptable manner, then difficulties are faced when he or she is trying to negotiate for things from the staff.

Ashley-Cameron and Dickinson (1979) found that nurses working with dying patients seem to be comfortable in a closed awareness context. Because nurses spend more time with patients in the hospital than do physicians, a closed awareness context may produce a more comfortable setting for the nurse. A recent study in England, however, produced a very different finding regarding awareness contexts. Open awareness was preferred in a study of 548 physicians and nurses (Seale, 1991). Eighty-one percent of these British practitioners said that they found it easier to work with dying patients if the patients were aware of their terminal condition. Perhaps the openness of the 1990s and/or the difference in cultures is reflected in these seemingly contradictory findings over different time periods.

Whether a person is a nurse or a physician, the ability to cope with terminally ill patients does not come easily. Society has not prepared one for such interaction. Some obviously react better than others. Working with patients with AIDS may be especially difficult due to the fear of acquiring AIDS. The key to good relations with dying patients is personally coming to grips with death. One should also have good communication skills. If the dying patient could be viewed as a person, not as the "lung cancer" case in Room 314, and treated as a human being, certainly the trauma of the dying process could be eased. In the end, medical personnel, the patient, and the family would benefit.

Family and Friends

It is most important that the family member or friend make contact with the dying individual, with the frequency of contact being somewhat dependent on the closeness of the relationship. Just being with the dying individual or making contact in other ways reminds the patient that you "care" and that the relationship is important to you. When you make contact, it is not as crucial what you say as it is that you "touch base."

Not only do family members and friends need to be cognizant of the ongoing human needs of the terminally ill patient, they need to continue to maintain relationships and continue to incorporate the dying person within the network of family and friends (Doka, 1993). It is important for family and friends to openly discuss their relationships with the dying person. If family members or friends have never verbalized to the dying person what she or he really means to them, this would be a good time to do that. If difficult to say in words, then write it down and share these feelings with the individual. Also, recall the good times you have experienced together. Reminiscing may provide meaningful interaction for the present and may mitigate subsequent grief.

Family and friends may participate with the dying individual in other ways in which they feel comfortable. For example, some may choose to be involved in personal care, monitoring medication, providing massages, or helping with grooming. Others may feel comfortable in social conversation, reading aloud, or watching television or listening to music together. Yet others may find meaning for themselves and the ill person by participating in religious rituals together. Family and friends should recognize the need to allow the dying individual the freedom to decide whether or not she or he wishes to pursue various activities, notes Kenneth Doka (1993).

As the family member or friend moves through the "process" of dying, assuming the death is not sudden and "unexpected," the caregivers become aware of physical changes and sometimes increased pain. Today, the typical dying trajectory is lengthening as the proportion of deaths from chronic illnesses is increasing. The degree of awareness of various family members and friends concerning the seriousness of the illness is important in relating to the sick person. That is to say, does everyone have the same degree of knowledge concerning the prognosis of the illness? The meaning(s) associated with knowledge of impending death make a significant contribution to the type of awareness displayed, and meanings are likely to be contingent on social context (Young, Bury & Elston, 1999).

Ira Byock (1997), former president of the American Academy of Hospice and Palliative Medicine, states that people who are dying also have a responsibility to their family and friends. It is very important for them to make an effort to reconcile strained relations. A dying person may wish to be "left alone," but that individual should recognize that family and friends may suffer because of the isolation from the dying person. It is the dying patient's right to be "left alone," since death is primarily a personal experience, but Byock notes that the friends and relatives may have their own personal experiences with regard to the patient's dying. Within the social context of dying, both the patient and family and friends should keep each other's feelings in mind. Byock (1997) thinks of family as a process: Family is marked by feelings of mutual connection, appreciation, and caring. He notes that individuals also have both relatives and friends who qualify as family in this sense of the word. He suggests that patients notice who shows up and who you miss when they do not. Even for individuals who have been utterly alone for years, it may be nurses or aides or volunteers who become family to the dying patient, observes Byock.

The Stress of Dealing With a Dying Family Member

As the family member/friend's health deteriorates and the individual moves into the terminal phase, family members and friends begin to cope emotionally and in other ways with the now-expected death and the ever growing burdens of care (Doka, 1993). While denial may exist with some friends and family members, most will probably recognize the possibility of death. These individuals will rally around the dying person, sometimes causing resentment between family members and/or friends.

A terminally ill person in the household places strain on family interaction. Primary caregivers are often fatigued, both mentally and physically, from lack of sleep and the strains of relating to a bedridden person—lifting, turning, changing diapers. Financial stress is also placed upon the shoulders of some family members. After awhile, family members and friends may wish (though they probably won't voice the thought) that the person would just "go ahead and die." A videotape entitled *Dying,* shown on national public television in the 1970s, presents Bill, a man in his early 40s who is dying from cancer, and his wife Harriet. Harriet actually says to Bill at one point, "Why can't you just go ahead and get this over with?" She shares a similar thought with her therapist. Harriet is frustrated over Bill's dying and feels that the man who said "I do" is leaving her "with two young boys to raise alone with drugs and everything." She notes that if he could go ahead and die now, she could then have time to find the boys another father, and therefore have help in raising them through the troublesome teen years! Besides anger and resentment, family members and friends may feel guilt about their own inability to be "perfect" all the time for the ill individual, notes Doka (1993).

How can family and friends find ways to reduce the stress of relating to a terminally ill person? Melodie Olson (1997, p. 212), nurse practitioner and professor, suggests that coping skills include "anything one can use to alter the relationship between the person and the environment to change the negative results of psychophysiologic consequences of stress." Family members and friends may find counseling and self-help groups useful, suggests gerontologist Doka (1993). They may also seek effective stress-reduction strategies: improving problem solving, increasing planning and communication, withdrawing from stressful situations or delegating, modifying unrealistic expectations of self, and examining lifestyle management such as good diet, regular exercise, social support, relaxation, and meditation. Certainly, the last suggestion of maintaining one's own health is perhaps the primary coping technique one can use to combat the stresses of caring for a dying person.

It is important for family and friends to allow themselves respite. An individual should not try to do it all, but call on other family members or friends to help. With a bountiful number of family members and friends, a division of labor can be established and a "share the care" plan can be devised. However, if family members and friends are few, having the ill person in hospice will help provide volunteers to give respite. Whatever the case, an individual should not try to care for a dying person 24 hours per day 7 days per week.

Helping Children Cope With a Dying Parent

When a parent is dying, families need to mobilize help from any source and make this help work for them, suggests thanatologist Phyllis Silverman (2000). Family and friends assist with the details of daily living such as being there when children come home from school and helping with daily routines of cooking, doing laundry, and other chores. Hospitals with policies that allow healthy children to visit and

THE FAMILY CIRCUS.   **By Bil Keane**

10-4
© 1996 Bil Keane, Inc.
Dist. by Cowles Synd., Inc.

"If somebody dies in the hospital,
angels move them to the
eternity ward."

REPRINTED WITH SPECIAL PERMISSION OF KING FEATURES SYNDICATE, INC.

WORDS OF WISDOM
I Noticed People Disappeared

Emily Dickinson

I noticed people disappeared,
When but a little child,—
Supposed they visited remote,
Or settled regions wild.

Now know I they both visited
And settled regions wild,
But did because they died,—a fact
Withheld the little child!

encourage the children to decorate the room provide greater opportunities for interaction and participation in the life of the dying parent. With these more liberal hospital policies, children "feel more connected to what is happening around them," according to Silverman (2000, p. 194).

Older school-aged children need more time to prepare for a parent's death than do younger school-aged children, observes social worker Grace Christ (2000) from her study of 88 families and their 157 children (ages 3 to 17) who experienced the terminal illness and death of one of their parents from cancer. Regularly updated information about the parent's condition seems to provide children with a helpful context. Visits with the dying parent can be meaningful, even when little verbal communication with the parent is possible. Visits provide "concrete evidence" for the reality that the parent is dying and give the children an opportunity to say goodbye, notes Christ. Being told about the parent's probable death in a timely way allows for **anticipatory grief** and helps the child to cope with the period of mourning after the death.

Children like to draw; thus, drawing pictures is an excellent way for them to express themselves. Elisabeth Kübler-Ross tells the story of a family in which the mother was terminally ill with cancer. Dr. Kübler-Ross was invited to their house to talk to the children. The father and mother were afraid to tell them and felt that the children knew nothing of the approaching death. Kübler-Ross sat down with the little girl, gave her some crayons and paper, and asked her to draw a picture. The little girl drew a dining table with four chairs, one of which was leaning against the table, rather than upright like the other three. Kübler-Ross asked her to explain the drawing. The little girl said that the three upright chairs were for her father, her brother, and herself. "The 'turned-down' chair is for Mommy. She is dying in the hospital and won't be coming back home." This was a child who *supposedly* was in the dark about the situation of her mother's dying. By asking her to draw a picture, she was able to explain freely her perception of what was happening. The little girl was not in a closed awareness context but was functioning in a mutual pretense environment.

PRACTICAL MATTERS
Signs of Approaching Death and What to Do to Add Comfort

Hospice exists to support the family's desire to aid a dying loved one in familiar surroundings. This time period is a very difficult one for families. The following was devised to help alleviate some of the fears of the unknown. The information may help caregivers prepare for, anticipate, and understand symptoms as patients approach the final stages of life. It is important to note that some symptoms may appear at the same time and some may never appear.

Symptom: The hospice patient will tend to sleep more and more and may be difficult to awaken.

Action: Plan activities and communication at times when he or she seems more alert.

Symptom: You may notice your loved one experiencing confusion about time, place, and identity of people.

Action: Remind your family member of the time, day, and who is with him or her.

Symptom: Loss of control of bowel and bladder may occur as death approaches, as the nervous system changes.

Action: Ask the hospice nurse for pads to place under the patient and for information on skin hygiene. Explore the possibility of a catheter for urine drainage.

Symptom: Arms and legs may become cool to the touch, and the underside of the body may become darker as circulation slows down.

Action: Use warm blankets to protect the patient from feeling cold. Do not use electric blankets since tissue integrity is changing and there is danger of burns.

Symptom: Due to a decrease in oral intake, your loved one may not be able to cough up secretions. These secretions may collect in the back of the throat causing noisy breathing. This has been referred to as the "death rattle."

(continued)

A child whose parent is dying often finds it useful to talk with other children who are having the same experience. Support groups such as "I Count Too" can provide an opportunity for children to express themselves in the presence of other children who are experiencing a similar situation in their own families. "I Count Too" provides counselors who sit in on the sessions to give guidance and advice to the children. Children experiencing the dying of a parent need intervention, because they typically have no history of such a process. Counselors can provide "tools" through which children can work toward coping with the dying of a parent. Without such help, these children are like a child sitting in a sandbox with no toys with which to play. We must provide the toys!

(continued from previous page)

Action: Elevate the head of the bed (if using a hospital bed) or add extra pillows. Ice chips (if the patient can swallow) or a cool, moist washcloth to the mouth can relieve a feeling of thirst. Positioning patient on his or her side may help.

Symptom: Hearing and vision responses may lessen as the nervous system slows.

Action: Never assume the patient cannot hear you. Always talk to the patient as if he or she could hear you.

Symptom: There may be restlessness, pulling at bed linens, having visions you cannot see.

Action: Stay calm, speak slowly and assuredly. Do not agree with inaccuracy to reality, but comfort with gentle reminders to time, place, and person.

Symptom: Your loved one will not take foods or fluids as the need for these decreases.

Action: Moisten mouth with a moist cloth. Clean oral cavity frequently. Keep lips wet with a lip moisturizer.

Symptom: You may notice irregular breathing patterns, and there may be spaces of time when no breathing occurs.

Action: Elevate the head by raising the bed or using pillows.

Symptom: If your loved one has a bladder catheter in place, you may notice a decreased amount of urine as kidney function slows.

Action: You may need to irrigate the tube to prevent blockage. If you have not been taught to do this, contact the hospice nurse.

Hospice in the Home Program, Visiting Nurse Association of Los Angeles.

DYING WITH DIGNITY

Just what is meant by "dying with dignity?" For the cowboy in the "days of the old West," dying with dignity might have meant dying with "one's boots on" while "riding the range." For a soldier in war, to die "in the trenches" fighting for a cause or one's country would probably have "dignity." For some early-day Eskimos, if an elderly member of the hunting party whose health was failing could not make it back to the camp, then he would be abandoned to die "with dignity." After all, for him to hold the hunting party back would not be appropriate. Thus, he would probably be remembered as having "died with dignity." An individual who, after going in for the fifth time from to rescue children from a burning building, did not return, would have "died with dignity." Most of us today, however, are not cowgirls/cowboys, soldiers, hunters gathering food for the family, or individuals rescuing children from burning buildings; thus, such deaths would not be fitting for us, either with or without dignity.

Cross-culturally, what might be considered dying with dignity may vary. The following story was told to George Dickinson in the spring of 1999 by a nurse in England. An elderly Chinese patient with cancer appeared to the hospital staff to be in pain. Because she spoke no English, the staff, via communication with the woman's daughter, confirmed their suspicion. Thus, they placed a shunt in her arm and administered pain medication. Soon after this action by the staff, however, the daughter removed her mother from the hospital, took her home, and ripped out the shunt. Though her mother was indeed in pain, the daughter had incorrectly communicated with the hospital staff. The daughter did not intend for the staff to control the pain, because her mother's religion would have her die *in pain* as part of the "appropriate" dying process for her culture. Pain was necessary for her mother to "die with dignity" and, therefore, have a good afterlife. Thus, cultural variations produce different "appropriate" ways of dying, and to die in pain is not a prerequisite for "dying with dignity" in most parts of the world.

John Stephenson (1985) states that dying is a human activity that is carried out in a normative manner. Each individual learns from society the meaning of death and the proper or "good" ways to die. One hopes to die what one's culture considers to be a good death. For example, anthropologist Margaret Mead seemed to close out her life in the way that she had lived it. A recorded interview with her on public television a few months before her death revealed a woman who gave the last moments of her life as energetically, as intelligently, and as openly as she had given the early years of her life (Cuzzort & King, 1995). Her conversation was animated—no brooding soul turned inward by the thoughts of impending death. At the end of the program she briefly revealed her attitude toward life and work and death. Then she arose and began to walk backward slowly, into the shadowy and dark recesses of the stage setting, still facing the camera. She smiled a charming smile, waved her cane, and said, "Goodbye, goodbye, goodbye"—knowing that it was the final farewell as she stepped into the darkness. As R. P. Cuzzort and E. W. King (1995) note, it was an act, yet a wonderful act. It summed up the way that she lived, wrote, and felt about people.

Though there is no good role model to follow in dying, perhaps the late Senator Hubert H. Humphrey of Minnesota came close by maintaining some control over final events in his life. Even in his dying, he kept a most positive attitude and frequently made telephone calls to individuals to wish them well. Calling former President Richard M. Nixon, for whom many in Humphrey's position would have had limited affection, to wish him a happy birthday epitomizes Humphrey's behavior during the final days of his illness. Because an **appropriate death** is generally consistent with past personality patterns, Senator Humphrey died as he had lived. Likewise, Jacqueline Kennedy Onassis spent her last few days of life as she had lived, with poise and dignity.

Avery Weisman (1993) outlines four characteristics of an appropriate death: (1) *Awareness* of impending death comes at about the time as learning that nothing else can be done to "save" the patient. (2) *Acceptance* of death may depend on the person speaking with a patient and with whom a patient is ready to speak.

DEATH ACROSS CULTURES
Dying in Scotland

Rory Williams examines the issue of a good death with elderly residents of Aberdeen, Scotland. He asks, "How is one defined as dying?" There is dying as defined by the doctor and confirmed by the event of death. Secondly, there is the situation where someone died before they were recognized to be dying, so that the recognition had to be reconstructed retrospectively. Then there is the situation where somebody recognized as effectively dying failed in fact to die. These three sorts of dying have a definite relation to ideals about good and bad deaths. To die "the proper way" is on the one hand to go quickly, easily, quietly, and unconsciously. Yet, this idea clashes with the idea that the dying should be cared for. Too quick a death might be "good for the one that's gone but bad for those left behind." Thus, the "properness" of the death depends on whether it is in the interest of the living or that of the dying. Thus, two broad ideals of dying well exist: going as quickly and unconsciously as possible and going only after an affectionate reunion with kin. However, both ideals of death are threatened from the third sort of dying when people lived on who were considered as good as dead and who were thought to be better off dead. Thus, there are two kinds of "good dying" and one kind of "bad dying."

From A Protestant Legacy: Attitudes to Death and Illness Among Older Aberdonians *(pp. 98–100) by Rory Williams, 1990, Oxford: Clarendon Press.*

---

(3) *Propriety* refers to the nonmedical features of illness that distinguish a good death from something more objectionable and refers to what a patient decides is right and proper based on community expectations and standards. (4) *Timeliness* is the question of when is the best time to die.

The concept of an appropriate death has been suggested as an alternative goal for those working with the dying (Hooyman & Kiyak, 1988). An appropriate death means that the individual dies as he or she wishes to die. The individual maintains as much control over dying as possible to make it meaningful. Although an appropriate death is usually only partially achieved, those working with the dying can assist them in exerting such control.

Anne Hawkins (1990), in an article on pathographies about dying, notes various "constructions of death" in the literature, including those of de Beauvoir, Conley, and Wertenbaker. Simone de Beauvoir's model of death (1965) describes her mother's death from cancer as "an easy death," one in which suffering is avoided. Herbert Conley (1979) describes death as one's "finest hour," when death is accepted as a part of life with avoidance of self-pity, concentrating on the needs of others and maintaining dignity and self-composure. Conley's description of death fits Elisabeth Kübler-Ross's concept in *Death: The Final Stage of Growth*

(1975). Lael Wertenbaker (1957) offers the model of a "heroic death," a "manly" death in which pain is endured as a test of bravery, courage, and heroism. In an Australian study of 100 dying individuals, Allen Kellehear (1990a) found that most of the terminally ill people engaged in some form of personal preparation for death (funeral arrangements, for example) and seemed to have an implicit conception of the **good death** that emphasized looking after the needs of the survivors in a practical way. A "good death" appears to be one that is "appropriate" at a particular time and place.

According to the Institute of Medicine (1997), a "good death" is one that is free from avoidable distress and suffering for patients, families, and caregivers, is in general accord with patients' and families' wishes, and is reasonably consistent with clinical, cultural, and ethical standards. Hospice coordinators, studied by D. C. Smith and M. F. Maher (1993), reported that dying individuals who experienced a "good death" tend to have the following qualities: having a sense of control, discussing the practical implications of dying, exploring an afterlife, talking about religious/spiritual issues, reviewing the past, having a sense of humor, not avoiding painful truths, taking an interest in personal appearance, benefiting from the presence of significant others, and participating in physical expressions of caring. As Beverly McNamara, C. Waddell, and M. Colvin (1994) note, the good death may be viewed as a complex set of relations and preparations, rather than as relating to a fixed moment in time. Thus, from this perspective, the good death is not a single event, but a series of social events.

In their article entitled "The Promise of a Good Death," Ezekiel and Linda Emanuel (1998) conclude that developed countries have the capacity to make a good death the standard of care. They argue that (1) society is focused on ensuring a good death, (2) physicians have more powerful medications and other interventions to alleviate pain than ever before, (3) clinicians are recognizing the multidimensional aspects of dying and the importance of attending to concerns other than pain, (4) hospice is widely available and increasingly used, (5) advance-care planning is strongly endorsed, (6) and medical schools, hospitals, and professional organizations are committing themselves to training physicians to improve the care of the dying. If indeed societies place more emphasis on a "good death," hopefully death fears can be reduced significantly.

THE DYING CHILD

According to Robert Kavanaugh (1972), a psychologist and former priest, "little people" enjoy the same human rights as their bigger counterparts. They have a right to know if they have a fatal condition. When kept in ignorance, children, like adults, will rarely grow beyond the initial stage of denial and isolation. When this occurs, children are robbed of the peace and dignity that can be theirs in the final stage of acceptance and resignation.

PRACTICAL MATTERS
The Dying Person's Bill of Rights

- I have the right to be treated as a living human being until I die.
- I have the right to maintain a sense of hopefulness, however changing its focus may be.
- I have the right to be cared for by those who can maintain a sense of hopefulness, however challenging this might be.
- I have the right to express my feelings and emotions about approaching death in my own way.
- I have the right to participate in decisions concerning my care.
- I have the right to expect continuing medical and nursing attention even though "cure" goals must be changed to "comfort" goals.
- I have the right not to die alone.
- I have the right to be free from pain.
- I have the right to have my questions answered honestly.

- I have the right not to be deceived.
- I have the right to have help from and for my family in accepting my death.
- I have the right to die in peace and dignity.
- I have the right to retain my individuality and not be judged for my decisions which may be contrary to the beliefs of others.
- I have the right to expect that the sanctity of the human body will be respected after my death.
- I have the right to be cared for by caring, sensitive, and knowledgeable people who will attempt to understand my needs and will be able to gain some satisfaction in helping me face my death.

From Cancer Care Nursing *(p. 33). by M. Donovan and S. Pierce, 1976, New York: Appleton-Century-Crofts.*

Relating to the Dying Child

The prognosis of death should be made known to children as soon as it is clear and final (Kavanaugh, 1972). Physicians, nurses, and families obviously need time to bring their own emotions under control. We now know how to treat the dying child kindly. Knowledge is kindness; ignorance is cruelty. The child is the patient whose life is being lost and whose concerns are preeminent. Kavanaugh (1972) notes that when children have known the truth about their condition and were allowed to talk about it openly, they have been as brave as any adult.

Reasons for the preceding advice given by Kavanaugh (1972) are based on the following observations:

First, the dying child is no ordinary child. The ordinary process of maturing quickens through lengthy illness with confinement, suffering, and deprivation. Children ill for a long time usually exhibit a maturity beyond their calendar years.

Second, children's consciences are more tender and concerned than most adults are in a position to know. Children on their deathbeds who are uninformed of their fate will often have guilt for the sadness and poorly veiled tears that they witness

Good communication channels are important to establish with children. With rapport in place, discussing serious matters like dying and death will then be easier.

around their bed. They can sense the phoniness around them, and they may believe that they are being punished for something evil that they have done. Their isolation is heightened in their heavy concerns. Nothing is sadder than a dying child learning of his or her fate from playmates.

Finally, moderately aware and normally alert children know what is predicted for them in the signs that they see—recognizing their plight in memories of dying scenes on television. They may ponder why the physician comes so often, why everyone is so nice to them, and why they are receiving all of the gifts. A review (O'Halloran & Altmaier, 1996) of studies on death awareness among children who are healthy, chronically ill, and terminally ill reveals that children with life-threatening diseases demonstrate increased understanding of death. In contrast, healthy and chronically ill children appear to require a certain age, cognitive development level, or intelligence threshold to understand these concepts. As anthropologist Myra Bluebond-Langner (1989) observes from her work with dying children that all terminally ill children become aware of the fact that they are dying before death is imminent. Yet, Bluebond-Langner notes that for these children the acquisition and assimilation of information are a prolonged process. Either a child

---

WORDS OF WISDOM
"Listening" Is the Key to "Telling" About Dying

How do we tell a child about his or her dying condition? Kavanaugh states that to adults brave enough to listen, this is not a valid question. The child will do the telling, if we create an atmosphere in which the child can make all appropriate deductions. The child's talk will flit in and out of the awful revelation.

Who should do the telling or serve as a catalyst for it? Kavanaugh says that anyone strong enough to take the consequences by being a regular visitor, a trusted confidant, and a patient listener can be the catalyst. Many adults cannot qualify.

From Facing Death *(pp. 142–143)*
by R. E. Kavanaugh, 1972,
Baltimore: Penguin Press.

---

shares what he or she knows about dying, or the final weeks and months become a lonely vigil, a sentence to fear and guilt, confinement, and confusion.

From her research with hospitalized terminally ill children between the ages of six and ten, Eugenia Waechter (1985) agrees with Kavanaugh's conclusions. Her findings indicate that, despite efforts to shield children from knowledge of the seriousness of their illness, the anxiety of those close to the children is likely to alter the emotional climate in the families. The emotional climate may be altered to such a degree that the children will develop suspicions and fears about their condition. Often they will feel that awareness of their condition is knowledge that they are not supposed to have, so the silence of those nearby isolates children from needed support.

The question of whether children should be told is meaningless, notes Waechter. No curtain of silence should exist around children's most intense fears. Support must especially be made available for them during and after actual encounters with death on pediatric wards. Support is needed to allow introspective examination of attitudes and fears related to death in general and to the death of children in particular.

In working with children with cancer, Yehuda Nir (1987) concludes that children's ability to cope with the stress of cancer has an impact on the way that they deal with the illness itself and with the treatment. Feelings of helplessness and vulnerability dominate, often leading to regressive behavior. Withdrawal and refusal to participate in the treatment are seen almost immediately at the onset of the disease in young children. Because these types of behavior interfere with the medical management of the illness, resistance to treatment is circumvented by methods such as the use of general anesthesia and hypnotic relaxation techniques. Although these interventions are medically justified, they intensify the children's feelings of passivity and helplessness.

Nir (1987) further notes that regression can take the form of acute separation anxiety. When the parents are unable to stay overnight in the hospital, they often speak with the children by phone. After talking with their parents, young children sometimes fall asleep with the telephone receiver in their laps, as if to maintain contact with what has become elusive.

Families who seem best equipped to cope are those who develop a therapeutic alliance with the **pediatric oncologist** (Nir, 1987). This alliance buffers and protects the child from much of the pain and stress of the daily treatment routines. Other parents who seem successful in reducing stress are those with strong religious beliefs, notes Nir.

By observing and talking to children with leukemia aged 3 to 9 in a hospital, Bluebond-Langner (1978) concluded that most of them knew not only that they were dying, but also that this was a final and irreversible process. Children may wish and need to be open about their condition, but adults are often traumatized by children's openness and honesty. Bluebond-Langner notes that the age of the children is not as important in their self-awareness of their dying as is their experience with the disease and its treatment.

The founder and executive director of Children's Hospice International, Ann A. Dailey, says (Gamarekian, 1987, p. 19):

> Kids intuitively know when they are dying and they tend not to have the fears that adults have. But they want to know what it is going to feel like. Will someone be with them to hold their hand? Will a grandmother who died last year be there to greet them? Families need to cry together, to tell how much they are going to miss each other.

Though one may not know what to say to dying children or may not feel "skilled" in this area, it is imperative that we show support to them and let them know that we care. Support of others is important throughout life, whether relating to a terminally ill person or otherwise. This point is illustrated in the following story told by Rabbi Harold Kushner (1985).

> A little boy had gone to the store, but was late in returning home. His mother asked, "Where were you?" He said, "I found a little boy whose bicycle was broken, and I stopped to help him." "But what do you know about fixing bicycles?" his mother asked. "Nothing," the little boy replied, "I sat down and cried with him."

Many times, we may not know how to "fix the bicycle," but like the little boy in the preceding story, we can give support to the individual in other ways.

Helping the Child Cope with Dying

Children sometimes consciously keep their feelings and responses secret in an attempt to protect their parents (Robinson & Mahon, 1997). Such secrecy could make it difficult to help a child cope with her or his dying. Gerry Cox (1998) advocates the use of humor, art, and music to help children express their feelings related to dying.

Adults who keep family secrets are less inclined to use humor. Humor and laughter are the opposite of secrets, notes Cox. To laugh together is a positive form of sharing and social support. Laughing is typically done in the company of others, not when alone. Laughing is contagious, as is crying. A more relaxed person, not one always taking self too seriously, is better able to laugh at one's self. Laughing together is viewed as caring and being supportive, observes Cox.

How do you make a dying child laugh? To make the child laugh it is useful to know what she or he particularly enjoys. Watching various cartoons on television might provoke laughter in some children. For others, play therapy, using a favorite toy, might produce an enjoyable, even humorous, situation, as the child's mind forgets the illness for a short while and goes into the never-never world of play and creativity. For some children, a visit from a clown to the hospital or home can produce a "funny" setting. Other children might enjoy relating to a puppy or kitten and find joy in the playfulness of the small animal.

Music is another form of expression that does not require oral fluency, says Cox (1998). Music is a therapeutic tool. Joyful music can encourage laughter and perhaps move one's emotions from despair to happiness. Though words are not essential for music to be helpful, the use of words can be powerful. Catchy tunes that repeat themselves such as a round (for example, "Row, Row, Row Your Boat" or "Three Blind Mice") can create a "happy" environment for children. Small children simply enjoy having someone sing to them, as they like having someone read to them. Anything to divert their thinking from the unpleasantness of their disease can only enhance the situation for the dying child.

The use of humor, art, and music allows children to remove the sense of distance between themselves and others and to enhance their self-esteem and lower death anxiety, notes Cox (1998). These therapies are effective ways to provide social support for dying children. Various therapeutic techniques, from story-telling to humor, exist for aiding these children. Dying children need to say goodbye by writing a letter, making a picture, or sending up a balloon with a message, suggests Cox. And, perhaps most importantly, adults need to support these efforts.

Parents of the Dying Child

Parental involvement with dying children often precedes the period of diagnosis, as parents and other family members may notice symptoms (Doka, 1993). Parents may have to cope with considerable anxiety and uncertainty. In the diagnostic period, parents learn what they and the child must face—the possible threat of a sustained illness and even death. The family's financial state is likely to be adversely affected by the costs of the illness. During the chronic phase of the illness, continued stress with occasional points of crisis occur. Family life tends to revolve around the disease, sometimes producing neglect for other children in the family.

Though parents can never be really prepared for the end, as the dying child continues to lose ground, they reach a point where resignation begins to replace hope for survival, and the inevitable end slowly comes into focus (Knapp, 1986). The character of hope changes from hoping for survival to hoping for a full range of living in the time available and for a comfortable, pain-free death. When parents consciously accept the fact that death is imminent, they become totally absorbed in the life of that child and try to make each day a memorable occasion. At this point, parents have to make decisions about where the child will die, at home or in the hospital or in hospice (Silverman, 2000). They are often faced with decisions about continuing life support and using antibiotics.

It is difficult for the parents to finally accept that their child's death is forthcoming. This out-of-order aspect of the child dying *before* the parents is difficult to comprehend. The unnaturalness of the child dying before the parents may also be complicated by death resulting from the child dying suddenly and unexpectedly in an accident. Parents' dreams of the child growing into an adult and relating to them as "big people," are shattered, as the child lies dying. Parents often wonder what their child would have been like as an adult. What would he or she have done for a living? Would he or she have married? Would their child have had children? Where would he or she have lived?

Parents tend to adjust best to their child's dying when they have a consistent philosophy, theology, or cosmology of life (Pine & Brauer, 1986). Such consistency seems to help the family accept the diagnosis and cope with its consequences. Well-adjusted parents also have a viable significant other for support during the course of the child's illness.

Parents of a dying child often take a protective approach toward communication in which they attempt to shield the child from the implications of the illness, notes Kenneth Doka (1993). Such a strategy may not work since children have access to a wide range of information about their condition: internal health cues, external treatment cues, information from books and television, and input from ill peers. Doka notes that the question from a dying child "Am I going to die?" may not be a request for information but a call for reassurance; thus, it is important to understand what the child is really asking. Visual art can be helpful in a child's communicating with an adult. Very young children can use different crayon colors to illustrate what they may be thinking. They can draw sad or happy faces to express themselves. Play therapy may also be appropriate to use in working with dying children. By giving the child a doll, she or he can "play" doctor and release frustrations of being stuck with a needle by doing the same to the doll.

Siblings of the Dying Child

Everyone within the nuclear family will probably be affected by the dying child's illness (Doka, 1993). Avoidance of the child is one way that some family members might react. Siblings might be prohibited by parents from playing with the ill child. Siblings of the dying child may definitely feel neglected by parents and other family support members, because their energies are being directed toward the ill child, not the siblings. Parents are not as free to play with the other children and go places with them. Even if free, the parents probably have little energy left to spend with the other children, because they may have been up most of the night with the ill child.

Individuals outside the immediate family or extended family members can play a significant role in giving attention to the siblings of the dying child by taking them to a movie or to the park, anything to show attention and divert the mind for a few hours from the "situation at home" or in the hospital. Teachers of siblings of a dying child should be notified of the situation at home, so that they can give special attention to the sibling and better understand why he or she may not be acting "normally." The sibling at school, for example, might "throw a fit" or hit another child, when normally he or she would not act in this way. "Look at me," the sibling is probably saying. "I exist too and would like some attention."

Depending on their ages, however, siblings of a dying child can play a key role in helping within the family setting. As Phyllis Silverman (2000) notes, children need to be involved and seen as active members of the family, as helpers, and as grievers. They can assist with chores and relate to the dying sibling by playing with her or him or reading to her or him. Parents should encourage siblings of the ill child to participate in and contribute to the situation. I. M. Martinson and R. G. Campos (1991) found that siblings who were allowed to participate in caring for a sibling who was dying of cancer felt pride and pleasure about their ability to help.

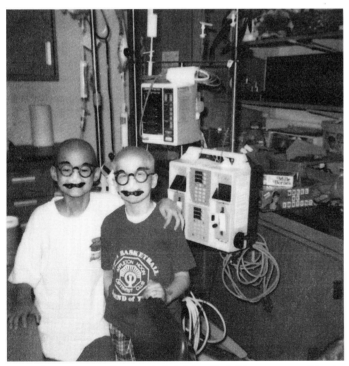

Mutual support can be very helpful for ill children, as noted in the child's own words:"My good friend from the hospital, who also had cancer, is Michael. He had a brain tumor. As you can see, we both had a good sense of humor, even in the toughest times. Sometimes when we didn't feel good, we would watch movies together. Another time we skipped school and played video games instead. I'm happy to say that Michael is in remission too."

Other children in the family need to know what is going on so that they can cope with the changes as the family adjusts to these new circumstances (Silverman, 2000). Physical, as well as behavioral, changes need to be anticipated. For example, siblings need to be forewarned if their dying sibling is receiving chemotherapy that will cause hair loss. Skin color may vary as the disease progresses. The dying child may have swelling of various joints or places in the body, depending on the nature of the illness.

The terminal illness of a child in the family often changes the course of the sibling relationship, bringing the siblings closer together or reducing tensions between them (Silverman, 2000). Children may mature rapidly and develop an understanding beyond their years as they learn to ask questions and take in what is going on. They are pushed to confront issues and feelings that would not ordinarily be expected of children their age. Indeed, with a dying child in the household, siblings may "grow up" quickly in numerous ways.

CONCLUSION

Because the thought of dying is stressful, it seems appropriate that we be aware of dying and death. With more awareness, we hope that a greater acceptance of dying and death will result. Through a better understanding of death meanings, our coping with dying and death should be enhanced. It is difficult to relate to the dying if we ourselves have not been sensitized to our own death.

We often play games in communicating with a dying patient. We know, the physician and nurse know, and the patient knows (probably whether told or not) when the condition is terminal, but we often exist in anything but an open awareness context. No one lets the other know that he or she knows. Death talk remains taboo.

SUMMARY

1. The meaning of dying depends upon the social context in which it takes place.

2. Time, space, norm, role, value, self, and situation are important components of the meaning of our dying.

3. One may go through stages in accepting his or her terminal self-meaning.

4. Dying a "good death" makes it easier on everyone.

5. According to Glaser and Strauss, four different awareness contexts exist between medical personnel, patients, and the patients' families: open, mutual pretense, suspicion, and closed.

6. The dying child needs support, as do the siblings of the dying child.

7. Societal disengagement refers to society withdrawing from an individual.

8. Kübler-Ross divides the dying process into five stages: denial, anger, bargaining, depression, and acceptance.

9. Socialization of physicians regarding communicating with dying patients and their families historically has been very limited.

10. For family and friends relating to a dying person, it is imperative that they first of all try to maintain their own health to better cope with the stress of taking care of the dying individual.

11. Primary caregivers to a dying individual must share the responsibility of care with others to reduce stress.

12. Music, humor, and art can be positively used as therapy with terminally ill children.

DISCUSSION QUESTIONS

1. What is meant by this statement: The meaning of dying will depend upon the social context in which the dying takes place?

2. Would you prefer to live with a person with a terminal disease or a person who is chemically dependent? Discuss the advantages and disadvantages of each.

3. How could steps be taken to overcome the diminished social and personal power of the hospital patient? Are such limitations on patients necessary for an orderly hospital?

4. Discuss this statement: The terminally ill eventually come to view death as a blessing.

5. You have just been told that you have inoperable cancer. Discuss how you think you would react. In what ways would you change your life?

6. How do you think you would relate to a dying child, if you were a parent?

7. Discuss Glaser and Strauss's four awareness contexts. Which do you think most often exists in a medical setting with a dying patient?

8. How do you think a physician should go about telling someone that she or he has a terminal illness? How would you prefer to be told this?

9. Discuss how "societal disengagement" applies to a dying person.

10. Discuss Kübler-Ross's five stages of the dying process. If you know of someone who is dying, try to "fit" her or him into the appropriate current stage.

11. How does research cited in this chapter show similarities or differences in how male and female physicians relate to dying patients and their families?

12. If you have experienced the death of someone close to you, what was the "awareness context" during most of that individual's illness? Would you have preferred another context?

13. Suggest ways in which art, humor, and music can be therapeutically used with dying children.

GLOSSARY

Anticipatory Grief: Experiencing grief before a death actually occurs; griefwork aimed at loosening attachment to the dying, making loss less painful when it occurs.

Appropriate Death: A person dying as he or she wished to die. The death is generally consistent with past personality patterns.

Awareness Context: What each interacting person knows of the patient's defined status and his or her recognition of the others' awareness of a personal definition.

Cadaver: A dead body. Human cadavers are used in medical schools for the purpose of dissection to learn the parts of the body.

Dying Trajectory: Perception about the course that dying will take.

Good Death: An "appropriate" death at a particular time and place.

Master Status: The status (position) most important in establishing an individual's social identity.

Norm: A plan of action or expected behavior pattern thought to be appropriate for a particular situation.

Pediatric Oncologist: A medical doctor practicing in the field of medicine related to children with tumors (often malignant).

Role: Specified behavior expectations for persons occupying specific social positions.

Scapegoat: A person, group, or object upon whom blame is placed for the mistakes of others.

Sick Role: A set of characteristic behaviors that a sick person adopts in accordance with the normative demands of the situation.

Situational Adjustment: The process by which an individual is "molded" by the group into which he or she is seeking acceptance. The person learns from the group how to continue successfully in a situation.

Societal Disengagement: A process whereby society withdraws from or no longer seeks the individual's efforts, as distinguished from social disengagement in which the individual withdraws from society.

SUGGESTED READINGS

Bethel, E. R. (1995). AIDS: *Readings on a global crisis*. Boston: Allyn and Bacon. Discusses the ecology of AIDS, the epidemics of AIDS in the United States, Africa, Latin America, Great Britain, and Asia, and programs for behavioral change.

Buckman, R. (1992). *I don't know what to say: How to help and support someone who is dying*. New York: Viking Press. Discussion of many issues that evolve in the process of watching someone die, including the topic of how to communicate with a dying person, from the author's clinical experiences as an oncologist.

Charmaz, K. (1991). *Good days, bad days: The self in chronic illness and time*. Rutgers, NJ: Rutgers University Press. Description from a sociologist of three stages through which chronic illness can progress: The illness disrupts life; it intrudes nearly every day and holds no promise of abating; and it engulfs the individual, becoming the paramount feature of life.

Christakis, N. (1999). *Death foretold*. Chicago: University of Chicago Press. "Giving bad news" to patients, from the perspective of the author, a physician and sociologist.

Doka, K. J. (1993). *Living with life-threatening illness*. New York: Lexington Books. A book to help patients, families, and caregivers in relating to a life-threatening illness.

Field, D., Hockey, J., & Small, N. (1997). *Death, gender and ethnicity*. New York: Routledge. Examines ways in which gender and ethnicity shape the experiences of dying and bereavement.

Frank, A. (1991). *At the will of the body: Reflections on illness.* Boston: Houghton Mifflin. How illness affects us and what it can teach us about life. The author is a medical sociologist who had a heart attack at age 39 and cancer at age 40.

George, R., & Houghton, P. (1997). *Healthy dying.* Bristol, PA: Taylor & Francis. Examines a wide range of issues surrounding the terminally ill, including difficulties faced by caregivers.

Goldman, L. (1996). *Breaking the silence: A guide to help children with complicated grief.* London: Taylor & Francis. Use of visualization, photography, artwork, clay, toy figures, punching bags, tape recordings, and story telling.

Hafferty, F. W. (1991). *Into the valley: Death and the socialization of medical students.* New Haven: Yale University Press. Using participant observation, describes the experiences of a class of first-year medical students as they are exposed to dying patients and gross anatomy laboratory.

Keizer, B. (1997). *Dancing with Mister D: Notes on life and death.* New York: Doubleday. Probes the intensity of the American preoccupation with dying and death as he shares his extraordinary experiences among the terminally ill. The author is a Dutch doctor with training in philosophy as well as medicine.

Kellehear, A. (1990). *Dying of cancer: The final year of life.* Chur, Switzerland: Harwood Academic Publishers. A sociological account of the social behavior and experiences of 100 individuals who were dying of cancer.

Konner, M. (1987). *Becoming a doctor: A journey of initiation in medical school.* New York: Penguin Books. Takes the reader through hospital rounds, in various specialty areas, and discusses encounters with death and with dying patients. The author is an anthropologist and physician.

Nuland, S. B. (1994). *How we die: Reflections on life's final chapter.* New York: Alfred A. Knopf. Accounts of dying that reveal not only why someone dies but also how from a surgeon and professor at Yale Medical School.

Proctor, R. N. (1995). *Cancer wars: How politics shapes what we know and don't know about cancer.* New York: Basic Books. Discusses how government regulatory agencies, scientists, trade associations, and environmentalists have managed to obscure the issues and prevent concerted action around the fight against cancer.

Riska, E., & Wegar, K. (1993). *Gender work and medicine.* London: Sage Publications. Collection of articles providing a critical assessment of the division of labor in medicine, setting current practice in its historical context.

Rosen, E. J. (1990). *Families facing death: Family dynamics of terminal illness.* New York: Lexington Books. Guidance from a family therapist to the wide range of disciplines that interact with families facing death.

Silverman, P. R. (2000). *Never too young to know: Death in children's lives.* New York: Oxford University Press. A guide for coping with dying and death of a child in the family.

Smith, J. M. (1996). *AIDS and society.* Upper Saddle River, NJ: Prentice-Hall. Discusses various aspects of AIDS: demographics, prejudice, discrimination, morality, institutional barriers, and economic and international issues.

Spiro, H. M., Curnen, M. G. M., & Wandel, L. P. (Eds.). (1996). *Facing death: Where culture, religion, and medicine meet.* New Haven: Yale University Press. Discussion of the current clinical setting for dying by physicians. Also, Christian, Judaic, Islamic, Hindu, and Chinese perspectives on death and mourning rituals are presented.

Webb, N. B. (1993). *Helping bereaved children: A handbook for practitioners.* New York: Guilford Press. Good suggestions for using drawings with dying children.

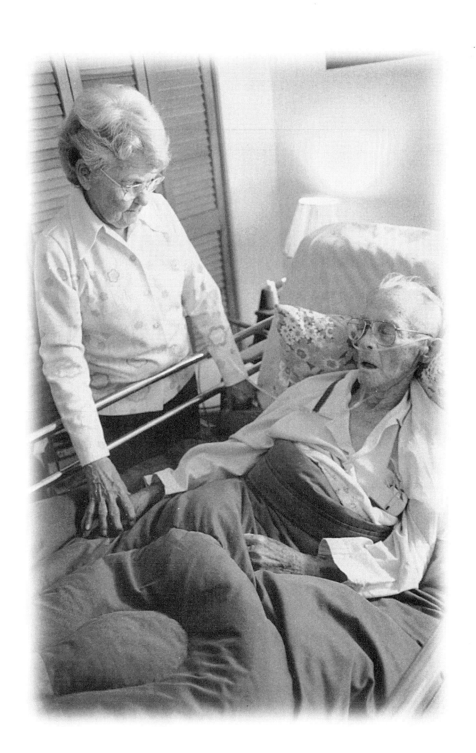

CHAPTER 6

LIVING WITH DYING

"I'm afraid of the pain."
"I don't want to be alone when I'm dying."
"I'm afraid of a long, protracted period of suffering."
"I don't want to die in a hospital. Let me die at home."
"I'm not afraid for myself, but I am worried about the effect of my death on those I love."

> —Frequent responses to the question: Does dying frighten you?

Many individuals are living for years with a serious illness. The chances of an illness going into remission or even "going away" are more likely this year than last, and the same will be true of next year because of scientific and medical advancements each year. In the early stages or remission phase of a chronic illness, where one is asymptomatic or experiences minor symptoms and little pain, it may be possible to continue to maintain a good quality of life, even one that approaches the standard of life before the diagnosis (Doka, 1993).

With a population better educated on health issues, the stigma of cancer (the "big C" disease) and other illnesses has diminished considerably in recent years. This allows people with life-threatening illnesses to live openly with their illnesses. Without the stigma, individuals can acknowledge their illness to others, making it easier to deal with their impending death. In this phase of relative health, an individual can begin to plan for the inevitable decline, researching long-term care options, making financial arrangements, and completing a will.

WORDS OF WISDOM
The Challenge of Living (While Dying)

The challenge of the dying process is the challenge of *living while dying* rather than *dying while living*.

From The Grace in Dying
by Kathleen D. Singh, 1999,
Dublin, Ireland: Newleaf.

As discussed in chapter 5, sociologist Arthur Frank (1991) observed that to seize the opportunities offered by illness we must live actively and must *talk* about the illness. In that way, as individuals and as a society, we can begin to accept illness fully. Only then can we learn that it is nothing special. Do not curse your fate, but rather count your possibilities, advises Frank (1991). Being ill is just another way of living, but by the time we have lived through illness, we are living differently. As a friend who was dying of lung cancer told George Dickinson, "I look at a sunrise and sunset and watch the tide come in and go out in a different way than before my illness." Indeed, in dying, one is living differently; dying changes the perspective on events in life.

It is amazing how positive an outlook many individuals living with life-threatening diseases have. Many of them continue to hold jobs, go to and from treatments, and basically live life to the fullest. In some cases, the person who is ill needs to help friends deal with the illness. A former colleague of George Dickinson's, who was dying of cancer, decided to continue teaching. Rather than continuing as if nothing was happening, he decided to be open about his condition and active in encouraging his friends and colleagues to be part of the process of dying and to *talk* about it. This communication helps both the patient and friends cope with the dying. He sent a letter to friends to come see him ("keep me company," as he put it), and noted that if "you don't know what to say," come anyway. He further said, "You don't have to avoid talking about health matters: My situation is real, and I'm not ignoring it. But I *am* determined to be upbeat and positive." Many individuals avoid visiting someone who is seriously ill for fear that they will not know "what to say" or, worse yet, will say the wrong thing. However, it is very important that you make contact with a significant other who is seriously ill. Even if you do not know what to say, go anyway to let the person know that you care. Dying of ALS, Professor Morrie Schwartz said to his former student (Albom, 1997, p. 36), "I may be dying, but I am surrounded by loving, caring souls."

Recently George Dickinson was asked to visit a 76-year-old woman who was in the latter stages of cancer. Her breast cancer had spread throughout her body, and the prognosis was not good. She wanted him to assist in planning her funeral and writing her obituary so that her adult children would not have to "worry themselves" with these matters. The day he met this delightful, cheerful individual, Dorothy Sutch, her mobility had decreased to the point that she was confined to a wheelchair. They talked for about an hour. When the assignments were completed, she threw her hands into the air and, with a smile on her face, said, "Now, I'm ready!" Her preparation for death was complete. She then talked about how dying was "like a journey." "It's happening so fast," she said. "It's exciting, unfolding before my eyes." He visited her two more times that week, and by the end of the week, she was dead. Indeed, it happened "so fast." Though he had not had the pleasure of knowing Mrs. Sutch, Dickinson later learned from her many friends that she had lived as one who cared and tried to make life enjoyable for those around her. In her dying, she again had been cheerful and caring. She died as she lived, as we all basically will.

We can learn much from persons with life-threatening diseases. As noted in chapter 1, physician Elisabeth Kübler-Ross says that working with the terminally ill has made her appreciative of life. Each day that she awakens she is thankful for the potential of another day of life. George Dickinson learned much about living and dying from his 1 week's acquaintance with Dorothy Sutch.

In dying, one often lives life to the fullest. It has been said that some individuals live more in their final months of life, knowing that the end is near, than they had lived in their previous years. Living with dying can be a most meaningful experience, both for the patient and for those with whom she or he comes into contact.

UNDERSTANDING AND COPING WITH THE ILLNESS

Though we may have limited experience with and may be initially uncomfortable in relating to someone who is dying, simply being aware of what that person may be going through should enhance our understanding of that individual's situation. Through better awareness, we hope we can better cope with the illness.

The Loss of Physical Functions

Depending on the particular illness, an individual with a terminal diagnosis will begin to lose bodily functions. Morrie Schwartz, dying of ALS, in Albom's *Tuesdays with Morrie* (1997), so well describes his gradual inability to dance, a favorite "leisurely activity" for him. Then he has difficulty simply walking and soon he is bedridden. Then comes the loss of the use of his arms and his ability to swallow and the capability to urinate and defecate without help—all such "every-day" events for us, *when healthy*. His speech became impaired. Simply turning over in bed became impossible without help. As small children, we are socialized to be independent; then in dying, we again become dependent. The sheer physical and psychological strain on caregivers of the dying can indeed be stressful.

If the patient has lost physical functions and is at home, various modifications within the home environs need to occur. A hospital bed, which can be adjusted up and down, will help the bedridden individual to get into different positions. A wheelchair and/or walker should be available, if the patient has some mobility. Oxygen may be required. A bedpan will be necessary, if the patient does not have tubes to eliminate waste. As noted elsewhere in this book, a financial drain on the family may occur. In addition, the physical aspects of turning and lifting the patient will take their toll on family members after a while. The psychological strain involved in relating to a dying individual will be stressful over time for caregivers. The patient will have health hygiene needs with which others will need to assist. The list of the needs of one who has lost physical functions is almost endless.

The Loss of Mental Capacity

With some illnesses, such as a stroke, an individual may lose the ability to think. Or medications and treatments may cause a very sick person to be "confused" and unable to think. If the patient has brain damage, as is sometimes the situation with a stroke, the brain may be reduced to that of a small child. The ability to speak, to write, or to respond in any meaningful way may be lost. In extreme cases, the individual is "alive" but may have virtually no mental capacity. It is, therefore, almost impossible for the individual to express specific wants and needs. Health-care providers must "guess" what the patient needs, and guesswork is not an exact science.

A disease affecting an estimated four million Americans today is **Alzheimer's disease (AD)**. AD is a chronic, degenerative, dementing illness of the central nervous system and involves personality change, emotional instability, disorientation, memory loss, loss of verbal abilities, and an inability to care for oneself (Weiss & Lonnquist, 2000). AD is difficult to determine and is often not diagnosed until late in the illness. Some patients with AD are violent and thus are difficult individuals for whom to care. Awareness of AD has increased significantly in recent years, and AD research is well underway today.

Dying of AIDS and Cancer

Though heart disease is the leading cause of death in the United States today, as noted in chapter 1, we tend to talk more about deaths by AIDS and cancer. Since AIDS was first diagnosed in 1981, the media has saturated us on this topic. Medical research is concentrating on various diagnoses of cancer; thus, cancer also stays at the forefront of media coverage. Both AIDS and cancer tend to reveal manifestations of the disease, if not from the disease itself then from side effects of the treatments. Thus, let us take a look at dying from AIDS and cancer.

AIDS

"Living with dying," when it is of **AIDS**, can have a certain stigma to it. Not only does the individual have a terminal illness, she or he has to sometimes put up with others' responses "because of how she or he may have acquired AIDS." A "morals" issue may enter into reactions to AIDS. One of the biggest difficulties facing the majority of persons with AIDS is the reversal of normal life development (Werth, 1995).

Acquired Immunodeficiency Syndrome (AIDS) Cases by Age and Sex   TABLE 6-1
(reported through June 1999)

Age at Diagnosis (years)	Male	Female
Under 5	3,364	3,308
5–12	1,064	860
13–19	2,134	1,430
20–24	18,266	6,944
25–29	77,384	18,179
30–34	134,150	26,679
35–39	133,954	25,164
40–44	99,416	16,916
45–49	58,257	8,606
50–54	30,779	4,460
55–59	16,860	2,648
60–64	9,267	1,595
65 or older	7,657	2,000
TOTAL	592,552	118,789

Source: Centers for Disease Control and Prevention. National Center for HIV, STD, and TB
Prevention. (1999, November 29). *HIV/AIDS Surveillance Report* (Vol. 11, No 1). Washington, DC:
U.S. Department of Health and Human Services, Public Health Service.

Nearly 90 percent of persons with AIDS are between the ages of 20 and 49 (CDC,
1999). Though far more deaths in the Third World are caused by diseases of childhood (respiratory disease, diarrhea, and measles), AIDS is rampant among young
adults. This situation is especially tragic in developing countries that rely on their
young adult population to lead them out of poverty ("Poor Man's Plague," 1991). It
is sometimes said that AIDS is a disease of the poor, and in the developed world this
is becoming the case. In the United States, it is in the inner cities that AIDS is spreading fastest. In poor countries, however, AIDS is often a disease of the relatively rich.

The age group with the greatest percentage of AIDS cases is 30 to 39 years (45
percent). In every age category of AIDS, males outnumber females (Table 6-1). AIDS
is affecting blacks proportionately more than it is affecting whites: Blacks make up
12 percent of the total population of the United States, yet approximately one third
of those with AIDS are African Americans. Andrew Sullivan (1990) noted in talking to minority men with AIDS that it was hard to avoid the impression that the
level of denial is measurably greater, the pain more intense, and the isolation more
complete than among whites. One gay black man said, "By being black, I'm separated from the white gay community, and by being gay, I'm separated from the
African American community."

Perhaps the "good news" about AIDS today in the United States is that the rates
for deaths from AIDS began dropping in the mid-1990s and dropped 48 percent

from 1996 to 1997 and again in 1998, according to data released in October 1999 (U.S. Department of Health and Human Services, 1999). The age-adjusted death rate from HIV infection in the United States declined an estimated 21 percent to a rate of 4.6 deaths per 100,000 in 1998. HIV mortality has declined more than 70 percent since 1995. However, HIV remains the leading cause of death among African American men aged 25 to 44 and the third leading cause of death among African American women in the same age group.

The "news" regarding AIDS on a global basis is not so good, however. In fact, the worldwide AIDS epidemic is considerably worse than had been estimated (Knox, 1997). Almost 31 million people are infected with HIV—8 million, or 30 percent, more than previous global estimates. The AIDS virus is now spreading at the rate of 16,000 new infections per day, up from 10,000 new cases given in previous estimates. AIDS has now surpassed malaria as the world's number one killer (Knox, 1997). Perhaps the most striking disparity between AIDS in developed and developing nations is in the number of children being infected by HIV. Each day in developing nations, 1,600 children under 15 acquire HIV; in the United States, preventive drug treatment has reduced infections among newborns to fewer than 500 a year.

Not only is AIDS a devastating disease, but for individuals in the gay community, support is literally dying out. Recently, George Dickinson served as a panelist at a regional sociology meeting discussing AIDS. One of the other panelists, who was gay and HIV positive, noted that of his immediate group of 40 friends, all but 6 had died of AIDS within a short time. Not only was he saddened, but he had also few friends left with whom to share his grief. A similar account is noted in a study of the psychosocial impact of multiple deaths from AIDS (Viney et al., 1992, p. 152) by one of the participants:

> I've been counting them up. You know, I've lost 50 people, friends and acquaintances, since this AIDS epidemic started. I've been to that number of funerals in the past few years. And now my lover is ill.

The psychosocial functioning of people who are coping with these problems is similar to that of people who are coping with any severe illness: higher than average levels of anxiety, depression, directly and indirectly expressed anger, and helplessness (Viney et al., 1992). Thus, for some young adults today, AIDS is a reckoning force bringing death into focus through personal experience.

What efforts are being made to accommodate individuals with AIDS in the United States? In 1992 a project in Seattle, Washington, called the Bailey-Boushay House was initiated to create a large home in which somebody was going to die every 30 minutes (Eagan, 1992). This project was one of the first opportunities for persons with AIDS to have a residence of their own. The house looks more like a small European spa than an institution. Inside, there is a greenhouse on the third floor and solariums at the end of every hallway. The greenhouse reflects the belief in the healing power of making things grow. Sofas are in every room for friends or relatives to stay overnight. Sun decks provide views of the Cascade Mountains. Beds are provided for 35 residents with day-care space for another 35 adults. Costs at the home are kept to a minimum. The facility is run by a nonprofit group. The average stay of a patient is 45 days.

The health-care challenge of the twenty-first century will be the increasing numbers of patients with AIDS and AIDS-related complex (ARC). Unlike other terminal illnesses, AIDS is a communicable disease involving extremely labor-intensive care for patients who are often without family support systems. Furthermore, the physical, psychosocial, and emotional complexities of AIDS often require especially sensitive and humane approaches by health-care providers of every community.

Cancer

The term **cancer** refers to a group of diseases that are characterized by an uncontrolled growth and spread of abnormal cells (Weiss & Lonnquist, 2000). Some cells are noncancerous (benign); others are cancerous (malignant). About one person in three now living in the United States will be diagnosed with cancer at one point (this does not include skin cancer). As noted in chapter 1, cancer is the second leading cause of death in the United States, claiming over 550,000 lives per year.

Causes of cancer are largely chemicals in the air, the water we drink, and the food we eat. We basically obtain cancer because of bad habits, bad working conditions, and bad luck— of the genetic draw, notes Robert Proctor in *Cancer Wars* (1995). The use of tobacco is a "bad habit" that contributes to cancer. G. L. Weiss and L. L. Lonnquist (2000) note other "causes" of cancer. Too much sun contributes to most of the one million cases of nonmelanoma skin cancer diagnosed each year. Excessive alcohol consumption is the main cause of cirrhosis of the liver, which makes one at high risk for developing liver cancer. Excessive exposure to radiation can contribute to cancer. Exposure via occupational hazards and environmental pollution correlates with cancer. Besides the above-mentioned cancer etiologies, heredity does play a role in some cancers, though most cancers are not inherited.

Cancer rates and causes vary (Proctor, 1995). The leading cancer killer in the United States is lung cancer, directly related to smoking; in Japan it is stomach cancer (perhaps caused by use of charcoal in preparing food); in India, cancer of the mouth is common, perhaps from chewing betel nuts and tobacco leaves; and in China moldy bread contributes to esophageal cancer. Religion may play a part in cancer rates as Mormons have lower death rates from cancer than non-Mormons (Mormons do not smoke). Cervical cancer is rare in nuns, but common in prostitutes.

The rate of cancer cases and deaths for all cancers combined declined slightly between 1990 and 1997 in the United States (National Cancer Institute, 2000). This was the first reported drop in cancer rates since a brief downturn in the 1930s. The greatest decline in cancer incidence rates has been among men, who generally have higher rates of cancer than women. Another trend is the increasing rate of survival for persons diagnosed with cancer. Individuals are living longer *with cancer*.

The top four types of cancer—lung, prostate, breast, and colorectal—account for slightly over half of all new cancer cases and are also the leading causes of cancer deaths for every racial and ethnic group. Cancer incidence and mortality rates are higher for blacks than for whites. Types of cancer that reveal increased incidence rates are non-Hodgkin's lymphoma among women and melanoma for both sexes. For males, the prostate, lungs, and colon are the most common sites for cancer, but lung cancer is by far the most lethal. For females, the breast, lungs, and colon are the most common cancer sites, but lung cancer is the most fatal.

Rates of cancer are probably going down because of better education and thus more frequent screening and earlier detection. Recent court decisions regarding smoking, followed by a media emphasis on antismoking education, have no doubt contributed to lower cancer rates, particularly those of lung cancer. Smoking is also associated with cancer of the mouth, pharynx, larynx, esophagus, pancreas, uterine cervix, kidney, and bladder.

Patients with cancer may experience fever, followed by chills and sweats as the body attempts to regulate internal temperature (CancerNet, 2000). The major causes of fevers in patients with cancer are infection, tumor-associated factors (e.g., acute leukemia, Hodgkin's disease, lymphomas, bone sarcomas, and hypothalamic tumors), allergic or hypersensitive reactions to drugs, and allergic or hypersensitive reactions to blood component therapies.

Anorexia, the loss of appetite or desire to eat, is the most common symptom in patients with cancer that may occur early in the disease or later as the cancer grows and spreads. The patient may also suffer from cachexia anorexia, a wasting condition in which the patient has weakness and a marked and progressive loss of body weight, fat, and muscle. Maintenance of body weight and adequate nutritional status can help patients feel and look better and maintain or improve their performance status. It may also help them to better tolerate cancer therapy (CancerNet, 2000).

Treatment options for patients with cancer generally include surgery, chemotherapy, and radiation. Sometimes all three options are chosen. Side effects of treatment could be nausea, vomiting, oral complications (mouth sores, dry mouth, and bleeding gums), delirium (confused thinking and hallucinations), fatigue, fever, constipation (impaction and bowel obstruction), swelling caused by blocked lymph nodes, diarrhea, sleep disorders (insomnia and disturbed sleep cycle), itching, and hair loss.

TREATMENT OPTIONS

Evaluating Treatment Options and Symptoms

George Dickinson recalls a story told to him by a sociologist a few years ago. After examination and testing, his neurologist informed the sociologist that he had a malignant tumor in his brain. The patient immediately said to his doctor, "That means I will die, doesn't it, Doctor?" The physician replied, "Not necessarily. We have surgery, **chemotherapy,** and **radiation therapy.** We will try all three, if needed." These treatment options suggested to the patient that indeed there were alternatives from which to choose to fight this cancer. The doctor's encouragement of treatment options gave hope to the patient and indeed a willingness to fight the disease.

PRACTICAL MATTERS
Steps Toward Cancer Prevention

1. Ban smoking.
2. Impose stiffer taxes on tobacco sales.
3. Initiate stiffer supervision of pesticides and alternatives to petrochemical agriculture.
4. Better regulate loopholes that allow nonfood pesticides to be used on crops we eat.
5. Empower OSHA to deal with indoor pollutants such as radon and secondhand smoke.
6. Allow the FDA to limit the substances added to tobacco in manufacturing cigarettes.
7. Modify trade practices at home that encourage cancer abroad.
8. Curtail the export of asbestos.

Taken from Cancer Wars *by R. T. Proctor, 1995, New York: Basic Books.*

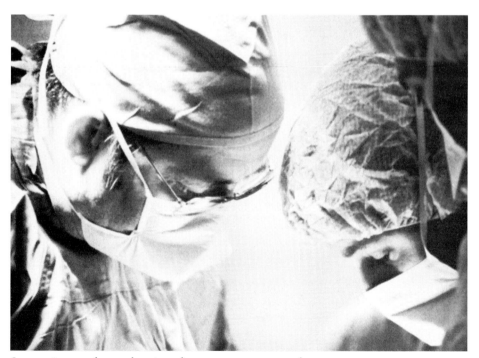

Surgery is one of several options for cancer treatment today.

After a diagnosis of a life-threatening disease, individuals will need to make decisions about treatment options (Doka, 1993). Treatment plans differ according to the illness. Some treatment plans will be relatively unobtrusive and may require little monitoring, yet others may cause lifestyle modifications. Heart disease, for example, may cause an individual to take daily medication, go on a special diet, and basically simplify the lifestyle and try to reduce stress. Yet, with other conditions, such as cancer, options, as noted above, for treatment may be surgery, chemotherapy, and/or radiation. The order in which these treatments are given may vary, depending on the particular health status of the patient. Certainly any one of these treatments or a combination of them can contribute to lifestyle modifications and limitations. Perhaps the good news about treatment options for basically all illnesses is that medical advancements are occurring annually; thus, one's chances of survival are extended as the onset of a disease is delayed.

Whatever therapies are available, feelings and fears about them should be explored, notes Kenneth Doka (1993). Surgery may be feared because of anesthesia, loss of consciousness, postoperative pain, visible damage (scars, for instance), loss of functional abilities, or financial considerations. The procedures of chemotherapy and radiation may cause anxiety and concern over side effects.

Men traditionally have not been good about going to physicians regarding potential health issues and thus would be less likely to have various treatments, compared to women. A recent Harris poll (Marquis, 2000) revealed the reluctance of men in seeing a doctor. Three times as many men as women had not seen a doctor in the previous year and one in three had no regular doctor, compared with one in five women in the 1998 survey. One fourth of the men said they would wait "as long as possible" before seeking help for a health-care problem.

Today, many women with a tumor in the breast choose to have the entire breast removed, even though study after study has shown that women with small- to moderate-sized tumors fare just as well with a lumpectomy—removal of just the tumor, leaving the breast intact (Proctor, 1995). Fear of breast cancer has led some women to have their breasts removed, thinking this will increase their chances of survival. Men have not yet begun to fear their prostate glands, though early signs of such fears are on the horizon, according to Proctor in his book on politics and knowledge about cancer.

Another treatment option for various illnesses is **immunotherapy** (treatment by stimulation of the body's own immune system). The immune system includes a complicated system of organs, tissues, blood cells, and substances to fight off infections, cancers, and other illnesses. Alternative forms of health care include **chiropractic medicine,** which relates to health problems based on blockages of the flow of vital energy caused by malalignments of the vertebrae. **Homeopathic medicine** (the use of natural drugs to treat patients) is built around the belief that if a small amount of a natural substance will precipitate symptoms in a healthy person, a similar amount in a sick person will aid in recovery. **Acupuncture** (using needles along meridians in the body to redirect energy flow) is a traditional Chinese medical treatment going back nearly 5000 years and is a holistic system that understands health in the context of the relationship between the human body and nature. **Faith healing** is the use of the power of suggestion, prayer, and faith in God to promote healing. Such healing involves mind over matter, whether the healing is from snake-handling groups in Appalachia or members of the Christian Science community. **Folk healing** (the use of "folk remedies" to heal the sick) exists in part because of a dissatisfaction with professional medicine and may include "treatments" such as ginger tea, honey, sassafras, salt, and butter. Folk healers include those referred to as root doctors, spiritualists, or voodoo practitioners. Some Native Americans practice folk healing by believing that diseases are caused by soul loss or spirit possession and use herbalists and bone setters to treat the sick.

Treating Drug Side Effects

Some anticancer drugs cause nausea and vomiting because they affect parts of the brain that control vomiting and/or irritate the stomach lining. The management of nausea and vomiting caused by chemotherapy is an important part of care for patients with cancer. There is no single best approach to alleviating these symptoms in all patients. Therapy must be tailored to each individual's needs, including factors such as the type of anticancer drug administered, the general condition of the patient, and age.

For treating loss of appetite and weight loss for patients with cancer, changes in diet and consumption of foods, taken through a tube or orally, can help. Researchers can evaluate the effect of the drugs alone or in combination on nausea and vomiting and assess differences in the quality of life among patients. Toxic effects related to the use of drugs can also be assessed.

Though controversial, the use of marijuana for controlling chemotherapy-induced nausea and vomiting in patients with cancer is generating much interest. The U.S. Food and Drug Administration (FDA) has approved the use of marijuana for treatment of nausea and vomiting associated with cancer chemotherapy in patients not responding to "standard" drugs (NCI Fact Sheet, 1999).

Though scientists say far more research needs to be done before marijuana can be considered a viable treatment, governmental reviewers have all but shut down clinical trials, basing their decisions on politics rather than on science, according to some researchers (Guterman, 2000). As of June 2000, only one clinical trial of the plant's medical effects was under way in the United States. Previous research (Guterman, 2000) before the government's war on drugs kicked into high gear, however, had shown that marijuana's major psychoactive ingredients could treat a number of conditions: relieve pain, reduce vomiting, and stimulate appetite in people with diseases, such as AIDS and cancer, that cause them to lose weight.

No matter how effective it is, however, smoked marijuana is unlikely to be approved as a medicine in the United States because of public resistance, based on longstanding social concerns, notes L. Guterman (2000). Those opposed to legalized use of marijuana for health purposes argue that the slippery slope curve could occur here: legalization of medical marijuana might open the door to recreational use and might even lead people to more harmful substances such as cocaine or heroin. Though not solidly supported by research, there are arguments against marijuana use grounded in physiology. Marijuana interferes with short-term cognition, causes the heart to beat faster, which could have negative effects for persons with heart disease, and could damage the lungs because it is smoked.

A nonsmoked medication has a better chance of being accepted by the medical establishment and approved by the government, observes Guterman (2000). The FDA has approved the drug, Marinol, taken as a pill, for appetite stimulation and relief of vomiting in patients with AIDS and cancer. The pill can be difficult for one with severe nausea to keep down, and Marinol takes hours to take effect, whereas smoked marijuana's instant hit of cannabinoids in the bloodstream allows patients to monitor their own dose, say the supporters of legalized medical marijuana. And the politics of medicine continues!

Pain and Symptom Management

The care of the dying is an art that should have its fullest expression in helping patients cope with the technologically complicated medical environment surrounding them at the end of life, notes Sidney Wanzer and his medical associates (1989). The concept of a "good death" requires the art of deliberately creating a medical milieu that allows a peaceful death. The physician must seek a level of care that optimizes comfort and dignity and needs to formulate a flexible care plan, tailoring treatment to the patient's changing needs as the disease progresses.

In today's medical environment, more and more doctors are integrating pain management into their treatment of dying patients. This pain and symptom management approach first took hold within hospice settings, which developed in response to the medical model approach to dying. Pain control must be a top priority of physicians, observe Wanzer and associates (1989), because one of the major causes of anxiety among patients and their families is the perception that physicians' efforts to relieve pain are badly deficient. As sickness progresses to death, dying patients may require palliative care of an intensity that rivals even that of curative efforts. Indeed, death education in medical schools must strive to help the student know how to make the dying patients' last few days more comfortable. As in natural childbirth—the other occasion on which the role of the physician and nurse is not to cure but to make the patient comfortable—natural death involves demedicalization to the extent that death is defined as a natural rather than a medical event (Walter, 1995). Nurses and doctors cannot conquer death, but they can become "midwives of the dying," assisting through this natural transition (Albery, Elliot, & Elliot, 1993).

Traditional medical care is often based upon a **PRN** (for the Latin *pro re nata*) approach, which means that medication is given "as the situation demands." In practice, this means that often a person must first hurt and ask for relief before pain

DEATH ACROSS CULTURES
Pain in Different Ethnic Groups

The subjective experience of pain is powerfully contoured by culture. In the early 1950s, a study of responses to pain among men in a veteran's hospital in New York revealed that these individuals accepted pain without complaint. These Americans had lived in the New World for many generations so that they no longer had ties with their countries of origin and thus were "Old" Americans. Irish Americans in this study also accepted pain without complaint, but the two groups differed greatly in how they experienced the meaning of pain.

There is a difference between private and public pain. Left alone, the Old Americans might collapse into tears, but never in public. In the presence of nurses and doctors, to admit to pain was permitted because the professional situation transformed complaints into purposeful discussion. They had faith in their doctors and were fairly optimistic about their outcomes.

On the other hand, the Irish-American patients differed. They lacked the optimism of Old Americans. In public, an Irish patient masked pain, but this was also done when talking to medical practitioners. The Irish-American felt helpless, guilty about becoming ill, and very pessimistic about the future.

Italian-American and Jewish-American veterans tended to display highly emotional responses and groaned and cried and complained and talked in detail about how they suffered. They shared a cultural trait that permits vivid expressions of pain. The Jewish-American patient tended to experience a future-oriented anxiety, and pain was taken as a frightening warning of ultimate possible doom. This patient needed reassurance from the physician, yet this same patient was skeptical toward the doctor. Jewish pain tended to be associated with powerful existential concerns and ultimate eschatological issues.

The Italian-American patient was equally vocal, yet the patient showed great trust in doctors and hospitals. Instead of a future orientation, the Italian-American patient experienced a present-oriented apprehension and begged for strong analgesics to quiet the pain.

Though this research by anthropologist Mark Zborowski on pain was conducted a half century ago and methodologically has some flaws, the findings at least hint at the impact of culture on behavior. Learned behavior indeed varies culturally, thus producing different outcomes.

From Magic, Science, and Health: The Aims and Achievements of Medical Anthropology *by Robert Anderson, 1996. Fort Worth, TX: Harcourt Brace College Publishers.*

management can be administered. This approach is responsible for much of the suffering endured by the terminally ill. British physician Phil Hammond observes (Hammond & Mosley, 1999) that orientation to pain management in medical school is "often pitiful" and that many students become doctors with "quaintly stoical views about pain" and a hesitation to use pain management.

In 1958 Cicely Saunders (1992, p. 20) developed an alternative method of pain control that is being used today, especially in hospice. She writes:

> Here at St. Joseph's Hospice, as elsewhere at that time, one saw people "earning their morphine," and it was wonderfully rewarding to introduce the simple and really obvious system of giving drugs to prevent pain happening—rather than to wait and give them once it had occurred. Here too there was the potential for developing ideas about the control of other symptoms, and also for looking at the other components of pain. But first of all I must salute the Sisters of St. Joseph's and the compassionate matter-of-factness of their dedicated care. Together we began to develop the appropriate way of caring, showing that there could be a place for scientific medicine and nursing. We could illustrate an alternative approach to the contrast between active treatment for an illness (as if to cure it were still possible) or some form of legalized euthanasia.

Pain Management

A pain management approach is based on the belief that a patient should not hurt at all. Regular medication is, therefore, given before the pain begins. The aim is to erase the memory of the pain that has been experienced and to deal with the fear of pain in the future. Pain medications are standardized to the needs of the patient. The aim is to control the pain and other symptoms without sedating the patient. Every symptom is treated as a separate illness because only when each symptom is under control can a patient begin to find fullness and quality of life. Pain is a subjective experience (Trauger-Querry & Haghighi, 1999). C. C. O'Callaghan (1996, p. 43) quotes the International Association for the Study of Pain's definition of pain as "an unpleasant sensory or emotional experience associated with actual or potential tissue damage. . . ." One's sensation of pain is a response to physical stimuli combined with the "meaning of such events," notes O'Callaghan. However, whatever the origin (response to psychological upheaval or a known physical precursor), people's reports of pain are real, and effective treatment must address the psychological and physical causes, says O'Callaghan.

Pain among Americans seems to be accepted, whether the individual is terminally ill or not. For example, a recent Gallup poll (Goldenberg, 2000) of 2,000 people found that regular pain was a pervasive part of American life and that few people sought medical attention for it. Sixty percent of those surveyed agreed that pain was "just something you have to live with." Of those adults surveyed, 42 percent reported experiencing pain every day. Only half said they had seen a doctor in the past 3 years for help with their pain.

On the other hand, the United Kingdom seemingly has a good grasp on the issue of pain and pain control. For example, in the spring of 1999 George Dickinson was riding with a British palliative medicine physician in South Yorkshire, England. In the course of the conversation, the physician asked Dickinson's opinion of physician-assisted suicide. Dickinson said he felt that in certain circumstances, such as a patient in the latter stages of AIDS, such action was justified. He said, "especially if the patient's pain is uncontrollable." At this point the palliative medicine doctor

LISTENING TO THE VOICES
Enduring Pain

An elderly male patient with cancer would not take his pain medication. The nurse asked to talk to him. She learned that he was in the landing of Normandy in World War II and endured much pain and hardship. Many of his comrades died in the invasion. Out of his feelings for those comrades and those memories of the landing,

he told the nurse that he could/would endure the pain. The pain brought back memories, and he must "bear it" out of respect for those fallen fellow soldiers.

*Told to George Dickinson by
a nurse in the spring of 1999
in Sheffield, England.*

almost wrecked his car and snapped back with, "But why should the patient be in pain?" To this British palliative medicine physician, there is no excuse for anyone to have pain. So much for Dickinson's argument of occasionally favoring physician-assisted deaths! Indeed, the physician is correct. There is no excuse for pain in the 21st century, in most cases. Adequate doses of opiates and appropriate adjuvant medications can control pain in 75 to 90 percent of patients with cancer without undue side effects (Schug, Zech, & Dorr, 1990). As Hammond (Hammond & Mosley, 1999, p. 184) states, "Morphine is a brilliant painkiller and has been used for centuries, and yet many doctors either refuse to give it or don't give enough for fear of overdosing or causing addiction." However, the political issue in the United States of using drugs, upon which one might become addicted (What should it matter if one is dying?), continues on, whether the drug is morphine or marijuana.

Anxiety and depression are part of the chronic nature of pain. The patient is anxious about the pain returning—anxious because of the meaning of pain. This is not the acute pain that doctors are taught to treat, unlike the pain of a toothache or childbirth or appendicitis—pains with a foreseeable end, notes S. A. Lack (1979). Acute pain may serve a purpose by warning of a malfunction. Pain for the patient with cancer has a sinister meaning. When awakening in the morning with a stiff neck, the patient with cancer assumes **metastases** (spread of the disease) whereas the patient without cancer assumes that she or he slept in a draft. The degree of perceived pain is totally different in these two situations, concludes Lack (1979). One must take into account the anxiety these patients suffer, along with the chronic depression caused by chronic pain.

Most cancer pain responds to pharmacological measures, and successful treatment is based on simple principles promoted by the World Health Organization (1986). Oral administration (by mouth) of analgesic drugs is preferred, and they are given regularly to prevent recurrence of pain (Hanks et al., 1996). A step-by-step approach is used, beginning with a nonopioid analgesic such as aspirin. At the second step a weak opioid such as codeine is added, and when this proves inadequate a strong opioid is substituted for the weak opioid. Morphine is often the recommended strong opioid analgesic. If pain returns consistently before the next regular dose is due, the regular dose is increased. If patients are unable to take drugs orally, alternative routes are rectal and subcutaneous, note G. W. Hanks and associates (1996).

Bone metastases are the most common cause of cancer-related pain (Bomanji et al., 1995) and have been found in up to 83 percent of patients with carcinomas of the prostate, breast, or lung at the time of death. People with painful bone metastases may survive a number of years and can experience chronic severe pain and impaired mobility; thus, pain control has a significant effect on their quality of life (Day, 1999). Strontium therapy, administered by intravenous injection, is used in pain relief for patients with bone metastases from tumors in the prostate and breast. Carried by the bloodstream, a single injection of strontium treats all painful sites simultaneously and delays the development of new pain.

In addition to traditional medical treatment of pain, art and music therapies can work to help control pain. In the medical model, it is common to address physical problems first. In a pain management approach, pain is the first thing to be treated. Pain lets us know that something is out of balance—pain is not an enemy. Part of the role of the therapist in pain management is to help patients and their families understand the complexity of pain and the numerous ways of addressing it. Art and music therapists offer interventions to assist the patient in balancing the focus of attention away from the physical sensation of pain. The therapist engages the patient using art or music, refocuses energy from the pain, and ultimately makes the person more balanced by the decrease of the perception of pain, note B. Trauger-Querry and K. R. Haghighi (1999).

Managing Dehydration

There are some symptoms associated with dehydration, but each is easily treated. Thirst, experienced in a small number of dehydrated patients at the end of life, is easily palliated with lip moisturizer, sips of water, ice chips, and hard candies on which to suck (Hoefler, 2000). In a small percentage of patients, electrolyte imbalance may lead to neuromuscular irritability and twitching, both of which are easily treated with sedation. Nausea, sometimes a symptom of dehydration, is treated with antiemetics. Normally, however, dehydration actually reduces nausea, vomiting, and abdominal pain. In addition, decreased urine output from dehydration may decrease the incidence of urinary tract infection. Pulmonary secretions also decrease with dehydration; thus, the patient experiences less coughing, congestion, choking, and shortness of breath. Dehydrated terminally ill patients often report less pain and discomfort than patients receiving medical hydration; thus, their need for pain medication is less (Hoefler, 2000). Dehydration also reduces swelling in the body, thus improving the person's sense of well-being, and reduces pressure on tumors, if they exist.

Organ Transplantations

Many individuals today are "living with dying," while waiting for an organ transplant, yet others are living because of organ transplants already received. Those individuals waiting for donors may indeed die while waiting. Thus, they probably live rather anxious lives. Organ and tissue donations have prolonged many individuals' lives, while improving significantly the quality of their lives. Particular tissues and/or organs in one's body may simply cease functioning and would cause the death of the individual, if not for new organs or tissues taken from someone else or some other animal.

Some organs and tissues come from individuals who have been legally pronounced dead, while others come from living persons. Approximately 25 different kinds of tissues and organs are used for transplantation including eyes, skin, bones, tendons, bone marrow, kidneys, livers, pancreases, blood vessels, lungs, and hearts. With advances in medical technology and preservation techniques, vital organs may be procured and transported hundreds of miles to a recipient center for transplantation. Hearts and lungs can be preserved for up to 6 hours before transplantation, and kidneys can be preserved up to 72 hours (United Network for Organ Sharing, 1997).

Since their beginnings in the mid-twentieth century, organ transplants have allowed thousands of individuals to live normal lives. Rapid means of transportation have enhanced the process of getting an organ to the recipient.

The first modern human organ transplant was a kidney transplant in 1954 (Kutner, 1990). The first liver transplant was done in 1963, the first pancreas transplant in 1966, and the first heart transplant in the United States—though not in the world—in 1968. Transplantation received especially wide coverage in the media with the first transplantation of a human heart, by Dr. Christiaan Barnard in South Africa in 1967. There is an increasing tendency today for individuals to donate at death their organs and tissues to the living. By 1996 there were a total of 16,845 transplants from living (2,789) and recently deceased donors (14,056). This includes approximately 10,000 kidneys, 3,500 livers, 2,000 hearts, 1,000 pancreases, and 700 lungs (United Network for Organ Sharing, 1997).

The National Conference on Uniform Laws developed the Uniform Anatomical Gift Act to help answer questions concerning transplantations, and by 1971 all 50 states had adopted the act with only minor variations (Sade, 1999). The act permits a person older than 18 years of age to donate part or all of his or her body upon death. If the body or body part has not been donated before death, a family member may agree to do so, with permission based on the priority of the relationship. The priority is as follows: spouse, adult son or daughter, parents, adult sibling, legal guardian, or other party responsible for disposition of the body.

Various problems affect human transplantations. Whether or not the donor and donee tissues are compatible can be a problem. There is often a shortage of donors, thus producing a difficult decision in selecting among potential recipients. The lack of a legal definition of death in some states prevents surgeons from removing healthy organs when brain activity has stopped but the heart and lungs are still functioning. The lack of a nationwide communications network to coordinate information presents another problem for human transplantations.

Other problems include the question of who will pay for expensive transplantation procedures. A few public insurers have decided in favor of coverage for heart and liver transplants. Kidney transplants have been covered since 1973 under Medicare's end-stage renal disease program. Though the costs of solid organ transplants are high, the benefits are substantial when one considers that they may be life-saving.

As medical technology has progressed and transplantations have become more successful, the demand for transplantation has greatly increased. In 1996 more than 50,000 patients were waiting for organ transplants, and even after nearly 17,000 successful transplants were performed, 4,000 Americans died while waiting for organs to be donated (United Network for Organ Sharing, 1997). If we add to this problem the fact that increasingly health maintenance organizations (HMOs), insurance companies, and other third-party providers will not pay for transplantation procedures, we face a situation in which those who receive organ and tissue transplantations are primarily those who can afford such surgery. Such an unequal distribution of resources is a good example of the conflict perspective, discussed in chapter 1.

> ## PRACTICAL MATTERS
> ### Uniform Donor Card
>
> Uniform Donor Card of (name of donor). In the hope that I may help others, I hereby make this anatomical gift, if medically acceptable, to take effect upon my death. The words and marks below indicate my desires. I give:
>
> a. _____ any needed organs or parts
>
> b. _____ only the following organs or parts—specify the organs or parts—for the purposes of transplantation, therapy, medical research, or education:
>
> _____
>
> c. _____ my body for anatomical study if needed
>
> Signed by the donor and the following two witnesses in the presence of each other:
>
> _____   _____
> Signature of donor   Date of birth of donor
>
> _____   _____
> Date signed   City and State
>
> _____   _____
> Witness   Witness

PALLIATIVE CARE

Separating **palliative care** (controlling pain and considering the patient's needs other than physical, such as social, psychological, spiritual, and cultural) from hospice is not easy, since palliative care is what hospice is all about. Palliative care seeks to satisfy the needs of patients and their families in several domains including the physical, psychological, social, and spiritual (Mino, 1999). Palliative care helps the person live out his or her final days in as comfortable and dignified a manner as possible, observes the director of a humanities program in a U.S. medical school (Mangan, 1998). However, palliative care is not restricted to individuals in a hospice program and indeed is much broader than that. Though much of this chapter entitled "Living With Dying" concentrates on hospice, it is because the "hospice way" of palliative care is an excellent model to follow throughout the health-care system as we move into the 21st century. Such *caring* treatment needs to be diffused throughout medical care. Pioneers of the modern hospice movement have always

---

WORDS OF WISDOM
Comfort: An Age-Old Goal

To cure sometimes,
To relieve often,
To comfort always.

Anonymous physician, 16th century.

---

maintained that the principles of palliative care, so successfully employed in the management of patients with cancer, can and indeed should "not only be facets of oncology but of geriatric medicine, neurology, general practice and throughout medicine" (Derry, 1997). However, Sally Derry notes that there is still a long way to go before the principles of palliative care become widely used in the management of patients with diseases that are acknowledged to be chronic and progressive.

Evidence-based practice (the conscious, explicit, and judicious use of current evidence in making decisions about the care of individual patients) is an important aspect of palliative care (Higginson, 1999). Evidence-based practice does not mean ignoring clinical experience or patient and family wishes, which are common components of palliative care. Rather such practice means integrating clinical expertise and findings with the best available external evidence from systematic research, notes I. J. Higginson. The evidence about effective palliative care services is rather like a sieve—there are gaps and holes in our knowledge, says Higginson. Patients with advanced illness are among the most difficult to study, and there is no second chance to get care right. Thus, Higginson observes that it is imperative to use existing research through appropriate systematic reviews to maximize the value of these data for the care of patients and their families. Phil Hammond (Hammond & Mosley, 1999) in his advice-giving, somewhat tongue-in-check book entitled *Trust Me (I'm a Doctor)* states that evidence-based medicine should be an international concept, with hard science "trumping" irrational cultural beliefs every time. But it is not, notes Hammond. He states, for example, that if you have a stomach ache, in France you get a suppository, in Germany you visit a health spa, in the United States they cut your stomach open, and in Britain they put you on a waiting list.

Palliative day care is a rapidly growing aspect of palliative care, especially in England (Goodwin et al., 2000). During the past two decades (Copp et al., 1998), there has been a significant increase in the provision of day-care services for patients requiring palliative care in the United Kingdom. Since the first palliative care day-care unit was established in 1975 in Sheffield, the number of units has increased to 230 in the United Kingdom, note Copp et al. Such day care is not unlike child day care or geriatric day care except that these individuals need special care due to the various manifestations of their illnesses. Positive aspects of a palliative day-care facility include the patient's getting out, engaging in various activities, and associating with others. Services received by the patients include the presence of a physician, nursing care, bathing, occupational therapy, arts and crafts, aromatherapy, massages, hairdressing, and relaxation. Clearly, the services provided by day-care centers for patients and their families are of considerable value and have been acknowledged as an integral and vital aspect of palliative care (Copp et al., 1998).

Palliative medicine, while currently not recognized as a medical specialty in the United States, is now recognized in five countries. The leader in palliative medicine is the United Kingdom. Palliative medicine in the United Kingdom has been recognized as making a significant contribution, not only to the practice of cancer medicine, but also to the care of those with terminal disease (Smith, 1994). Patients treated by a team with skill and expertise in the care of the dying will value such care whatever the primary specialty of the medical team. These patients can gain from pain relief, symptom control, communication, and family support—the skills of palliative medicine practitioners (Smith, 1994). Perhaps the health-care system in the United States will begin to see the benefits of palliative medicine and palliative care and more rapidly move in this direction.

THE HOSPICE MOVEMENT

The **hospice movement** shares all the goals of palliative care. **Hospice** has developed as a response to fears related to the dying process and the institutionalized ways in which death is typically handled in institutional settings. The three primary patient concerns that hospice care directly addresses are the problems related to symptom and pain control, the apprehension caused by having others in control of one's life, and the anxiety about being alone at the time of death (Magno, 1990). The primary goals of hospice care are to promote patient-family autonomy, to assist patients in obtaining pain control and real quality of life before they die, and to enable families of patients to receive supportive help during the dying process and the bereavement period. The National Hospice Organization, which represents nearly 2,500 hospice care programs in the United States, in February of 2000 changed its name to the National Hospice and Palliative Care Organization (NHPCO), demonstrating the sharing of core values and philosophies and the national evolution in end-of-life care (National Hospice Organization and Palliative Care Organization, 2000). The hospice movement proclaims that every human being has an inherent right to live as fully and completely as possible up to the moment of death. Some traditional health care, emphasizing the *curing* but not so much the *caring*, of the patient at any cost, has ignored that right.

Many physicians have been trained, for example, to focus on restoring the patient to health. Accordingly, many patients are subjected to a series of operations designed to prolong life, even though a cure is sometimes impossible, as in the case of a rapidly progressing cancer. Most hospice patients have had some surgery, chemotherapy, or radiation treatments. Some continue these even while they are hospice patients because of the pain-relieving nature of the treatments (radiation

Saint Christopher's Hospice in London, England, is the model hospice for hospice programs around the world. Besides having 66 beds for hospice patients, Saint Christopher's is an educational center for teaching health care professionals about palliative care.

may reduce the size of a tumor and, therefore, reduce the discomfort). There comes a point, however, if quality of life is a goal, when one should refuse further surgery, seek ease of pain without curing, and attempt to live qualitatively rather than quantitatively. In hospice, cure goals for patients are changed to comfort goals, and every patient has a significant role in all health-care decisions.

Because of the emphasis on quality of life, hospices pay attention to many facets of pain reduction, discussed earlier, including, but not limited to, physical pain. Hospice medical professionals have spent considerable time in developing a variety of methods of pain control that subdue not only what the patient describes as pain, but also the symptoms related to the illness.

Much of this emphasis on pain control has developed despite the practice by some professionals of sedating patients in pain. Quality of life cannot be achieved if the patient is "knocked out" or has become a "zombie." Physicians try to find the point at which the pain is managed but sedation has not occurred. Such pain management has necessitated considerable retraining of health-care leaders.

Hospice people also deal with social, psychological, financial, and spiritual pain. Terminally ill patients may experience social abandonment or personal isolation that comes when friends and acquaintances stop visiting them because of an inability to cope with issues of death, a lack of knowledge about what to say or do, or simply a lack of awareness of what the patients are experiencing. According to C. L. Chng and M. K. Ramsey (1984–1985), "in too many cases family members may inadvertently 'reject' the patient when confronted with the reality of death, while the professional staff may distance itself to avoid becoming too emotionally involved." It is ironic that the dying patient's need for social support and companionship comes at a time when he or she may often be more alienated than at any other time in life.

Financial problems are also experienced by patients and their families who face large hospital and medical bills at a time when family income may be diminished. Finally, there is a spiritual pain that people experience when they seek answers to existential questions and ultimate meaning and purpose in the face of suffering. "Why did God allow this to happen to me?" and "Why do bad things happen to good people?" are questions frequently asked by patients and their families.

Critical questions are: What constitutes quality of life? What do people most want to accomplish before they die? What do they most want to do? When one of Michael Leming's hospice patients in Northfield, Minnesota, was asked that last question, he said that he had always wanted to take a helicopter ride. With the help of the local NBC television station, they made this dream a reality. Like this patient, almost everyone has unfinished business in life. Some may wish to renew relationships with friends or family members. Others may desire to put their own affairs in order, to write their memoirs, to plant a garden, to watch the sunset, or to plan their own funeral service. Robert Kavanaugh (1972) tells the story of Elaine, who, in her last months of life, studied for her real estate license examination, passed the test, and with the help of her husband sold two houses. Thus, at age 37, Elaine found her first job, *while dying.*

The Hospice Team

Hospice care is provided by an interdisciplinary **hospice team,** with each discipline having something to contribute to the whole. All disciplines work together, each in its own area of expertise, and each interdisciplinary team includes several layers or levels of care. At the center of the team is the *patient and his or her family.* The hospice movement emphasizes the need for people to make their own decisions with the supportive help of health-care professionals and other trained persons. A vital part of the process is the *patient's own physician*—the professional who will continue to be in charge of the care of the patient and write medical orders when necessary.

The next layer of the team includes the hospice's professional caregiving staff. This consists first of *physicians,* required to direct medical care. *Nurses* comprise the next layer. Registered nurses are responsible for coordinating the patient's care. Licensed practical nurses and nurse's aides are also included—especially in inpatient settings.

The *hospice social worker* constitutes an important part of the team. The social worker spends considerable time working with families, thus enabling family members to communicate with each other. Although family members may be aware that the patient is dying, they may never have discussed it with each other or with the patient. The social worker also spends time dealing with social problems, such as alcoholism and marriage problems, and working with the children or grandchildren of patients. All too often in modern society children have been shielded from participating in events focused on the death of a family member.

Pastoral care is a basic aspect provided by the team. A larger hospice may employ a *chaplain,* who will direct pastoral care to patients and their families, counsel other members of the caregiving team on spiritual issues, and try to involve clergy of the community in the care of their own people. In smaller hospice programs all of the care may be provided by local clergy who work closely with the hospice staff.

Financial counseling is a significant service provided by the hospice team. Because patients and families have often exhausted their financial resources at the time of care, attention is given to seeking other forms of third-party reimbursement, such as those provided by Medicare or Medicaid or private insurance companies and to seeking other programs for which the patient may be eligible.

The next layer of the hospice team includes a variety of health-care professionals or other key leaders in the community whose help may be called upon during the illness. A *psychiatrist* or *psychologist* may be needed to provide expert counseling help. *Nurses, home health aides,* and *homemakers* employed by public health nursing agencies—such as visiting nurse associations—may be needed to provide special continuing health care or to share in the provision of patient care. *Physical* and/or *occupational therapists* may be needed to work with the patient to ensure maximum daily functioning. Finally, the services of a *lawyer* and/or *funeral director* may be required to help the patient settle personal affairs and provide for the needs of survivors after his or her death.

Artists are increasingly recognized as important members of the hospice team. The Connecticut Hospice pioneered the development of an arts program that considers the arts as a means to help patients find meaningful fulfillment during their last days. In many programs artists in areas such as metalwork, photography, pottery, drama, dance, and music work with patients interested in such self-expression.

Trained *volunteers* comprise an essential part of the hospice team. Medicare reimbursement is predicated on the requirement that volunteer time represents 5 percent of all patient care—no hospice can exist for long without a strong volunteer component. NHPCO-affiliated programs recruit their volunteers from many sources: churches (94 percent), civic groups (80 percent), social groups (70 percent), professional organizations (65 percent), business and industry (47 percent), colleges (34 percent), and secondary schools (12 percent).

Hospice volunteers include people from the following groups: women not employed outside the home, students, retired persons, and professionals such as social workers, psychologists, teachers, gerontologists, members of the clergy, and architects (Chng & Ramsey, 1984–1985). Many volunteers have experienced the death of someone close and find that this provides them with an opportunity to serve others. Some volunteers are retired health-care professionals such as physicians or nurses. Others are nonprofessionals who are deeply interested in the needs of dying patients and their families. Each volunteer has skills and experiences that can greatly enhance the life of the terminally ill patient, note Chng and Ramsey.

Before volunteers begin a hospice program's extensive training program, they are interviewed by the volunteer coordinator and may be asked to complete specially designed questionnaires that assess their feelings and sensitivity toward dying persons. The National Hospice Organization (1997) claims that, on average, hospices require 22 hours of training before a volunteer is allowed to work directly with a patient or family. In addition to the initial volunteer training program, every hospice program has regular in-service training to maintain and update the volunteers' skills. The average volunteer provides services for 3 years, and 50 percent of hospice volunteers stay 6 or more years (National Hospice Organization, 1997).

Some volunteers work in patient-care tasks such as providing transportation, sitting with a patient to free family members to get out of the house for a while, carrying equipment, or providing bereavement counseling to family members after the death of the patient. However, according to Chng and Ramsey (1984–1985), hospice volunteers perform primarily three roles: companion/friend, advocate, and educator. The average hospice annually receives 3,300 hours of service from volunteers. Nationwide, that translates into more than 5.25 million hospice volunteer hours annually—approximately two thirds of which pertain directly to patient care (National Hospice Organization, 1997).

A national study of volunteers by the NHPCO determined that 87 percent of volunteers were female and that 58 percent of these were 60 years of age or younger, compared with 53 percent of male volunteers. The vast majority of these volunteers had experienced the death of a significant other (National Hospice Organization, 1993).

In a study of volunteers at the Mercy Hospital Hospice Care Program of Urbana, Illinois, M. Patchner and M. Finn (1987–1988) discovered that 86 percent of the hospice volunteers felt capable of performing all of their duties involved in hospice work and that they were very satisfied with volunteering but were most satisfied with assignments that involved direct contact with patients and families and palliative care work within the hospital. Volunteers were least satisfied with the following instrumental tasks: completing forms, attending volunteer meetings, and performing clerical duties. These volunteers were involved in hospice work primarily because they felt that they had something to offer others and wished to be of service to others, observed Patchner and Finn.

The unpaid volunteer has the double benefit of being identified by the patient and family as being knowledgeable but not having the professional status that can create a social distance. By being a stranger who provides a "listening ear," without emotional involvements or professional entanglements, the volunteer can support the patient and family members as can no other participant in the social network of dying—"stranger and friend" at the same time.

The volunteer also functions in an advocate role by acting on behalf of the dying patient and family. Sometimes patients and their family and friends are afraid to challenge or ask questions of physicians and other medical personnel. The volunteer, who has become a trusted friend and confidant, can often speak up for patients and their families and make their needs known to those responsible for their care. Michael Leming once served as the primary volunteer for a male patient who was undermedicated. When the patient complained to his nurse regarding his pain, he was told to "brave it out." Knowing the medical system, Leming was able to contact the appropriate individuals, who were, indeed, able to have his pain medications reevaluated. The words of Chng and Ramsey (1984–1985, p. 240) provide good advice at this point:

> To be truly effective, the suggestions of the volunteer have to stem from knowledge and understanding of the intricate patient-family-institutional configurations. Under the careful guidance of professionals, the volunteer can serve a significant role as ancillary to professionals.

The final role served by volunteers is that of educator. Most individuals in our society have not had many personal experiences with death. The hospice volunteer can learn from each experience in working with dying patients and pass on insights that may be helpful to patients and families. The volunteer can help dying persons and their family and friends understand that the dying process is usually complex, stressful, and disordered. In addition, most patients and families have a strong need to have their feelings and experiences validated. Patients and families who have a difficult time understanding their feelings, emotions, and experiences during the dying process should be assured that they are "quite normal."

Within the community at large, a number of influences either assist with patient care or help to make it possible. Family members or friends are urged to participate in the patient's care as much as possible. When family members cannot provide as

PRACTICAL MATTERS
Sharing the Journey: A Ministry in Hospice

Janet White

As a trained hospice volunteer, I provided vital nonclinical services such as transportation, companionship, respite care, and light housekeeping. My weekly visits also kept the patient and family in contact with the world beyond their own home, preserving the tie between them and the community, and preventing further isolation during the last phases of illness.

My first patient was like a grandfather to me. When I took my weekly turn in caring for Tom, I always received more than I gave. His attitude reflected an inner joy and peace, even though his frail body was wrecked by the perpetual coughing and shortness of breath that left him weaker every day.

"Please open the curtains," he would whisper first thing in the morning. "I want to see the sunshine."

His reply was always the same when I accomplished the task. "Thank you, my dear, that's grand." Oh, how he loved the sunshine. He reveled in each warming ray.

Tom loved music, and had sung in the church choir for most of his 89 years. Sometimes I brought him Christian cassettes to listen to in the long hours alone. He played the "Hallelujah Chorus" again and again.

"Soon I'll be singing Hallelujah right to Jesus," he struggled to say with a weak smile. "I'll have a new voice and plenty of breath to praise the Lord." After his funeral, I cried every time I heard that beloved classic, as I thought of Tom lying in the hospital bed in his living room. The brevity of life had helped both of us focus on what was truly important and of eternal value.

Adapted from "Sharing the Journey:
A Ministry in Hospice"
(pp. 24–25), by J. White, 1992,
The Lutheran Journal, 60.

much care as may be needed at certain times, hospice personnel will try to meet the patient's needs by exploring all possible options to do so (National Hospice Organization, 1988). The patient-family support system is the most significant factor in the dying process for many patients, but elements of the system also include numerous close or distant relatives, friends, neighbors, members of local churches, and/or other civic groups.

Hospice programs depend on a high degree of community interest and support. Bringing this about requires a planned program of public information. The concept of hospice must be sold to the medical community and to members of the larger community. Specific activities require not only financial support (especially while the hospice program is developing), but also a willingness to testify before regulatory agencies about the granting of hospice accreditation, Medicare certification, and/or approval to begin offering services to the people of the area.

The Patient-Family as the Unit of Care

One of the distinguishing features of the hospice concept of care is that, whenever possible, hospice enables patients to make decisions about how and where they want to live their lives. **Patient-centered** care is nonjudgmental, unconditional, and empowering.

One of the patients of the hospice program where Michael Leming volunteers provides an example of this philosophy of care. This patient had adult children in town but lived as a single person with his dog. He had lung cancer and desired to die at home alone. He was also a smoker and heavy alcohol drinker. The hospice program agreed to honor the patient's desires whenever it was possible. Therefore, a hospice nurse visited his house every 4 hours, and members of the police department checked in on him every hour from 10 p.m. to 7 a.m. The patient's pain was kept under control without sedation. A hospice volunteer (who also happened to be a licensed vocational nurse) visited the patient two to three times each day, and she, along with friends and family members, met the patient's requests for liquor and cigarettes. Visits from all members of the hospice team never lasted longer than 10 minutes. The patient died as he wanted—in his home, free from pain, and in control of his own care.

Although not every member of the hospice team, or the patient's family, would have chosen to die as this patient did, everyone respected his right to make decisions regarding his care. The hospice philosophy states that patients and their families have the right to participate in decisions concerning their care and that they should not be judged because their decisions are contrary to the beliefs of their caregivers.

Traditional health care has concentrated on the patient and ignored the family. Perhaps many health-care workers would say, if given an opportunity to state their opinions confidentially, that they would prefer family members to stay away. Traditional ratios of physicians, nurses, social workers, or chaplains to those needing care have been based on an assumption that only the patients need attention. Although hospice staff, to be sure, are not given the responsibility to meet physical needs of family members, they do have tremendous concern for the social, psychological, and spiritual needs of the family.

Hospices challenge the health-care system to provide an adequate ratio of professional staff members to patients. For example, in the state of Connecticut the public health code, in its regulations for hospice licensure, stipulates that at all hours of the day or night there must be at least one registered nurse for every six patients and at least one nursing staff member (licensed practical nurse or nurse's aide and a registered nurse) for every three patients.

Family care, however, involves much more than numbers of staff. It requires that health-care workers know how to cope with the fears, worries, tears, and turmoil of family members and when to speak, when not to speak, and what to say. It requires that they take time to listen, to determine how they may be most helpful.

Hospice care is costly care due to the number of staff people involved. It challenges society as a whole to give priority to such care because of the right of the dying to quality of life. A harried nurse in a traditional hospital setting, trying to meet

Dame Cicely Saunders, founder of the modern hospice movement, in her office at Saint Christopher's Hospice in London, England (with George Dickinson). With degrees in nursing, social work, and medicine, indeed Dr. Saunders is well qualified for her work!

the needs of perhaps a floor of patients at night, is not being granted the time required to sit with a dying patient for whom night is especially fearful. Neither does this nurse have the time to be of assistance to husbands, wives, or children struggling with grief.

The interdisciplinary team supports the staff person within each discipline by enabling the resources of the entire team to come into play in meeting family needs. For example, a night-shift nurse who is asked questions relating to spiritual care might wish to give an answer at the time that the question is asked. This nurse will, however, also have the resources of the chaplain to determine the best methods to meet patient needs. In hospice care the patient-family unit is involved in decision making. This poses crucial questions to caregivers who may be accustomed to making decisions and having everyone go along with what they have decided.

The Cost of Hospice Care

Even though hospice care is personalized to meet the needs of each of its patients (involving an entire team of professional and volunteer caregivers), it is also very cost effective because more than 90 percent of hospice care hours are provided in patients' homes, thus substituting for more expensive multiple hospitalizations. A study released in 1995 by Lewin-VHI and commissioned by the National Hospice

Organization (1997) showed that for every dollar that Medicare spent on hospice, it saved $1.52 in expenditures. Furthermore, during the last year of life, hospice patients incurred $1,786 less in costs than those not in hospice care. These savings totaled $3,192 in the last month of life as hospice home-care days often substituted for expensive hospitalizations. Medicare rates for fiscal year 1997 were around $94 per day for home care and $419 per day for general inpatient care (National Hospice Organization, 1999).

Ideally, no prospective hospice patient should be turned away because of lack of money. Even though only 18 percent of all hospice programs were for-profit organizations in 1998, each year these programs provide some care for those patients who cannot afford to pay, and most poor, underinsured, and uninsured families do receive some financial assistance (National Hospice Organization, 1999). According to the National Hospice Organization (1999), sources of payment for hospice services are as follows: Medicare, 74 percent; private insurance, 12 percent; Medicaid, 7 percent; and "other" (donations, grants, private pay), 7 percent.

Hospice care is insured by the Medicare Hospice Benefit enacted in 1982, provided that the hospice is Medicare certified. A hospice program must undergo a vigorous evaluation of the services that it provides to become Medicare certified and must agree to directly provide the following services: nursing care, medical social services, physician services, counseling, and volunteer services. As of 1998, 80 percent of all hospices were Medicare certified (National Hospice Organization, 1999). According to the National Hospice Organization, the daily, per-patient payment rates in 1997 made to Medicare-certified hospices (before local wage adjustments) were:

Routine home care $94

Continuous home care $550

Inpatient respite care $97

General inpatient care $419

To the preceding payments there is a $13,974 annual per-patient program cap (this cap is computed on an aggregate basis for all Medicare patients in the hospice program). Furthermore, Medicare specifies that not more than 20 percent of hospice care provided by a certified hospice program may be provided within inpatient facilities (National Hospice Organization, 1997). In 1994 Medicare spent $1.2 billion on hospice services. Additionally, hospice became an optional benefit under state Medicaid in 1986. In 1999 state Medicaid hospice benefits were available in 43 of the 50 states (National Hospice Organization, 1999), and annual payments exceeded $150 million for hospice services.

The General Electric Company was the first major employer in the United States to provide a hospice benefit for its employees. Presently coverage for hospice is provided to more than 80 percent of employees in medium and large businesses. Furthermore, the majority of private insurance companies offer a comprehensive hospice care benefit plan, and major medical insurance policies, provided through

---

PRACTICAL MATTERS
How to Live With a Life-Threatening Illness

With the help of hospice you can:

1. Talk about the illness. If it is cancer, call it cancer. You can't make life normal again by trying to hide what is wrong.

2. Accept death as a part of life. It is.

3. Consider each day as another day of life, a gift from God to be enjoyed as fully as possible.

4. Realize that life never is going to be perfect. It wasn't before, and it won't be now.

5. Pray, if you wish. It isn't a sign of weakness; it is your strength.

6. Learn to live with your illness instead of considering yourself dying from it. We are all dying in some manner.

7. Put your friends and relatives at ease. If you don't want pity, don't ask for it.

8. Make all practical arrangements for a funeral, will, etc., and make certain your family understands them.

9. Set new goals; realize your limitations. Sometimes the simple things of life become the most enjoyable.

10. Discuss your problems with your family, including your children if possible. After all, your problems are not individual ones.

From O. Kelly, Make Each Day Count *newsletter.*

---

insurance companies and offered to employees as part of a benefit package, also underwrite hospice coverage in most instances. However, many hospice programs still rely on grants, donations, and memorials to meet the needs of their patients and families that are not covered by Medicare, Medicaid, and insurance reimbursements (National Hospice Organization, 1997).

Hospice leaders hope to make it possible for any person of any age suffering from a terminal illness to be eligible for the coverage of costs related to hospice care. They are also firm in their conviction that such care saves considerable money in the long run. Many patients currently hospitalized would not need hospitalization if hospice services were available for patients and families. A basic societal question is whether as Americans we believe enough in quality of life for the dying to be willing to make it possible.

Though hospice care requires a higher ratio of staff to patients than that usually provided in health-care programs, it should be remembered that the cost is nonetheless lower than that for other forms of care. Because the majority of hospice patients are able to remain at home for much, if not all, of the illness, the costs of patient home care, when compared with any forms of inpatient care, are proportionately low. Because of the level of services provided, hospice **inpatient care,** however, will normally be higher than that provided in a nursing home, but lower than inpatient care in a general hospital setting.

Public Attitudes

The hospice movement began at a time when public consciousness of dying and death issues had reached an all-time high. It afforded an opportunity to do something tangible for other people, and many took advantage of the chance to volunteer for an active role. At the same time, increasing public awareness of dying and death gave rise to considerable publicity in the media. This helped provide public support when regulatory agencies held hearings on granting approval for hospice services.

Public attitudes toward care of the terminally ill persons and their families will play an increasingly important role in the future. These attitudes will help to determine whether health-care professionals will, in fact, broaden the scope of care to encompass the family and strengthen their skills in dealing with dying patients. Patients and families are, after all, consumers. In this age of consumer awareness it is becoming increasingly evident that those who purchase services can control to some extent the types of services available. Health-care professionals are becoming increasingly responsive to desires of their clients. The most important factors causing caregivers to seek improvement of skills will be the desires of those they serve. At the same time, especially in areas of competition among hospitals, consumer awareness will play an important part in encouraging such institutions to humanize the care that they give.

Many physicians, nurses, social workers, clergy, and other personnel at hospitals and nursing homes have heard about hospice care and have taken the initiative to secure specialized training and to incorporate the hospice philosophy into routine treatment of their patients. When any people's movement arises, an immediate question is whether it will become institutionalized to such an extent that the original spirit will be lost as it adjusts to the reality of regulation, control, and payments of costs. The hospice movement is currently at that juncture. There is every cause for hope that one of two things will happen: Either hospices will continue to provide the specialized care for dying patients, or the health-care system itself will change to incorporate many of the improvements represented by the hospice movement. Either outcome should result in the betterment of medical care for patients and their families.

A Gallup Poll in 1996 revealed that 9 of 10 adults would prefer to be cared for at home, if terminally ill with 6 months or less to live. Nearly 90 percent of adults believe it is the family's responsibility to care for the dying. Thus, hospice programs in the United States seem to be giving that which the majority of Americans desire—an opportunity to die at home under the care of family.

Yet, sometimes people are reluctant to use hospice programs (Doka, 1993). Many may not be familiar with the hospice philosophy or even be aware of the existence of hospice in their community. Caregivers might be reluctant to discuss hospice with the family for fear the topic might upset them. To enter a hospice program may be viewed as "giving up." Such misinformation might contribute to one's failing to enter a hospice program when indeed the individual could benefit from hospice, notes Kenneth Doka. Entering a hospice program very late in the illness denies the patient counseling services and social support that hospice can provide to increase the quality of the patient's remaining time.

Evaluation of Hospice Programs

Although evaluations of hospice programs have not flooded the literature, some studies have taken a look at the hospice approach. An early in-house evaluation of the Connecticut Hospice (Lack & Buckingham, 1978) reported less depression, anxiety, and hostility among terminally ill patients receiving home care compared with others not receiving these services. In addition, it was found that the overall social adjustment of family members was improved—individuals were better able to express thoughts and feelings with less distress.

Colin M. Parkes (1978) reports that only 8 percent of those who died at St. Christopher's Hospice in London suffered unrelieved pain; this compares with 20 percent who died in hospitals and 28 percent who died at home. John Hinton (1979), in comparing the attitudes of terminally ill patients at a British hospice with those in hospital wards, noted that hospice patients appeared less depressed and anxious while preferring the more open type of communication environment available to them.

Janet Labus and Faye Dambrot (1985–1986) compared the experience of 28 patients with cancer who died in an independent, voluntary, community-based hospice program with 29 nonhospice patients with cancer who were treated in a small community hospital in northeastern Ohio. Whereas all of the nonhospice patients died in the hospital, 50 percent of the hospice patients died at home. Furthermore, hospice patients at the time of admission were less likely to be ambulatory than hospital patients. According to Labus and Dambrot:

> The hospice group was younger, had more people living with them, and had been ill for a shorter period of time prior to hospice admission than the hospital group. Hospice patients spent more time in the hospice program after admission prior to death. The hospital patients spent fewer days in the hospital prior to death. Hospice patients were referred to other agencies significantly less frequently than hospital patients.

Labus and Dambrot concluded that this hospice program was succeeding in fulfilling the needs of its patients and in containing health-care costs through the multidisciplinary team—only one outside referral had to be made, and 50 percent of the patients died at home.

In the evaluation of hospice programs, another important concern is the effect of these programs upon professional caregivers. A study of hospice nurses conducted by Pamela Gray-Toft and James Anderson (1986–1987) sought to explain burnout and rapid turnover in the nursing staff. They found that hospice caregiving is significantly more stressful for nurses for the following reasons:

1. Traditional nurses often resent the special roles and treatment received by hospice nurses within institutional settings. Many hospital-based hospice nurses report feeling socially isolated from other hospital nurses.

2. Hospice nurses are often regarded by others as being "weird" or "different" because they choose to work with the dying.

3. The hospice concept of care creates increased (and often unrealistic) patient expectations; consequently, hospice nurses experience inordinate demands placed upon them by the patients and families with whom they work.

4. Because hospice often creates long-term relationships between dying patients, families, and nurses, nurses often experience strong feelings of loss and grief after the death of their patients. Many times hospice nurses are not given an opportunity to grieve after a patient's death because of the demands of their other duties.

5. Hospice nurses are overexposed to death, which can sometimes contribute to a morbid view of life.

As a result of these findings Gray-Toft and Anderson created an experimental staff-support program for hospice nurses. Based on the initial success of this experiment, the group-support program has now expanded to include the professional contributions of a chaplain and medical social workers.

Many emerging reports on hospice are emphasizing the cost of hospice care compared with that of traditional hospitalization. Hopefully, the question of cost will not gain undue prominence and become of greater concern than the caring process itself.

There is concern that hospice care is becoming increasingly bureaucratized and susceptible to the forces of rationalization and routinization and that this is compromising the movement's founding ideals (Bradshaw, 1996). Yet, it is also believed that the process of routinization and bureaucratization in hospices is necessary and inevitable for their survival. With health-care workers perhaps no longer attracted by a charismatic ideal—a calling—they are increasingly people who are motivated to care for the terminally ill as employment or for career purposes.

Even though AIDS accounts for only 4 percent of hospice patient deaths, in 1996 NHO-affiliated hospices provided terminal care for one third of all patients who died of AIDS in the United States (National Hospice Organization, 1997). There are problems in applying the hospice model uncritically to patients with AIDS, notes Martha Downey (1997): In contrast to patients with cancer, for whom the transition between curative and palliative care is relatively easily marked, making it easier to decide when to offer hospice care to the patient and family, patients with AIDS often experience health "highs" and "lows." The next few years will bring new challenges for those providing palliative care to patients with AIDS. Jeannee Parker Martin (1988), a member of a health-care team of the Visiting Nurses and Hospice of San Francisco, describes her hospice's treatment of patients with AIDS. She exalts the appropriateness of the interdisciplinary team approach. Professional caregivers and volunteers in hospice must be trained to deal with the special needs of patients with AIDS. Hospice programs must review all policies and procedures, such as admission criteria and infection-control policies, as they reach out to those infected with the AIDS virus. Martin concludes that from a societal point of view it is imperative that home care and hospice services be available to all persons with AIDS and that when home care is no longer a viable alternative, other long-term care options should be available.

It was the conclusion of Edward Crowther (1980), over two decades ago, that the United States is not ready for hospice. He noted that it is not a problem of training or funding, but more a problem of being unable or unwilling to change our attitudes toward proper care for a dying patient. Crowther cites typical hospice development committee problems such as power struggles, personal conflicts, and different opinions on approaches to take. The biggest problem, however, seems to be attitudinal. He notes that the medical community remains to be convinced that current methods of treating the terminally ill must be changed. Dying persons cannot be treated in the same way as those who can expect to be cured. Physicians must learn the difference between the hospice goal of "care" and the traditional goal of "cure." The care concept admits that the patient is dying. Medical personnel must put their knowledge and efforts into relief of distressing symptoms and into human understanding. If this change in attitude does occur, Crowther believes, terminal care consistent with the ideals of the hospice movement can be provided in the United States.

This notion is supported by Josefina Magno (1990), the first executive director of the National Hospice Organization, when she claimed that the biggest problem of the hospice movement in the 1990s would be the lack of involvement of the medical profession. Physicians are most likely to wait to refer a patient to hospice until the last few days of life when they are absolutely sure that the patient is dying, rather than at the time of the terminal diagnosis, when the hospice concept of care would be more beneficial to the patient and family. Such an assertion is supported at the end of the 1990s by C. B. Johnson and S. C. Slaninka (1999) who cite studies which confirm that early access to hospice is impaired in part due to lack of awareness and lack of referrals.

In conclusion, despite various concerns cited above, hospice today appears to have established itself fairly well within the overall health-care establishment, notes famed American thanatologist Robert Kastenbaum (1998). Many health-care professionals and administrators have become persuaded that hospice does what it claims to do and does so in a rational, accountable, and cost-effective manner, notes Kastenbaum.

LISTENING TO THE VOICES
Hope at the Hospice

In October, 1985, I had been treated for 6 months for HIV. I was in complete remission, but I was wasting away. I had no appetite. I lost about 25 percent of my body weight. I had no energy. I was sleeping uncomfortably about 20 hours a day. I couldn't stand up without blacking out. My whole body ached all the time.

The doctor finally figured out that I was suffering from adrenal insufficiency caused by the medication. She called at 7:30 one evening after she'd seen me and said I had to come into the emergency room immediately. "You're likely to be dead by morning. You need cortisone NOW!" I had my neighbor drive me in, and she dropped me at the emergency room door.

I was eventually put in triage, where all my valuables were taken away and I was hooked up to monitors. I remember being vividly conscious of everything that was going on around me, even though I had no ability to do more than blink my eyes.

The ER staff took vial after vial of blood, but kept delaying the cortisone my own doctor had ordered. One resident thought he saw a shadow in my lung x-ray, and wanted me tested for PCP. A doctor told him the shadow was my heart.

A nurse took my blood pressure and told me it was 50 over 30. She suggested I say my prayers. I prayed with every breath.

As they were taking one more vial of blood, my blood stopped flowing, and someone said, "Pump your hand." I remember thinking, "Why isn't he doing what they're telling him to?"

Then I didn't care anymore. I was floating free of my body, and for the first time in a long time, I felt no pain. I felt perfect peace, the "peace that passes all understanding." I felt whole in a way I'd never felt before: I finally understood all those things I'd never understood about myself. I felt completely surrounded by love, just as if every person who had ever loved me was right there with me, holding me, and caring for me. I had been terrified of dying alone, and I discovered that fear is totally irrelevant. We are not alone, even in dying.

Then I was conscious of being back in my body, and I was angry. It had been so restful, so perfect, and now I was in pain again.

I came away from this experience no longer afraid of dying or of death. I know now that what awaits us after we leave our bodies is perfect peace, understanding, and love. There is nothing to be afraid of.

This is why I believe hope is possible even at the hospice. It is through faith, but also through experience, that I know hope can be a reality for people who are dying. Hope is theologically correct, even in facing death. Faith gives us the hope we need to live, even as we die.

Adapted from "Spirituality Column for the Body," by Reverend A. S. Pieters, March 15, 1996.

CONCLUSION

A diagnosis of cancer or heart disease or AIDS today does not necessarily mean a death sentence. With medical advancements in recent years, many individuals in our midst are "living with dying." Though they may have a terminal illness, they indeed may "live" many years with this disease. We have made progress in being more open about the "C" word (cancer) and AIDS. Such diseases are being viewed more as attacking individuals in all walks of life. Medical treatment options are more numerous, thus perhaps offering more encouragement to patients. The ability to transplant more than two dozen tissues and organs also contributes to the options for one with a terminal diagnosis.

As the American way of life has changed from a primary group orientation to a more secondary, impersonalized orientation, so has dying shifted from the home to the hospital or nursing home setting—away from kin and friends to a bureaucratized setting. The birth of the hospice movement in the United States might be considered a countermovement to this shift. As we seek out primary group relations in our secondary-oriented society, we seek to die in the setting of a familiar home rather than in the sterile environment of a hospital. Perhaps we are evidencing a return to a concern for each other—a dignity to dying may be on the horizon.

Hospice is a return to showing care and compassion. It is a revival of neighbors helping neighbors—a concept so often lost in our urbanized society. Hospice consists of professionals literally going the extra mile and coming to one's home when needed—medical personnel actually making house calls. Hospice, for example, encourages children younger than age 14 to be present with the terminally ill person rather than making them wait in the hospital lobby. Hospice is a grassroots movement springing up in small communities, as well as in larger urban settings, to provide better health care. To paraphrase the words of Robert Kavanaugh (1972), the hospice concept of care helps us to unearth, face, understand, and accept our true feelings about death and provides us with the opportunity to live joyfully and die as we choose. In short, hospice is a movement that transforms our awkwardness in death situations into a celebration of life.

With federal money now covering most hospice expenses and with rigid government requirements for approval of hospice programs, it is important that every effort be made to prevent hospice programs from being strangled by the bureaucracy from which they receive financial assistance. Hospice programs must also continue to make the patient-family unit the focus of their care and treat these clients in a nonjudgmental and unconditional manner, thus empowering them as autonomous human beings.

Summary

1. Hospice is a specialized health-care program that serves patients with illnesses such as cancer during the last days of their lives.

2. Palliative care concerns pain control but also is being sensitive to social, psychological, spiritual, and cultural aspects of the dying person's needs.

3. Pain perception various with cultural background.

4. The hospice team is made up of various health-care professionals and the primary caregiver and volunteers.

5. The hospice movement supports the inherent right of every human being to live as fully and completely as possible up to the moment of death.

6. Though the majority of Americans today die in hospitals and nursing homes, the majority of individuals enrolled in hospice programs die at home.

7. Much of the cost of hospice care is covered by third-party reimbursements: Medicare and/or Medicaid.

8. Hospice programs are attractive to patients with AIDS because they provide a unique opportunity for patients to be cared for humanistically and because the special individual needs often required by patients with AIDS can be addressed within the hospice concept of care. However, AIDS also presents special challenges for hospice programs.

9. Treatment options today are more numerous than in the past, thus tending to give patients hope that if the first one does not work, there is now something else to try.

10. Because of the sensitive nature of the issue, human organ transplantation has not been traditionally widely discussed in our society.

11. Persons with terminal illnesses have treatment options other than surgery, chemotherapy, and radiation. These alternatives include immunotherapy, homeopathic medicine, chiropractic medicine, acupuncture, faith healing, and folk healing.

12. Pain *can* be controlled, but political issues get in the way, thus preventing widespread use of morphine in the United States. Legalizing marijuana for medical purposes is also controversial, yet marijuana can be useful in preventing nausea caused by certain cancer treatments.

DISCUSSION QUESTIONS

1. What is hospice care? How does it differ from the treatment given by most acute-care hospitals? Identify the major functions of a hospice program.

2. Discuss issues related to the family as the unit of care in hospice programs. How do hospices try to achieve quality of life for each of the patients they serve?

3. Discuss the pros and cons of legalizing marijuana for medical purposes.

4. Pain control is not a goal of all dying patients. Discuss the cultural ramifications of controlling pain.

5. Do you feel that bereavement care should be offered to the families of the terminally ill even after the individual has died? Justify your answer in terms of medical, emotional, and financial considerations.

6. If you were terminally ill, would you consider entering a hospice program? Explain your answer and refer to specific reasons such as cost, family burden, and imminent death.

7. Discuss the special opportunities and challenges in providing hospice services to patients with AIDS.

8. Discuss some of the problems affecting human organ transplantations from both a societal and a personal point of view.

9. Should society make eligible for a transplant only those persons who can afford it, or should the potential societal contribution of the individual be taken into account?

10. Discuss some of the problems involved with dying individuals such as the loss of physical and mental functions.

11. Have you ever been around someone who had Alzheimer's disease? If so, discuss your reactions. How can you relate to a patient with Alzheimer's disease, if he or she is in the latter stages of the illness?

GLOSSARY

Acupuncture: Involves the use of needles to redirect the flow of energy within the body to treat illness.

AIDS: *Acronym for acquired immunodeficiency syndrome,* a lethal syndrome caused by a virus that damages the immune system and weakens the body's ability to fight bacteria.

Alzheimer's Disease: A chronic, degenerative, dementing illness of the central nervous system.

Cancer: Refers to a group of diseases that are characterized by an uncontrolled growth and spread of abnormal cells.

Chemotherapy: Treatment of cancer with certain chemicals that attack and destroy certain types of cancer cells.

Chiropractic Medicine: A therapeutic approach to healing that involves a hands-on manipulation of bones in the spinal column to relieve pressure on nerves.

Evidence-Based Practice: The conscious, explicit and judicious use of current evidence in making decisions about the care of individual patients.

Faith Healing: Uses the power of suggestion, prayer, and faith in God to promote healing.

Folk Healing: Primarily uses "folk remedies" passed down orally from generation to generation and common ingredients found with a particular group of people to treat illness.

Homeopathic medicine: Involves the use of natural drugs to treat patients.

Hospice: A specialized health-care program that serves patients with life-threatening illnesses, such as cancer, during the last days of their lives.

Hospice Movement: A response to fears related to the dying process and the institutionalized ways in which death is typically handled in institutional settings.

Hospice Team: An interdisciplinary team of professionals and volunteers who work together to contribute their expertise to the quality of patient care.

Immunotherapy: Treatment by stimulation of the body's own immune system.

Inpatient Care: The type of institutionalized care that is required, for example, as an illness progresses and that may be provided in a hospice facility.

Metastases: The spread and invasion of cancer cells to other organs or tissues. When this happens, the disease is said to have metastasized.

Palliative Care: Care designed to give the patient as pain-free a condition as possible. In addition to physical needs, the patient's social, psychological, cultural, and spiritual needs are considered.

Palliative Day Care: A facility, not unlike child day care or geriatric day care, for patients with special needs—sort of like hospice on a day-by-day basis.

Patient-Centered Care: A distinguishing feature of the hospice approach, which enables patients to make decisions about how and where they want to live their lives. Patient-centered care is nonjudgmental, unconditional, and empowering.

PRN, or *pro re nata*:   A traditional medical approach which means that medication is to be given "as the situation demands." In practice, it means that patients must first hurt and ask for relief before pain management can be administered.

Radiation therapy: Treatment using x-rays to destroy cancerous tissues.

SUGGESTED READINGS

Aldridge, D. (Ed). (1999). *Music therapy in palliative care—New voices.* London: Jessica Kingsley. A collection of work from music therapists in different countries.

Clark, D., & Seymour, J. (1999). *Reflections on palliative care.* Buckingham: Open University Press. Philosophy and development of palliative care, especially in the United Kingdom. Written from the perspectives of sociology and social policy, this book also focuses on policy issues.

Connor, S. R. (1997). *Hospice: practice, pitfalls, promise.* Bristol, PA: Taylor & Francis. Describes hospice operations and problems, examines the goals of hospice, and also looks at the business side of hospice.

Katz, P. (1999). *The Scalpel's Edge: The culture of surgeons.* Boston: Allyn and Bacon. An in-depth picture of how surgeons in a particular hospital work, think, make decisions, manage their emotions, and communicate with their patients and other doctors.

Picardie, R. (1998). *Before I say goodbye.* London: Penguin Books. A 34-year-old woman talks about her dying.

Ray, M. C. (1997). *I'm here to help: A guide for caregivers, hospice workers, and volunteers.* New York: Bantam Books. Discusses and delineates the communication skills that are necessary in the hospice environment to keep it peaceful and comfortable and presents new ideas on how to facilitate communication between hospice workers and their patients and among all of those involved in the process.

Reoch, R. (1997). *Dying well: A holistic guide for the dying and their carers.* London: Gaia Books. A guidebook to therapies, exercises, and practical information for the dying.

Singh, K.D. (1999). *The grace in dying.* Dublin, Ireland: Newleaf. An in-depth psychological, religious, and spiritual analysis of the experience of dying.

Spiro, H. (1998). *The power of hope.* New Haven: Yale University Press. A physician explores how patients and caring doctors can help lessen suffering when illness occurs.

CHAPTER 7

DYING IN THE AMERICAN HEALTH-CARE SYSTEM

I left medical school. . . . with no idea how to break bad news or manage the sick patients I was about to be confronted with.

—Phil Hammond, MD

It is a great thing to die in your own bed, though it is better still to die in your boots.

—George Orwell

Within the American health-care system, most individuals believe that the sole responsibility for their terminally ill family members lies with the physician. Yet, Chinese individuals believe that the family *and* the physician are responsible for the patient's treatment (Tanner, 1995). The American way of life emphasizes individual-centeredness of autonomy, assertiveness, and independence within a youth-oriented perspective, whereas Chinese values display generational continuity, family solidarity, respect for elders, and situation-centeredness—individual impulses are subordinate to the will of the family as a group, observes Jane Tanner (1995). In China, when the patient, medical staff, and family members know that the patient is dying, they tend to practice "mutual pretense," not unlike in U.S. hospitals. In the Chinese culture, mutual pretense is the predominant context in some hospitals, notes D. Lin (1992).

Often traditional Chinese medicine (acupuncture and herbal treatments) is used in conjunction with Western medicine, and all Chinese physicians today receive some training in both. Chinese hospitals tend to always be overcrowded and have inconsistent quality (Hu, 1984). Due to the expense of Chinese hospitals, family members may do considerable labor such as washing floors, bringing in food, and taking the patient to laboratories or for x-rays. Such assistance may be supported by giving the family member time off from work (with pay), according to Schneider (1993).

Because China has a socialist medical system, health care is a state-provided public service; thus, one's medical expenses while dying would be paid for by the government, certainly a rare case in the United States. Whereas dying in the United States primarily occurs in an institutional setting, in general, dying at home has been the traditional arrangement for most Chinese families. Dying at home allows more latitude for the Chinese to handle dying in culturally appropriate ways (Lee, 1995). Also, there is a Chinese belief that people who die away from home will become lonely spirits, ghosts in the wild. In addition, in dying at home, the dying individual is often integrated into the round of daily life and is involved in living while dying. The Chinese feel that a symbolic completion of their shared relationship and solidarity results when the person dies at home. Thus, dying varies significantly in different cultures, yet there are similarities. Let's now take a more detailed look into dying within the American health-care system.

THE MEDICAL MODEL APPROACH TO DYING

The **medical model** in the United States is basically the idea that, when sick, we go to a physician to be "made well." If we are terminally ill, however, the doctor cannot "make us well." Thus, dying does not "fit" the medical model of being made well and is often reacted to by avoidance. Western medicine focuses on living and keeping individuals alive, not dying and death. This model ignores the fact that illness may be part of dying, and in treating the illness, we are trying to cure the individual. With people living longer today, we are more likely to die of illness. Modern medicine tries to cure the illness and not allow the individual to die.

By ignoring the "terminalness" of an illness within the medical subculture and failing to accept that dying is normal within the life cycle, the medical model views the dying patient as a deviant. When one is labeled "deviant," an entire interactional framework is created within which the "normals" relate to the deviant. Regardless of whether or not the individual is responsible for the deviant label, the stigmatized individual is still discredited and treated with less respect than normal persons (Larson-Kurz, Wheeler, & Dickinson, 1979).

Talcott Parsons, one of the best-known structural functionalists in the United States and one of the early medical sociologists (1951), observed that the physician does not view the patient in a holistic sense, but rather views her or him as someone with a particular disease or ailment that needs to be treated. Thus, the physician's goal is instrumental—use medical expertise to cure the patient's disease and return her or him to health and normal social functioning. Illness is a rejection of the societal value of health, noted Parsons (1958). A structural-functional perspective assumes that most individuals are healthy most of the time, and those who are ill must be "made well" by the medical system, so that society can function smoothly.

Dying as Deviance in the Medical Setting

As we have discussed, the medical profession's attitudes toward patients, and in particular dying patients, are functional to the reinforcement of the view of the "physician as healer" and functional with respect to maintenance of order in the medical subculture. The dying patient is a deviant in the medical subculture because death poses a threat to the image of the "physician as healer" (Wheeler, 1973). George Dickinson recalls submitting a manuscript on physicians' attitudes toward terminally ill patients to a medical journal a few years ago and the editor writing back, stating that dying patients cannot be called "deviants" (not a good word to use within a medical setting!).

In observing patients within a hospital setting, Daniel Chambliss (1996) argues that the patient is first separated from home, work, and even from biography. For most, illness (terminal or otherwise) is abnormal in that it is an unusual breakdown in a normally healthy daily routine. The patient is dressed in a gown. He or she becomes an object of looking and talking. Hospital life for patients is an endless round of being looked at, listened to, touched, and poked and prodded. The patient's rights to privacy are forfeited. Chambliss (1996) suggests that medicine needs the object it can treat—defined problem, curable disease, and the grateful patient who believes in the doctor, the nurses, and the hospital. Patients not fitting this sort of object (who have a chronic or incurable illness or who are noncompliant) challenge more than medicine's effectiveness. They challenge medicine's entire worldview. They are indeed deviant to the medical system.

Death also creates embarrassing and emotionally upsetting disruptions in the scientific objectivity of the medial social system. Thus, the disruption caused by death in the medical social system, if not controlled, could lead to a great deal of conflict.

Labeling Theory

A major school of thought explaining deviance is **labeling theory.** The perspective of this school of thought does not focus on the act or the actor, but rather on the audience observing. Erving Goffman (1963) describes the stigmatized person as one who is reduced in the minds of others from a "whole and usual" person to a "tainted and discounted" one. Therefore, the key to the identification of deviance is found in the audiences labeling the individual as deviant. Thus, in analyzing the dying patient as deviant in the medical subculture, the medical audiences who interact with and participate in the labeling of the patient must be examined.

With such labeling, an entire interactional framework is created within which the "normals" relate to the "deviant." Given the idea that all dying patients are deviant, imagine the stigma attached to persons dying of AIDS. An individual with AIDS has a double deviant label—*dying with AIDS*. Many HIV/AIDS-related issues are cast as moral ones (Smith, 1996). Such characterizations occur because HIV infection is spread through unpopular and/or illegal behaviors, such as homosexual relations or intravenous drug use. Because HIV infection is spread through sex and is considered fatal, it evokes certain base-level fears in people. Those affected by the disease are subjected to moral pronouncements and may indeed be discriminated against, notes J. M. Smith (1996). Because discrimination disenfranchises people from meaningful participation in society, groups typically discriminated against may not have any incentive to comply with social norms (Becker, 1963).

A former student of George Dickinson's, who has AIDS, desired volunteer work at a funeral home because of his fascination with death. His request was denied. He later, rather jokingly, said that he did not know if the refusal came because he is gay or because he has AIDS. As W. C. Earl, C. J. Martindale, and D. Cohn (1992) report from their study of coping with HIV infection, how individuals become infected is frequently of more concern in the evaluation of positive adaptation than is the stage or progress of the ailment. Thus, whether the patient acquired AIDS through sexual activity or intravenous drug use or through some other means, the stigma of deviance tends to apply.

According to Elliot Freidson (1972), when a person is labeled deviant, the stigma interferes with normal interaction. Although other individuals may not hold the deviant person responsible for her or his stigma, they are nonetheless "embarrassed, upset or even revolted by it," says Freidson. Therefore, the assumption can be made that the deviant person elicits certain aversive attitudes from the audience with whom interaction occurs. These aversive attitudes may be of sufficient strength to elicit attempts to manage them and to decrease aversion through avoidance behavior.

Deviance Results in Punishment

The primary reaction to deviance of any type is punishment of some sort. French sociologist Emile Durkheim (1961) observed that the primary purpose of punishment is not to punish the deviant himself or herself, but rather to affirm in the fact of the offense the rule that the offense would deny. Thus, one could suggest that illness is a rejection of the societal value of health or that the terminally ill patient to the physician is the antithesis of healing and getting well taught in medical school. For the patient with AIDS, **homophobia** may be present in the community (Schofferman, 1988). The patient may feel extreme guilt or shame for being ill or may blame himself or herself for "getting what was deserved."

Robert Kavanaugh (1972, p. 8) characterizes American deification of physical health, the hospital, and the physician in the following way:

> Our universal deity, Physical Health, is the major god currently worshipped. His demands for untold dollars in tribute are incessant. Every day it costs more to worship at his shrine (the hospital), yet the devout seem only too willing to scrape and to pay. They sit uncomplaining for endless hours in the offices of His high priests (physicians) and will purchase any drug or pill or lotion the priests prescribe, in any combination. In His shrine, the dying are excommunicated, the dead are damned.

The dying person can seldom assume normal role functioning, although he or she views the illness as undesirable and has tried to cooperate with medical personnel to get well. The dying person, therefore, is permanently cast into a deviant role due to the inability to respond to treatment and get well.

Normalization of Dying in the Medical Setting

The process of dying is no less normal than the process of living, for they coexist in the same world (van Eys, 1988). Dying must be considered a normal state in and of itself—*Death: The Final Stage of Growth*, as Elisabeth Kübler-Ross entitled one of her books. It is the world of patienthood which is artificial. Modern medicine, with its technology, has created industrial complexes called hospitals, where the environment is so artificial that both patient and staff must work to normalize the dying process. **Normalization of dying** refers to maintaining roles, relationships, and identity, though dying. In dying, both for the patient and significant others, living a life as normal as possible is a real challenge. Roles and relationships may have to be modified and an acceptance of bodily changes in the patient are in order. The ill person may look somewhat differently, because of the disease and/or side effects from treatments; thus, self-image may be affected. Others may tend to be overly protective of one with a terminal illness, notes Kenneth Doka (1993).

Normalization of the life of the dying person is not the same as is normalization of the life of a patient, notes van Eys. To be dying is normal and that is determined by the patient. In attempting normalization for the patient the staff tries to keep the patient from slipping into an exaggerated state of patienthood. In normalization for the dying, the dying person must teach the staff that there is a normal process going on.

Hospitals are places in which dying could be allowed, notes van Eys. There are many things patients need at times of their dying to make their time left more tolerable. There is room in hospitals to allow for the dying. Normalization is not really based on ethnic preferences, social patterns, and familial behavior. Normalization targets self-esteem, goal orientation, and abatement of loneliness. It generates a community that allows patients to continue to have full participation in human commerce.

The dying person redefines her or his community. The dying person is the high priest of his or her own temple, says van Eys. It is as if life has a holy of holies where only those initiated into that inner sanctum are allowed. It transcends the human construct of hospital and the secular state of patienthood. Hospitals will be able to accommodate dying persons only if the staff recognizes this transcendent element of the dying process. The current view of the human body as a machine that can be fixed without the spirit leaving the driver's seat makes it difficult to accommodate dying.

While trying to normalize dying, as living is normalized, nonetheless hospitalized patients may find themselves at a loss as to how they can handle events surrounding their illness (Schroepfer, 1999). In the depersonalized setting of a hospital, controls over tasks that are normally performed themselves are relinquished to health-care professionals. A power differential is also created between the patient and physician because doctors possess the information, knowledge, and skills

---

LISTENING TO THE VOICES
The Sickness of Dying

Poison drips into our veins. Radiation scars our lungs. Bureaucracy diminishes us. Impersonality angers us. We are irritable, wet, cold, seasick, vomiting. We are bad company. Shivering, suffering. Self-centered.

From Seeing the Crab *(p. 80),*
C. Middlebrook, 1996. New York: Basic Books.

---

required for dealing with patients, thus again affecting the patient's sense of control, states Schroepfer.

Dying in a Technological Society

A cultural value that makes it difficult to decide to let nature take its course and allow death to happen is the **technological imperative** (Freund & McGuire, 1999), a prevalent idea in most Western societies, especially the United States. Such a concept urges that, if we have the technological capability to do something, we should do it. This idea implies that action in the form of the use of an available technology is always preferable to inaction.

Barbara Koenig (1988) noted that the technological imperative in American and Western European medicine is sustained through social processes of routinization in which experimental procedures or new technologies quickly come to be considered "standard" in clinical practice. For example, by the early 1970s, the mechanical ventilator or "breathing machine" was standard equipment in medical centers and community hospitals in the United States. The technology was available and thus should be used. The "breathing machine" contributed to the creation of the intensive care unit and immediately was considered essential technology there (Kaufman, 2000).

Progress, as enabled by science and manifested in technology, is an enduring feature and primary value in modern medicine, observes Sharon Kaufman (2000). The most powerful form of progress in contemporary clinical practice is the technological imperative, criticized for decades for being a means without an end, an activity carried out in the absence of reflexive guidance. The technological imperative shapes the goals-on-the-ground of medical practice and frames its gaze, notes Kaufman. Yet clinical medicine has ends—to save lives and manage the course of disease. That moral imperative also guides and rationalizes medicine's specific practices, especially reliance on and commitment to uses of technology. These means and ends exist as basic assumptions and give meaning both to medical practitioners' actions and public perceptions of progress and "the good" in medical practice.

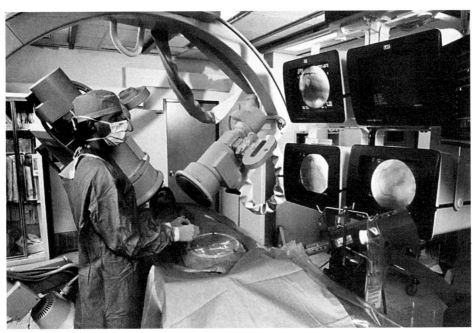

Medicine in the twenty-first century does indeed involve advanced technology. A surgical facility today resembles a Star Wars movie set, unlike an operating room in the early twentieth century.

The technological imperative is embedded in institutional responses to health crises, observe P. E. S. Freund and M. B. McGuire (1999). Institutional decisions to invest in technology are often driven by financial incentive—government subsidies, for example. Most hospitals have created several high-technology wards (e.g., coronary and neonatal intensive care units) equipped with high-tech equipment. Such expensive high-tech equipment requires costly specialized support workers and thus has caused the costs of health care to escalate significantly.

Sociologist and physician Nicholas Christakis (1999) notes that the explosive growth in both the amount and sophistication of technology deployed by physicians to combat disease has given them unrealistic expectations about their own abilities. Thus, physicians tend to regard death as a personal failure. Christakis says that it is not surprising that the technological forces arrayed to treat serious illness, in an effort to control death by postponing it, have in recent years come to be focused on controlling death by managing and predicting it. Such "management" is expressed in the increasing technicalization of euthanasia, for example. With various medical technologies emerging today, the relevance of prognosis for physicians has increased. For example, obstetrical ultrasound may reveal information about the internal anatomy of a baby that would not otherwise be known until it is born. Also, the advent of genetic testing technology provides yet another important new arena

---

WORDS OF WISDOM
The Routinization of a Hospital

Since the introduction of modern technology, the hospital has lost its aura of being a place of comfort and has instead become an establishment resembling a factory, where illnesses are taken care of, rather than human beings.

From "Human Values in the Medical Care of the Terminally Ill" (p. 19–30), by H. H. Heinemann, 1972. In B. Schoenbert, A. A. Carr, D. Peretz, & A. A. Kutscher (Eds.), Psychosocial Aspects of Terminal Care, New York: Columbia University Press.

---

for prognostication for physicians. Analysis of a person's genes may reveal relevant medical outcomes years or decades in advance of manifestation.

The business of dying over the past half century has shifted from the moral to the technical order (Cassell, 1975). The moral order has been used to describe those bonds between individuals based on sentiment, morality, or conscience that describe what is right. The technical order rests on the usefulness of things, based on expediency and not founded in conceptions of "right." The moral and technical order that E. Cassell refers to is identical in meaning to Talcott Parsons' pattern-variables of particularism-universalism (1951). Particularism, noted Parsons, is the judgment by the actor of a physical or social object in accordance with its uniqueness by criteria peculiar to the individual object. Universalism, on the other hand, is the judgment by the actor of a physical or social object according to criteria applicable to an entire range of objects.

Some of the contributing factors in this shift from the moral to the technical order, noted by Cassell (1975), are the advances in medical technology in combating death. In addition to technological advances in medicine, there has been a change in the place where death occurs. Early in the 20th century, death took place in the home where the dying person was surrounded by family and friends. Today in the 21st century, death more likely takes place "offstage" in institutions—hospitals, nursing homes, and other extended care facilities. Cassell also notes the widespread acceptance of technical success itself. The belief is that death need not occur in the foreseeable future and that death is a reversible event like actors on the stage who are "killed" but reappear in subsequent productions. The technology of cryonics, discussed earlier in chapter 2, illustrates this point well.

When people began to question the price society seemed to be paying for medical progress (hopelessly ill or brain-damaged patients being kept alive by respirators, feeding tubes, and pharmacologic maneuvers) that could not be halted without court orders, living wills were created in 1967, as discussed in chapter 12 (Grady, 2000b). The documents were an effort (the success of which might be questioned) to wrench dying from technology's grip. Nonetheless, the 21st century is a time of high-tech "everything"; thus, modern medicine is in vogue with its high-tech emphasis, whatever the consequences might be.

LISTENING TO THE VOICES
Low Status for Hospitals and Physicians

Writing in 1950 from his own experiences in European hospitals, George Orwell observed that the previous 50 years or so had brought a great change in the relationship between doctor and patient. Prior to the 20th century, a hospital was popularly regarded as much the same thing as a prison, and an old-fashioned, dungeon-like prison at that. A hospital was a place of filth, torture and death, a sort of antechamber to the tomb. Only the destitute would have thought of going into such a place for treatment. The whole business of doctoring was looked on with horror and dread by ordinary people. From the 19th century one could collect a large horror-literature connected with doctors and hospitals. The dread of hospitals probably still survives among the very poor.

From "How the Poor Die"
by George Orwell, 1950. In Shooting
an Elephant and Other Essays,
New York: Harcourt Brace.

The Environment of the Dying Person

Choosing a place to die, if indeed one has such a "luxury," is not a difficult choice for most Americans. The overwhelming majority of U.S. citizens prefer to die "at home," as noted in chapter 1. Let's take a brief look at the environs in which one will probably die: hospital, home, nursing home, and hospice.

Hospital

More than 75 percent of adults are hospitalized at some point during the year before they die, and almost 60 percent of adults see a physician at least five times during the last year of their life (Christakis, 1999). Thus, hospitals and doctors become frequent visiting places for persons toward the end of life.

From a societal perspective, as Talcott Parsons and Renee Fox (1952) suggest, hospitalization both protects the family from many of the disruptive effects of caring for the sick in the home and operates as a means of guiding the sick and injured into medically supervised institutions where their problems are less disruptive for society as a whole. A hospital is a bureaucracy characterized by specialization, rationalization, development of power through expert and specialized knowledge (knowledge secretly protected), and depersonalization. As sociologist Max Weber (1968) observes, the more the nature of bureaucracy is developed, the more the bureaucracy is dehumanized. The more the hospital succeeds in eliminating purely personal and emotional elements from its daily operation, the closer the bureaucracy comes to perfection (Moller, 1996).

Modern societies are inherently bureaucratic societies, argues Weber. As dying becomes bureaucratized, it takes place in specialized institutions and the social role

---

WORDS OF WISDOM
The Socioeconomic Stigma of Dying

In 1967, sociologist David Sudnow, in comparing a private hospital with a public hospital, offered the following advice: If you anticipate having a critical heart attack, keep yourself well-dressed and your breath clean. The response of the emergency room staff may depend to a certain extent on their interpretation of striking social characteristics—age, moral character, and clinical teaching value.

From Passing On: The Social Organization of Dying, *by David Sudnow, 1967, Englewood Cliffs, NJ: Prentice Hall.*

---

of "formal caretakers of the dying" emerges, notes D. W. Moller (1996, p. 25). With such bureaucratization, the responsibility of the community and family to care for the dying person is basically a thing of the past. As dying is prolonged by an active process of medical treatment, a hostility is established between medical technology and death. In such a "medicalized" situation, death becomes transformed into an enemy to be defeated. The depersonalization is also found in physicians' being socialized to remain emotionally neutral and undisturbed in the presence of dying and death (Moller, 1996).

David Sudnow (1967) reveals how medical staff respond to death in a standardized, routinized manner. Thus, the bureaucratization of death within an institutional setting contributes significantly to an impersonal way of dying. The "normality" of dying within an institutional setting, therefore, is basically redefined to fit the model of bureaucracy.

Dying in a hospital traditionally gives patients very little control over the circumstances of their daily lives and the course of their dying, notes Jack Kamerman (1988). The alienation of dying patients in the United States today is a "self-exacerbating blend of naturally and societally occurring powerlessness" (Moller, 2000, p. 126). We all know that we must someday die, since that is the nature of life, yet death becomes perceived through societal orientation as a thief that robs us of our most precious possession. The awareness contexts of suspicion, closed awareness, and mutual pretense of Barney Glaser and Anselm Strauss, discussed in chapter 5, often exist in a hospital setting.

Home

In both community and clinical samples, a majority of individuals consistently express a preference to die at home, according to studies cited by Judith Hays and colleagues (1999). Adults sampled in Australia and Italy favored home death over hospital death by three to one, and in a multiethnic Los Angeles sample, Anglo-American and Japanese-American adults preferred home death to hospital death by three to one, African Americans by two to one, and Mexican Americans by five to three. Clinical studies of preference for place of death have been conducted among terminally ill patients in the United Kingdom, the United States, Canada, and Japan and a preference for home death was reported in 54 to 74 percent of dying patients when the only alternative was a hospital setting. When inpatient hospice care was also an alternative, 53 to 58 percent of those dying of cancer and 32 percent of those dying of AIDS expressed a preference for a home death (between 15 and 29 percent preferred a hospice death).

Though "home" is the preferred place to die, as noted earlier, the majority of Americans die in an institutional setting, such as a hospital or nursing home. The majority of individuals dying in a hospice setting in the United States (less than 15 percent die in a hospice setting) do indeed die "at home," under the care of hospice. Yet, not everyone who dies at home is enrolled in a hospice program. Dying at home with hospice as a choice has only been available since the early 1970s.

The first hospital was established in the United States in Philadelphia in 1713. By 1873 there were only 178 hospitals in the United States, but by the early 1990s, the number had increased to over 5,600 (Cockerham, 1998). Thus, for most of the history of the United States, dying in a hospital was not an option, since they barely existed. Even with hospitals proliferating in the 20th century, many rural areas still did not have adequate access to a hospital; thus, dying at home was about the only option. Also, without the life-support equipment, which only came into existence in the latter half of the 20th century, in many cases hospitals could do nothing more for the dying person than could be done at home. Thus, dying at home was indeed the way individuals died in early America. Nursing homes were not available "on every corner" until late in the 20th century; thus, again the home was "the place" where individuals died.

George Dickinson can remember his great-grandmother dying at home back in the 1950s. There were no nursing homes in the community, hospice did not exist, and she did not need hospital care. She was dying of "old age" in her 90s. Thus, the "option" was to die at home. Different relatives took turns sitting up with her and caring for her various needs. The doctor made "house calls." She died in familiar environs and surrounded by her family. In a way, this was "the good old days."

Nursing Home

Two thirds of persons who consider a nursing home their usual place of residence will remain in the nursing home until death (Hanson, Henderson, & Rodgman, 1999). Death indeed is an everyday occurrence in nursing homes and retirement communities. The major difference in dying between these two is the degree to which residents have control over the course of their dying, observed Victor Marshall (1976) in a comparison in the 1970s. Dying in the nursing home was structured by a formal administrative policy—dying patients were moved to the "dying room" in the infirmary. In the retirement community, however, the residents informally worked out their own approach to handling dying and death.

Jaber Gubrium (1975) used **participant observation** in studying residents in a nursing home. His **ethnography** was entitled *Living and Dying at Murray Manor*. Gubrium notes that nursing home residents view the nursing home as their *final* place of residence. When sent from home to the hospital, there is at least the hope of returning home—not the case when sent to a nursing home. The situation in a nursing home is not unlike that in a hospital—very regimented and bureaucratized. The daily "routine" is set by the administration, and it is followed. Residents tend to define their futures in terms of death. Vivid signs of dying in Murray Manor include physical crises such as heart attacks, being transferred to the third or fourth floor "to die," and daily reports from those with dying spouses to other residents in the home. After death, the bodies are taken out through the front lobby at night— by day, they are taken to the basement via the elevator to the service ramp.

Hospice

Hospice care makes the meeting of social, psychological, financial, and spiritual needs a major priority in patient care. By doing this, hospice provides an alternative to the health care found in most medical treatment centers. However, none of the social, psychological, financial, or spiritual needs of patients (or their families) can be met until all health-care professionals are comfortable with discussing death-related issues. If a physician, for example, is afraid of death or chooses to ignore it, it will be difficult for him or her to enable the patient to deal with the issues involved.

Although different types of hospice institutions exist, all hospice programs are unified by the general philosophy of patient care. In 1996 approximately 78 percent of hospice patients in the United States had cancer, 10 percent had cardiovascular diseases, 4 percent had AIDS, 1 percent had renal or kidney disease, 1 percent had Alzheimer's disease, and the remaining 6 percent had a variety of other diseases (National Hospice Organization, 1997).

Because 77 percent of American hospice patients die at home (National Hospice Organization, 1999), one of the questions frequently raised by family members is what to do if an emergency develops in the middle of the night or on a holiday. Although many physicians and other health-care professionals do not make house calls, hospice personnel do. Home care for hospice patients is made viable by the fact that a physician and a nurse are on call 24 hours a day, 7 days a week. This gives patients and families confidence that they can manage at home.

Community physicians continue to be involved in the care of their patients, and usually remain primary caregivers, while the patient is receiving home care. Such community involvement relates the hospice program to the area in which it is located and tends to give hospice care greater visibility than is sometimes true of health-care programs.

Hospice Inpatient Care

Inpatient care in hospice usually becomes necessary for one of three reasons. The first is that in order to bring a patient's pain and symptoms under control, a stay of a few days in an inpatient facility may be necessary or helpful. The second is that the family taking care of the patient at home may become exhausted and need a few days' rest while the patient is cared for elsewhere. The third reason is that home care may be inappropriate at a given stage of the illness due to the patient's condition or home situation. It is hoped that upon admission to an inpatient hospice facility, patients will be able to move back and forth from home care to inpatient care at various stages of the illness.

When admitted to hospice inpatient care, special efforts are made to make the patient feel as much "at home" as possible. When taking a tour of St. Luke's Hospice in Sheffield, England, in the spring of 1999, the medical director was showing George Dickinson the admittance area in which the patients were brought into hospice. The gurney was awaiting the arrival of the new patient. Sticking up under the top sheet was a bed warmer! The patient coming into the hospice within the next few minutes would indeed receive a "warm" welcome. Psychologically, how fitting for hospice. They are all about *care.*

Whereas traditional medical care in recent years has tended to concentrate care in specialized hospitals or in nursing homes, hospice care returns the focus to the family. Because the family is the unit of care within the inpatient hospice facility, sufficient space must exist for a large number of family members to congregate. In addition, such care requires a home-like environment—the idea is to make the facility as much like a home away from home as is possible. Patients are encouraged to bring with them favorite possessions such as pictures, a favorite chair, or plants.

No arbitrary visiting restrictions are placed on those wishing to see hospice patients. One may visit at any time of day or night. Visitors of any age, including young children, are not restricted in their visitation. Furthermore, family pets, such as dogs or cats, may come as well. The goal of an inpatient hospice facility is to provide a homelike environment where the patient and his or her family can appreciate the joys of social relationships.

The inpatient facility of the Connecticut Hospice in Branford illustrates the preceding principles. The family room is off limits to staff and is provided solely for the comfort of family members. Hospice care places considerable emphasis on the tastiness, attractiveness, and nutritional value of food prepared for patients. The Connecticut Hospice employs a gourmet chef who trained in Paris to supervise its food preparation. Kitchens containing a refrigerator, a microwave oven, a stove, and a sink are also available for use by families. Washing machines are likewise maintained for their use. Large living rooms with fireplaces are available. Ten rooms with four beds in each help patients develop social support systems among family groups. There are four single bedrooms, also. Spacious corridors next to patient rooms contain plants and areas for family gatherings. Beds may be moved around as desired—on a nice day these beds are often outside on patios. A common room and chapel are used not only for religious services but also for presentations by various kinds of artists. Operated by volunteers, a beauty parlor is available to help patients feel better about themselves. A preschool exists for 3- and 4-year-old children of staff, volunteers, and people in the community. When the patient dies, he or she is taken to a viewing room for the family members.

LISTENING TO THE VOICES
Dying in a Hospice Environment

After a week in the hospital and a few days at a regional medical center, we found out that my father had a rare type of cancer. It only took 10 weeks to take his life, but he did put up a fight. He tried chemo twice, but it was not working. Toward the end of my father's illness, we checked him into hospice. My mother and I could not take care of him the way he needed anymore. It was just becoming too much. The people at the hospice program could not have cared for my father any better. They were like angels sent down from heaven to make his last days comfortable. They would read to him, talk to him, etc. They let my whole family stay with him night and day.

The hospice program is wonderful in the help that they provide for the patient and the family. They helped us cope with the loss, and still call to see how we are doing. Actually, a few of the employees came to my father's service.

From a letter received on November 12, 1997 from Courtney Schomp, a former student in George Dickinson's class entitled "Death and Dying."

The Connecticut Hospice in Branford, Connecticut—the first hospice in the United States—is a facility for the terminally ill in their last few weeks of life. The emphasis is on care, not on cure.

DEATH EDUCATION IN MEDICAL SCHOOLS

As noted in chapter 1, medical education has historically offered only limited assistance to the medical student encountering death for the first time (Dickinson, 1985). From 1975 to 1985 the number of full-term death and dying courses in medical school in the United States increased from only 7 to 14 (out of a total of 113 and 128 medical schools, respectively). By 1990 the number of courses had increased to 18 (Mermann, Gunn, & Dickinson, 1991); yet by 1995 the number had dropped to 9 (Dickinson & Mermann, 1996). Eighty percent of medical schools in 1975 and 1985 offered death education in the form of an occasional lecture or minicourse; however, this number had increased to 90 percent by 1995 (Dickinson & Mermann, 1996). The average number of years of these offerings in 1985 was 8. Thus, an emphasis on dying and death in medical schools has a brief history and remains somewhat limited today, although the exposure for students is increasing.

The University of Massachusetts Medical School integrates sessions on dying and death into the gross anatomy courses. These are typically taught during the first few weeks of the course (Marks & Bertman, 1980). During this course, students are encouraged to identify and articulate feelings about death and about the experience of dissecting a cadaver. Without this discussion, the students tend to become desensitized to the human dimension of death.

In a study (Dickinson et al., 1997) of first-year medical students in the gross anatomy laboratory, pre- and postsurveys of the 84 students revealed that 54 percent showed less death anxiety after completing gross anatomy, 29 percent had increased fear of death, and 17 percent experienced no change. Thus, the gross anatomy course tends to desensitize students to death. While the reduction of death anxiety is good, this is often part of a general "numbing" that occurs during medical school that may affect doctors' ability to relate to dying patients as people. To address this problem, some medical schools are trying to treat cadavers as people rather than objects. For example, several medical schools routinely hold a memorial service for the cadavers that are dissected. Through such a dedication service the students give thanks to those who in death taught them about life.

Developing a Sensitivity to Social/Psychological Needs

The medical training of most physicians historically seems to be primarily concerned with the patient's physical state rather than with the patient's social-psychological needs. Granted, the patient wants (demands?) a physician knowledgeable of her or his physical state, yet such knowledge is only part of the picture. Some changes are occurring, however, as evidenced by Dartmouth Medical School's interest in "facilitating the medical students' understanding of the patient as a bio-social-psychological being" (Nelson, 1980). In Dartmouth's program, the medical students are required to learn from the terminally ill patient what it is like to be dying. Thus, the

---

WORDS OF WISDOM
More Emphasis in Death Education Needed in Medical School

Physicians are not being taught to recognize when patients are reaching the final stages of life, to manage symptoms associated with dying, or to attend to the special needs of dying patients. Medical students learn to fight disease and are taught to view death as a failure. They are trained to respond to illness with aggressive care, to subject patients to invasive tests, procedures, and machines, until death is recognized as imminent. The dying process and the dying person are largely ignored. Such practices will persist if the system for educating doctors and other health-care workers continues to emphasize the basic sciences and does not teach students how to care for the dying.

Adapted from "Medical Education Must Deal With End-of-Life Care" (pp. A56–A57), by F. Cohn, J. Harrold, & J. Lynn, May 30, 1997, Chronicle of Higher Education.

---

patient is used as "teacher." The inclusion of a humanistic emphasis in death education should help both the dying patients and the medical students.

At the Yale School of Medicine, Alan Mermann (Mermann, Gunn, & Dickinson, 1991) offers a seminar for first- and second-year medical students on problems in communicating with dying and seriously ill patients. Professor Mermann also uses very sick patients as teachers. Approximately one third of the first-year students take this elective course. Goals for students in the course are to (a) learn to talk with, and listen to, sick persons; (b) learn to establish a professional relationship without the intrusion of friendship; (c) ascertain the meaning of compassion without sentimentality and the need for humility in the context of the physician's ignorance; (d) learn of our common frailty as human beings, the finality of death, and the need that we all have for companionship when death is near; and (e) enrich the students' understanding of those in their care.

Developing Communication Skills

A survey of more than 600 physicians (Dickinson, 1988) in 1986 revealed that the majority (78 percent) agreed that more emphasis in medical school should be placed on communication skills with terminally ill patients and their families. A study of medical students in England (Field & Howells, 1986) found that, of the 77 percent of students expressing worries about the prospect of interacting with dying patients, communicating with such patients was their most frequent concern. Results from another study of 350 family physicians in South Carolina (Dickinson, 1988) in 1987 found that the majority felt that their medical education was inadequate in helping them relate to terminally ill patients and their families.

In a Harris survey (Richardson, 1992), 91 percent of 1,501 practicing dentists, nurses, pharmacists, physicians, and veterinarians said that teaching students how to communicate effectively with patients and their families is very important, yet 36

Good communication skills for health-care professionals should be emphasized in their medical education, just as technical skills are stressed. Rapport with patients is very important in helping them cope with illness.

percent gave their schools a poor rating in this area. Seeing the need for better communication skills with patients, the Association of American Medical Colleges in 1991 added two 30-minute essays to the admission test to help medical schools evaluate communication skills (Altman, 1989). Similarly, the American Medical Association (Montgomery, 1996) announced in 1996 a major effort, beginning in 1997, to teach doctors how to aid the dying by helping patients and their families plan for dying, by providing effective ways to reduce suffering, and by treating psychiatric complications.

In 2000, the American Medical Association concluded that treating terminally ill patients with kindness, tact, and good medical judgment may come more naturally to some doctors than to others, but the AMA decided that the skills can be learned (Grady, 2000a). A 2½-day course entitled "Education for Physicians on End-of-Life Care" has been developed. As of June of 2000, the course had been given seven times in different cities to groups of 100 to 150 doctors each time. The doctors learn details of pain control that were not taught in medical school, as well as how to use drugs to ease shortness of breath, nausea, seizures, agitation, and other problems. They are urged to encourage family members to lie in bed with a dying patient if they wish. They are taught how to turn off a respirator in a way that does not leave the patient gasping for air. They learn how to help prepare a patient's family for the physical changes that take place as a person dies.

In a study of 441 family medicine practitioners in South Carolina (Durand, Dickinson, Sumner, & Lancaster, 1990), those who reported having been taught concepts concerning terminally ill patients and their families while in medical school had a more "positive" attitude toward death than did those who received no instruction. Other studies (Bugen, 1980; Dickinson & Pearson, 1981) have reported attitudes that are more favorable, less fearful, and less avoiding with respect to dying and death as a result of death education.

Physicians themselves may be of therapeutic value to their patients, simply by talking regularly to them and assuring them that they will not be abandoned as death approaches (Cohn, Harrold, & Lynn, 1997). Medical students must be trained to understand that this role is important and involves skills that can be taught, including the management of care as a patient moves among different settings (hospital, home, nursing home, and hospice, for example), and how to offer spiritual and emotional support.

Treating a dying person may involve a period of "mixed management" in which a patient continues to need some aggressive care, even while the doctor places increased emphasis on palliation, note F. Cohn, J. Harrold, and J. Lynn (1997). Physicians can help patients by raising and discussing issues and working out ways of incorporating visits from family members, members of the clergy, or counselors into the plan of care.

Ethicists Cohn, Harrold, and Lynn (1997) suggest that medical education requires the development of specific competencies to bring about better end-of-life care. For example, when students learn to conduct patient interviews, they must learn to address emotional and spiritual issues as well as medical concerns. Students should learn not only about the existence of living wills and "durable power-of-attorney" forms, but also how to discuss them with patients. Students should be taught to ask and appreciate what is important to the patient and must learn how to care for themselves as they face the task of caring for dying people.

A good start in reorienting medical education should be to increase students' contact with dying patients, recommend Cohn, Harrold, and Lynn (1997). As discussed earlier in this chapter, Professor Mermann at Yale Medical School assigns students to a dying person in his seminar on dealing with very sick patients. In suggesting more contact with dying patients, Professor Mermann (1997) notes that what may be missing in a medical school lecture is the unique individuality of each of us, especially when confronted by the prospect of disaster, with its accompanying anxiety and fear, and the possibilities of pain and suffering. Students could also accompany nurses and social workers on home visits to dying patients. Students could take the hospice training and volunteer to work on a weekly basis with a dying person—such an assignment is common in England but somewhat rare in the United States.

Good rapport with dying patients might be an intrinsic quality of certain medical students that is reinforced by experience, or it might be something in which medical students are given instruction. Though values, ethics, and communication

---

LISTENING TO THE VOICES
Presence of the Physician in the Dying Situation

Every physician has a professional mandate to help other human beings in need. But the inevitability of a patient's death may challenge this code, unless the physician is willing to add a new dimension to the patient-doctor relationship, that of helping the patient deal with the experience of dying. The presence of the doctor in this last phase of life may be crucial for a peaceful death. It can allow the patient not only to die with self-respect, but also to feel less lonely.

From "The Dying Patient, the Physician, and the Fear of Death" (p. 1729), by E. Seravalli, 1988, New England Journal of Medicine, 319.

---

skills may be presented and "learned" in medical school, there is no guarantee that they will be carried over into clinical practice. Perhaps the results would be known if only the majority of medical schools in the United States offered a course on relating to terminally ill patients and their families. With the current trend since 1975, however, this is not likely to occur in the near future.

Ira Byock (1997), former president of the American Academy of Hospice and Palliative Medicine, says that care for dying patients is still inadequately taught. He notes that the culture of curative medicine is entrenched within medical universities with attention directed toward preserving life at all costs. The values of caring are subordinate. Considerations of money, physical comfort, human dignity, and the quality of life experiences are, at best, secondary, according to Dr. Byock. The problem is not that medical professionals are heartless, but rather that the goals of end-of-life care are often not well considered, if at all. Though the situation is improving at the beginning of the twenty-first century, M. Lattanzi-Licht and S. Connor (1995) noted that the most training in palliative care an American medical school student could expect in the mid-1990s was approximately one lecture during four years of schooling. "Small wonder that the care of the dying in the United States is still generally poor and occasionally horrible," conclude Lattanzi-Licht and Connor.

PUBLIC POLICY AND HEALTH NEEDS
OF THE CRITICALLY ILL

Many Americans assume that health care should be a guaranteed right of all citizens who desire and can afford to pay for it. For those unable to pay, the government provides basic health care through Medicaid. Although passive euthanasia would

---

WORDS OF WISDOM
Difficult Questions With No Easy Answers

Should the artificial heart be considered an inalienable right of any needy patient? Should the public at large pay the extra cost of saving premature babies? What is to be gained by extending the lives of comatose patients a few more years? It is clear that our society faces a danger: as medical technology advances and we spend more and more for less and less, we edge ever closer to what might be called the tragedy of the medical commons. A tragic end can be prevented only by saying NO! But what should we limit? The kinds of defects and malfunctions eligible for treatment? The age beyond which a patient may not be treated with heroic medicine? These are good candidates. Every point in our medical and legal systems seems to be biased in favor of compassion. At some point compassion must yield to principle.

From "Crisis on the Commons"
(pp. 21–25), by Garrett Hardin,
September–October 1985, The Sciences.

---

affirm the right of those who choose to reject medical intervention, many believe that those who wish to live have a legitimate claim to the resources of society in providing the medical assistance necessary for living. According to *The Economist* ("Euthanasia: What Is the Good Death?," 1991):

> The terminally ill now account for 20 to 30 percent of hospital expenditures, yet no amount of money is going to make them better. Doctors tend to shy away from principles of "utility" and cost worthiness. Instead, they argue a principle of "justice."

Americans are beginning to reevaluate universal access to health care. This reevaluation of public policy is motivated by a utilitarian concern for distributive justice.

Utilitarianism is the belief that social policy should be directed toward providing the greatest good (maximum benefits) for the greatest number of people. If providing an artificial heart for one patient who will live 3 months inside a hospital costs more than a million dollars, utilitarians believe that the same amount of money might be more appropriately spent on prenatal care for 10,000 women. Furthermore, in the United States, approximately 50 percent of the medical expenditures that an individual will incur during his or her lifetime will be for medical services delivered in the final 6 months of life. Because all of us are involved, to some extent, in paying for these medical costs (either through government subsidy or health insurance premiums), it is easy to question whether or not these expenditures are a worthwhile investment for such a limited return.

These types of cost-benefit analyses deemphasize the health needs of particular individuals in favor of the health needs of the entire society. Garrett Hardin (1985), an ecology professor at the University of California at Santa Barbara, presents this argument with the following statement: "Easy access to health care is exhausting America's medical resources." According to Hardin, whenever meeting medical needs becomes an obligation of the entire community, utilization will always benefit individuals, whereas costs will be shared by all members of the system. Therefore, individuals can never be expected to limit their use, and health-care costs will naturally rise to such a level that the system will collapse—the end of health care for all.

A recent survey (Jankel, Wolfgang, & Hoff, 1994) of Georgia state legislators asked the respondents to rank the importance of funding for specific health-care services. The legislators assigned the highest priority score to providing care to pregnant women and their unborn children, and the lowest score to organ transplants. It seems that, for at least some elected officials, in the allocation of government health-care dollars the goal is to maximize the state's return on its investment.

Dr. Daniel Callahan, director of the Hastings Center, which deals with end-of-life issues, has advocated that persons older than 80 years be ineligible for Medicare reimbursements for cancer surgery, intensive care, and heart bypass surgery (Callahan, 1987). Callahan says:

> As a nation, we must set a hard line against the endless prolongation of life. We must direct our medical energies only to the relief of pain and suffering in the elderly, and to help them prepare for a graceful death after a full life.

Yet, there are more radical proposals that even the most utilitarian of persons would have a difficult time accepting. K. K. Fung (1993) proposed that the hopelessly ill voluntarily elect physician-assisted death and thereby save society's resources. To ensure that the saved resources would not be rechanneled to additionally futile treatments for other terminally ill patients, those who chose active euthanasia would be allowed to determine how the saved resources would be redeployed. A related, but significantly more modest proposal by Howard Curzer (1994) argues that justice sometimes requires increasing a patient's risk of death by shifting scarce health and financial resources to patients who need them more. Although these decisions are very difficult, the cost and availability of life-extending medical procedures financed primarily by the government and third-party providers force us to admit that there cannot be a guaranteed right to health care in the United States.

As Nicola Clark (1992) observes in looking at the high costs of dying, perhaps public health policy has become so focused on interventional strategies for the individual that we have lost sight of the broader needs of the community. Government-funded health care is a zero-sum game, notes Clark. One patient's use of health-care resources does restrict the resources available to others. The costs become very clear when discussing end-of-life care.

DEATH ACROSS CULTURES
Health Care in Sweden

Sweden has demonstrated that a socialized system of health-care delivery can be effective in a capitalist country through the formation of a national health service. The Swedish National Health Service (NHS) is financed through taxation. Taxes in Sweden have been the highest in the world, but Sweden remains one of the world's most egalitarian countries. Universal health insurance would eliminate the worries of finances for one dying in Sweden. George Dickinson had a friend visiting in Sweden who had to have major surgery. Her bill for the entire hospital stay and surgery was less than $5. She had to pay for a long-distance telephone call from her hospital room!

Physicians in Sweden are paid according to the number of hours worked rather than the number of patients treated. Sweden is one of the leaders in the world in palliative medicine; thus one's pain in dying would largely be eliminated. General hospitals are owned by county and municipal governments. Drugs are either free or inexpensive. Sweden remains committed to universal and equal access to health services paid by public funding.

Taken in part from Medical Sociology *(pp. 289–291), by* William C. Cockerham, 1998, Upper Saddle River, NJ: Prentice Hall.

THE COST OF DYING

Before he died last month, 90-year-old Patient M spent 41 days in the intensive care unit of New York City's St. Luke's Roosevelt Hospital (Clark, 1992). Patient M suffered from severe bilateral pneumonia, a fractured hip, and gradual respiratory, kidney, and heart failure. From the outset, his doctor did not believe he would survive. Still, Patient M received aggressive care. In the end, the cost of keeping him alive in his final days—the respirator, the heart monitor, feeding tubes, antibiotics, radiographs, blood tests, and sedatives amounted to, in one preliminary total, $93,111. Was it worth it? Most Americans still feel every person, regardless of age, income, or level of insurance, deserves life-prolonging care. But one of every seven health-care dollars spent each year is used during the last 6 months of life. Is this really the most effective way to allocate health-care resources?

The cost of health care in the United States is approximately one trillion dollars annually (Matcha, 2000); thus, approximately 14 percent of the U.S. gross domestic product (GDP) is spent on health care. The United States has the most expensive health-care system in the world. Predictions for health-care costs early in the 21st century are 18 percent of the GDP. D. A. Matcha notes that 41 percent of all the money spent in the world on health care in 1990 was spent in the United States.

Costs associated with illnesses are varied but can be characterized as direct and indirect (Smith, 1996). **Direct costs** typically include those for a wide variety of

ZIGGY

types of medical care from hospital to home-based care (e.g., services of the medical staff, medication, and use of the hospital facility). **Indirect costs** may include overhead costs of running the hospital (e.g., postage, utilities, and upkeep of the building) or costs associated with lost productivity (e.g., money lost from one's not having a job).

The United States' method of health-care financing can only be described as a shell game, according to former Secretary of Health, Education and Welfare Joseph Califano (1986). The pea represents the cost of health care and the shell represents the vendors (e.g., federal government and insurance companies) who provide funding for health care. Ultimately, the American people pay all health-care costs, whether for a terminal or a chronic illness, but health-care dollars are directed from the people to the providers in three main patterns: (1) direct payment by the consumer, (2) private insurance, and (3) government taxation at various levels (Freund & McGuire, 1999).

In 1996, over half (53 percent) of all health services were paid for by private sources, most coming from private health insurance companies, with the rest being paid out-of-pocket by individual patients and their families (Weiss & Lonnquist, 2000). The first private insurance, Blue Cross, was established in 1930. With a company like Blue Cross (later Blue Shield was added for nonhospital services), an individual (or her or his employer) pays a premium, usually monthly. When the individual becomes ill, she or he receives medical treatment and then the bill is sent to the insurance company, which pays most (or all in some cases) of the medical bill. This **third-party payment** is very popular in the United States today.

In the latter part of the 20th century, **managed care organizations** entered the scene—this form of private insurance controls what they spend on health care by screening prospective utilization of insured medical care. They also manage expenditures by limiting coverage to care provided by doctors and health facilities. Demands of patients and their advocates are balanced against the price purchasers are willing to pay (Wholey & Burns, 2000).

The public welfare system, funded via government taxation, for health purposes includes **Medicare** and **Medicaid**. In 1996, 34 percent of all health services were paid for by Medicare and Medicaid (Weiss & Lonnquist, 2000). Because 70 percent of all people who die in the United States each year are 65 years of age or older (Clark, 1992), they are probably covered under Medicare. Medicare was created by legislation enacted in 1965 to finance acute medical care mainly for elderly Americans (Freund & McGuire, 1999). Part A of Medicare is hospital insurance; Part B is supplementary medical insurance for physicians' services, outpatient care, laboratory fees, and home health care. Chronic illness and long-term care are not part of the Medicare plan; thus, many individuals do not have adequate medical insurance to cover their expenses. Such a void in medical coverage in our society has been a major political issue for several years.

For individuals with low incomes in the United States, Medicaid provides financial assistance with health-care costs regardless of age. Medicaid was created at the same time as Medicare, but functions differently. Medicaid is perceived as "charity" rather than "deserved aid," note P. E. S. Freund and M. B. McGuire (1999). Each state operates its own Medicaid program but receives federal contributions. Despite Medicare and Medicaid, an estimated 43 million Americans (16 percent of the population) have no insurance to cover medical expenses. Further, an estimated 61 million have no health insurance coverage for part of the year. Many individuals have no health insurance coverage simply because they cannot afford it, some individuals do not "believe in" health insurance (pay as it comes with no regard for what will happen if a *major* illness occurs or "God will take care of you"), and others have no insurance because they may have just graduated from college and are no longer included on their parents' policy, but have limited money and perhaps no job to cover insurance.

In the United States health care is closely linked to employment status—medical insurance often comes with a job as a fringe benefit. Blacks and Hispanics are particularly likely to be uninsured because they are disproportionately represented among the working poor, having jobs that pay no insurance benefits and in living in regions of the country where labor is less commonly unionized and Medicaid insurance is unavailable (Ginsberg, 1991). Looking at young adults, irrespective of ethnic origin, those without private health insurance are 35 percent more likely to die than those with insurance (Rogers, Hummer & Nam, 2000). Thus, having private health insurance coverage, as opposed to none, enhances one's chances of survival. Mortality risks are reduced when income is used to purchase health insurance and to buy goods and services that will promote health and prevent disease. There is most definitely a correlation between income and mortality.

Because one's major health-care costs normally occur during the last few months of life, the process of dying can indeed be expensive. The current health-care cost crisis has caused many people to become paupers simply because they are incurably ill and are not dying quickly enough (Byock, 1997). Driven by Medicare and the insurance industry to reduce expenditures associated with the last year of life, hospitals, clinics, and health maintenance organizations (HMOs) have instituted an array of cost-containment measures reflecting a de facto strategy of "doing more with less," notes Ira Byock. Costs of care are therefore being shifted onto the backs of the patients and families in need. Byock cites a survey of the impact on the family of caring for a seriously ill person: 20 percent of families reported that a member had to quit work, delay his or her own medical care, or make another major life change to provide the needed care; 29 percent of families experienced the loss of most, or all, of their major source of income; and 31 percent reported the loss of most or all family savings

Byock (1997, p. 242) further states, "To be terminally ill or elderly in America today is to be reminded frequently that you are a drain on the nation's resources." Byock (p. 242) also says that the message to the elderly and the incurably ill is: "Limit your use of resources and get out of the way to make room for those who are younger, vigorous, and still able to contribute to society." Our society places

ZIGGY

value on "producers" and not so much on those who are strictly "consumers." Terminally ill and elderly persons are high consumers but are often unable to be productive, and thus they are considered less worthy recipients of medical dollars.

CONCLUSION

Indeed, while one may think that the cost of living is expensive, the cost of dying is *not* inexpensive. Should our society have more Jack Kevorkians to help reduce the high costs of dying by shortening the period of the dying process? Should the government simply stop funding medical care costs for the very sick elderly persons? The answer to both of these questions for most Americans would probably be an emphatic "no." With escalating health-care costs in the United States, we have to take a careful look at expenditures at the end of life, so that we can better "manage" these expenses.

It is encouraging that American medical schools are leaning more toward a humanitarian approach to health care in the socialization of their students. Yet, palliative care must continue to be better integrated into our medical schools to encourage more treatment of the whole person—physical, social, psychological, and spiritual. We must work toward a point in medical treatment where the dying patient is indeed a "normal" aspect of the medical subculture and is treated warmly as a human being, not just the "abdominal cancer case in Room 614."

SUMMARY

1. The process of dying is not extraordinarily different from the process of healthy normal living.

2. Dying patients are considered "deviants" in the medical subculture because they do not fit the "norm" of sickness—they cannot be "made well."

3. The focus of labeling theory is on the audience doing the labeling, not the person being labeled.

4. Medical schools have traditionally offered very little death education.

5. In a few medical school offerings on relating to seriously ill patients, a dying patient is used as teacher.

6. Medicare and Medicaid are programs to assist the elderly and poor, respectively, with health-care needs.

7. The cost of health care in the United States is approximately one trillion dollars annually.

8. Approximately 70 percent of all people who die each year in the United States are elderly (65 years of age or older).

DISCUSSION QUESTIONS

1. Discuss why the process of dying is not really different from the process of healthy normal living.

2. How does "labeling theory" tend to **make** one a "deviant"?

3. Why is an person with HIV/AIDS a "double deviant?"

4. Why is a dying patient considered to be a "deviant" in the medical subculture?

5. Why do you think that medical schools have traditionally offered very little death education?

6. Discuss why a terminally ill patient might be very effective in socializing medical students on how to relate to seriously ill patients.

7. Discuss the concept "utilitarianism" in relation to health-care policy.

8. Does the United States have a utilitarian policy in dealing with health-care issues of the critically ill? Why or why not?

9. Discuss whether or not elderly patients should be prevented from having costly treatments paid for by the federal government, when that money could apply toward the health-care needs of children in poverty pockets in the United States.

GLOSSARY

Direct Costs: Usually includes costs for services of care such as the medical staff, the use of the hospital room/surgery room, medicine, and use of equipment.

Ethnography: The systematic description of a culture based on firsthand observation; literally, *writing* about a *culture*.

Homophobia: An intense dislike of or prejudice against homosexuals.

Indirect Costs: Typically includes overhead expenses such as utility bills for the hospital or costs of billing and may also include lost wages for the time the patient cannot work.

Inpatient Care: The type of institutionalized care that is required as an illness progresses. Hospice-type facilities include a free-standing hospice or a wing or designated area within a hospital or nursing home.

Labeling Theory: The theory based on the idea that whether other people define or label a person as deviant is a critical determinant in the development of a pattern of deviant behavior.

Managed Care Organizations: Groups that use administrative processes or techniques to influence the quality, accessibility, utilization, costs, and prices or outcomes of health services provided to a defined population by a defined set of providers.

Medicaid: A federal and state program that uses general revenues to fund health care for the poor. Eligibility is tied closely to economic status.

Medical model: When sick, an individual goes to a physician to be made well.

Medicare: A federal program of health insurance for persons 65 years of age and older.

Normalization of dying: Maintaining roles, relationships, and identity, though dying.

Participant Observation: A method of data-gathering, particularly popular in cultural anthropology, in which the researcher "lives with" and observes the group that she or he is studying.

Technological Imperative: A concept urging that if we have the technological capability to do something, we should do it. Action in the form of the use of an available technology is always preferable to inaction.

Third-Party Payment: A health-care payment scheme in which the patient pays a premium into a fund and the doctor or hospital is paid from this fund for each treatment provided to the patient.

Utilitarianism: The principle of evaluation based upon the goal of providing the greatest good (maximum benefits) for the greatest number of people.

SUGGESTED READINGS

Byock, I. (1997). *Dying well: The prospect for growth at the end of life.* New York: Riverhead Books. Informative book on care at the end of life by a former president of the American Academy of Hospice and Palliative Medicine.

Chambliss, D. F. (1996). *Beyond caring: Hospitals, nurses, and the social organization of ethics.* Chicago: University of Chicago Press. A documentation of the contemporary hospital, its nurses, and their moral and ethical crises.

Christakis, N. A. (1999). *Death foretold: Prophecy and prognosis in medical care.* Chicago: University of Chicago Press. A sensitive study by a physician/sociologist of a difficult topic. Dr. Christakis explores doctors who must predict lifespan and make pronouncements of mortality for terminally ill patients.

Diamond, J. (1998). *C: Because cowards get cancer too.* London: Vermillion. Experiences in the medical system of a journalist as he fights his battle with cancer.

Gerlach-Spriggs, N., Kaufman, R. E., & Warner, S. B. (1998). *Restorative gardens: The healing landscape.* New Haven: Yale University Press. Examines the history of restorative gardens and their therapeutic role in six modern medical centers.

Hammond, P., & Mosley, M. (1999). *Trust me (I'm a doctor).* London: Metro Books. A consumer's guide to how the British system of medical care works.

Lowenstein, J. (1997). *The midnight meal and other essays about doctors, patients, and medicine.* New Haven: Yale University Press. Presents humanistic approaches to learning and practicing medicine and explores the importance of being a caring doctor.

Mermann, A. C. (1999). *To do no harm: Learning to care for the seriously ill.* Amherst, NY: Prometheus Books. A study of the ways people experience serious and life-threatening illnesses, the types of suffering they endure, and how we can come to understand their lives. Dr. Mermann's course on teaching medical students at Yale Medical School to relate to seriously ill patients is discussed.

Moller, D. W. (2000). *Life's end: Technocratic dying in an age of spiritual yearning.* Amityville, NY: Baywood Publishing. Addresses dying in a technological society as it relates to spiritualism.

Proctor, R. N. (1995). *Cancer wars: How politics shapes what we know and don't know about cancer.* New York: Basic Books. Tells the story of how government regulatory agencies, scientists, trade associations, and environmentalists have managed to obscure the issues and prevent concerted action in the fight against cancer.

Rogers, G. R., Hummer, R. A., & Nam, C. B. (2000). *Living and dying in the USA: Behavioral, health, and social differentials of adult mortality.* San Diego, CA: Academic Press. A guide to the health and mortality landscape of modern America.

Selwyn, P. A. (1998). *Surviving the fall: The personal journey of an AIDS doctor.* New Haven: Yale University Press. Memoir of the first decade of the AIDS epidemic written by a physician whose encounters with dying patients allowed him to come to terms with his own losses.

Timmermans, S. (1999). *Sudden death and the myth of CPR.* Philadelphia: Temple University Press. Documents the failure of modern medicine to deal with sudden death.

Wiener, C. L. (2000). *The elusive quest: Accountability in hospitals.* Hawthorne, NY: Aldine de Gruyter. An examination of the costly enterprise that has been developed to hold hospitals accountable for improved quality of care. The author places the accountability arena in its historic, economic, and sociological context.

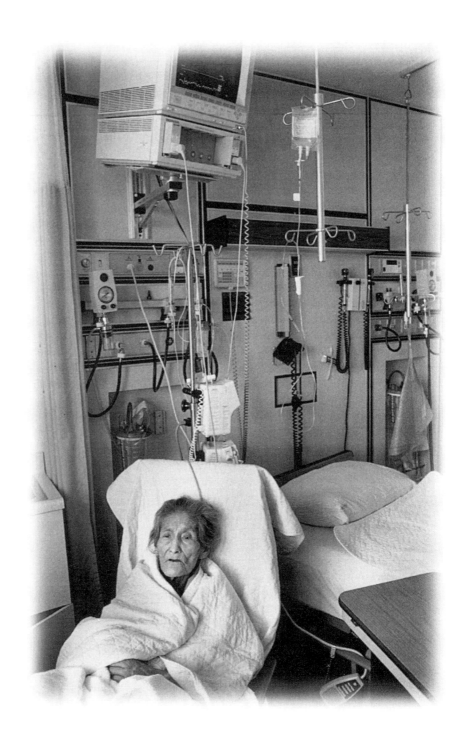

CHAPTER 8

BIOMEDICAL ISSUES AND EUTHANASIA

It is silliness to live when to live is torment;
And then have we a prescription to die when death is our physician?
— William Shakespeare, *Othello*

All substances are poisons; there is none which is not a poison.
The right dose differentiates a poison and a remedy.

— Paracelsus

Recently, a 90-year-old woman came to the emergency room of a New York City hospital in critical condition with severe chest pain due to a serious heart attack confirmed with an EKG (Jauhar, 2000). The children of the woman were told that if nothing were done to help their mother she might die. Angioplasty (a catheter is threaded through an artery in the groin to open the blockage cutting off blood flow to the heart) probably offered her the best chance of surviving this crisis, but it was risky due to her age, especially in an emergency. Their mother could die on the table or suffer brain damage or end up on a ventilator. The outcome was uncertain. She could be treated with a thrombolytic drug to dissolve blood clots, but this simple injection carried a risk of bleeding into her head. So what should they do, they asked? The children could not decide. Perhaps, they suggested, they should ask their mother, since she was awake and seemingly competent. The physicians could keep her alive, but only she should decide whether the effort would be worth it, the doctors and family thought. When the options were explained to the patient, she wanted assurances they could not give. In the end, she looked out at the room of

---

PRACTICAL MATTERS
Making Life and Death Decisions

1. A newborn is brain-dead. Should the baby receive life support, as the religious mother wishes?

2. A baby has massive internal defects. Doctors recommend treatment after treatment, but the parents say, "Enough is enough."

3. A teen with a painful, fatal disease is ready for death; her parents aren't.

4. An old woman makes an adult child her power of attorney for health care. When that child says to let Mom die, another child says no.

Adapted from "When Decisions Are Life-and-Death" (p. 26), by R. L. Guyer (1998, February 6–8), USA Weekend.

---

white coats, slowly shook her head, and said she did not know. Since neither the family nor the patient could make a decision, the patient was taken to the cardiac care unit for observation. Who decides and what is decided for such life and death decisions? Such questions are not easy for anyone to answer.

Physicians, depending on their specialty, are confronted daily with these kinds of situations regarding life and death matters. Perhaps it is no wonder that physicians generally have shorter life spans than do other professionals, given the life and death nature of the decisions they must make. How long should specialists work to save the life of a premature baby who weighs slightly more than 1 pound at birth? Should a teenager be kept alive with life support equipment after a skull-shattering auto accident? What is the "right" decision regarding a 90-year-old adult receiving life support? Even the most medically savvy people find dilemmas such as these wrenching and bewildering (Guyer, 1998).

With the technical transformation of medicine, it is now possible to extend the life of critically ill patients, restoring some to their former health, allowing others to live in a severely disabled state, and prolonging the dying process for still others (Mebane, Oman, Kroonen, & Goldstein, 1999). Such an ability to extend life raises ethical issues about quality of life and resource allocation, issues complicated by a lack of consensus in our diverse society. If the use of technology merely prolongs the life of a severely ill patient with no hope for cure, the costs to society are enormous, with very little or even negative return to the individual patient, note E. W. Mebane and colleagues (1999). Yet, for religious reasons, some individuals may feel compelled to keep the individual alive at all costs. By keeping the person alive, a "cure" may be developed and the patient may then have a good quality of life. Some may also ask, "Who am I to say that the person should be allowed to die?"

Many factors are included in decision making regarding **biomedical issues.** The mere financial costs of medical procedures developed through scientific research often present a dilemma. According to Donald Joralemon (1999), biomedicine has been associated with some negative issues such as little impact on some health problems afflicting the majority of the world's population, limited access to biomedical care for some due to a scarcity of medical personnel, and severe constraints on public health budgets. On the other hand, however, biomedicine has had positive outcomes (e.g., increased lifespan, eradication of smallpox, organ transplants, and effective treatments for many other diseases). This chapter will address biomedical issues with which we are faced in the 21st century. The issues are usually not clear-cut but fall in the grey zone, so who decides and upon what grounds?

ETHICAL BEHAVIOR

Regarding ethical behavior, we must realize that medical ethics cannot really do more than lay out the issues clearly for all to see and argue about. If we expect medical ethics to give answers, this would be cause for disappointment, observes Arthur Zucker (2000), law and ethics editor of *Death Studies*.

What Is Ethical Behavior?

Ethical behavior refers to a conscious reflection on moral beliefs and seems typically to be applied to specific cases in a setting as opposed to the nature of the setting itself (Chambliss, 1996). In professional settings such as hospitals, ethical behavior is usually based on the codification of moral principles by an occupational group and may reflect the group's long-range self-interest in its image as a servant of the community. In medicine, the use of bioethical language has made moral debates more abstract, rights-driven, individualistic, and centered on discrete cases. Ethical behavior is acting after trying to answer the question, "What should be done?" More practically speaking, the question is, "What can be done?" Ethics often assumes that people are autonomous decision makers sitting in a fairly comfortable room trying logically to fit problems to given solution-making patterns, notes D. F. Chambliss. But that is not the real world where decisions have to be made; that is the hypothetical world.

Inside hospitals, decisions are driven not by academic problem-solving techniques but by the routines of life in a professional bureaucracy. Too often the response to the patient is strictly based on the rules of the hospital without taking into consideration the individual situation. For example, the elderly mother of a friend of George Dickinson's had been comatose for several months when an emergency health situation occurred. This 95-year-old woman was rushed from the nursing home to the emergency room in the hospital and had major surgery performed on her. Her "living will" was ignored, and the hospital staff did what they had *learned* to do, according to the "medical rules." The surgery "successfully" kept her alive. She lived another 3 weeks and died in the intensive care unit. If students were exposed to "real settings" ("live scenarios") they would find the "exercise" less academic and more of a real-life situation and would respond to "What *can* be done?" and not "What *should* be done?" In the case of the friend's mother, *no action*, other than keeping her comfortable, would have been appropriate.

Ethical behavior in health care is inseparable from the organizational and social settings in which ethical problems arise, observes Chambliss (1996). He notes that ethical problems are not random or isolated failures of the system; they are often in fact fundamental, if unintended, products of that system. People work in organizational and professional roles and settings, and these shape their behavior; ethical decisions are not made in some hypothetical "free choice zone." For example, nurses and physicians have very different professional roles, though both treat patients. Doctors have vastly more power, give the orders, are rewarded for scientific expertise, and hold the legal responsibility for the patient's well-being. Nurses, on the other hand, carry out the doctors' orders, are largely rewarded for organizational skills, and handle much of the day-to-day work of patient care. Because of training, physicians tend to lean toward aggressive treatment and are often reluctant to stop treatment, yet the nurses may be getting frustrated dealing with a patient whose condition is not improving and who is begging to be left alone. Thus, a moral dilemma between the more powerful and the "ones carrying out the orders," notes Chambliss.

George Dickinson recalls back in the 1960s when kidney dialysis machines were scarce, not the situation of abundance of machines that we have today. A medical intern lamented to him that dialysis treatment for a young woman in his hospital had been stopped, which meant that she would die. Her dialysis was stopped because they had only three machines but had four persons in need of dialysis; thus **triage** was practiced because of limited sources. A "bureaucratic decision," based on the "policy" of the hospital had deemed that she must "come off," and thus would die. Stopping kidney dialysis will lead to uremic poisoning and death. It was "someone else's turn" to have the machine. As Chambliss says, an organization makes the decision—very impersonal in what is a most personal situation.

Bioethicists

Patients today believe that they have a *right* to take part in their own care—they expect their wishes to be a determining factor in doctors' decisions, observes physician Eric Cassell (1991). The right of patients to refuse or discontinue treatment is now an acknowledged aspect of American medicine. The appropriate allocation of scarce resources (deciding who will get treatments or organs when there are not enough to meet existing needs) is a current issue in this time of budgetary constraints.

How does one make these difficult decisions regarding various moral problems posed by modern medical science and technology? In our very specialized society today, we indeed are not alone in decision making regarding biomedical issues. There is a specialist to assist with these difficult medical decisions—a **bioethicist**. Such a specialist is a recent addition to the world of medicine (DeVries & Subedi, 1998). Hired by hospitals and academic medical centers, bioethicists are called on for their philosophical and legal expertise to help with difficult decisions about medical treatments and end-of-life care. This medical consultant meets privately with the patient, family, and medical staff to collect facts and beliefs about the patient's illness and wishes and to discuss how each individual sees the situation (Guyer, 1998). Often this fact-gathering search ends in the "most justifiable decision," and the case is ended. However, it is not a perfect world, and in other situations the bioethicist gathers all parties together to talk. If this fails, the hospital ethics committee will conduct an official review and make formal recommendations. Bioethicists have no medical or legal authority, notes Ruth Guyer.

Guyer (1998) describes what a bioethicist is. She notes that a bioethicist is someone with a bioethics degree (23 universities in the United States offer master's degrees in bioethics and 13 offer Ph.D.'s in philosophy or religion with an emphasis on bioethics). The bioethicist is trained in philosophy, law, psychology, religion or humanities, and social science disciplines and learns to "talk the talk" of medicine when working in hospitals. A bioethicist can also be a doctor, nurse, or other clinician who has studied moral philosophy. George Dickinson recalls meeting an individual working in a hospital in Boston in the early 1980s who introduced himself as a "moral philosopher." After talking with him, it was obvious that he was what today is called a bioethicist.

Many individuals hope this new bioethical presence in health-care settings will make medicine more humane and more ethical. However, a sociological view suggests that bioethicists might begin with good motives, but that their position in the structure of health care may come to color their decisions. If we consider the interests of those who are represented by bioethicists, most bioethicists claim that they represent the patient, protecting her or his autonomy against the power of medicine. However, journalist Ruth Shalit (1997) comes to a very different conclusion. She notes that a swelling number of bioethicists who are cashing in on their ethical expertise and marketing their services to managed care executives are eager to dress up cost-cutting decisions in labels and lofty principles.

As R. DeVries and J. Subedi (1998) observe when looking at the organizational location of bioethics, the presence of bioethicists in medical institutions leads to an affinity between bioethicists and other professionals there. They liken the role of the bioethicist to a public defender in the American legal system. The formal role of each is to represent the interests of a client in a large and confusing bureaucracy, but, like public defenders, bioethicists must maintain good relationships with other members of that bureaucracy, many of whom are working against the interest of their clients. Given this organizational situation, bioethicists will be inclined to represent the interests of medical professionals and medical institutions over those who are merely passing through—the patients and their families—say DeVries and Subedi.

USE OF THE BODY IN MEDICAL RESEARCH AND TRAINING

The debate in the United States about the marketing of fetal tissue for medical research and drug production is an example of how culturally specific conceptions of the person influence where the boundary is drawn between alienable and inalienable body parts (Joralemon, 1995). Perhaps transplanting organs together with genetic engineering, artificial reproduction, therapeutic cell lines, and mechanical implants will be enough to overwhelm the intuitively unambiguous connection we feel to our bodies.

Embryonic stem cell research is providing an exciting potential treatment for a host of diseases, but the controversial research has left many conservative lawmakers hesitant about its use (Nelson, 2000). A 4-year ban on public funding of such research has effectively transferred most current efforts to the private sector. In June 2000, Congress was to debate a bill that would overturn the ban. The more conservative argument questions the federal government's sponsoring the "killing of human embryos." This approach concludes that all humans should be treated as persons and that research must not deliberately harm any humans, including embryos—at the moment of conception a new and unique individual comes into being and needs protection and nourishment to become an adult person (O'Mathuna, 1996). A more "liberal" argument, such as that of Joseph Fletcher, in his well known book *Situational Ethics* (1966), views every situation as ethically unique so that we can never say that a certain act is always right or always wrong. Fletcher argues that it is ethical to treat different humans in different ways.

Embryo stem cells are desirable because of their ability to be transformed into any cell type, which establishes a blank slate waiting to be filled in by the appropriate instructions. Theoretically, the cells could serve as a ready supply of replacement tissue for damaged or missing cells in patients with such diseases as diabetes, Huntington's chorea and Alzheimer's disease, or in those who have suffered spinal cord injuries; thus, they have a tremendous potential for the medical field. The National Institutes of Health researchers have discovered a method of generating virtually unlimited supplies of specialized neurons from mouse embryonic stem cells. The most promising source of these human stem cells is leftover embryos discarded from in vitro fertilization clinics.

The cells have also been derived from aborted fetuses, raising a number of objections, notes B. Nelson (2000). Fetal stem cells share many properties with their embryonic counterparts, though technically there are some differences. Opponents of using discarded embryos to harvest stem cells have based their arguments on their belief that life begins at the moment of conception. Thus, researchers who harvest cells from frozen 4- or 5-day-old embryos, therefore destroying them, are then committing the act of abortion.

The individuals who perform the tasks of removing body parts are called "technicians" and are employed by companies that retrieve body parts, also known as "harvesters," such as the Anatomic Gift Foundation in Illinois (O'Meara, 1999). Because of a federal law prohibiting the sale of human tissue or body parts, the traffickers have worked out an arrangement to expedite the process and remain "legal." The harvesters receive the fetal material as a "donation" from the abortion clinic. In return, the clinic is paid a "site fee" for rental of laboratory space where technicians perform the dissections. The harvesters then "donate" the body parts to the researchers and, rather than pay for them, the researchers "donate" the cost of the service with a formal price list. No laws are broken, since no body parts from aborted fetuses are sold. Normally, the fetus is dissected, and the body parts are shipped to their destination.

Nearly 75 percent of women choosing abortion agree to donate the fetal tissue (O'Meara, 1999). Anti-abortionists agree that if women think that something good is coming out of the abortion, it is easier for them to make the decision to abort. Thus, the ethical issue is difficult to separate from the scientific issue.

Human cadavers have been used for training medical students in the United States since the 19th century. Initially, such usage of "dead bodies" was illegal, but the necessity of using cadavers for gross anatomy laboratories in medical schools eventually won approval. While using cadavers for "learning purposes" may not be agreed to by all individuals, dead bodies are not supposed to be commodities. Donors usually will their bodies with the thought of advancing science.

Willed-body programs with surplus cadavers are not allowed to sell them for a profit, only at cost. However, with universities drawing up price lists—$350 for a full leg, $500 for a torso—and private companies vying for the specimens, the competition for cadavers is heating up (Kowalczyk & Heisel, 1999). The family giving the body to science trusted the medical school to act ethically and use the body for students to learn; yet donors sign contracts allowing medical schools to use their bodies for any legal purpose. Normally, medical schools have enough cadavers and sometimes have "leftovers." Computer software for learning gross anatomy is becoming more popular in medical schools; thus the day *may* come when medical schools need fewer human cadavers.

Cadavers are also used by the Department of Transportation to measure the effectiveness of various crash protective devices in automobile collisions (Schiedermayer, 1994). The use of cadavers in crash testing has been significantly curtailed in recent years, however, and they are now used primarily to calibrate the dummies. Only a few dozen cadavers are used annually for testing crash protection devices in car collisions.

Though not familiar to the public, doctors in training in emergency rooms sometimes perform an unnecessary invasive procedure, just for practice and without informed consent, on patients for whom resuscitation efforts appear to be failing, according to a recent article in *The New England Journal of Medicine* (Kaldjian et al., 1999). Though the practice does no physical harm and one third of the doctors in training say they approve of it, the procedures are indeed questionable to some. Physicians opposed to the practice argued that it was medically unnecessary and violated patient's rights or dignity, and the procedure could be learned in other settings. According to L. C. Kaldjian and colleagues, the practice reduces patients to "mere objects of use in education" and may reinforce unethical attitudes among interns and residents, especially since it is done without the patient's consent. Such "practice procedures" involve threading a tube into the femoral vein in the groin, a procedure best learned on a "live

---

WORDS OF WISDOM
Teaching Surgery Without a Patient

To avoid some of the ethical dilemmas of using "live patients" or "dead ones," surgical residents can perform "operations" on machines rather than patients in a growing number of teaching hospitals. The cost is more than a million dollars just in simulation equipment. The system is designed to train residents in laparoscopy or minimally invasive surgery performed through small incisions. The simulator makes the surgeon feel like she or he is actually cutting into tissue. They observe what they are doing on a television screen. Residents can also practice life-support techniques on a life-like simulated patient who has a chest that rises and falls with each "breath."

Some physicians are leery of these simulation techniques and are not convinced that a computerized system, no matter how realistic, can teach residents the skills they will need in a real operating room. Yet, such simulated procedures do not carry the anxiety of working against the clock in a real setting on a live patient. With our highly computerized society, who knows what the 21st century will bring to medicine in the operating room. Stay tuned!

From "Teaching Surgery Without a Patient" (pp. A49–A50), by K. S. Mangan (2000, February 25), The Chronicle of Higher Education, 45.

---

patient," since pulsating vessels are used as landmarks to guide the insertion of the tube. Though using the human body and its parts in medical research and training may be beneficial to society, such usage lacks the approval of all segments of society.

ORGAN TRANSPLANTATION

Attitudes toward organ transplantation illustrate different people's cultural constructions of the sacred, notes British sociologist Clive Seale (1998). In the United States, rationality about the body under the tutelage of medicine has gone very far, and the transplantation of body parts is accepted as the rational solution to otherwise certain death for some individuals. Likewise, the use of animal organs is acceptable, though in some countries, for example, Israel and Denmark, this is resisted.

 In Japan, there is great resistance to human organ transplantation, deriving from Japan's equivocal position as a society which on the one hand seeks to break free from religious and traditional forms, but at the same time is suspicious of the wholesale import of Western approaches to matters of life and death, observes Seale. In Japan, only one heart transplant had ever been performed up to 1995, and the doctor responsible was prosecuted for murder (Lock, 1995). The Japanese have religiously inspired concerns about the wrath of ancestors whose bodies have been violated. The Japanese resistance to organ transplantation can also be equated with earlier opposition to life insurance in the United States and European countries and involves similar issues about what is to be considered as sacred and thus beyond human calculation.

Some of the ethical issues involved in organ transplantations are entitlement criteria for acceptance to waiting lists such as age, ownership of cadaver organs, criteria for allocation of scarce organs, surgery on healthy people as live donors of organs, purchase of organs from the poor to serve the rich, and abortion for viable replacement tissue. If solid organs from different animals were available for transplantation purposes, called **xenotransplantation,** many of these ethical hazards could be diminished (Koshal, 1994). Besides the rejection problem with many animal tissues, the use of animals for transplant purposes has its ethical problems also. Do animals have rights? If so, are these rights overridden by the rights of humans? Should xenotransplantation, involving the intentional death of an animal, be considered before exhausting all the possibilities of human organ replacement? Are there ethical issues in using healthy animals for their organs and killing and consuming dead animals for food?

In the past 2 to 3 years there has been a significant surge of interest in xenotransplantation. Attention has been focused on the pig, which is already sacrificed in large numbers for food consumption. Pig heart valves have been used for the last 25 years. Current efforts are underway to breed pigs specifically for organ donation purposes. A recent survey in the United States suggests that 51 percent of those polled would be willing to receive an animal organ if needed (Koshal, 1994).

The initial donor policy in the United States was one of pure voluntarism—donation was made legal and it was hoped that volunteers would come forward (Weiss & Lonnquist, 2000). Courts ruled that competent adults could voluntarily donate organs to relatives, and minors could donate only with parental and judicial consent. Yet, the demand for organ donations soon exceeded the supply, and a more assertive organ donation policy of encouraged voluntarism evolved.

This has led to some serious ethical issues. With the current demand for human organs for transplantation purposes being much greater than the supply, a tendency exists in some parts of the world to sell certain organs. Though such selling is unlawful in the United States, a company in Germany routinely sends a form letter to all persons listed in the newspaper as having declared bankruptcy offering $45,000, plus expenses, for a kidney, which is then sold for $85,000 (Weiss & Lonnquist, 2000). Is it okay for one to sell body parts for money? Is it appropriate for a small number of people to benefit from public financing of an expensive technology when a larger number of people could benefit from expenditures on a greater diversity of less expensive problems? What criteria should be used to select organ transplant recipients, and are such criteria consistent with the values of a democratic society? What are the medical and quality-of-life outcomes associated with organ transplantations? There are probably more questions than answers on the ethics of organ donations, since numerous opinions exist regarding these questions.

There is the question of "ownership" as it applies to the corporeal part of the person, notes anthropologist Donald Joralemon (1995). Legal scholars have observed that the courts and public opinion have come a long way toward recognizing property rights and commercial value in some of what the body produces

SOURCE: DON WRIGHT/*THE MIAMI NEWS*

during life—e.g., blood, hair, semen, and ova. Exactly which body parts are considered to be integral to personhood varies by culture. For example, selling hair may be a legitimate economic venture in America, but in some societies (e.g., where voodoo is practiced) it would subject the individual to the risk of sorcery according to a more inclusive notion of bodily integrity, states Joralemon.

To take organs from a dead body, when is the person actually "dead" so that the organs can then be removed? As discussed in chapter 2, the definition of "dead" is not so easy to determine, yet the "brain dead" definition seems acceptable in many states in the United States and in many European countries today for organ donation purposes. Research (Callender, 1987) in minority communities on attitudes toward organ donation, however, reveals widespread suspicion that brain death might be declared prematurely to provide access to organs for transplant.

In 1992, the University of Pittsburgh Medical Center adopted a controversial plan that added a new category of potential donors—"non-heart-beating cadavers" (Joralemon, 1999). This category refers to persons whose lives will probably end when mechanical support is withdrawn, but will not meet the criterion of "brain death" as long as life support is maintained. If the patient or next of kin requests that life support be withdrawn and if there is a request that the organs be donated for transplantation, these procedures would be followed: (1) Take the person to an operating room where the life support is withdrawn; (2) administer medication to minimize any pain; (3) declare the person dead; and (4) procure organs immediately after death.

The National Organ Transplantation Act of 1984 authorized the creation of a national organ and transplantation network. This legislation led to the formation of the United Network for Organ Sharing (UNOS), a nonprofit organization and a

---

DEATH ACROSS CULTURES
A Ban on Selling Kidneys in Egypt

In an effort to end the sale of human organs by the poor, Egyptian physicians have banned all kidney transplants from living donors other than between relatives. The physicians are setting up an organ bank and have begun to transplant kidneys from cadavers, rather than live donors. Egypt, like India and other third-world countries, has a big business in the sale of organs, especially kidneys. Foreigners wanting to buy Egyptian kidneys must come to Egypt, since there are no provisions for the preservation of transplant organs. Egypt is somewhat unique in having large numbers of poor people willing to sell their organs.

The sale of organs evolved into an organized business in Egypt in 1987. Private laboratories act as brokerage houses and send out recruiters to the slums of Cairo to entice prospective donors in for tissue tests. Kidneys sell for $10,000 to $15,000. The ban by physicians will put more pressure on obtaining kidneys, since about 10,000 Egyptians suffer kidney failure annually but only 2,000 receive kidney dialysis treatment because of the high cost and the shortage of equipment. Setting up the organ bank will require a change in the Constitution, which stipulated that the dead must be buried immediately in accordance with the Islamic law. Thus, such a change involving ethical issues will not take place overnight.

From "Egypt's Doctors Impose Kidney Transplant Curbs," by C. Hedges (1992, January 23), New York Times International.

---

system of regulations governing the distribution of organs from donors. According to these regulations, developed by the U.S. Department of Health and Human Services, organs are distributed in local areas first, even if there are sicker patients elsewhere. Transplant surgeons and the UNOS object to government agencies setting the rules, saying allocation of organs is a medical decision that should be made by medical experts ("U.S. to Release Rules on Use of Donor Organs," 1999). The UNOS lobbied Congress to eliminate these rules on distribution. In response, Congress imposed a 1-year moratorium on adherence to the regulations. However, that moratorium expired in October of 1999, and the old regulations are back in effect.

In addressing the issue of donor shortages of organs, surgeon Robert Sade (1999) suggests that the most common reason for nondonation is denial of consent by the donor's family. Sade recommends that patients, before their deaths, or families, after the deaths, could select a category of recipients or a specific recipient for donated organs. This would be similar to the process involved in estate planning, where assets are distributed to designated groups or causes. By designating organ recipients, the donor could feel comfortable that the donation was consistent with her or his personal beliefs and values.

Going a step further, British ethicist John Harris (Jenkins, 1999) suggests that bodies should become public property on death, allowing surgeons to harvest usable organs without going through the traumatic process of asking grieving relatives for consent. Harris' idea fits the Western cultural construction of donation of blood and body parts as a gift of life. The idea is that individuals live within a community of faceless, statistically defined, risk-bearing individuals whose obligations to each other are based on an abstract sense of common humanity, rather than particular ties of blood or kinship, notes Clive Seale (1998).

Harris also recommends that a commercial market in live organs could be set up and that people should be able to trade their body parts to cut down on waiting lists. Perhaps husbands and wives who divorce in the near future will make organ donations a part of their divorce settlement (Rumbelow, 1999). The monetary value of a previously donated organ to a spouse would figure into the settlement.

Government officials argue that if organ transfers were left to the free market, evil results would ensue (Brown, 1996). The worst-case scenario is that people would begin murdering others and stealing their body parts, in order to profit from the sale of their organs. Though the idea of people buying and selling human organs in a free market is repulsive to many individuals, what right does the government have to interfere, suggests a conservative argument on this issue.

To encourage more organ donors, Pennsylvania plans to begin paying the relatives of organ donors $300 toward hospital and funeral expenses (Krauthammer, 1999). This would be the first effort in the United States to reward organ donation. Such a "reward" might be in violation of a 1984 federal law that declares organs a national resource not subject to compensation. An argument against this plan is that the reward money will be more appealing to the poor than the rich; thus a disproportionate number of the poor will succumb to the incentive and provide organs.

Regarding social class variations in the receiving of organs, the whole idea of considering a future where the rich have the ability to secure organs from the poor is indeed problematic. Paying $15,000 may be considered a reasonable price to a wealthy individual and would be almost irresistible to a potential donor whose annual salary equals that amount. Thus, the whole question of who would receive an organ poses ethical questions. Is it a mere coincidence that the former New York Yankees baseball great Mickey Mantle or "I Dream of Jeannie" and "Dallas" television star Larry Hagman received liver transplants "almost immediately" when needed, whereas individuals who are not celebrities died while on waiting lists for this same organ? In our very political world who one is can indeed often make a difference. Money and influence can definitely be a factor in the distribution of organs today. The question of ethics remains, and it is one that will have to be addressed in the years to come.

EUTHANASIA

The meaning of **euthanasia** today is not the same as the original Greek meaning of "a good death" or "a gentle death," because the current meaning is the administration of death to the dying, a hastening or advancing of death. Today, patients and their families are demanding increasing control over their health care and finalization of life. A desire for more personal control reflects the trend in Western society. People in Australia, Canada, the United States, the United Kingdom, and other parts of Europe and the world fear that they will not be able to die in this gentle easy way, not "in control." Indeed, recent public opinion polls suggest that euthanasia is favored: 67 percent in the United States, 76 percent in Australia, 77 percent in Canada, and 82 percent in the United Kingdom favor some form of euthanasia (Clark et al., 2000). Yet, the rapid development of technology has threatened individuals' sense of control (Kelner & Bourgeault, 1993). Whereas new technologies have made it possible to sustain life longer, patients often have limited say in initiating these complex treatments. If an individual, for example, is in very bad health, has a poor medical prognosis, and desires to die, should the medical profession have the option of assisting that individual with death?

Distinguishing between killing and letting a person die seems easy, at least on the surface, notes Margaret Battin (1994). *Killing* involves intervening in an ongoing physiological process that would otherwise have been adequate to support life, whereas *letting die* involves not intervening to aid physiological processes that have become inadequate to support life. Yet, there are ambiguous cases. For example, removing a respirator may seem to be letting the patient die or it might be viewed as killing the patient. Despite gray areas, to grant that there is "a difference" between killing and letting die that is adequate to support assertions against killing is not to grant that there is or must be a moral difference.

A now rather famous article by James Rachels (1975) in *The New England Journal of Medicine* challenged the conventional conception of the moral difference between killing and letting die with this case: Smith and Jones will each inherit a considerable fortune from their respective 6-year-old cousins, should the cousins die. One evening, while his cousin is taking a bath, Smith sneaks into the bathroom and drowns him. Meanwhile, Jones is also planning to drown his own cousin, who is also taking a bath, but as Jones sneaks into the bathroom the child hits his head and slips under the water, and Jones does nothing to save him. Now both children are dead. Smith has killed his cousin; Jones has merely allowed his cousin to die. Clearly a conceptual and causal difference in what the two men did exists, yet is there a moral difference? Not really, since they both acted despicably, and it is no excuse for Jones to say that he did not kill his cousin, he "merely" let his cousin die.

Richard Trammell (1975), however, argues that there is a difference between Jones' killing and Smith's letting die. He says that the difference is like trying to taste the difference between two fine wines when they are mixed with green persimmon juice. There is a difference between the wines, but in those overpowering circumstances one cannot see what it is. Just so with the drowned cousins because the behavior of both men is so repugnant it has a "masking" effect and we cannot see the difference in what they did.

Battin (1994) says that in some cases killing is right—the coup de grace granted a mortally wounded, dying soldier who cannot be saved on the battlefield or abortion to save the life of the mother. The **double effect** is another example when "killing is right." It must be recognized that the distinction between killing and letting die does not succeed in carrying the moral weight often placed on it. Even if one could always draw clear conceptual lines between the two, the conceptual distinction almost always brings along with it an unjustified moral distinction. To describe a procedure as "mercy killing" could infer that because it involves killing it is wrong, thus taking the moral baggage along with the conceptual distinction, though the inference does not follow. Each case needs to be argued on its own merits, suggests Battin.

Sanctity-of-Life Versus Quality-of-Life Debate

Occasionally, a newspaper article will say that the survivor of a serious automobile accident is in a "vegetative state." Has the patient died? Is the patient no longer human? Now consider the situation in which an individual is assaulted and put into an irreversible coma. Has the assailant committed murder or assault and battery? If this same patient is disconnected from life support equipment or **intubation** (tube feeding) and dies, who caused the death—the person who committed the crime or the medical personnel who withdrew the life support or feeding device? In general, people respond to such questions about medical conditions from one or two orientations concerning the meaning of life. The first orientation emphasizes the **sanctity of life,** whereas the second emphasizes a **quality of life.**

Sanctity-of-Life View

Euthanasia from a sanctity-of-life orientation would contend that all "natural" life has intrinsic meaning and should be appreciated as a divine gift. As a consequence, human beings have the obligation to prolong life. Hessel Bouma and colleagues (1989) note that the Hippocratic tradition in the medical profession recognizes that physicians' responsibility to terminally ill patients is to simply "mitigate their suffering while allowing them to die." Furthermore, the Hippocratic Oath, reacting against an earlier practice of actively and intentionally hastening the deaths of terminally ill patients, forbids the giving of "a deadly drug" to dying patients. According to the Hippocratic Oath, part of the essential qualification for being a doctor is an absolute rejection of killing the patient (Jeffrey, 1993). The following two quotations illustrate the sanctity-of-life orientation:

> The quality of all our lives suffers, it insists, unless every human life is considered inviolable because of the very fact of its existence. A dying patient's relationship to those about him or her symbolizes the relationship of all individuals to one another. To practice direct euthanasia, even at the request of the patient, is to weaken the claim of each one of us to the right to have others respect and not violate us. (Weber, 1981, p. 49)
>
> No horror against life is impossible once we have allowed anyone but the Creator to usurp sovereignty over life. Whom the gods would destroy they first make mad. Legalized euthanasia is such madness. (Morriss, 1987, p. 149)

Recently Mebane et al. (1999) examined racial differences in patient preferences for end-of-life treatments and physician-assisted deaths and report that black patients tend to request more life-sustaining treatments, view physician-assisted deaths more negatively, and place a higher value on longevity than white patients. Likewise, physicians share similar preferences for end-of-life treatments with their patients: Black physicians are more likely than white physicians to request aggressive medical treatments for themselves in the persistent vegetative stage and organic brain disease situations. The African Americans in the Mebane et al. study tend to be taking a sanctity-of-life perspective. They prefer living "at all costs."

Religion *may* be a factor in the sanctity-of-life view. R. Kalish and D. Reynolds (1976) concluded that African Americans are reported to be more religious than whites; thus black patients and physicians might feel guilty if treatment were stopped, based on religious reasons. Mebane and colleagues (1999) argue that longevity is valued as an intrinsic good, more for blacks than whites, based partly on the history of oppression experienced by older blacks and the idea that suffering may be an expected part of life rather than a reason to terminate life.

Quality-of-Life View

From a quality-of-life perspective, the concept of "quality" is even more difficult to define. Who should decide when life has quality? *Quality* is a relative term, and its meaning often changes as one progresses through her or his life cycle and as a result of social circumstance. Would not your quality of life be diminished if you were paralyzed, blind, and/or deaf? Yet, many people with these disabilities experience a life of quality filled with purpose and meaning.

The quality of life orientation holds that when life no longer has quality or meaning, death is preferable to life. What may be quality to one may not be to another. Some individuals might prefer death over life, for example, if they were in a permanently paralyzed state, yet others may find quality in such a situation and consider themselves to have a "quality of life." Others might consider the inability to walk to have no "quality" and would choose death to life. To accept any frailty in life might be more than some individuals can bear, and they might consider such a life unbearable with no "quality" and would therefore prefer death.

The research on perceptions by race of end-of-life issues reported by Mebane and colleagues (1999) suggests that whites in their study follow more the "quality of life" view. They found that white physicians are more likely to want physician-assisted death for themselves in the persistent vegetative state and organic brain disease scenarios. Black doctors are less likely to view tube feeding as "heroic" and less likely to report positive attitudes toward advanced care planning than white physicians (advanced care planning would limit treatments).

The hospice movement takes a quality-of-life view in that their goal is for the patient to "live with dignity," even while dying. Hospice philosophy suggests that the individual have a "good death," free of pain and a sense of control of her or his life.

Though advanced medical technology aids in prolonging life, the quality of life may be limited; thus modern technology is a curse in the eyes of some quality-of-life supporters. Individuals supporting the sanctity-of-life view should see advanced medical technology as a positive development in that life can be extended, no matter the quality. However, Mebane and colleagues (1999) found that African Americans, who are sanctity-of-life advocates according to their findings, are more likely to have

---

LISTENING TO THE VOICES
Not a Good Quality of Life

Any reasonable definition of life must include the idea of its being self-supporting. I will spend the rest of my life in a hospital, but while I am there, everything is geared just to keeping my brain active, with no possibility of it ever being able to direct anything. As far as I can see, that is an act of deliberate cruelty.

The best part of my life was, I suppose, my work. The most valuable asset that I had for that was my imagination. It's just a shame that my mind wasn't paralyzed along with my body. Because my imagination—which was my most precious possession—has become my enemy. And it tortures me with thoughts of what might have been and of what might be to come. I can feel my mind slowly breaking up.

I am filled with absolute outrage that you, who have no connection with me whatever, have the right to condemn me to a life of torment because you can't see the pain. There's no blood and there's no screaming so you can't see it. But if you saw a mutilated animal on the side of the road, you'd shoot it. Well, I'm only asking that you show me the same mercy you'd show an animal. But I'm not asking you to commit an act of violence, just take me somewhere and leave me. And if you don't, then, you come back here in 5 years and see what a piece of work you did today.

Statements by Claire Harrison, the central character in Whose Life Is It Anyway? *(pp.73–74), a play by Brian Clark (1981).*

---

negative perceptions about hospital care and have some anxiety about life-support equipment being stopped prematurely *because of* their race. Thus, some aspects of life-support technology are *not* viewed positively among the African Americans in the study. Looking at the overall impact of technology on death, Ivan Illich (1976, pp. 207–208) illustrates the dilemma:

> Today, the man best protected against setting the stage for his own dying is the sick person in critical condition. Society, acting through the medical system, decides when and after what indignities and mutilations he shall die. The medicalization of society has brought the epoch of natural death to an end. Western man has lost the right to preside at his act of dying. Health, or the autonomous power to cope, has been expropriated down to the last breath. Technical death has won its victory over dying. Mechanical death has conquered and destroyed all other deaths.

The Right to Die

If we conceive of the right to die as being like most other rights (e.g., the right to property, the right to an education, the right to travel, the right to go about our daily business in peace and so on), then we should, as in all these other cases, have a right to die and be able to call upon the resources of the community to help us to fulfill that right, notes Scottish psychologist John Beloff (1989).

Although the overwhelming majority of those who are terminally ill fight for life to the end (Hendin, 1997), some individuals, whether terminally ill or otherwise, may wish to die and feel that it is their right to die. H. M. Chochinov, K. G. Wilson, & M. Ennis (1995) reported that 44 percent of 200 terminally ill patients reported occasional wishes that death would come soon. Indeed, does an individual have a "right" to die? The media have highlighted the actions of Jack Kevorkian, the retired pathologist, as he has assisted individuals who wished to die. In fact, Kevorkian was a champion of the right to die issue throughout the 1990s and is currently serving time in prison for his actions. In 1996 the Ninth Circuit Court of Appeals in the state of Washington ruled that, based on the Fourteenth Amendment, an individual has a right to determine when and how she or he will die, as this court determined that it was unconstitutional for the state to prohibit physicians from prescribing life-ending medication for use by terminally ill patients.

The whole question of the right to die was in the news in the United States and throughout the world during the 1990s and the issue is continuing into the 21st century. The difficult decision to forgo (withhold or withdraw) life-sustaining treatment, however, has been lacking in the medical education of physicians (Faber-Langendoen & Bartels, 1992). A major emphasis in medical education is placed on the diagnosis and treatment of diseases, yet traditionally there has been

Jack Kevorkian has advocated for severely ill people's right to commit suicide and claims to have assisted over 100 individuals with their deaths. He is currently in prison for practicing AVE.

limited guidance given for physicians regarding which treatments to discontinue, once a decision to forgo treatment has been decided. Medical education should teach students and residents to grapple with the complexity of forgoing treatment, note K. Faber-Langendoen and D. M. Bartels (1992).

Recent federal legislation now allows one to rightfully refuse medical treatment. **The Patient Self-Determination Act (PSDA)** is a federal law affecting all health-care agencies that receive any federal funding (Timmermans, 1999). The federal government now mandates that health-care providers inform patients about their right to refuse and accept treatment, even when they have lost decision-making capacity. The goal of the PSDA is to motivate people to make their wishes about treatment known in advance, but not to force them to do so. The PSDA has been criticized because the timing of informing them—as they enter the hospital for treatment—is bad. Thus, the right to die is being wrestled about in the courts, in legislative bodies, in ballot boxes, and in the media. This is a topic that probably will not go away anytime soon.

As will be discussed in chapter 12, a **living will** provides a vehicle by which individuals make their intentions known concerning the withholding of medical treatment. In allowing a patient to die, palliative care becomes of utmost importance. Medical personnel must try to keep the patient comfortable and attempt to respond to psychological, social, and spiritual needs. Palliative care personnel should frame chronic pain in strictly behavioral terms by developing an appreciation for the cognitive and evaluative aspects of the experience, suggests Joseph Kotarba (1983) in researching chronic pain management. The physician should inquire into the patient's cognitive resources for normalizing suffering. If the patient is a member of a religious group, for example, the usefulness of that membership should be stressed. If such membership is lacking, the individual should be encouraged to seek a compatible source of meaning, says Kotarba.

Many palliative care professionals cite weakness and fatigue as some of the most disabling symptoms of terminal illness because there is little that can be done to alleviate this state of powerlessness, observes anthropologist Beverly McNamara (1997). These manifestations of terminal illness are just a few of many symptoms that will influence the dying person's conception of self. Nausea, vomiting, breathlessness, constipation, diarrhea, edema (swelling), smell from infections and wounds, confusion, and pain are also associated variously with disease and with intervention procedures. Knowledge of approaching death further complicates self knowledge and the capacity to act to change circumstances. Some individuals view euthanasia as a dignified death, whereas others believe it is unlawful killing. Does dying "suddenly" mean that people would rather not have control over the circumstances of their death? asks McNamara.

The terminally ill person's access to power should be seen in contrast to that of the health professionals who care for them. A doctor, for example, who is dying of cancer can rather easily negotiate her or his own medical management with the aid of colleagues, but an Australian Aboriginal person is unlikely to benefit from very basic forms of support, notes McNamara. The degree to which terminally ill persons engage in decisions about their dying and forthcoming death will be mediated by their access to power and to the place and time of their death. In the last days of life individuals will be better able to maintain some degree of authority if resources are channeled into facilities which support their needs and wishes.

Many dying people fear pain, but also feel dissatisfied with the debilitating side effects of pain-controlling medications such as morphine. A dilemma is posed for the health professional: How can terminally ill persons articulate their needs and desires if their cognitive functions are altered by medications, yet how can they focus on autonomous decision making if they are incapacitated by pain? asks McNamara. A cautionary note about the reliance upon technology's capacity to provide answers to suffering is that overuse of medical technology will not only anesthetize people to pain, it may also anesthetize them to the act of dying.

If an individual wishes to die "naturally" and not be kept alive by extraordinary means, then *allowing* that person to die constitutes passive euthanasia. As noted earlier, going to the hospital today is not a "place where you go to die" but rather is a "place where you go and they will *not let you die*." Passive euthanasia is *letting* the patient die. On the other hand, if the patient wishes to be *helped* to die through physician-assisted death or active voluntary euthanasia, this is active euthanasia and is illegal in the United States, except for Oregon, where physician-assisted death is legal. Let's look more closely at these two types of euthanasia.

The Trent Palliative Care Center in Sheffield, England, attached to Saint Luke's Hospice, conducts extensive research in end-of-life issues and has an active educational program for health care professionals.

Passive Euthanasia

The action of allowing a patient to die is often referred to as **passive euthanasia**. Passive euthanasia may involve withholding medical treatment or removing life-support equipment. Such action is supported by people who affirm both sanctity-of-life and quality of life perspectives. Such action is indeed legal in the United States and other countries around the world. The patient is "being allowed" to die naturally. She or he was only being kept alive because of the life-support equipment and/or the medication. Though action was involved to remove the support, the patient still died on her or his own accord.

Allowing a patient to die has not historically been easy. There have been several situations in which court action was needed to "allow the individual to die." In the 1970s such a well-publicized case was that of Karen Ann Quinlan, a college student who mixed alcohol with other drugs and went into a comatose state. After several months of a negative medical prognosis, the courts ruled that she could be "allowed to die" by removal of life-sustaining equipment.

Distinguishing Between Ordinary and Extraordinary Measures

If one accepts the legitimacy of passive euthanasia, as set forth in the living will, it is clear that the use of extraordinary measures is not required of attending physicians for terminally ill patients and members of their family. However, differentiating between ordinary and extraordinary measures is not easy. **Ordinary measures** of preserving life are all those medicines, treatments, and operations that offer a reasonable hope of benefit for the patient and that can be obtained and used without excessive expense, pain, and other inconvenience (Bouma et al., 1989). **Extraordinary measures** of preserving life are all those medicines, treatments, and operations that cannot be obtained without excessive expense, pain, or other inconvenience or that, if used, would not offer a reasonable hope of benefit.

These measures introduce other ambiguous terms—"reasonable hope," "excessive expense," "excessive pain," and "excessive inconvenience." What is reasonable and excessive? Who among us can make decisions using these highly subjective concepts? Which of the following groups should society allow to determine these matters: politicians, physicians, lawyers, judges, members of the community at large, and/or patients and their families?

Many within the medical community have operationalized ordinary measures to be any procedures that are "usual and customary" medical practice and extraordinary measures to be any procedures that are exotic, innovative, and medically experimental. However, it is clear that this distinction provides us with little help because technology has a way of imposing itself upon modern society in an amoral fashion and what has been considered innovative and experimental becomes usual and customary with time. Consider the situation in which an infant with Down's syndrome starves to death because parents will not permit a physician to perform a simple operation that would allow food to be digested by the child. The fate of this infant is a classic example of what sociologists call **cultural lag** (not all aspects of a culture change at the same pace; one aspect "lags behind" another)—a condition created by rapid medical technological development and relatively static definitions of biological life and death.

Medical practice today can prolong life (or postpone death) without having precise definitions of a point at which meaningful living has stopped and death has occurred. This condition would be of little significance if medical science were unable to maintain some biological functioning of the individual by artificial means, but such is not the case. Compared to the horrors of senility and extended terminal treatment because of modern medicine's ability to prolong life, the terminally ill patient may indeed welcome death.

Furthermore, even if one were able to distinguish between ordinary and extraordinary measures, would it be acceptable for patients (or someone acting on their behalf) to request the withdrawal of ordinary life support such as food and water? This is an issue that tends to divide persons who affirm a sanctity-of-life perspective and favor passive euthanasia.

Nutrition and Hydration

Withholding and withdrawing artificial nutrition and hydration is a perplexing and emotional issue in end-of-life care, observe Z. Huang and J. C. Ahronheim (2000) in their update on these issues. Since the Karen Ann Quinlan case in 1976, as mentioned earlier, in which the court authorized termination of life support for a patient in a persistent vegetative stage, courts have realized an important distinction between killing and letting die. This has enabled the patient's wishes to be upheld without the patient or care provider being held criminally liable.

Commentators have questioned the practice of giving intravenous fluids to terminally ill patients and have explored whether dehydration causes distressing symptoms in dying persons (Huang & Ahronheim, 2000). Yet, studies on artificial nutrition and hydration at the end of life suggest that tube feeding is the rule rather than the exception (Hoefler, 2000). Many elderly and infirm patients spend their last days in long-term and tertiary-care facilities, and many of these institutions find tube feeding to be more cost efficient than hand feeding patients who have trouble feeding themselves. In addition, reimbursement rates are higher for tube-fed patients (Hoefler, 2000). Caregiver anxiety about withdrawing food and fluids also is a factor, at least partly because of the important role that food plays in our culture. Physicians also get caught up in the cross-cutting currents about whether to feed hopelessly ill and demented patients in their care or not, especially when family members are clamoring for the physicians to "Do something!" In addition, physicians' concern with legal liability may cause anxiety. Many physicians also seem to be confused about medical-ethical issues associated with hydration and artificial nutrition—many agree that tube feeding can be withdrawn from patients in a vegetative state at the request of a surrogate, but are reluctant to do so, thinking that they would either be violating professional norms or their action would lead to increased patient discomfort (Hoefler, 2000).

Nutrition and hydration problems call for expertise in dealing with terminally ill patients. From an ethical perspective, at what point is the patient not given food and water? Nutritional deficiencies usually result from nausea, mouth soreness, or anorexia (Gentile & Fello, 1990). However, a person can live quite a while without food. Hydration is more critical. A human being cannot live for any length of time without fluids. Intravenous (IV) therapy questions are often raised by families. If a patient has been ambulatory but is weakening and is dehydrated, 1 or 2 liters of fluid can enhance the quality of life temporarily, if that is what the family and patient desire. If, however, the patient is in an extremely debilitated state, is bedridden, and is barely conscious, starting IV fluid replacement could be considered invasive and will probably not add to the quality of life, note M. Gentile and M. Fello.

When food and fluids are withheld or withdrawn, patients die of dehydration, not starvation (Hoefler, 2000). Death by starvation is a long, arduous process involving major weight loss and body wasting. Bodily functions begin to shut down compromising the production of white blood cells, thus weakening the immune system. Toward the end of the starvation process, the intestines begin to fail leading to uncontrollable diarrhea. Eventually, the heart muscle gives out, and the person dies of cardiac arrest. The process of starvation can drag out for weeks.

Unlike starvation, death via dehydration is relatively quick and painless—dehydration tends to be the direct cause of all "natural" deaths (not associated with violent trauma or acute infection). According to Gentile and Fello (1990), slow dehydration caused by ingestion of less and less fluids, and finally no fluid, does not usually create much discomfort. Research on dehydration suggests that patients who stop taking in food and fluids slowly sink into a state of unconsciousness over the course of several days, and die peacefully shortly after that (Hoefler, 2000). Usual symptoms of the slow dehydration process are slight elevation of body temperature and drying of the mucous membranes such as the nose and mouth. Rapid dehydration, however, which is usually caused by prolonged vomiting, severe diarrhea, or mechanical evacuation of stomach contents, can be significantly more uncomfortable. Without IV fluid replacement, a person will weaken and die quickly. Some patients choose to forego IV therapy knowing full well the eventual outcome, observe Gentile and Fello.

To let a person die from dehydration or lack of nutrition is not agreeable to all groups. For example, some Orthodox interpretations of Jewish law conclude that death cannot be hastened through withdrawing or withholding artificial nutrition and hydration or other treatments that may prolong life (Rhymes, 1996). Some patients and family members may have religious concerns based on a misunderstanding of the requirements of their religion.

In the end, however, according to James Hoefler's thorough study of the topic (2000), there is little doubt that the prevailing sentiment among those who have had experience with death and dehydration is that the benefits of dehydration outweigh the burdens for certain patients, namely, those who are terminally ill, irreversibly unconscious, and severely demented.

CPR Versus DNR

CPR (cardiopulmonary resuscitation) follows a sanctity-of-life view in that the idea is to prolong life—or restore the person to life. On the other hand, **DNR** (do not resuscitate) would follow a quality of life view for an individual in a terminal condition.

In the early 1960s, a leading group of resuscitation researchers agreed that CPR—the combination of securing an open airway, mouth-to mouth ventilation, and chest compressions—was the most promising resuscitation technique. CPR addressed the often fatal problem of an obstructed airway (Timmermans, 1999). Before CPR was introduced in 1960, health-care providers alternated between resuscitating aggressively and preparing for impending death on the basis of the patient's social viability (e.g., age and "moral character"). In contrast, legislators and biomedical researchers have now designed an emergency system in which care providers are required to save lives whenever medically indicated (Timmermans, 1999). Now, only people who choose not to be resuscitated should be exempt from a reviving attempt. No longer should anyone systematically withhold potentially beneficial care from groups with a low social value. To do so would mean passive euthanasia for some and aggressive but futile lifesaving efforts with needless suffering for others, whose presumed social value is higher, notes S. Timmermans (1999).

Timmermans, in his book on the myth of CPR, asks whether resuscitation protocols and legal protections, instituted since the 1960s, have in fact eliminated the social inequality of emergency room care. As a sociologist using participant observation of CPR, Timmermans notes that the initial impression of the emergency room staff is based on the circumstances of the CPR, the social characteristics of the patient, the biomedical indicators, and the treatment. The team members then ask if the patient has a reasonable chance of survival and how long they will work on the patient. When the patient arrives in the emergency room (ER), the physician may declare her or him **"DOA"** (dead on arrival). Legal changes since the 1960s have mandated that when someone calls 911, a resuscitative effort begin and be virtually unstoppable until a physician sees the patient in the ER. Unless the patient shows obvious signs of death, the **EMTs** (emergency medical technicians) or paramedics need to begin CPR as prescribed by their protocol. Upon arrival at the ER, the physician cannot legally stop the lifesaving attempt until all protocols are exhausted. Stopping sooner would qualify as negligence and grounds for malpractice. Timmermans observes, however, that the ER staff do not always aggressively revive all clinically viable patients, nor do they give up on all biologically dead patients. Thus, there is a lack of consistency, depending on the circumstances.

We are in danger of abusing the procedure of CPR today (Rahman, 1989). At some point, we must stop adding to the suffering of dying patients by pounding on their chests and possibly breaking bones, just to extend their lives by a few days or weeks. With the routine use of CPR today and the estimated survival rate of 1 to 3 percent of victims who have cardiac arrests out of hospitals, the efforts really appear to be rather futile (Cummins, Oranto, & Thies, 1991). Why batter the sternum of a drastically debilitated, 85-year-old woman who is suffering from pneumonia or try to revive with CPR a patient who has been sick for years with diabetes, hypertension, heart disease, or kidney failure and who has just had an operation for a perforated intestine? The idea of beating on the chest of elderly sick persons such as these is morally and physically repugnant, notes Rahman. Yet, the procedure of CPR continues to be used indiscriminately. When the question of using CPR, in the event it is needed, is not raised beforehand, under the duress of the moment, it is difficult for families to make a quick decision; thus the physician simply follows the routine of performing CPR.

Whereas CPR is a method for sustaining life, DNR (do not resuscitate) is a formal order, written on the patient's chart, which indicates that, if the patient's heart stops, no attempt will be made to resuscitate the patient. Approved nationwide guidelines explicitly allow doctors to write DNR orders. Officially approved procedures can be adhered to, and all parties involved are legally protected (Chambliss, 1996). Such orders are quite common, and on any given day in a large hospital a number of patients will have "DNR" or "No Code" orders on their charts. Approximately 70 percent of patients who die in the hospital have a DNR order in place (Council on Ethical and Judicial Affairs, 1991).

Officially, physicians are responsible for signing DNR orders. A DNR order states that a particular patient whose breathing and circulation fails should not be revived. By making such a call, and then following it through, the medical personnel are "allowing the patient to die." Once the DNR order is signed by the doctor, the nurse is organizationally responsible for monitoring the death. He or she must sit and watch the patient turn blue and listen to the final gurgling sound of the patient "drowning in his or her own juices" (Chambliss, 1996).

Physicians, following the Hippocratic Oath to "prolong life," sometimes have difficulty with DNR orders. Such resistance is especially true for young doctors, interns, and residents, observes D. F. Chambliss. In general, it seems that everyone involved with a DNR order has some desire to sidestep responsibility for this awesome decision. Families may prefer to have the doctors decide, because they are the "experts," yet staff will argue that families should make the decisions themselves. A good example of this difficulty with DNR orders is found in journalist T. M. Shine's *Fathers Aren't Supposed to Die: Five Brothers Reunite to Say Good-Bye* (2000). Shine and his four brothers are having problems with the fact that their dad is dying and are debating the issue of the DNR decision. One of Shine's brothers suggests that, instead of signing a statement "Do not restart this man's heart under any circumstances," the statement could read "At least try once." Indeed, both family and medical staff are reluctant about the DNR order.

CPR, on the other hand, fits the medical model in U.S. hospitals to "prolong life" and "make the patient well." Yet, CPR can be controversial if the patient preferred death, and death is denied to her or him. Outside of a hospital setting and in an emergency situation, however, EMT personnel will routinely apply CPR, if the individual's condition requires such.

Active Euthanasia

The previous section discussed the issue of allowing the patient to die. Such a procedure involves passive euthanasia and is indeed legal in the United States and most

---

PRACTICAL MATTERS
Ways to Avoid Excessive CPR

Suggestions to avoid excessive use of CPR include the following:

1. The public and professionals must be educated to honor a peaceful death, if so desired.

2. Patients should sign living wills.

3. Patients with incurable diseases should be tactfully approached by their families and physicians about the use of aggressive measures.

4. If a patient is opposed to resuscitation, the primary physician must write clearly on a patient's chart, "DNR."

Adapted from "Routine CPR Can Abuse the Old and Sick" (p. 9A), by F. Rahman (1989, February 27), Minneapolis Star and Tribune.

---

countries of the world. We now turn our attention to **active euthanasia** whereby the individual is helped to die. Such "help" can come through a physician giving advice about how the patient can take her or his life and/or giving a prescription for a lethal dosage of a medication to end the person's life. This action is typically referred to as **physician-assisted death (PAD,** sometimes called physician-assisted suicide or PAS, because in essence the individual takes her or his own life). A more aggressive form of "helping" an individual to die is called **active voluntary euthanasia (AVE)** whereby, upon request by the patient, the physician ends the patient's life through direct action (e.g., injecting a lethal dose of a drug into the patient).

There is an obvious analytical distinction between active and passive euthanasia. However, after one accepts the legitimacy of letting people die by terminating life-sustaining treatment, it is an easy step to accept the legitimacy of assisting terminally ill patients to commit suicide or to perform active euthanasia. A telephone survey (Huber, Meade-Cos, & Edelen., 1992) of 200 Midwestern respondents determined that 90 percent favored some kind of personal control over death circumstances (e.g., passive and active euthanasia).

Clive Seale (1997) summarizes the current causes for the high support today of AVE and PAD: (1) As medical technology has increased, so have fears of inappropriate life-sustaining treatment evolved; (2) the rise in support is an aspect of the decline in religious faith and the emergence of rationalistic approaches to matters of life and death; (3) as social bonds weaken, people opt for more individualistic solutions to problems; and (4) the population has aged, resulting in more lingering suffering than in the past. Seale notes that increasing secularization and rising levels of education are likely to lead to even more increased levels of public support in the future.

Organizations Supporting Active Euthanasia

In the United States, the **Hemlock Society,** founded in 1980, assists terminally ill individuals in the act of **self-deliverance.** In 1981 the Hemlock Society published a

DEATH ACROSS CULTURES
Euthanasia in the Netherlands

In November 2000, the lower chamber of Parliament in the Netherlands voted 104 to 40 to allow doctors to help patients die, under certain restrictions. The bill became law in April 2001. Before this vote, AVE was not legal, but it was "decriminalized" in the Netherlands, as long as the physician followed certain guidelines. Since this meant that an act of PAD or AVE would not be prosecuted, the country was identified as one in which the patient had the "opportunity" to have a physician help end a life. The new law sets out strict guidelines, not significantly different from the "decriminalized" ones, which are as follows:

1. The request must be voluntary and well considered.

2. The patient must be facing unremitting and unbearable suffering, but does not have to be terminally ill.

3. The patient must have a clear and correct understanding of the situation and prognosis.

4. The physician and patient must reach the conclusion that there is no reasonable alternative acceptable to the patient. The decision to die must be the patient's own.

5. The physician involved must consult at least one other independent doctor who has examined the patient.

6. The physician must end the patient's life in a medically appropriate manner.

7. The physician must report the cause of death as euthanasia or suicide.

8. The patient must be an adult. Young people from 12 to 16 must have parental consent. Children under 12 are not eligible under the present guidelines.

Following are some of the facets of Dutch society that have influenced attitudes toward end-of-life care. Ninety-nine percent of the population has medical coverage, and all have coverage for catastrophic illness. Malpractice suits are almost unheard of in the Netherlands because of high trust of and esteem for physicians. Forty percent of patients in the Netherlands die at home, compared to 15 percent in the United States. Dutch churches have less political power than those in the United States. The Dutch often act together for the collective good. Most all Dutch citizens have a general practitioner, many of whom make house calls.

Studies show that about half of the physicians in the Netherlands have performed euthanasia. Nonetheless, they seem to have approached it circumspectly, almost reluctantly. Out of an estimated 25,000 requests for information, 9,000 patients actually make a formal application to their physician, and 3,000 have it carried out. Many patients are reassured to find their doctors receptive to ending their suffering, but in the end decide to die a natural death. The fact is that despite the social acceptance of euthanasia, these deaths only comprise 2 percent of the 135,000 annual deaths in Holland.

(continued)

(continued from previous page)

Part of the reason for this relatively small percentage of physician-assisted and AVE deaths may be a reluctance to file reports of the cases. Between 1990 and 1995, more than half of the occurrences of AVE and PAD went unreported. Reasons given for not reporting by the doctors were "too much paperwork," objections to the "criminal" overtones of the investigation that follow the reporting, and reluctance to put the grieving family "through the additional burden of interrogation" that is a part of the investigation. Thus, the reporting structure for PAD and AVE was changed in November of 1998: a three-person panel composed of a physician, a lawyer, and an ethicist, reviewed all cases and made recommendations regarding investigation. Under the 2000 legislation, physicians will continue to be accountable to a panel of peers, including legal, medical, and ethical experts, no change from 1998. The hope has been that this change will result in greater compliance with reporting.

The Netherlands have also been criticized for their presumed lack of palliative care and the existence of only very few hospices in the country. From a sociological point of view, one may be tempted to interpret the shift toward autonomy-based requests for euthanasia as a byproduct of a liberal society, with its emphasis on self-government, control, and rational choice. Others will say that more emphasis on patient autonomy fits perfectly into the process of emancipation of the patient than has been going on since the beginning of the 1970s.

J. B. Vander Veer, after completing his study of euthanasia in the Netherlands, concludes that the Dutch system would not work in the United States because the societies are too different. The United States lacks universal health insurance and coverage for catastrophic illness, and the "slope seems too slippery" in the United States. With euthanasia now being legal in the Netherlands, not "just *decriminalized,*" it will be of interest to observe if the rate of physician involvement with patient deaths changes.

Adapted from "Euthanasia in the Netherlands" (pp. 532–537), by J. B. Vander Veer, (1999, May), Journal of the American College of Surgeons, *188(5); "Slipperly Slopes in Flat Countries— A Response" (pp. 22–24), by J. J. M. van Dalden (1999, February)* Journal of Medical Ethics, *25; and "Dutch Becoming First Nation to Legalize Assisted Suicide" (p. A3), by Marlise Simons (2000, November 28),* New York Times.

book written by its cofounder and director, Derek Humphry, entitled *Let Me Die Before I Wake: Hemlock's Book of Self-Deliverance for the Dying.* The first 10 chapters of this book provide detailed case studies of individuals who chose self-deliverance as an alternative to passive euthanasia (often with the help of family members and friends). The remainder of the book provides a bibliography and an extended discussion of personal and legal issues related to the decision to take one's life. Needless to say, this book was controversial in some segments of the American society.

A year earlier, the English counterpart of the Hemlock Society—**EXIT** (founded in 1935 as the British Voluntary Euthanasia Society)—published *A Guide to Self-Deliverance* (1980). This book was unique because it provided the reader with detailed instructions to successfully and painlessly commit suicide. In the United States this book would have been construed as an attempt to intentionally aid and/or solicit another to commit suicide—a felony in some states. In 1991, Derek Humphry published *Final Exit: The Practicalities of Self-Deliverance and Assisted Suicide for the Dying.* The most controversial chapter was entitled "Self-Deliverance via the Plastic Bag." This chapter provides detailed instructions for self-deliverance including inserting one's head into such a plastic bag, "firmly tied around the neck with either a large rubber band or a ribbon," after one takes a sufficient number of pills to induce unconsciousness. The instructions came without a guarantee, but Humphry could predict satisfaction by stating that "in the vast majority of cases, the drugs work." After *Final Exit* gained wide public attention through *The Wall Street Journal,* it began an 18-week journey on the *New York Times* bestseller list and eventually sold more than a million copies (Nuland, 1998).

Protesters in wheelchairs, wearing shirts with the words "Not Dead Yet" demonstrate in front of Dr. Jack Kevorkian's home in Michigan in 1996. Dr. Kevorkian's assisting in the deaths of individuals indeed did not set well with many Americans.

Until the publishing of *Final Exit,* the Hemlock Society had not been willing to be explicit in providing suicide information to its members. However, with public sentiment turning more favorably toward active euthanasia, the Hemlock Society began to take a more active role in promoting self-deliverance. In 1991 the Hemlock Society became a strong advocate of Washington State's Initiative 119, which would have legalized PAD had the bill passed, and contributed over $300,000 in support of the bill (Gabriel, 1991).

After Humphry retired as executive director of the Hemlock Society in 1992, he created the Euthanasia Research and Guidance Organization (known as **ERGO!**). This nonprofit educational corporation sought to improve the quality of background research in physician-assisted deliverance for persons who are terminally or hopelessly ill and who wish to end their suffering. As well as conducting opinion polls, ERGO! develops and publishes guidelines—ethical, psychological, and legal—for patients and physicians to better prepare them to make life-ending decisions. The organization supplies literature to and does research for other right-to-die groups worldwide and also briefs journalists, authors, and graduate students who are interested in right-to-die issues. In addition, ERGO! is willing to counsel dying patients (and their families) provided that they are competent adults and are at the end stage of a terminal illness. Persons suffering from unbearable mental illness are referred by ERGO! to alternative forms of skilled help.

The Hemlock Society and ERGO! have been careful to differentiate suicide from self-deliverance and self-deliverance from mercy killing. Suicide is often condemned socially and religiously as a selfish act or as an "overreaction of a disturbed mind." Suicide is considered by many to be an irrational act that is a permanent solution to what is often a temporary problem. In contrast, self-deliverance should be considered a positive action taken to provide a permanent solution to long-term pain and suffering for the individual and her or his significant others faced with a terminal condition.

Self-deliverance is a completely voluntary act on the part of the patient, whereas mercy killing involves other people's behavior that may or may not be sanctioned by the patient. U.S. laws do not distinguish between murder and mercy killing, and therefore mercy killing is predominantly treated as a criminal offense regardless of the motivations of individuals involved. The defense (other than for a physician) sometimes used in criminal court for a mercy killing is "temporary insanity."

The Hemlock Society, ERGO!, and EXIT wish to eliminate the stigma attached to self-killing and to extend to all individuals the right of active, rational, and voluntary euthanasia whenever the dying process promises only unrelieved pain and a life devoid of dignity, meaning, and purpose.

The Slippery-Slope Argument # 6

Persons who oppose active euthanasia often do so on the grounds of what they call the **slippery-slope-to-Auschwitz argument** (Wilkinson, 1995; Potter, 1993). They allege that the justification for withholding some treatments also applies to withholding any and all treatments from those whom society considers unworthy. It is further argued that if a society wishes to extend to any of its members the right to active euthanasia, none of its members is protected from being killed. Such was the case under Hitler when German society's "unproductive," "defective," and "morally and mentally unfit" persons—Jews, homosexuals, and handicapped—were sent to Auschwitz and other places of confinement to be "euthanized" (murdered).

The slippery-slope argument alleges that society moves toward a disregard for human life in predictable steps. The first step is the acceptance of passive euthanasia when medical technology and surgical procedures are withheld from (or rejected by) chronically and/or terminally ill patients. The second step occurs when ordinary or customary procedures are withdrawn—such procedures would include intravenous delivery of food and water. We arrive at the third step when society accepts the legitimacy of suicide (self-deliverance). The fourth step is when these "rights" are extended to patients with nonterminal illnesses. In the fifth step, the medical system will assist patients in self-deliverance or active euthanasia (as is presently the situation in Holland), like a beloved animal that is "put to sleep." In the sixth step all of these provisions are extended to the non-terminally ill persons and/or persons acting on their behalf. Finally, we arrive at the seventh step—"Auschwitz"—when society is willing to accord these "privileges" to the unwilling, incompetent, and/or undesirable.

To prevent society from reaching such a level of desensitization, people who embrace the slippery-slope argument argue that no type of euthanasia should be considered socially acceptable—in other words, to be safe, society should "stay off the slope entirely!" However, other biomedical scholars who favor AVE and PAD (Ogden, 1995; Weijer, 1995; Logue, 1994; Hill, 1992) argue that if patients voluntarily permit and/or cooperate with their physicians to cause a premature death, their rights and dignity as persons will be enhanced, and a Nazi-like program of extermination will be unlikely. However, Robert Twycross (1996) notes that once the "slipping" begins, it simply continues. He observes that the Dutch have already slipped a long way down the slope in that both voluntary and involuntary euthanasia exist in the Netherlands and that both physical and psychiatric disorders are "grounds" for euthanasia.

Physician-Assisted Death

As previously noted, the issue of active euthanasia and physician-assisted death was highlighted in the media during the 1990s. Dr. Jack Kevorkian's "death machine" in Michigan was the subject of numerous headlines. In 1991, as noted earlier, Derek Humphry's *Final Exit,* providing detailed information on how to "self-deliver," went on the market and became a best-seller. *The New England Journal of Medicine* published Dr. Timothy Quill's experience of assisting his patient "Diane" by prescribing barbiturates after she refused treatment options for leukemia in 1991 (see abridged version in box). Voters went to the polls and courts made decisions regarding PAD throughout the 1990s. Today in the 21st century, this issue is not going away.

Legal Measures.   In 1991 the voters of Washington defeated Initiative 119 (54 percent to 46 percent), which would have legalized PAD. Likewise, in 1992 California voters failed to pass Proposition 161 (52 percent to 48 percent), an initiative similar to Washington's Initiative 119. Active euthanasia was legalized in the Netherlands (as of April 2001), and physicians there are now allowed to help patients take their own lives under certain circumstances (Switzerland allows PAD in carefully controlled situations). The Supreme Court of Canada, in 1993, refused (by a vote of five to four) to allow a woman with amyotrophic lateral sclerosis (ALS) to have her physician legally assist in her death. In 1994 Measure 16 passed (51 percent to 49 percent) in Oregon, legally allowing physicians to hasten death for the terminally ill. The results, however, were tied up in court, and Oregon voted again on this issue in November of 1997 and it passed again (by a vote of 60 percent to 40 percent). Thus PAD is now legal in the state of Oregon. Since the passing of the Oregon law in 1997, 16 patients died in 1998 from PAD and 27 patients died in 1999 from PAD (Sullivan, Hedberg, & Fleming, 2000).

In 1996 the Ninth Circuit Court of Appeals in Washington ruled that the Washington statute prohibiting physicians from prescribing life-ending medication for use by terminally ill, competent adults who wished to hasten their own deaths is unconstitutional because it violates the Due Process Clause of the 14th Amendment of the United States Constitution. Also, in 1996 the Second Circuit Court of Appeals in New York reached a similar ruling that expanded on the Ninth Circuit Court's ruling in also finding that allowing terminally ill patients to die through physician removal of life-sustaining treatments, such as mechanical ventilators or artificial hydration and nutrition, while not allowing such patients to die by taking physician-prescribed medication, is a violation of the equal protection guarantee of the 14th Amendment. The U.S. Supreme Court ruling in 1997 declared that terminally ill patients do not have a fundamental constitutional right to assisted death and upheld the New York and Washington laws that make it a crime for physicians to give life-ending drugs even to mentally competent terminally ill adults who have made such a request. However, states retain the right to enact laws that legalize assisted death; thus, the matter was thrown back by the U.S. Supreme Court to the states. Physicians still maintain the right to provide terminally ill people with adequate pain relief medication even if this shortens life.

In the Northern Territory of Australia in 1996 the world's first law permitting voluntary euthanasia was tested when the first person died after it was enacted. Six months later in 1997, however, after four acts of voluntary euthanasia were reported, the law was revoked. Currently, as noted earlier, the state of Oregon is the only place in the United States where PAD is legal, but other states have similar initiatives on their ballots for voters to decide in the near future. However, the Oregon ruling, as of this writing, is in jeopardy of being overridden by the U.S. Congress. If Congress passes the Pain Relief Promotion Act (PRPA), the federal Controlled Substances Act would be violated when physicians administer drugs to hasten a patient's death.

LISTENING TO THE VOICES
My Patient's Suicide

I have long been an advocate of the idea that an informed patient should have the right to choose or refuse treatment, and to die with as much control and dignity as possible. Yet there was something that disturbed me about Diane's decision to give up a 25 percent chance of long-term survival in favor of almost certain death. Diane and I met several times that week to discuss her situation, and I gradually came to understand the decision from her perspective. We arranged for home hospice care, and left the door open for her to change her mind.

Just as I was adjusting to her decision, she opened up another area that further complicated my feelings. It was extraordinarily important to Diane to maintain her dignity during the time remaining to her. When this was no longer possible, she clearly wanted to die. She had known of people lingering in what was called "relative comfort," and she wanted no part of it. We spoke at length about her wish. Though I felt it was perfectly legitimate, I also knew that it was outside of the realm of currently accepted medical practice and that it was more than I could offer or promise. I told Diane that information that might be helpful was available from the Hemlock Society.

A week later she phoned me with a request for barbiturates for sleep. Since I knew that this was an essential ingredient in a Hemlock Society suicide, I asked her to come to the office to talk things over. She was more than willing to protect me by participating in a superficial conversation about her insomnia, but it was also evident that the security of having enough barbiturates available to commit suicide, if and when the time came, would give her the peace of mind she needed to live fully in the present. She was not despondent and, in fact, was making deep, personal connections with her family and close friends. I made sure that she knew how to use the barbiturates for sleep, and how to use them to commit suicide. We agreed to meet regularly, and she promised to meet with me before taking her life. I wrote the prescriptions with an uneasy feeling about the boundaries I was exploring— spiritual, legal, professional, and personal. Yet I also felt strongly that I was making it possible for her to get the most out of the time she had left.

(continued)

Public Opinion and Physicians' Attitudes.   Both U.S. medical ethics literature and U.S. medical associations have traditionally condemned physician-assisted death (Jecker, 1994). The American Medical Association's Council on Ethical and Judicial Affairs stated that, although life-prolonging medical treatment may be withheld, "the physician should not intentionally cause death" (1992). In addition, the American Geriatrics Society (1994) has stated that physicians should not provide interventions that will intentionally cause the death of patients.

(continued from previous page)

The next several months were very intense and important for Diane. Her son did not return to college, and the two were able to say much that had not been said earlier. Her husband worked at home so that he and Diane could spend more time together. Unfortunately, bone weakness, fatigue, and fevers began to dominate Diane's life. Although the hospice workers, family members, and I tried our best to minimize her suffering and promote comfort, it was clear that the end was approaching. Diane's immediate future held what she feared the most: increasing discomfort, dependence, and hard choices between pain and sedation. She called her closest friends and asked them to visit her to say good-bye, telling them that she was leaving soon. As we had agreed, she let me know as well. When we met, it was clear that she knew what she was doing, that she was sad and frightened to be leaving but that she would be even more terrified to stay and suffer.

Two days later her husband called to say that Diane had died. She had said her final good-byes to her husband and her son

that morning, and had asked them to leave her alone for an hour. After an hour, which must have seemed like an eternity, they found her on the couch, very still and covered by her favorite shawl. They called me for advice about how to proceed. When I arrived at their house we talked about what a remarkable person she had been. They seemed to have no doubts about the course she had chosen, or about their cooperation, although the unfairness of her illness and the death were overwhelming to us all.

Diane taught me about the range of help I can provide people if I know them well and if I allow them to express what they really want. She taught me about taking charge and facing tragedy squarely when it strikes. She taught me about life, death, and honesty, and that I can take small risks for people I really know and care about.

Adapted from "Death and Dignity: A Case of Individualized Decision Making" (pp. 32–34), by T. Quill (1991, May), Harper's Magazine. *Originally published in* The New England Journal of Medicine, 324, *March 7, 1991.*

Though there is opposition to physician-assisted death, acceptance among health professionals is beginning to occur (Dickinson, Lancaster, Summer, & Cohen, 1997). For example, few people publicly criticized the case in which Dr. Timothy Quill assisted "Diane" in ending her life in 1991. In 1996 the 30,000-member American Medical Student Association filed a brief before the Supreme Court supporting physician-assisted suicide death.

Public opinion of physician-assisted death is also becoming more favorable in the United States and in other places, as noted earlier. In 1950 only 34 percent of people in the United States supported active euthanasia for incurably ill patients if they and their families requested it (Blendon, Szalay, & Knox, 1992). By 1973, however, a Gallup poll in the United States found that 53 percent of those interviewed said that physicians should have the legal right to painlessly end the life of a person with a terminal illness if the patient and family request it (Nagi, Puch, & Lazerine, 1978). In 1977, as many as 60 percent supported active euthanasia. By 1991, 63 percent of the U.S. population supported active euthanasia, and a majority (64 percent) favored allowing PAD (Blendon et al., 1992). By 1997, 67 percent in the U.S. supported PAD (Angell, 1997). In Australia in 1996 (Hassan, 1996) 76 percent agreed that a doctor should comply with a terminally ill patient's request for a lethal dose; the Netherlands have shown a progressive increase in public support with 92 percent of Dutch people supporting mercy killing in 1998 (Onstad, 2000); a Canadian survey (Edwards & Mazzuca, 1999) found that 77 percent support euthanasia in patients with terminal illness and severe pain; and in the United Kingdom 82 percent agreed in 1994 (Donnison & Bryson, 1995) that doctors should be allowed by law to end the life of a patient if such a request is made with a painful, incurable disease.

Why is there increased support for PAD and AVE? Possible answers to this question include (1) better educated populations, (2) more autonomy for individuals (an additional choice), (3) more technology contributing to more fear of misuse, (4) more secularized populations (a decline in religion) and thus a more rational approach, (5) an aging population contributing to more lingering suffering, and (6) the emergence of AIDS having an impact on younger, urban dwellers.

Because physicians would be involved with active euthanasia, we must wonder about their views on this topic. A comprehensive survey in 1993 of 938 physicians in Washington indicated that they are polarized: A slight majority favor legalizing physician-assisted death and euthanasia in at least some situations, but most would be unwilling to participate in these practices themselves (Cohen et al., 1994). In a **replication** study of the Washington State survey, George Dickinson and colleagues (1997) found that 587 South Carolina physicians had attitudes remarkably similar to those of the physicians in Washington: Their attitudes were sharply polarized, with fewer than 15 percent giving a neutral response to the questions of euthanasia and assisted suicide. Another survey (Watts, Howell, & Priefer, 1992) of 727 geriatricians on their attitudes toward assisted suicide among patients with dementia found that 21 percent would consider assisting in the suicides of competent, nondepressed patients. H. G. Koenig (1993) states that increasing support within the medical community for physician-assisted death comes from a recent decision by Michigan physicians to reverse their stand against the practice, preferring that it not be considered a felony. A study (Meier et al., 1998) of 1,902 physicians in the United States revealed that 24 percent would perform AVE and 36 percent PAD *if* it were legal. Regarding attitudes of both black and white physicians in the United States (Mebane et al., 1999), of the 502 physicians responding, 37 percent of white physicians and 27 percent of black physicians found PAD "an acceptable treatment alternative."

Physicians in other countries have also voiced their opinions about AVE and PAD in recent years. A study in Italy (Grassi, Magnani, & Ercolani, 1999) of 396 general practitioners found that only 15 percent gave responses "somewhat" or

"strongly" favoring AVE and PAD. A Canadian study (Verhoef & Kinsella, 1996) showed that physicians' agreement on legalizing AVE and PAD decreased over a 3-year period (29 percent versus 15 percent and 50 percent versus 37 percent, respectively). In the United Kingdom (Clark et al, 2000) responses from 333 geriatric medicine doctors revealed strong opposition to legalization of AVE and PAD (80 percent agreed AVE is never ethically justified and 68 percent agreed that PAD is never ethically justified).

Why is there such strong opposition to PAD and AVE on the part of physicians? Some answers given in the literature include the following: (1) Legalization would have a negative effect on palliative care and on physician/patient relationships. (2) Legalization of AVE would follow legalization of PAD. (3) Involuntary euthanasia would follow AVE, thus the slippery-slope argument. (4) Sanctity of life is emphasized in the Hippocratic Oath. (5) A patient might be deterred from seeking medical advice, if a part of the physician's role was to "kill" patients.

A problem noted with PAD when the patient is acting without the physician present is the likelihood of serious accidents inherent in the patient's giving himself or herself an overdose. Vomiting may occur as the patient slips into a coma, with aspiration of the vomitus. If isolated patients change their minds, they may nevertheless choke to death in a panic, or, if rescued, die of pneumonia. With states of confusion resulting from morphine, barbiturates, or other compounds, some patients may experience terror or panic (Report of the Committee on Physician-Assisted Suicide and Euthanasia, 1996).

WORDS OF WISDOM
A Good Death

A good death does honor to a whole life.

Petrarch (1304–1374)

As noted above, public opinion polls regarding PAD and AVE and the opinions of physicians on these issues generally do not agree. Whereas the public overall seems to have the attitude that patients should have a right to have assistance from their doctors in ending their lives, physicians generally are not in favor of such action and in some cases are strongly against this practice.

Guidelines for Assisted Suicide.   With Oregon being the only state in the United States where physician-assisted death is currently legal, their guidelines in the list below could be the model for similar states in the future, pending voter approval:

1. The patient must request a physician's assistance in suicide three times, the last in writing, with the statement dated and signed by the patient in the presence of two witnesses.

2. Two physicians must determine that the patient has a life expectancy of 6 months or less.

3. The physician must wait at least 15 days after the initial request, and at least 2 days after the final request, before writing the prescription for the lethal drugs.

4. The physician must determine that the patient is not suffering from a psychiatric or psychological disorder or depression causing impaired judgment.

5. At least one of the two witnesses to the patient's written request for the lethal prescription must be a person who is not a relative of the patient, does not stand to benefit from the estate of the patient, and is not an employee of the institution where the patient is being treated.

With widespread public interest in assisted suicide and euthanasia, whether legalized or not, the medical profession should work to improve the care of terminally ill patients through more effective control of pain and other symptoms. However, whether patients are terminally ill, pain-free, or not, rational suicide may be the answer for some individuals.

CONCLUSION

Physicians in the United States often press for more treatment than either a family or a terminally ill patient desires, yet some families, on the other hand, often insist, even in hopeless situations, that "everything possible" be done for the patient. Though these motives are kind, the results are often not, notes ethicist and physician Charles Culver (1990). Because of our ever-increasing technological sophistication, dilemmas and disagreements about allowing patients to die are frequent and will probably increase in the future. Bodies can be kept alive for a long time. There are cases where further treatment would not be totally pointless, but the patient, were she or he competent to decide, would not desire further treatment. Or if the patient were truly altruistic, she or he might opt to die and donate organs so that others could live a quality life. But then deciding when one is dead and to whom to give the organs, once the individual is declared dead, poses other questions. Thus, the saga continues.

SUMMARY

1. Euthanasia, or "good death," must be defined within the context of one's understanding of the meaning of life.

2. The sanctity-of-life perspective of euthanasia contends that all "natural" life has intrinsic meaning and should be appreciated as a divine gift. As a consequence, human beings have the obligation to enhance the quality of life as it may exist.

3. The quality-of-life perspective of euthanasia contends that when life no longer has "quality," death is preferable to living a life devoid of meaning.

4. Individuals try to facilitate a "good death" by two methods—passive and active euthanasia. Passive euthanasia involves a protocol whereby no action or medical intervention hastens death. Passive euthanasia usually involves the removal of medical technology or the refusal of medical intervention. Passive euthanasia is supported by people who affirm both sanctity-of-life and quality-of-life perspectives.

5. Active euthanasia requires a direct action to bring about death. Those who favor active euthanasia prefer to use the term *self-deliverance* when the life of a terminally ill patient is ended. Self-deliverance should be considered a positive action taken to provide a permanent solution to the long-term pain and suffering of the individual and his or her significant other who are faced with a terminal condition.

6. Persons opposing active euthanasia often do so with the "slippery-slope-to-Auschwitz" argument. They allege that the justification for withholding some treatments also applies to withholding any and all treatments from those whom society considers unworthy. The slippery-slope argument asserts that society moves toward a disregard for human life in predictable steps.

7. Because of the sensitive nature of the issue, human organ transplantation has not been widely discussed in society. Difficult decisions need to be made about what qualifies a person to be eligible to receive a transplant from a donor.

8. Physician-assisted death has been a major issue in the United States and throughout the rest of the world in the 1990s and will continue to be an issue into the 21st century.

DISCUSSION QUESTIONS

1. Compare and contrast passive and active euthanasia. What difficulties does implementing each of these approaches have for current social policy dealing with health care for the terminally ill?

2. If your parent were dying from an extremely painful and incurable form of cancer and decided that life was not worth living, what role would you be willing to play in assisting him or her in self-deliverance? Would you be unwilling to assist your parent? If not, what actions would you take in discouraging your parent's action?

3. Provide arguments for and against the removal of food and water from a terminally ill patient in a coma.

4. Discuss some of the problems affecting human organ transplantations from both a societal and a personal point of view.

5. Should society make eligible for a transplant only those persons who can afford it, or should the potential societal contribution of the individual be taken into account? Are there some medical procedures that should never be financed by insurance benefits or government funds because they are too expensive?

6. Discuss your reaction to Dr. Timothy Quill's assisting his patient with leukemia, "Diane," with her death.

7. Physicians are sworn to "prolong life and relieve suffering," yet physician-assisted death runs counter to "prolonging life." The majority of people in the United States favor physician-assisted deaths. Discuss the pros and cons of physician-assisted deaths.

GLOSSARY

Active Voluntary Euthanasia or AVE: Upon the request of the patient, a physician directly contributes to his or her death (e.g., by injecting a lethal dosage into the patient).

Active Euthanasia: A direct action that causes death in accordance with the stated or implied wishes of the terminally ill patient. Such action can be voluntary (patient requested) or involuntary (not requested by the patient).

Biomedical Issues: Issues concerned with molecular biology to the extent that social, psychological, and behavioral dimensions of illness are disregarded.

Bioethicist: A professional person who assists in decision making between biomedical practitioners and their patients regarding values and standards within a medical setting.

CPR: Cardiopulmonary resuscitation—the act of using mouth-to-mouth resuscitation and chest compressions to try to restore one's breathing.

Cultural Lag: A situation in which some parts of culture change more slowly than others. A typical situation occurs when technology changes faster than the values of the culture.

DOA: "Dead on arrival"—a classification given to a patient upon arrival at the emergency room if the physician declares the patient dead. Thus, no resuscitation efforts would follow.

DNR or Do Not Resuscitate: An order required by some hospitals for heroic care or other resuscitative measures to be withheld.

Double Effect: A physician giving a patient pain control medicine and increasing the dosage, as the pain increases. Too much medication such as morphine, for example, may indeed "kill" the patient, and this is the "double effect" (i.e., the pain is being controlled, but a side effect of too much medication might cause the death of the patient).

EMTs: Emergency medical technicians.

ERGO!: Euthanasia Research and Guidance Organization.

Euthanasia: Literally, a "good death."

EXIT: A British society that promotes active euthanasia.

Extraordinary Measures: All of those medicines, treatments, and operations that cannot be obtained without excessive expense, pain, or other inconvenience or that, if used, would not offer a reasonable hope of benefit.

Hemlock Society: An organization in the United States that supports active euthanasia.

Intubation: A medical procedure whereby the patient is fed through a tube placed in the stomach.

Living Will: A document stating that one does not want medical intervention if the technology or treatment that keeps one alive cannot offer a reasonable quality of life or hope for recovery.

Ordinary Measures: All of those medicines, treatments, and operations that offer a reasonable hope of benefit to the patient and that can be obtained and used without excessive expense, pain, or other inconvenience.

Quality of Life: The perspective that when life no longer has quality, death is preferable to living a life devoid of meaning.

Passive Euthanasia: The withholding of treatment, which, in essence hastens death and allows the individual to die "naturally."

Patient Self-Determination Act (PSAD): Went into effect on December 1, 1991, and requires all health-care providers who receive federal funding to inform incoming patients of their rights under state laws to refuse medical treatment and to prepare an advance directive.

Physician-Assisted Death or PAD: A physician "giving advice" to a patient about taking her or his life and/or prescribing a lethal dosage of medication whereby the individual can take her or his own life (sometimes called "physician-assisted suicide" or PAS).

Replication: The repetition or duplication of the procedures followed in a particular study in order to determine whether or not the same findings are obtained at another time and/or place.

Sanctity of Life: The perspective that all natural life has intrinsic meaning and should be appreciated as a divine gift. As a consequence, human beings have the obligation to enhance the quality of life as it may exist.

Self-Deliverance: A rational and voluntary act of taking one's life; an alternative to the terms *suicide* and *mercy killing*.

Slippery-Slope-to-Auschwitz Argument: The belief that if a society extends to any of its members the right to active euthanasia, none of its members is protected from being killed. Auschwitz was a German concentration camp where people were killed during World War II.

Triage: A system of assigning priorities of medical treatment. Organ transplants are sometimes an example of triage. Who receives the organ and on what basis is the decision made?

Xenotransplantation: The transplantation of organs between different species.

SUGGESTED READINGS

Baker, R. B., Caplan, A. L., Emanuel, L. L., and Latham, S. R. (Eds.) (1999). *The American medical ethics revolution: How the AMA's Code of Ethics has transformed physicians' relationships to patients, professionals, and society*. Baltimore, MD: Johns Hopkins University Press. Provides important historical and evaluative insights into the ethics activities of the AMA.

Battin, M. P. (1994). *The least worst death: Essays in bioethics on the end of life*. New York: Oxford University Press. Offers insight into the controversial topics of withdrawing and withholding care, euthanasia, and suicide.

Chambliss, D. F. (1996). *Beyond caring: Hospitals, nurses, and the social organization of ethics*. Chicago: University of Chicago Press. Offers a sobering revelation of the forces shaping moral decisions in hospitals in the United States today.

DeVries, B. (Ed.). (1999). *End of life issues: Interdisciplinary and multidimensional perspectives*. New York: Springer Publishing Company. Examines the ways in which individuals at the end of life are influenced by aspects of their interpersonal and social environments.

DeVries, R. A., & Subedi, J. (Eds.) (1998). *Bioethics and society: Constructing the ethical enterprises*. Englewood Cliffs, NJ: Prentice Hall. An anthology of sociological articles concerned with biomedical and bioethical issues.

Hoefler, J. M. (1997). *Managing death*. Boulder, CO: Westview Press. Examines the medical, legal, ethical, and clinical aspects of right-to-die issues that rise over decisions about life-sustaining medical treatment.

Joralemon, D. (1999). *Exploring medical anthropology*. Boston: Allyn & Bacon. Though a book on medical anthropology, the author has some interesting sections on medical ethics.

McKhann, C. F. (1998). *A time to die: The place for physician assistance*. New Haven: Yale University Press. Discussion of many aspects of physician-assisted dying by a specialist in cancer surgery. Dr. McKhann explains why he thinks it should be made legally available under certain circumstances.

Randall, F., & Downie, R. S. (1996). *Palliative care ethics: A good companion*. Oxford: Oxford University Press. Discusses the importance of palliative care in being good companions to the dying and their relatives.

Seale, C. (1998). *Constructing death: The sociology of dying and bereavement*. Cambridge, UK: Cambridge University Press. Draws on sociological, anthropological, and historical studies to demonstrate the variability that exists in human social constructions for managing mortality.

Thomasma, D. C., & Kushner, T. (Eds.) (1996). *Birth to death: Science and bioethics,* Cambridge, UK: Cambridge University Press. End-of-life topics presented from the perspective of a scientific or historic viewpoint of the available technology and its achievements, and then other contributors discuss the ethical implications that arise from each.

Timmermans, S. (1999). *Sudden death and the myth of CPR.* Philadelphia: Temple University Press. A sociologist shares his observations of happenings in emergency rooms regarding CPR.

Tong, R. (1996). *Feminist approaches to bioethics: Theoretical reflections and practical applications.* Boulder, CO: Westview Press. Confronts the moral diversity that reveals feminism in bioethical issues.

C H A P T E R

SUICIDE

Suicide is a permanent answer to a temporary set of problems.
—*NBC Evening News*, March 12, 1987

There is but one truly serious philosophical problem, and that is suicide. Judging whether life is or is not worth living amounts to answering the fundamental question of philosophy.

—Albert Camus

Physician-assisted death is a medical response to the wishes of terminally ill patients. This action involves a discussion and consideration between a doctor and the patient. Unassisted suicide is a very different matter. The decision is usually made in isolation by an individual who is often suffering from depression. Whereas physician-assisted death is a relatively recent phenomenon, unassisted suicide has been around for as long as recorded history. Yet, the word *suicide* is not in the Bible or in the pamphlet by John Donne (1644/1930) on self-homicide. The *Oxford Dictionary* states that the word was first used in English in 1651 and is derived from the modern Latin *suicidium*, which stems from the Latin pronoun for "self" and the verb "to kill" (Farberow, 1975). Though the word *suicide* was not used, the earliest recorded suicides known to Western culture were the deaths of Samson and Saul around 1000 B.C. (Stillion & McDowell, 1991). Two other Old Testament individuals who took their own lives were Abimilech and Achitophel. Probably one of the more famous suicides was that of Socrates, by drinking hemlock.

Changing Attitudes Toward Suicide

Historically, attitudes toward suicide have been somewhat mixed. Early societies sometimes forced certain members to commit suicide for ritual purposes. For example, suicide was occasionally expected of the wives and slaves of husbands or masters who had died as an expression of fidelity and duty. In the ancient Roman Empire, suicide was socially acceptable until slaves started to practice it. Because the suicide of slaves caused a severe financial loss, however, suicide was declared a crime against the state (McGuire & Ely, 1984). Modern Judaism, although officially regarding suicide as a sin, rests upon a long tradition of honoring "heroic" suicide to avoid rape or forced slavery or idol worship (Curran, 1987). Until the middle of the 20th century, the Roman Catholic Church denied Catholic ritual and burial to those who took their own lives (*Codex Juris Canonici*, 1918). For Japanese kamikaze pilots in World War II, suicide represented the "great death." The Chinese regarded suicide as acceptable and honorable, particularly for defeated generals or deposed rulers (Ingram & Ellis, 1992). The New Testament records only the suicide of Judas Iscariot, but in no way is the act of taking one's own life condemned in the scriptures (Alvarez, 1971).

Attitudes toward suicide changed radically when St. Augustine, drawing heavily from the philosophy of Plato and Aristotle, laid down rules against suicide that became the basis for Christian doctrine throughout the succeeding centuries. Societal opposition to suicide in Christian communities continued throughout the Middle Ages, until the Renaissance broadened thinking on the subject. Public and private opinion, however, remained far from unanimous. Even today in the United States, attempted suicide is considered a felony in nine states: Alabama, Kentucky, New Jersey, North and South Carolina, North and South Dakota, Oklahoma, and Washington (Curran, 1987).

The Stigma of Suicide

Traditionally, suicide has been morally proscribed by Western culture. L. G. Calhoun and B. G. Allen (1991) note that surveys have suggested that families in which a suicide has occurred are expected to feel shame and that individuals who commit suicide have been viewed as "crazy," mentally ill, and psychologically stressed. Yet when respondents were asked to evaluate individuals they knew who had been bereaved by a loved one's suicide, the negative stigma was less. Knowing the individual may buffer the degree of negative social reactions. In addition to the negative feelings, surviving relatives of suicide victims are thought to be more grief stricken because their loved ones willfully took their own lives (Ingram & Ellis, 1992). Calhoun and Allen (1991) conclude that although recent data indicate that society rejects suicide less today under some specific conditions, such as a terminal illness, suicide continues to be viewed as an undesirable and abnormal act. The traditional idea in the United States is that the only legally, ethically, and morally correct response to suicide is intervention, with such ethical codes being found in many helping professions and their organizations, such as the American Psychological Association (1992). Action to prevent suicide is expected and required, based on the assumption that anyone who considers suicide is suffering from some sort of mental problem (Werth, 1995). Because suicide is still socially unacceptable, the family may lack the support systems normally available to those grieving other kinds of death—a traumatic loss is suffered, and the taboo act generates feelings of disapproval and shame (Dunn & Morrish-Vidners, 1987).

The difficulty of the stigma attached to a suicide for family members is pointed out in a study by J. A. Gibson, L. M. Range, and H. N. Anderson (1987). A vignette about the death (from either suicide or a rare disease) of an 11-year-old boy who came from either a single-parent home or a two-parent home was provided to 120 high school juniors and seniors. The suicide victim and his family were perceived more negatively than their rare disease counterparts, but knowledge that the parents were divorced did not correlate with the cause of death to affect the respondents' impressions. A study by L. M. Range and W. C. Goggin (1990) also found that people respond more negatively to the victim and the bereaved family if the death is by suicide rather than by viral illness.

The general attitude of the public toward suicide remains confused and sometimes contradictory (Ingram & Ellis, 1992). Extreme views on suicide range from total acceptance to total rejection of the right of an individual to commit suicide. At the heart of the debate is the question of whether people should be allowed to die without interference.

Defining Suicidal Acts

The meaning of suicide continues to be problematic. If we accept the definition of suicide as "any death resulting either from a deliberate act of self-destruction or from inaction when it is known that inaction will have fatal consequences" (Theodorson & Theodorson, 1969, p. 427), we could possibly classify the following persons as engaging in suicidal behavior:

1. A person who smokes cigarettes, knowing that the Surgeon General has determined that smoking is a major cause of lung cancer

2. A race car driver who races even though he or she knows that in any given race there is a good chance that someone will be killed

3. A person who takes a bottle of sleeping pills, hoping to call attention to self as one who has personal needs that are not being met

4. A person who refuses to wear a seat belt when studies show that wearing a seat belt reduces the death rate in automobile accidents

5. A person who continues to eat fatty foods after having suffered a heart attack

6. A person with advanced stage cancer who refuses chemotherapy or surgery

7. A person who overeats (or undereats) to the extreme that his or her health is directly affected

8. A person with high blood pressure who refuses to take medication to control it or fails to control food intake or to exercise

9. A person who ingests a mixture of alcohol and drugs

Probably most of us would not consider the preceding list of persons as suicidal. Rather, we would be concerned about whether the person in question was intentionally trying to kill himself or herself. In classifying deaths as suicides, it is difficult to determine the motivations of a person no longer living.

There is a qualitative difference between a "suicide gesture" and a "completed suicide." Suicide gestures are motivated by a need for aid and support from others, whereas completed suicides are acts of resignation. At this point, we have a problem of tautology—suicide gestures that mistakenly end in death are classified as "intentional suicides," and unsuccessful suicide attempts are considered "suicide gestures." In addition, suicides are sometimes classified as accidents or natural deaths as a favor to family members or as a method of "providing a more positive view" of the deceased person. Thus, there are deaths resulting from intentional acts of self-destruction that are recorded as "natural deaths," and there are suicidal gestures accidentally ending in death that are recorded as suicides. Obviously, there are problems in compiling suicide statistics!

In recent years suicide has attracted increasing interest as a large-scale social phenomenon (Battin, 1982). Awareness of the problem of suicide has caused both increased scrutiny of the phenomenon by sociologists, psychologists, and researchers in other disciplines and increased efforts to reduce its frequency by physicians, counselors, social workers, and the police. In fact, Lindsay Prior (1989) notes that during most of the 20th century, the study of death was the study of suicide and that in turn was the study of social causation. Efforts to seek causes and to reduce the frequency of suicide work together. Suicide research centers and suicide prevention services have developed in recent years, and the literature on suicide has increased significantly. Even a "cookbook" on ways to commit suicide, entitled *Suicide, Its Use, History, Technique and Current Interest* by C. Guillon and Y. LeBonniec, came out in France in 1982 and soon worked its way to the United States. In 1991 Derek Humphry's *Final Exit: The Practicalities of Self-Deliverance and Assisted Suicide for the Dying*, which tells people how to commit suicide, rose to number one on the *New York Times* bestsellers list. And throughout the 1990s physician-assisted death (discussed in chapter 8) was a major topic of the media, an issue at the ballot boxes, and a concern in the courts.

SOCIAL FACTORS INVOLVED IN SUICIDE

In the United States, nearly 30,000 individuals commit suicide each year—one person every 20 minutes (Marrone, 1997). The number of reported suicides represents only 10 to 15 percent of the people attempting suicide. Suicide is often associated with age, gender, marital status, and socioeconomic factors. For example, suicide is rare in childhood, rises sharply during adolescence and early adulthood, and peaks with elderly persons.

A study (Taylor, 1990) conducted by Cornell University Medical School revealed that men infected with AIDS were 36 times more likely than average men to commit suicide. The same study also found that patients with AIDS were far more likely to choose to end their lives than were men with other fatal diseases. Elisabeth Kübler-Ross (1987) suggested that the motivation of many people with AIDS to consider suicide is to avoid enduring their painful feelings of despair, isolation, and discrimination and the ravaging deterioration that they have seen among their friends. A study of 44 terminally ill patients (Brown, Henteleff, Barakat, & Rowe, 1986) concluded that patients with a terminal illness who are not mentally ill are no more likely than the general population to wish for premature death. People with AIDS, however, are different.

Age, Gender and Marital Status Factors

Typically, the suicide rate for males shows peaks during adolescence and especially during the elderly years, whereas the rate for females shows a gradual increase from young ages until ages 45 to 50, after which the rate gradually declines (McCall, 1991).

In relation to marital status, the suicide rate is lower for individuals who are married than for those who never married and is highest for those who are widowed, divorced, or separated. Individuals with children commit suicide less than childless individuals. These data on marital and parental status fit Emile Durkheim's theory (discussed later in the chapter) that suicide is a function of social integration.

Listed in decreasing order of frequency, the ways of committing suicide in the United States are shooting with firearms (especially handguns), overdosing with drugs (most often those prescribed by physicians), cutting and stabbing, jumping from high places, inhaling toxic gas, hanging, and drowning (Oaks & Ezell, 1993). For elderly persons, not taking medication or taking medication irregularly or simply not eating can be added to the list of ways of committing suicide. Sometimes elderly persons might "give up" psychologically, and this triggers biological changes that increase potential for disease. Whereas women attempt suicide more often, men more often complete suicide. Because men are more likely to use firearms in suicide attempts and women are more likely to use drugs, this obviously contributes to the higher suicide "success" rate for males.

In looking at cross-national comparisons of suicide among adolescents and young adults (aged 15 through 24) in 34 of the world's wealthiest nations, 34 percent of these suicides were firearm-related (Johnson, Krug, & Potter, 2000). Methods of committing suicide, however, vary with cultures. For example, the most common suicide method among the older than 65 age group of males in Italy is hanging/strangulation (De Leo, Conforti, & Carollo, 1997), followed by jumping from high places and submersion. For females in Italy, the most common method of suicide is jumping from high places. For young people in Italy, the most common method of suicide is hanging/strangulation.

Doonesbury   BY GARRY TRUDEAU

©1986 REPRINTED WITH PERMISSION OF UNIVERSAL PRESS SYNDICATE

Socioeconomic and Cultural Factors

Suicide rates tend to be highest at both extremes of the socioeconomic ladder. Individuals in certain professions such as physicians, dentists, and lawyers tend to be particularly suicide prone, whereas the suicide rates of teachers and members of the clergy are rather low. The high rates may be due to stress and the fact that physicians and dentists also have access to drugs with which to take their lives. Dentists do not always have a good image in the public's eyes and thus may suffer from low self-esteem. Just ask an adult to recall childhood images of trips to the dentist! The recollections are often not pleasant; thus, they have a negative association with dentists.

The suicide rate for Native Americans has been reported to be greater than the national average, 18.1 per 100,000 (McIntosh, 1995); however, rates have been found to vary greatly from reservation to reservation (Ellis & Range, 1989). Suicide among Native Americans tends more often to be alcohol related and violent methods (firearms and hanging) are more commonly used than in the mainstream U.S. population (May, 1990).

There is also a difference in suicide rates between blacks and whites in the United States. Suicide rates among blacks are substantially lower than among whites—9 blacks versus 16.7 whites per 100,000 population annually (McIntosh, 1995). A recent study (Gibbs, 1997) has identified the following as protective factors that reduce suicide risks among African Americans:

1. Church attendance or affiliation is associated with lower rates of suicide. The African-American church has traditionally served as a source of social cohesion, social support, and stress reduction. Religious beliefs have likely influenced the cultural beliefs of many blacks that "suicide is a white thing."

2. The central role of women in the African-American community may be a significant factor in their low suicide rates. Black women have learned to cope with high levels of stress, poverty, and discrimination by forming strong social networks, sharing resources, assuming flexible family roles, and turning to religion as a source of comfort for their anger and depression.

3. Suicide rates tend to decline with age among African Americans. Elders are treated with respect and dignity. They live independently or in their extended families rather than in nursing homes or retirement communities. In addition, many blacks look forward to retirement from low-status, low-wage jobs.

4. Suicide rates for blacks tend to be lower in the South and in areas of less racial integration. The cohesive social environment of ethnic neighborhoods has reinforced the role of the extended family. In communities where African Americans are less integrated, they might experience greater social cohesion. The social and psychological benefits of a cohesive black community have been considerably weakened by major changes such as integration, urbanization, deindustrialization, and secularization of the broader society. These changes may have contributed to a declining sense of community and fewer social supports for vulnerable African Americans.

5. Suicide rates for blacks tend to be higher among those who live alone and are not involved in family units. The extended family and kin networks have played a major role in the survival of African Americans in a frequently hostile society by fulfilling important functions such as economic support, emotional support, and social support. The extended family serves as a buffer between its members and the stresses of the external environment, giving them a sense of identity, security, and support, thus reducing the risk of suicide among rural and low-income blacks.

More suicides occur in heterogeneous urban areas than in more homogeneous rural areas. Though nostalgic times such as the latter part of December are thought to drive higher suicide rates, the "renewal of life" times such as spring have consistently had the highest rates for suicide. The second highest suicide rate is found in autumn, followed by winter and then summer. Days with the highest suicide rates are Monday and Tuesday, with Wednesday, Thursday, Friday, and Sunday having lower rates. The risk is lowest on Saturdays (McIntosh, 1995). The highest rate of suicide is found in Jews; Protestant faiths have higher rates than Catholics (Oaks & Ezell, 1993). Contrary to popular belief, suicide occurrence does not vary by lunar phase (Maldonado & Kraus, 1991).

THEORETICAL PERSPECTIVES ON SUICIDE

Why do human beings kill themselves? The personal and sociocultural factors of suicide, taken individually and in combination, are so complex that it would be difficult to arrive at a definitive explanation, notes David Moller (1996). From a medical model perspective, an attempt or an act of suicide is considered abnormal behavior, requiring clinical intervention. Yet, when suicide becomes a source of duty or even honor, as for a kamikaze pilot or for anyone sacrificing her or his life for the group or cause, motivations become more social and cultural.

Sociological Perspectives

The most significant contribution of sociologists to the study of suicide has been the sociological perspective itself—the insistence on seeing suicidal actions as in some way the result of social factors (Douglas, 1967). According to sociologists, suicidal behavior can only be understood within the context of shared social values, beliefs, and patterns of social interaction.

Durkheim's Theory

The noted French structural-functional sociologist, Emile Durkheim, proposed a comprehensive theory of suicide that is a classic sociological study. Durkheim's purpose in *Suicide: A Study in Sociology* was not to explain why individuals commit suicide but to explain why suicide rates exhibit such stability—the proportion of individuals committing suicide within a given population remains relatively stable over long periods of time, and a similarity of patterns exists across different societies (Hughes, Martin, & Sharrock, 1995). The most frequent interpretation of Durkheim's *Suicide* by American sociologists involves the following propositions (Douglas, 1967, pp. 39–40):

1. The structural factors cause certain degrees and certain patterns of social interaction.

2. The degrees and patterns of social interaction then cause a certain degree of "social integration."

3. "Social integration," defined as either states of individuals or as a state of the society, is then defined as the "strength of the individual's ties to society."

4. The "strength of ties" is then defined either in terms of egoism, altruism, and anomie, or else the "strength of ties" is hypothesized to be the cause of the given degrees of egoism, altruism, and anomie.

5. "Egoism" is defined as a relative lack of social or collective activity that gives meaning and object to life, "altruism" is defined as a relatively great amount of social activity, and "anomie" is defined as a relative lack of social activity that acts to constrain the individual's passions, which, without constraint, increase "infinitely."

6. And, finally, the given balance of the degrees of egoism (not well integrated into society), altruism (too well integrated into society), and anomie (not knowing societal expectations) is hypothesized to be the cause of the given suicide rate of the given society.

Thus, the three types of suicide—egoistic, altruistic, and anomic—are related to Durkheim's conception of social integration. Two recent time-series analyses of sociological and economic variables accounting for suicide rates in the United States give somewhat mixed findings compared to Durkheim's theory. One study of the United States from 1940 to 1984 by Bijou Yang (1992) concluded that economic growth seemed to have a beneficial impact on male suicide rates by lowering the rates, yet a detrimental impact on female suicide rates by raising the rates. This means that the suicide rates did not increase for everyone during the economic booms and busts as predicted by Durkheim. The impact depended on the social group involved. Durkheim had posited that because economic prosperity and depression bring about less social integration and less social regulation than do normal economic situations, the suicide rate rises accordingly. Another time-series study by Bijou Yang, D. Lester, and C. Yang (1992) correlated suicide rates in the United States with economic and social variables. The study revealed that, from 1952 to 1984, the economic variables of gross national product per capita and the unemployment rate contributed significantly to the suicide rate, as did the social variables of divorce rate and female participation in the labor force. Thus, social processes that have been presumed to decrease social integration appear to be associated with a higher suicide rate.

Durkheim's main point was that, although we think of suicide as a supremely individual and personal act, it occurs within a social context that influences individual acts. For example, different types of society produce different rates of suicide. **Egoistic suicide** occurs when persons are inadequately integrated into the society, such as intellectuals or persons whose talents or stations in life place them into a special category (celebrities or stars) and who therefore are less likely to be linked to society in conventional ways.

Altruistic suicide happens when individuals are overly integrated into the society (an exaggerated concern for the community) and are willing to die for the group, such as the Japanese kamikaze pilots of World War II or terrorists of today. A rising concern is developing in Japan today over a phenomenon known as **karoshi** (death from overwork), a sort of altruistic suicide. This obsession with overwork is found primarily in middle-aged men who work exceedingly long hours (12 hours per day on the job plus another nearly 2 hours commuting to and from the job) and become so exhausted that they literally die, usually from cardiovascular complications (Sakano & Yamazaki, 1990). Through their hard work, these men are demonstrating their loyalty to the company, being willing to stay and work after hours—overly integrated into the group. Many feel guilty when they take holidays and are away from their jobs. Also, many of these workers tend to think of their jobs even when they are not at work. They die from stress and overwork "for the good of the company."

Anomic suicide results from the lack of regulation of individuals when the norms governing existence no longer control those individuals (feel let down by the failure of social institutions), such as people in business who commit suicide after a stock market crash or middle-aged individuals who kill themselves after suddenly losing their jobs because of downsizing.

A fourth type, called **fatalistic suicide,** was later introduced by Durkheim and has recently received more attention. This type suggests that a person may receive too much control by society and feel oppressed by its extremely strict rules. Fatalistic suicides are a response to the despair engendered by a society allowing little opportunity for satisfaction or individual fulfillment. Both the altruistic and the fatalistic suicides involve excessive control of the individual by society (Kastenbaum, 1995). Extreme examples of such control would be living in slavery or under totalitarian rule. A less-extreme example would be rules and regulations in a bureaucracy that

A problem with many workers (primarily men) in Japan is that they are so "married" to their jobs that they literally kill themselves from overwork. They "overdo it" for the good of the company. Several Japanese businesses are currently trying to address this problem, called karoshi (death from overwork).

make no sense to some employees; nonetheless, "they are the rules" and one must abide by them.

Conflict Theory

From a conflict perspective, not everyone within a society has an equal access to power and material objects. Thus, some individuals are in a more favored position than others. One who feels "left out," mistreated, or not in control of a situation may feel somewhat alienated. According to Karl Marx, **alienation** is a common state of consciousness in a capitalist society where humans are more concerned about the products they create than the people who produce them.

Capitalism, Marx noted, separates laborers from nature, from their work, and from each other. In addition, workers are not allowed any creativity and can never realize their full potential. Thus, alienation results in a loss of self or a lack of self-realization. A worker on an assembly line, for example, performing the same task over and over again for 8 hours, is likely to feel alienated. Whereas a craftsperson, who takes a raw material in hand and creates a finished product, probably would not be inclined to feel alienation.

These constant feelings of alienation may lead over time to a feeling of power-lessness, which in turn may lead to helplessness and hopelessness. Suicide may be the individual's response to these intolerable feelings caused by alienation in a capitalist society.

Dramaturgical Perspective

Derived from the general approach of symbolic interactionism, the **dramaturgical perspective** uses the metaphor of the drama to explain behavior. Rather than asking questions to find out about the motivations of the individual who committed suicide, someone using the dramaturgical approach would observe events related to the suicide. For example, after a suicide, family members and friends react in different ways. A dramaturgical analysis might examine those actions and interactions, paying close attention to nonverbal behavior. They would attempt to find the meaning of the suicide in these actions, not in interviews with the survivors.

Behavior is analyzed from the standpoint of the observer (**etic approach**) rather than from the subjective meanings of participants (**emic approach**). Whereas the dramaturgical analyst emphasizes action, the symbolic interactionist emphasizes intention. Both perspectives lead to an examination of meanings derived from interaction.

A good example of the dramaturgical approach is the film entitled *But Jack Was a Good Driver*. In this film, Jack, a high school student, has died in a single-car accident. As his two friends walk away from the graveside service at the cemetery, they begin to recall recent events involving Jack. They describe the actions of Jack in recent weeks. He had flunked a chemistry exam, broken up with his girlfriend, and given away his CD collection, all actions that were out of character for Jack. They conclude that Jack's "accident" was probably suicide (Jack was a good driver and should have been able to negotiate that curve) because his actions of recent weeks showed some evidence that he might have been planning for his death. The viewer learns about Jack's suicide through the conversation of his friends. The movie exemplified a dramaturgical approach.

Existentialist Perspective

Existentialist philosophy acknowledges that human existence is finite and that we all must face death. One must assume responsibility for his or her own actions and must be responsible for moral choices and actions. This anticipation of death affects daily actions, bringing about an understanding of how death and life are linked.

From an existentialist point of view, it is conceivable that the confrontation with death during a suicidal crisis may cause suicidal individuals to sense their aloneness and personal responsibility for their experience. The choice of death is considered as valid as the choice of life—both the result of reflection (Charmaz, 1980).

Existentialism is linked to the phenomenological method of inquiry. Using an emic approach, death is studied by asking a dying person about the meaning of death. In studying attempted suicide, one would conduct interviews with persons having attempted suicide to determine what they were thinking during their crisis (Charmaz, 1980).

SUICIDE THROUGH THE LIFE CYCLE

Our perception of life and death and the way we respond to stress changes as we go from childhood to adolescence to adulthood to old age. And these changes affect the way we view suicide. Likewise, suicide rates vary. Rates are high among adolescents and young adults and highest among elderly persons. Let's look at suicide through the life cycle.

Childhood Suicide

Suicidal deaths are virtually nonexistent before age 5 and are rare between 5 and 9 years of age. Suicide is similarly uncommon among children between the ages of 10 and 14 (Smith, 1985). In fact, the U.S. Division of Vital Statistics does not report any deaths of children younger than age 8 as suicides, regardless of what data may have been entered on a given death certificate—"other" and "unknown" and "unspecified causes" are the classifications used for these deaths.

Adults prefer to believe that children do not commit suicide, notes Israel Orbach (1988), possibly harboring the false perception that childhood is a carefree, happy time. For example, one U.S. county coroner stated that any death of a child younger than 14 should be regarded as accidental, even when a suicide note is left behind (Orbach, 1988).

As with many other suicides, child suicides are easily mistaken for accidents. The causes of death themselves involve an element of chance: traffic accidents (sometimes running into traffic), falls (jumping?) from high places, fatalities in handling guns, or drowning (Orbach, 1988). The fact that young children do not write suicide notes compounds this difficulty because notes are a chief category of evidence that coroners use to verify suicide (McGuire & Ely, 1984). The Suicide Prevention Center of Los Angeles, in its work with children and adolescents in crisis (Peck, 1980), estimates that 50 percent of the deaths reported as accidents in the groups they studied are, in reality, unreported suicides. Thus, the state of reporting suicides and suicide attempts of children and adolescents is filled with inconsistencies, confusion, and concealment.

The role of family seems to play a major part in suicide attempts of children. A study of 8- to 13-year-olds, including psychiatric inpatients and those who had attempted suicide (Asarnow & Carlson, 1988) found strong support for a link between suicide attempts in children and perceptions of low family support. Another study (Garfinkel, Froese, & Hood, 1982) of 505 children and adolescents who made 605 suicide attempts and were seen in an emergency room (compared with a control group of another 505 children and adolescents admitted and matched by sex and similar age) led to a similar conclusion. In this case it was family disintegration that seemed to be a key factor. Children with one or more family members who have committed suicide are at higher risk of attempting suicide than are children who do not have this family history; this is true regardless of the degree of closeness that the children had with the suicidal family members (Gutierrez, King, & Ghaziuddin, 1996).

Though the suicide rate among children is not high, a study of junior high school children (Domino, Domino, & Berry, 1987) indicates that more than 20 percent have "seriously thought about suicide" and that a slightly higher percentage have personally known someone who committed suicide.

Adolescent Suicide

A noticeable shift in suicide rate occurs at age 15, when it increases dramatically. In the early 1990s, the adolescent suicide rate stood at 12.9 per 100,000 (National Center for Health Statistics, 1995). Suicide is the second leading cause of death among adolescents and young adults from age 15 through 24. Adolescent and young adult suicide is not unique to contemporary times, but the rate has risen significantly since 1950, when it was only 4.5 per 100,000. These numbers could reflect a greater willingness to report suicide, as attitudes have changed since 1950. In recent years, copycat suicides (discussed later in this chapter) have certainly called attention of the media to suicide and have caused parents and educational personnel to be even more aware of suicide.

The suicide rates for adolescents and young adults between the ages of 15 and 24 doubled between 1960 and 1989 but leveled off between 1979 and 1984 (Bingham, Bennion, Openshaw, & Adams, 1994). However, between 1986 and 1991, while the rate remained stable for females (aged 15 to 19) in all ethnic groups, the suicide rate among black males was increasing at a faster rate than that among white males (Shaffer, Gound, & Hicks, 1994). While adolescent black males and females still have lower rates overall than whites, there has been a sudden and sharp recent increase among young black males (Gibbs, 1997).

Accidents and homicides are viewed by many researchers as disguised suicides (Coleman, 1987). Many drug overdoses, fatal automobile accidents, and related self-destructive food and alcohol disorders are probably uncounted teen suicides. Thus, the total number of adolescent suicides may be greater than reported. In N. Conrad's study (1992) comparing 11th and 12th graders who reported suicidal behavior with those who did not, 93 percent of those who reported suicidal behavior also knew a friend, family member, or acquaintance who had attempted suicide. On the other hand, only 50 percent of the students in the sample who did not report suicidal behavior knew someone who had attempted suicide. Likewise, P. Gutierrez, C. A. King, and N. Ghaziuddin (1996), in a study of adolescent attitudes about death in relation to suicide, concluded that adolescents experiencing the suicide of a friend or immediate family member or exposed to attempted suicide re ported a weaker attraction to life and a stronger attraction to death than did adolescents lacking this experience.

In taking their own lives, boys tend to use violent means such as shooting and hanging, though girls seem to be turning to violent means also (Brozan, 1986). Perhaps by using violent means these adolescents are suggesting that they indeed are serious about their "attempts."

David Moller believes that attempted adolescent suicide must be interpreted as more than a "cry for help" (Moller, 1996). It is a response to the awareness of painful life circumstances and a fatalistic view of them. It is an *absence* of hope, with hope being the ability to imagine a different future. Therefore a suicide attempt is an effort to restore a sense of personal autonomy and power in the face of perceived powerlessness. Moller notes that a completed suicide for the adolescent makes a final and dramatic assertion of personal strength. This final act of life is ironically life-affirming—motivated by the adolescent's thirst for a life that is meaningful, stable, and autonomous. But the suicide must be successful. A failed attempt is considered much less favorably, according to Daniel Bell (1977). Bell found that college students in the mid-1970s viewed suicidal peers as "cowardly, sick, unpleasant, and disreputable" and were much more negative in their attitudes toward peers who attempted suicide and lived than toward those who completed suicide.

Robert Kosky (2000) observes that it would be interesting to find out the prevalence of attempted suicide among adolescents in cultures where formal puberty rites of passage exist. His prediction would be that, in these societies, there is less need for young people to find their own way of testing identity themselves and, therefore, less likelihood of uncontrolled or personal self-injurious behaviors. Perhaps attempted suicide in adolescence may be modern Western society's form of the rite of passage rituals.

Warning Signs of Suicide

Any unusual behavior on the part of an adolescent *may* be an indication that his or her life is not right and *could be* an early warning of more bizarre behavior in the future, such as attempted or completed suicide. For example, if an individual begins taking long walks at 2 A.M., when she or he had previously not done so, this is probably a sign that something is bothering him or her. Suddenly acting in a strange way, such as offering to give away a CD collection, without an obvious reason, is not normal.

Any stressful situation, especially those involving a loss, can trigger suicidal behavior. Breakup with a boyfriend or girlfriend, moving to a new community and being the "new kid in town," death of a significant other, divorce in the family, or failure on an exam are all situations that could lead to a suicide attempt. Friends

---

LISTENING TO THE VOICES
The "Solution" Is Suicide

The following letter, written by a 17-year-old who took his life, apparently indicates that death was "the solution" to this adolescent's dilemmas of life. He lived with various problems but tended to keep his pain mostly to himself.

> Dear Mom, Dad, and everyone else,
>
> I'm sorry for what I've done, but I love you all and I always will, for eternity. Please, please, please don't blame it on yourselves. It was all my fault and not yours or anyone else's. If I didn't do this now, I would have done it later anyway. We all die some day, I just died sooner.
>
> Love,
>
> John

After John's body was found, people began to piece together his life and were filled with the inevitable "if onlys." This reconstruction of an interpretation after the event of suicide follows the dramaturgical approach, as discussed earlier.

From "Helping Suicidal Adolescents: Needs and Responses" (pp. 151–166), by A. L. Berman, 1986. In C. A. Corr & J. N. McNeil (Eds.), Adolescence and death *[Microfilm] New York: Springer Publishing Company.*

---

and family should be alert to the "warning signs" of suicide. These warning signs could include depressed mood, eating problems, and truancy and other conduct problems (Wichstrom, 2000). Other predictors include substance abuse, low self-worth, a restricted social network and a lack of social integration, suicidal ideation, and alcohol intoxication.

Sometimes an adolescent will commit suicide without having left any "visible" signs. Even if one is signaling warnings, they may not be obvious to those around. Since most of us are not trained in clinical psychology or psychiatry, we need to be alert to unusual behavior around us, so that we might be in a position to thwart an adolescent's suicidal efforts and help him or her seek professional counseling.

Explaining Adolescent Suicide

Adolescence is a time of frustration and confusion. Yet at the same time, because adolescents are often treated like children and feel scared and uncertain, it is no wonder that the suicide rate in this age group is so high. Suicide becomes an available solution to one's problems. When stress is experienced, a constricted view of future possibilities and a momentary fix on present escapes may prevail. Impulsive action and limited alternative solutions to problems may turn suicidal fantasies into suicidal behaviors. Suicide is an answer (though permanent) to a set of temporary problems. The problem is gone, but so is life.

Because adolescents are caught between childhood and adulthood, the often intense and conflictual task of separating from the world of family can be more difficult when family dynamics interfere with a child's move toward self-sufficiency (Berman, 1986). Parents of suicidal adolescents have been found to experience more overt conflict and threats of separation and divorce. Loss of a parent before the children's 12th year also seems to be a predictor of future suicidal behavior. Suicidal adolescents report receiving little affection, hold negative views of their parents, describe time spent in their family as unenjoyable, have deficient problem-solving skills, tend to have more prevalent use of drugs and alcohol, and see themselves as different from their parents. Socioeconomic adversity, poor parent-child attachment, exposure to sexual abuse, high rates of neuroticism, and novelty seeking also seem to be predictors of suicide attempts (Ferguson, Woodward, & Horwood, 2000). Psychiatric conditions that may also contribute to suicidal behavior include depression, anxiety disorders, substance abuse disorder, and conduct disorder. Suicidal behavior in childhood appears to be more closely associated with family turmoil and stress than is suicidal behavior in later life. Once a suicide attempt has been made, future attempts may be expected, if the first attempt occurs at a younger age. Also early pubertal development, not living with both parents, and poor self-worth may contribute to a second suicide attempt.

Active peer involvement may be the support that one needs to discourage suicidal behavior. Being involved with peers is negatively correlated with a depressive mood in adolescence. Suicidal adolescents tend to show greater withdrawal from, and less involvement with, the school milieu. On the other hand, suicidal adolescents tend to have more frequent and serious problems with peers, to be more interpersonally sensitive, and to be less likely to have a close confidant (Berman, 1986). Adolescents are especially vulnerable to rejections from their peer group (Rich et al., 1991). Suicidal adolescents are particularly sensitive and suggestible because they often lack reinforcing role models and may be low in self-esteem and ego strength. Thus, suicidal behavior by one may lead to an epidemic of followers.

Copycat Suicides

Friends of teens who commit suicide may be at risk for cluster or copycat suicides. A **suicide cluster** is an excessive number of suicides occurring in close temporal and geographical proximity and most closely approximates the concept of an "outbreak" in a particular community (Gould, Wallenstein, & Davidson, 1989). Philip May (1990), in compiling a bibliography on suicide among Native Americans, concluded that suicide clusters among particular groups of youths tend to take on epidemic form in many Native-American communities. Suicide clusters among teenagers have been noted in the media, and imitation is generally cited as the "cause." Cluster suicides account for between 1 and 5 percent of all youth suicides in the United States (Fulton & Metress, 1995). Teens who have already lost a friend to suicide see dying as glamorous and attention getting. Studies of clusters (Olson, 1997) confirm that teens do not commit suicide without thought, but suicide by a classmate or acquaintance may give them the idea of a dramatic way to solve some of their problems.

DEATH ACROSS CULTURES
Suicide Ideation Among Native-American Adolescents

Three distinct adolescent Native-American tribal groups were studied for differences in factors associated with **suicide ideation.** The groups were the Pueblo tribe (who are **matrilineal,** have an agrarian economy, and have a highly cohesive social structure that emphasizes strong controls over individual behavior), the Southwest tribe (who are characterized as matrilineal with a pastoral tradition and maintain a social structure that is slightly less cohesive than the Pueblo group and for whom to speak or think of death is highly inappropriate), and the Northern Plains tribe (who are **patrilineal,** are nomadic hunters and gatherers, and place great emphasis on individuality and self-sufficiency).

Findings from this 5-year study of 1,353 students in grades 9 to 12 reveal that the factors associated with suicide ideation among adolescents in each of these three groups were consistent with each tribe's social structure, conceptualization of individual and gender roles, support systems, and conceptualization of death. For the Pueblo youth, the identified correlates of suicide ideation were (1) reporting that a friend attempted suicide in the last 6 months, (2) lower perceived social support, and (3) depressed affect. For the Southwest adolescents the analyses indicated that coming from a home without both biological parents and experiencing more interpersonal stress contributed to thinking seriously about suicide. For the Northern Plains tribe youth, suicide ideation was associated with low self-esteem and higher levels of depressed affect. These youth were more likely to think seriously about suicide, had more negative thoughts about their abilities, and had a more depressive symptomatology.

Whereas considerable heterogeneity in the factors associated with suicide ideation emerged across these three Native-American tribes, no significant group differences were evident in the prevalence of suicidal ideation. This finding suggests that levels of critical suicide ideation may show limited variation across tribes, and a small percentage of youth in all of these tribes may experience serious and recurrent thoughts about suicide. Substance use is not a key correlate of suicide ideation in any of these tribes, though this finding does not agree with all studies of suicide about Native Americans. It may be that although substance use is not a key correlate of suicide ideation, it could instead be a critical contributor for acting upon such thoughts. Another factor could be socioeconomic adversity, a condition shared by all these groups. It also seems that social cohesion is not a significant factor in suicidal ideation.

From "Factors Associated With Suicide Ideation Among American Indian Adolescents: Does Culture Matter?" (pp. 332–346), by D. K. Novins, J. Beals, R. E. Roberts, & S. M. Manson, 1999, Winter, Suicide and Life-Threatening Behavior, 29(4).

Madelyn Gould and David Shaffer (1986) studied the variation in the numbers of suicides and attempted suicides by teenagers in the greater New York area 2 weeks before and 2 weeks after four fictional films with suicides were broadcast on television in the autumn and winter of 1984 and 1985. The number of suicide attempts in the

---

PRACTICAL MATTERS
How You Can Help in a Suicidal Crisis

1. Recognize the clues to suicide. Look for signs of hopelessness and helplessness. Listen for suicide threats and words of warning. Notice if the person becomes withdrawn and isolated.

2. Trust your own judgment. If you believe that someone is in danger of suicide, act on your beliefs. Don't ignore the signs of suicide.

3. Tell others. Share knowledge with parents, friends, teachers, employers, or other people who might help. If you have to betray a secret to save a life, do it. Don't worry about breaking a confidence if someone's suicidal plans are revealed to you.

4. Stay with a suicidal person. Don't leave a suicidal person alone if you think there is immediate danger. Stay until help arrives or the crisis has passed.

5. Listen. Encourage a suicidal person to talk. Don't give false reassurances that "everything will be okay." Listen and sympathize with what the person says.

6. Urge professional help. Offer to make an appointment for and go with the person for professional help, if that is what it takes. Call your community hotline or crisis number for suggestions.

7. Be supportive. Show the person that you care. Help to make the person feel worthwhile and wanted.

From Youth and Suicide *(p. 96),*
by F. Klagsbrun, 1976,
New York: Pocket Books.

---

period after the broadcasts was significantly greater than the number before the broadcasts, leading the researchers to conclude that some teenage suicides are imitative.

Another study, by David Phillips and Lundie Carstensen (1986), examined the relation between 38 nationally televised news or feature stories about suicide from 1973 to 1979 and the fluctuation of the rate of suicide among American teenagers before and after these broadcasts. The observed number of suicides by teenagers within 7 days after these broadcasts was significantly greater than the number expected. The more networks carrying a story on suicide, the greater was the increase in suicides thereafter. Teenage suicides increased more than adult suicides after stories about suicide. An extensive review of suicide clusters by Madelyn Gould, S. Wallenstein, and L. Davidson (1989) confirms that temporal and geographical clusters of suicides occur, that they appear to be multidetermined, and that they appear to be primarily a phenomenon of youth. Imitation and identification are factors hypothesized to increase the likelihood of cluster suicides.

With the suicide of Kurt Cobain, some individuals feared that large numbers of copycat teen suicides would occur (Waters, 1994). However, it was noted that those apt to copy the rock star were those who already had significant emotional problems. Those individuals with low self-esteem and self-identify tend to get hooked on heroes and are more likely to behave in the same self-destructive way.

Though the preceding findings certainly suggest a correlation between media broadcasts on suicide and increased suicide clusters among adolescents, this is not to suggest that such media exposure "causes" an overall increased rate of suicides. Indeed, viewers previously thinking of suicide might be persuaded one way or the other after watching television programs on suicide. Those who do kill themselves are more likely to have made a previous attempt at suicide, to have lost a close friend or relative to violent death, or to have suffered a recent breakup with a girlfriend or boyfriend (Fulton & Metress, 1995).

For nonsuicidal persons as a group, however, the impact may be somewhat different. For example, in exposing university students to films on suicide and violence, one study (Biblarz, Biblarz, Pilgrim, & Baldree, 1991) concluded that there was no reason to believe that nonsuicidal young people as a group will view suicide in a more positive light as a result of watching movies with suicidal or violent content. However, exposure to suicide films appears to have increased viewers' arousal levels significantly, if only temporarily.

Studies of the effect of the media on suicide have neglected the audience receptivity to suicide stories, notes Steven Stack (1992). Thus, Stack set out to test the symbolic interaction theory that the degree of media influence is contingent on audience receptivity. He developed a taxonomy of stories from the Great Depression using the years 1933–1939 (1933 being the first year for which suicide data were collected for the nation as a whole). Stack hypothesized that the economic context of the nation's worst

The suicide of Nirvana's Kurt Cobain in April 1994 shocked millions of young rock fans throughout the world and led to widespread grief and confusion. In Seattle, Washington, Cobain's hometown, a candlelight vigil was attended by 5,000 people, including these two fans.

economic depression of the 20th century would increase the effect of the media on suicide—the economic collapse would promote a suicidal mood of depression, disrupted family organization, and other psychosocial processes thought to underlie suicide risk. However, there was no overall impact of suicide stories on suicide. The only significant relationship found by Stack to be associated with increases in the suicide rate was publicized stories about suicides by political heroes.

Adult Suicide

Though suicide rates are highest among elderly persons and adolescents, adults in between are not immune to death by suicide. Contributing factors to suicide found in adults include feelings of hopelessness, personal losses such as health problems or bereavement, alcohol or drug abuse, and financial problems. In addition, unemployment, not unrelated to financial problems, is a major factor contributing to suicide among adults (Pirkis, Burgess, & Dunt, 2000).

As noted in chapter 3, middle-aged adults are in a stage of life where the years remaining yet to live are shorter than those already lived. Thus, individuals in this age category may begin to panic over the fact that "time is running out," and they will not accomplish all they had hoped. This age group begins to see deaths of peers from natural causes (heart disease and cancer) and is therefore reminded that life is not forever. Also, being the "sandwich generation," often having to deal with "hang-on" young adult children and aging parents at the same time, may prove too stressful for some. Suicide again may become a permanent solution for a temporary set of problems.

Some attempted suicides in adults may belong to a "culture of risk" in which the individual has not matured appropriately and may carry the adolescent developmental processes in an unresolved state (Kosky, 2000). Overall, one would expect that the older people get, the less influence the risk-taking culture of adolescence would have on them. Attempted suicide among adults would probably be failed suicide.

A recent study (Miller, Hemenway, & Rimm, 2000) of 50,000 middle-aged and elderly white males in the United States examined the relation between smoking and suicide from 1986 through 1994. A positive association between smoking and suicide was found among these men. Although we cannot infer causality, simply a correlation, the findings did indicate that the smoking-suicide connection is not entirely due to the greater tendency among smokers to be unmarried, to be sedentary, to drink heavily, or to develop cancers.

Most adult individuals who commit suicide tend to talk about their attempt before the act. Other than an outright statement, "I am going to kill myself," more subtle verbal clues might include statements such as "I'm not the person I used to be," "You would be better off without me," "I can't stand it anymore," "Life has lost its meaning for me," and "Nobody needs me anymore." Behavioral signs of suicide might include getting one's house in order as if preparing to depart, frequent crying without explanation, sudden irregular sleeping habits, loss of appetite, inability to concentrate, sudden changes in appearance, or sudden withdrawal from society. Though the "signs" may appear obvious after a completed suicide, at the time, we might not recognize them as signs of suicide.

Elderly Suicide

Throughout the rest of the world (Moscicki, 1995), as well as in the United States, a much greater percentage of elderly people commit suicide than of any other age group (Brown, 1996). It is estimated that 15 percent of the general population in Western societies are 65 and older, yet this group commits up to 30 percent of all known suicides (Adamek & Kaplan, 1996).

Suicide among the elderly may be even more prevalent than the statistics suggest, note researchers R. N. Butler, M. I. Lewis, and T. Sunderland (1991). Reasons cited are the following: (1) intentional overdoses of prescribed medications may go unrecognized or unreported; (2) older persons may commit "chronic" or "passive" suicide whereby they lose the will to live and simply stop caring for themselves; (3) they may cease taking medications, take increased health risks, or delay treatment for medical conditions; and (4) they may slowly stop drinking and eating until they become sick and die.

The suicide rate of persons aged 65 and older rose nearly 9 percent from 1980 to 1992, according to the *Morbidity and Mortality Weekly Report*'s "Suicide Among Older Persons" (1996). In the three decades before 1980, the suicide rate of elderly persons exhibited a decreasing trend and decreased 42 percent between 1950 and 1980 (Osgood & McIntosh, 1986). Some sociologists and psychologists have proposed that the increased suicide rate among elderly persons may be due to technological advances that extend lives and result in intolerable quality of life and/or medical treatments of ailing older patients that result in the depletion of family savings (Tolchin, 1989). In addition, the increased suicide rate among elderly persons may result from social isolation and society's growing acceptance of suicide ("Suicide Rate," 1996). Our society also places great value on youth, health, and beauty. This ageist view adds to the sense of worthlessness that some elderly persons experience (Devons, 1996).

Age by itself does not cause suicidal behavior. The person statistically most likely to commit suicide in the United States, however, is the white male older than age 85 (Devons, 1996). The traditional, assertive, achievement-oriented, middle-aged male may not be as adaptable to the retirement situation as the traditionally more passive, nurturant, middle-aged woman. Thus, the greater change in behavior and self-esteem necessitated in the older man may create more stress and depression if he is unable to find satisfaction in this new situation. A greater potential for suicidal behavior results (Aiken, 1991).

Gender Differences

In the United States, men commit 81 percent of suicides among the elderly ("Suicide Rate," 1996). Because gender is one of the most important predictors of suicide in the elderly, Silvia Canetto (1992) has addressed the question of why older women are less suicidal than older men. She suggests that women are less likely to die of suicide because they are more likely to seek professional help. Suicidal behavior in elderly persons has been attributed to poverty, yet older women are more likely to be poor but less likely to be suicidal than older men. Canetto surmises that for many women, limited access to financial resources has been a long-standing problem, whereas for many men limited access to financial resources is less common until retirement. Thus, financial difficulties for older men may be a new experience.

Though living alone is frequently assumed to be a precipitant of suicidal behavior among elderly persons, it seems to be a more significant factor for men than women. Canetto (1992) concludes that living alone may be more normative for older women than for older men who tend to be more isolated than older women living alone.

Though widowhood is a predictor of suicide in both women and men, the effect seems to be stronger for men than for women. Canetto (1992) observes that older men are more likely to depend on their wives for emotional support than vice versa because older women appear to have more emotional connections with friends than do older men. Older men depend on their wives for their personal care and the running of the home. Losing a spouse may thus disrupt social support and emotional functioning more for a man than for a woman.

Although older men have a much higher suicide rate than older women, between 1979 and 1992 data from the National Center for Health Statistics revealed an increasing trend in rates of suicide for women over the age of 65 (Adamek & Kaplan, 1996). During this period of time, shooting replaced poisoning as the most prevalent method of suicide by women 65 and older.

Causes of Suicide

The *threat* of suicide is uncommon in elderly persons—they simply kill themselves, and their attempts rarely fail. When an elderly person attempts suicide, he or she usually has a profound death wish and thus selects ways that are likely to fulfill that wish. Recent studies have revealed that suicide is rarely an impulsive act unheralded by any warning signs but rather as a rule is an act that can be anticipated. It has been estimated that 75 percent of elderly individuals contemplating suicide consult their physicians only shortly before committing the act (Achte, 1988). However, physicians apparently do not typically bring up suicide with elderly patients, as noted by a recent Gallup survey (1993). In a survey of 802 Americans aged 60 and older who lived independently, Gallup found that, during their last physical examination, only 6 percent had been asked about depressed, suicidal thoughts.

Kalle Achte (1988, p. 58) states that loneliness is the great tragedy of aging: Old age has been called "the era of losses." After each loss, an elderly person has to try to pull himself or herself together and to adjust in order to be able to retain psychological equilibrium. Losses for the elderly may include jobs, social status, health, independence, friends, and family (Osgood, Brant, & Lipman, 1991). These losses may result in stress at a time when the individual is least able to cope. If there is no possibility of communication with other people, loneliness can become extremely distressing for the older individual.

Robert Kastenbaum (1992) agrees that the reduced likelihood of open and substantive communication contributes to the possibility of suicide in the elderly. Kastenbaum notes that the following tend to reduce communication channels for elderly persons: (1) living alone reduces the opportunities to express and explain suicidal intent to a caring person; (2) residing in a low-income, transient urban area tends to make a person socially invisible; (3) the depressive state often associated with suicidal behavior frequently has the effect of reducing communication; and (4) the fact that most suicides among elderly persons are committed by men suggests that reluctance to seek help may be strongest among those elderly persons who are at greatest risk.

Though some danger signals for suicide among elderly persons have already been noted, a list of all such signals would include depression, withdrawal, bereavement (especially within the last year), isolation, expectation of death from some cause, retirement, loss of independence, physical illness, desire and rational decision to protect survivors from financial disaster, philosophical decision (no more pleasure or purpose in life), decreased self-esteem, and organic mental deterioration (Eddy & Alles, 1983).

A National Institute of Mental Health study (Hohmann & Larson, 1993) found that an elderly person with a severe mental health problem is more likely to seek the help of clergy than the aid of a mental health specialist. Because 80 percent of elderly persons are members of a church or synagogue and over half (52 percent) worship at least once a week (Gallup, 1994), both clergy and the larger religious community may have an effective means to help those elderly persons in distress (Weaver & Koenig, 1996). More than 50 studies in the past 15 years demonstrate a strong association between faith practices (for example, regular church attendance, Bible reading, and prayer) and mental health among older persons, according to Andrew Weaver and Harold Koenig (1996). Older persons who practice their faith frequently have lower rates of depression, alcoholism, and hopelessness (Koenig, 1994).

Joseph Richman (1992) says that there is a great publicity campaign in favor of the view that suicide is a rational response to growing old. Though many elderly suicidal persons have severe problems in accepting their age, many of them have revealed similar problems in self-acceptance throughout their lives, and many felt "old" at age 15, notes Richman. The decision to commit suicide does not arise overnight, but rather is part of a lengthy process during which the significance of the individual's entire existence has been called into question. In working with over 800 suicidal persons, many of whom were elderly, ill, or terminally ill, Richman (1992, p. 134) says, "I can state unequivocally that suicide is not a rational response to aging."

RATIONAL SUICIDE

There have been times in history when misery was so common and the outlook so bleak that suicide probably seemed a serious option. As Mark Twain once said "there are times in life when we would like to die temporarily." The Judeo-Christian tradition, however, has generally advocated life and strongly urged that life not be disposed of no matter what the temptations. For an elderly person, taking one's life tends to be a more rational or philosophical decision. Elderly individuals and others today seem to be asking if life is to be valued under all conditions because it has intrinsic value or if the value of life is relative to the circumstances.

Elderly persons are held in high esteem in many traditional societies because of the store of knowledge they have accumulated over the years. Perhaps not so coincidentally, suicides among the elderly in these cultures seem to be less common than in those of many modern industrialized nations.

The "Listening to the Voices" box, describing Mr. and Mrs. Saunders' final days, suggests that the value of life for them was relative to the circumstances.

Although suicide is usually treated as the product of mental illness or as a desperate cry for help used by someone who does not really want to die, suicide can be considered a rational act. James Werth (1995) describes **rational suicide** as having five components: (1) the person has a realistic assessment of the situation; (2) the person's mental processes are unimpaired by psychological illness or severe emotional distress; (3) the person has a motivation that would be understandable to a majority of uninvolved community members; (4) the decision is deliberated and reiterated over a period of time; and (5) if at all possible, the decision-making process should involve the suicidal person's significant others.

Certainly one could argue that the Saunders' act was a rational homicide/suicide. Jacques Choron's assertion (1972) that "rational" implies "not only that there is no psychiatric disorder but also that the reasoning of the suicidal person is in no way impaired and that his motives would seem justifiable, or at least 'understandable,' by the majority of his contemporaries in the same culture or social group" seems to "fit" the case of the Saunders.

Richard Brandt (1986) states that the person contemplating suicide is obviously making a choice between future world courses—the future world course that includes his or her demise an hour or so from now and several possible courses that

---

LISTENING TO THE VOICES
After 60 Years of Marriage, Couple Decides to Leave World Together

Julia Saunders, 81, had her hair done. Her husband, Cecil, 85, collected the mail one final time and paused to chat with a neighbor. Inside their mobile home, they carefully laid out a navy blazer and a powder-blue dress.

After lunch, the Saunders drove to a rural corner of Lee County and parked. As cows grazed in the summer heat, the couple talked. Then Cecil Saunders shot his wife of 60 years in the heart and turned the gun on himself.

Near the clothes they had chosen to be buried in, the couple had left a note:

"Dear children, this we know will be a terrible shock and embarrassment. But as we see it, it is one solution to the problem of growing old. We greatly appreciate your willingness to try to take care of us.

After being married for 60 years, it only makes sense for us to leave this world together because we loved each other so much."

On the floorboard of the car, Cecil and Julia Saunders had placed typewritten funeral instructions and the telephone numbers of their son and daughter.

In Philadelphia, a police officer stood by as the Saunders' son, Robert, 57, was told of his parents' death. His parents wanted no tears shed over their decision to die. The note they left for Robert and his sister, Evelyn, 51, ended with a wish: "Don't grieve because we had a very good life and saw our two children turn out to be such fine persons. Love, Mother and Father."

From Minneapolis Star and Tribune *(p. 12A), October 4, 1983.*

---

include his or her demise at a later point. Although one cannot have precise knowledge about many features of the latter courses, it is certain that they all will end with death sooner or later. The basic question for the person to answer to determine the best (or rational) choice is which course would be chosen under conditions of optimal use of information, when all personal desires are taken into account. It is not a question just of what is preferred now.

Certainly one's desires, aversions, and preferences may change after a short while. When one is in a state of despair, nothing but the thing that he or she cannot have—a love that has been rejected or a job that has been lost—may seem desirable. The passage of time is likely to reverse all of this. So if one acts on the preferences of today alone, when the emotion of despair seems more than one can stand, death might seem preferable to life. But if one allows for the preferences of the weeks and years ahead, when many goals might be enjoyable and attractive, life might be found to be preferable to death. Thus, Brandt is suggesting that a future world course might well be different than that of the present.

Brandt further suggests that one must take into account the infirmities of one's "sensing" machinery. Knowing that the machinery is out of order will not tell one what results it would give if it were working; thus, the best recourse might be to refrain from making any decision when one is in a stressful frame of mind. If decisions have to be made, one must recall past reactions, in a normal frame of mind, to outcomes like those under assessment. The future world course obviously did not look "enjoyable and attractive" to the Saunders after 80 years of life.

Individuals opposed to suicide outright would argue that a person does not have the right to commit suicide regardless of the circumstances. Others believe that circumstances may be so unbearable that suicide is understandable, but that this should be an individual act and not one encouraged by society or the legal system, note Ellen Ingram and John Ellis (1992) in their review of literature on suicidal behavior. Opponents with extreme views argue that life should be preserved regardless of circumstances even in the case of a terminal illness. They would argue that suicide is an ambivalent act and that every person who wants to die also wants to live. Such a view suggests that suicidal thoughts are only temporary and that through intervention the crisis often passes (Ingram & Ellis, 1992).

CONCLUSION

Just what constitutes a suicide is not clear today. Many persons display "suicidal behavior" by constantly taking risks and gambling with their lives; yet if death results from such behavior, it is often classified as "accidental" or "natural." Sociologists argue that suicidal persons construct their meanings of suicide and motivations for committing it out of collective values upon which the social structure rests. Meanings of suicide arise out of what people think, feel, and do.

Persons bereaved by a suicide may have a longer and more difficult grieving process due to their perceiving less social support than other bereaved individuals (Thompson & Range, 1992–1993). Some grief reactions, such as shame or rejection, may account for this lack of support. Shame may occur because of the stigma surrounding death by suicide (Silverman, Range, & Overholser, 1994–1995). Rejection may result because suicide implies that the deceased rejected not only life but also family and friends. Suicide survivors may also have more difficulties discussing the death with others; thus, they may try to deny the cause of death.

Suicidal persons are more likely to be adolescents or very elderly, male, and not married. Though age by itself does not cause suicidal behavior, the person most vulnerable to suicide is a white male older than age of 85. Suicidal persons tend to talk about suicide before the act and often display observable signs of suicide. Males are perhaps more "successful" at completing suicide than are females because males tend to use a more lethal weapon—firearms. For suicidal persons, this act becomes an easy "solution" to their problems. Suicide can be a rational act, and one does not have to be "insane" to take one's own life.

Below are some suggestions for dealing with the aftermath of suicide:

1. Be available to the survivors by taking the initiative to help. Do not hesitate to visit them, even if you do not know what to say. Just being present shows that you care.

2. Discourage guilt feelings. Suicide tends to result from long-term feelings; thus, the grievers should not blame themselves with "if only" feelings.

3. Be truthful and honest at all times.

4. Discourage self-pity and encourage the individual to become involved in activities.

5. Encourage the griever to express his or her real feelings. Help guide the individual toward thinking ahead to the future.

SUMMARY

1. Suicide has been around for as long as recorded history.

2. Suicide has historically been viewed in different ways in different places. Even today the general attitude toward suicide remains confused and sometimes contradictory.

3. The meaning of suicide is problematic. Suicide research centers have evolved in recent years to address the complex issues of suicide.

4. In the late 1800s French sociologist Emile Durkheim noted that suicide is related to one's degree of social integration. Different types of society produce different suicide types: egoistic, altruistic, anomic, and fatalistic.

5. The dramaturgical approach to suicide study uses the drama or action to explain the behavior.

6. The existentialist perspective on suicide may view the choice of death as it views the choice of life—both the result of reflection—and not view death as "the enemy."

7. If one feels alienated and powerless, suicide may appear to be a solution.

8. Though age by itself does not cause suicidal behavior, elderly males are the most vulnerable to suicide in the United States.

9. People who commit suicide tend to talk about their attempt before they act.

10. The most frequent method of completed suicide in the United States is by firearms.

11. The concept of rational suicide suggests that a person does not have to be mentally ill to take his or her own life.

12. The media have contributed to copycat suicides by teenagers, according to some researchers.

13. Suicidal deaths are virtually nonexistent before age 5 and are rare between 5 and 9 years of age. Adults prefer to believe that children do not commit suicide.

14. Social factors tend to be associated with suicide: age, gender, marital status, and socioeconomic level.

DISCUSSION QUESTIONS

1. Discuss why the meaning of suicide continues to be problematic today.

2. Give examples of Durkheim's four types of suicide.

3. How does the dramaturgical approach view suicide?

4. How does an existentialist perspective "explain" suicide?

5. Describe signs of suicide. What should you do if you observe some of these signs in a friend?

6. Discuss why the suicide rate tends to be high among adolescents and elderly persons.

7. The concept of rational suicide implies that one does not have to be mentally ill to take his or her own life. Discuss whether you feel that one can be "sane" and take his or her own life.

8. Discuss your reaction to the deaths of Julia and Cecil Saunders.

GLOSSARY

Alienation: Not feeling in control and not feeling a part of a situation. This often results in a lack of self-realization and a feeling of powerlessness and hopelessness.

Altruistic Suicide: Overly integrated into society and willing to die for the group.

Anomic Suicide: Results from a lack of regulation of the individual when the norms governing existence no longer control that individual.

Dramaturgical Perspective: Highlights the construction of action by using the metaphor of the drama to explain behavior.

Egoistic Suicide: Occurs when the person is inadequately integrated into the society.

Emic Approach: An analysis of behavior from the perspective of the participant (the one being studied). If an individual is studying suicide, for example, data could be gathered from one who has *attempted* suicide. What was this experience like from her or his point of view?

Etic Approach: An analysis of behavior from the perspective of the observer (the one doing the study, not the participant).

Existentialist Philosophy: Choice of death being like the choice of life. Death is not the enemy as viewed by other perspectives.

Fatalistic Suicide: Too much control over a person by society may lead to feelings of oppression under extremely strict rules.

Karoshi: Death by overwork, a current phenomenon of concern in Japan.

Matrilineal: A unilineal (not bilineal where kinfolk are counted both through the mother and the father's line of descent) descent or kinship through the mother's line.

Patrilineal: A unilineal descent or kinship through the father's line.

Rational Suicide: Suicide in which the individual is not insane and is aware of what he or she is doing.

Suicide Cluster (Copycat Suicide): An excessive number of suicides occurring in close temporal and geographical proximity.

Suicide Ideation: The formation or conception of suicide ideas by the mind.

SUGGESTED READINGS

Fuse, T. (1997). *Suicide, individual and society*. Toronto: Canadian Scholars Press. Includes discussions of the history of social thought on suicide, theories of suicide, culture and depression, and prevention, intervention and postvention.

Holinger, P. C., Offer, D., Barter, J. T., & Bell, C. C. (1994). *Suicide and homicide among adolescents*. New York: Guilford Publications. An examination of suicide and homicide in terms of the similarities and differences in the two forms of adolescent wastage. This book offers a comprehensive approach to adolescent violence with an emphasis on a public health approach.

Jamison, K. R. (1999). *Night falls fast: Understanding suicide*. New York: Knopf. Traces the reasons underlying suicide, looks into journals, drawings, and farewell notes of suicides, and discusses biological and psychological factors in suicide and new treatments for mental illness.

Kennedy, G. J. (Ed.). (1996). *Suicide and depression in late life*. New York: John Wiley and Sons. Written from a psychiatric perspective, this anthology covers three areas: critical issues in clinical science, therapeutic approaches, and a more-informed public policy.

Laufer, M. (Ed.). (1995). *The suicidal adolescent*. Madison, CT: International Universities Press. An easy-to-read work written primarily from a psychodynamic perspective toward suicidal attempts.

Minois, G. (1999). *History of suicide: Voluntary death in Western culture*. Baltimore: Johns Hopkins University Press. Follows the religious, philosophical, literary, and judicial debate for and against self-murder from antiquity to the end of the Enlightenment.

Pearson, J. L. and Conwell, Y. (Eds.). (1996). *Suicide and aging*. New York: Springer Publishing Company. Examination of how the aging process, cultural factors, cohort effects, personality, medical disorders, and the concept of an individual's life course trajectory may contribute to the problem of suicide.

Smolin, A., & Guinan, J. (1993). *Healing after the suicide of a loved one*. New York: Simon and Schuster. Directed toward the lay public and individuals suffering from the loss of a significant other by suicide.

Stilling, J. M., and McDowell, E. E. (1996). *Suicide across the life span*. Bristol, PA: Taylor & Francis. A thorough review of suicide literature from a developmental psychology perspective.

Werth, J. 1996. *Rational suicide?* Bristol, PA: Taylor & Francis. Presents rational suicide as a legitimate option for some individuals, which has implications for mental health professionals.

Williams, J. M. G. (1998). *Cry of pain: Understanding suicide and self-harm*. New York: Penguin. Helpful in dealing with the full range of emotions families are left to cope with for years after a death by suicide.

Zimmerman, J. K., & Asnis, G. M. (Eds.). (1995). *Treatment approaches with suicidal adolescents*. New York: John Wiley and Sons. A comprehensive and sophisticated program of suicide assessment, treatment interventions, and preventive programs for work with suicidal adolescents.

C H A P T E R 10

Diversity in Death Rituals

Some children love to watch the cremations. The skull is usually the last thing to be burned. Sometimes it collapses with a loud pop, like a balloon bursting. When that happens, the children clap their hands.

—Alexander Campbell, *The Heart of India*

Death occurs in all societies, yet it evokes an incredible variety of responses. At the moment of death, survivors in some societies remain rather calm, others cry, and still others mutilate their bodies. Members of some societies officially mourn for months, while others complete the ritual within hours. In many societies families are involved in preparing the corpse for the funeral ritual; in others, families engage professional funeral directors to handle the job.

The variety of responses to death is further noted by Richard Huntington and Peter Metcalf in *Celebrations of Death: The Anthropology of Mortuary Ritual* (1992). They state that corpses are burned or buried, with or without animal or human sacrifice; they are preserved by smoking, embalming, or pickling; they are eaten—raw, cooked, or rotten; they are ritually exposed as dead or decaying flesh or simply abandoned; or they are dismembered and treated in a variety of these ways. Funerals are times for avoiding people or holding parties, for weeping or laughing, or for fighting or participating in sexual orgies.

Whereas death **rituals** in the United States are generally subdued and rather gloomy affairs, some societies engage in rather spirited activities. The Bara of Madagascar, for example, engage in "drunken revelry" at a funeral—rum is consumed,

343

sexual activities occur, dancing takes place, and contests involving cattle occur (Huntington & Metcalf, 1992). Among the Cubeo of South America, simulated and actual ritual coitus is part of the mourning ritual (Goldman, 1979). The dances, ritual, dramatic performances, and sexual license have the purpose of transforming grief and anger over a death into joy.

In most non Western societies, death is not seen as one event, but rather as a process whereby the deceased person is slowly transferred from the land of the living to the land of the dead (Helman, 1985). The process is illustrated by rituals marking biological death, followed by rituals of mourning, and then by rituals of social death. The deceased person is often viewed as a soul in limbo during rituals of mourning, though he or she is still a partial member of the society (Sweeting & Gilhooly, 1992). For the Kota people of south India, for example, a person is not socially dead until after the dry funeral—the second funeral for the deceased, held annually and lasting 11 days (Mandelbaum, 1959).

UNDERSTANDING DEATH RITUALS

Ritual can be defined as the symbolic affirmation of values by means of culturally standardized utterances and actions (Taylor, 1988). People in all societies are inclined to symbolize culturally defined feelings in conventional ways. Ritual behavior is an effective means of expressing or reinforcing these important sentiments, and it helps make death less socially disruptive and less difficult for individuals to bear (Haviland, 1991). Rituals differ from other behaviors in that they are formal—stylized, repetitive, and stereotyped (Rappaport, 1974). Rituals are performed in special places, occur at set times, and include liturgical orders—words and actions set forth previously by someone.

Death Rituals as a Rite of Passage

Throughout life we will occupy social positions within the societies of which we are a part. Whenever our social position changes our identities change as well. These transitions in identity require of us, as well as of those who are significant in our life, the ability to adjust and adapt to the transitions. Human beings construct rituals as one means of acknowledging and adapting to change. Consider the following transitions: a child becomes an adult, a single person commits to another in marriage, a married couple become parents, a worker retires, and a person dies. In most societies, each of these transitions is marked by collective actions (or social rituals) that acknowledge a change in peoples' identities.

Death is a transition, but only the last in a long chain of transitions, or rites of passage, according to Richard Huntington and Peter Metcalf (1992). In many cultures and religions, being dead is another status change, replete with new roles and obligations. Often the dead are expected to give advice, cure illness, reward good deeds, and ensure a good harvest. In Ireland, the dead are called upon to cure the afflicted and to comfort the lonely (Enright, 1994). Thus, at funerals in Ireland people never say farewell because they fully expect to hear from their friends and loved ones again.

The moment of death is related not only to the process of afterlife, but also to the process of living, aging, and producing progeny. Death relates to life—to the recent life of the deceased and to the lives that he or she has procreated and now leaves behind. There is an eternity of sorts on either side of the line that divides the quick from the dead. Life continues generation after generation, and in many societies it is this continuity that is focused upon and enhanced during the rituals surrounding a death.

One of the best-known accounts of death as one of a series of such **rites of passage** through the life cycle comes from A. Van Gennep (1909/1960) in his treatment of funerals. Van Gennep compared funeral rituals with other rites of passage (such as marriage ceremonies and graduation exercises) and claimed that all rites of passage will have three subrites—rites of separation, rites of transition, and rites of reincorporation.

In the rite of separation, those whose identity is undergoing change will be perceived by the group as different or "other." At the graduation or wedding the graduates or the members of the "bridal party" dress differently. In the middle of the rite of passage the identity of the ones changing will experience some ambiguity—after the bride and groom have exchanged vows, are they married or single? Is the deceased person in the casket as dead as he or she will be after burial or cremation? Finally, for all rites of passage there will be a rite of reincorporation in which the community incorporates in a new way those who have taken on a new identity. These rites of reincorporation are often meals or feasts—graduation or wedding receptions and after-funeral meals.

Funeral or mortuary rites in many societies are critical because they ensure that the dead make the transition to the next stage of life or non-life. Although many cultures continue to have relationships with the dead, they only do so when the dead are safely in the next world. However, persons for whom no rites are performed

> . . . are the most dangerous dead. They would like to be reincorporated into the world of the living and since they cannot be, they behave like hostile strangers toward it. They lack the means of subsistence which the other dead find in their own world, and consequently must obtain them at the expense of the living. (Van Gennep, 1909/1960, p. 160)

The view of death rituals as a rite of passage is supported by a study of the similarities between the Irish funeral wake and the events staged to send immigrants to America in the nineteenth century (Metress, 1990). Both affairs involved public participation and an opportunity for the family and community to come together to grieve over their loss. Though sad, one could actually rejoice at the departed's rebirth to a new state: death and immigration freed one forever from the stark hopelessness of poverty. Music, singing, dancing, and wake games were usually a part of the affair.

Structural-Functional Explanations

As we discussed in chapter 1, the term *function* refers to the extent to which some part or process of a social system contributes to the maintenance of that system. Function means the extent to which an activity promotes or interferes with the maintenance of a system (Cuzzort & King, 1995). Rituals are a particularly effective means of promoting the maintenance of social systems.

The structural-functional perspective of sociologist Emile Durkheim (1915/1954) and anthropologists Bronislaw Malinowski (1925/1948) and A. R. Radcliffe-Brown (1964) reduces the ritual process to an equilibrium-producing system. The specific functions of rituals include validating and reinforcing values, providing reassurance and feelings of security in the face of psychological disturbances, reinforcing group ties, aiding status change by acquainting persons with their new roles, relieving psychological tensions, and reestablishing patterns of interaction disturbed by a crisis (Taylor, 1988).

Similarly, burial rite functions can be enumerated as follows. First, they give meaning and sanction to the separation of the dead person from the living. Second, burial rites help effect the transition of the soul to another, otherworldly realm. Third, they assist in the incorporation of the spirit into its new existence.

Finally, death rituals are emphasized as mechanisms that recreate social solidarity and reaffirm social structure. Death "disrupts" social networks, relationships, and patterns of interaction. Rituals performed at death help to restore order to that which has been disrupted. Individuals gain strength from the ritual affirmations of the community, and they are then better able to continue their lives.

The function of reaffirming social structure is often accomplished through the family reunion that occurs as an unintended consequence, or a **latent function,** of the funeral ritual. The intended consequence, or **manifest function,** of a funeral is to pay respects to the deceased and support the survivors. The latent function is to bring family and community together. This latent function is as important as the manifest function.

Ronald Barrett (1992) observes that African-American funerals are indeed "a primary ritual and a focal occasion with a big social gathering after the funeral and the closest thing to a family reunion that might ever take place." This is a particularly important restorative function for African Americans, whose history of slavery included families being routinely, and arbitrarily, broken apart. However, a study by Melvin Williams (1981) in Pittsburgh, Pennsylvania, reveals class differences in the way that the restorative function of the funeral operates. For the middle class, a funeral may establish, validate, and reinforce social status, whereas for the lower class, funerals tend to be rites of intensification (where the feelings of the bereaved are experienced more intensely) and solidarity—as people put aside feuds and squabbles for the moment.

This function of reaffirming the social order is also evident in some egalitarian tribal cultures where funerals involve elaborate feasting and gift-giving. The "Death Across Cultures" box illustrates how the egalitarianism of the Vanatinai is reinforced by funeral rituals.

DEATH ACROSS THE CULTURES:
Honoring the Dead

Maria Lepowsky (1994) has character-ized the matrilineal society of the Vanati-nai as being egalitarian; that is, there are no ideologies of male superiority or of fe-male inferiority. This egalitarianism is re-flected in the elaborate mortuary rituals. Mortuary ritual and the hosting of feasts and exchange of valuables that accom-pany it are the primary avenues to per-sonal power and prestige in the society. On Vanatinai, the deaths of men and women are marked equally by elaborate mourning and feasting, and the burden of mourning obligations is the same for men and women. Women are expected to ob-tain and present ceremonial valuables, and both women and men may strive to enhance their reputations as "big men" (*giagia*, literally "givers") by first accumu-lating these valuables and subsequently giving them away and by hosting or con-tributing a great deal to feasts.

After the burial there will always be a se-ries of feasts. The first feast, called *jivia*, oc-curs within a few days or weeks. During this feast valuables are exchanged and the de-ceased's kin ritually feed the mourning spouse or representatives of the father's line-age (Lepowsky, 1994). A *velaloga* feast 2 weeks or 2 months later releases a widow or widower from taboos against leaving the hamlet or bathing. There are a number of other feasts associated with mourning, but the final feast is the *zagaya*, which is the largest and most important of all. The *za-gaya* is held after about 3 years of intensive prepreparations. After the *zagaya*, all taboos are lifted from people and places, and the be-reaved spouses can again make themselves attractive, court, and remarry.

The relations of power revealed by cer-emonial exchange on Vanatinai do not separate men from women. Different forms of wealth are not associated exclu-sively with one or the other. Strength, wis-dom, and generosity are qualities looked for in both men and women. Only those with these qualities will be able to host a successful *zagaya* feast. The *gigia* are both male and female, and their power stems from their possession of wealth and their knowledge of magic.

From "Honoring the Dead" by Daniel G. Bates and Elliot M Fratkin, 1999. In Cultural Anthropology, *2nd ed., Boston: Allyn and Bacon.*

For some, the burial ground can serve as a symbolic representation of the social order (Bloch & Parry, 1982). Among the Merina of central Madagascar, for exam-ple, after death one returns "home" to the tomb, representing a regrouping of the dead—a central symbol of the culture and an underlying joy of the second funeral. By the entry of the new corpses into the collective mausoleum, the tomb and the re-united dead within it represent the undivided and enduring descent group and be-come the source of blessings and the fertility of the future. The force of this symbol of the tomb as the representation of the eternal undivided group can be sustained only by downplaying the individuality of the corpses that enter it.

In older American cemeteries, family solidarity was preserved in the "family plot," where many members of a single family were buried together in a space delineated by fences, headstones, and footstones. Shared tombstones for married couples often declared this family unity and solidarity with the epitaph "Together Forever."

MOURNING BEHAVIORS

The expression of emotions experienced by those who lose a loved one ranges from complete to amplified wailing, notes Effie Bendann's *Death Customs: An Analytical Study of Burial Rites* (1930). She states that after a death the aborigines of Australia and Melanesia indulge in the most exaggerated forms of weeping and wailing. They display other manifestations of emotional excitement seemingly because of grief for the departed. At the end of a certain designated time period, however, they cease with metronomic precision, and the would-be mourners indulge in laughter and other forms of amusement. Likewise, Hindus in India are encouraged to express their grief openly, even sometimes extravagantly, through shrieks of women mourners and floods of tears, yet no weeping is to occur during the cremation ceremony (Tully, 1994).

A study of **ethnographic** data from 78 societies (Rosenblatt, Walsh, & Jackson, 1976) to identify the essence of universal grief behavior showed that death is nearly universally associated with emotionality and that the most usual expression among the bereaved is crying. One could argue that people display the outward emotion of crying at the time of death because they are sorrowful. Perhaps it is not as simple as that. Anthropologist A. R. Radcliffe-Brown (1964) notes two types of weeping. There is reciprocal ritual weeping to affirm the existence of a social bond between two or more persons. It is an occasion for affirming social ties. Although participants may not actually feel these sentiments that bind them, participation in various rites will strengthen whatever positive feelings they do have. The second type of weeping—weeping over the remains of a significant other—expresses the continued sentiment of attachment despite the severing of this social bond.

Richard Huntington and Peter Metcalf (1992) observe that Radcliffe-Brown was strongly influenced by the French sociologist Emile Durkheim. Durkheim (1915/1954) argued that the emotions developed are feelings of sorrow and anger and are made stronger by participation in the burial rite, whereas Radcliffe-Brown argued that those participating in ceremonial weeping come to feel emotion that is not sorrow but rather togetherness. What Durkheim finds significant is the way that other members of society feel moral pressure to put their behavior into harmony with the feelings of the truly bereaved. Even if one feels no direct sorrow, weeping and suffering may result. Thus, to say that one weeps because of sadness may be too simplistic. From this perspective death rituals become rites of intensification whereby feelings and emotional states are intensified by ritual participation. It is not uncommon for people to say "I was doing quite well in 'dealing with my emotions' until the funeral began." Oftentimes the function of death rituals is to intensify feelings and emotions and then provide a means by which individuals can express their sentiments.

Even though expressions of sorrow seem to be pervasive in most cultures, there are other cultures in which excessive alcohol consumption, dancing, and spontaneous expressions of joy are encouraged as a socially acceptable response to death. Such experiences celebrate the new exalted status of the person who has died and proclaim that the deceased person is better off in this new situation. This is certainly evident in the Irish wake, where the merriment at wakes for the dead may seem disrespectful. From a cultural point of view, the wake actually demonstrates the strong belief that the physically or symbolically dead will continue to exist more happily elsewhere. The American wake reiterates the significance of transitional times for the Irish, whether changes in the seasons or major life changes such as birth, immigration, and death (Metress, 1990).

The traditional funeral for the Igbo of Nigeria includes males dancing to celebrate the life of the deceased.

Expressions of joy and happiness during funerals or wakes may also offer a socially acceptable method for escaping feelings of sadness and anger. As mentioned earlier, during funerals for the Bara of Madagascar, rum is consumed, sexual activities occur, dancing takes place, and contests involve cattle (Huntington & Metcalf, 1992). In a functionally equivalent manner the Cubeo of South America encourage simulated and actual ritual coitus as part of mourning rituals (Goldman, 1979).

The "Listening to the Voices" box provides an example of how dancing can be used to transform cultural grief and sadness into joy and celebration.

As we have observed, societies encourage both the expressions of sadness and joy in their mourning rituals. In the former, participants are given the opportunity to collectively express their grief. In the latter, individuals will dance, participate in dramatic performances, and even experience sexual license all with the purpose of transforming grief and anger over a death into joy.

Changes in Mourning Behaviors in the United States

In the days of the Puritans, elaborate and extended mourning was discouraged because it was thought to undermine the cheerful resignation to God's will that was essential to the Puritan experience. However, in the nineteenth century, when the Romantic influence's emphasis on emotions led to simple sentimentalism, mourning practices became much more dramatic.

For most of the nineteenth century, the main outlet for the grief of sentimentalism was the ritual of mourning, including the funeral, but extending beyond that. This ritual differed markedly from the simple Puritan rite as mourners immersed themselves in grief to become, through their expressive (and often excessive) emotions, the central feature of the ritual. It allowed many members of the middle class (especially the women, who were supposed to be creatures of the "heart") to indulge in grief as "therapeutic self-indulgence." Like other forms of sentimentality, the sentimental mourning ritual counterpointed "the real world" because it forced all mourners to consider the power of personal connections in their lives. It turned people from life to death, from the practical to the ceremonial, from the ordinary to the extraordinary, and from the banal to the beautiful (Taylor, 1980, pp. 39–48). This romantic emphasis led to the development of the beautiful funeral. Concerned with an etiquette of proper social relations, people used the funeral, which was designed to reflect "taste" and "refinement," to preserve appearances among their middle-class peers and distinguish themselves from the common folk (Farrell, 1980, pp. 110–111). Especially around the mid-1880s, the tastefully refined middle-class funeral was a dark and formal affair. After death, which still generally occurred in the home, the family members either cleaned and dressed the corpse, or, if possible, hired an undertaker to care for the corpse. If they had secured an undertaker, he would place a black badge over the doorbell or door knocker to indicate the presence of mourning and to isolate the family from the unwanted intrusions of everyday life. The family members would also close window shades and drapes.

LISTENING TO THE VOICES
Cuttin' the Body Loose: Dancing in Defiance of Death in New Orleans

"They gon cut the body loose!" One short brother with a mustache was running up and down the funeral procession explaining that they wasn't going to have to go all the way to the cemetery on account of they was going to cut the body loose. This meant that the hearse would keep on going and the band and the second liners and the rest of the procession was going to dance on back to some tavern not too far away. So we stood in a line and the second liners were shouting, "open it up, open it up," meaning for the people in front to get out the way so the hearse could pass with the body. After the hearse was gone we turned the corner and danced down to the bar.

. Like a sudden urge to regurgitate and with the intensity of an ejaculation, an explosive sound erupted from the crowd under the hot New Orleans sun. People spontaneously answered the traditional call of the second line trumpet.

"Are you still alive?"

"YEAH!"

"Do we like to live?"

"YEAH!"

"Do you want to dance?"

"YEAH!"

"Well damn it, let's go!"

The trumpeter was taunting us now, and the older people were jumping from their front porches as we passed them, and they were answering that blaring hot high taunt with unmistakable fires blazing in their 60-year-old black eyes. They too danced as we passed them. They did the dances of their lives, the dances they used to celebrate how old they had become and what they had seen getting to their whatever number years. The dances they used to defy death.

Nowhere else in this country do people dance in the streets after someone has died. Nowhere else is the warm smell of cold beer on tap a fitting conclusion for the funeral of a friend. Nowhere else is death so pointedly belittled. One of us dying is only a small matter, an occasion for the rest of us to make music and dance. Nothing keeps us contained. With this spirit and this music in us, black people will never die, never die, never.

We were all ecstatic. We could see the bar. We knew it was ending, we knew we were almost there and defiantly we danced harder anyway. We hollered back even that much louder at the trumpeter as he squeezed out the last brassy blasts his lungs could throw forth. The end of the funeral was near, just as the end of life was near for some of us but it did not matter. When we get there, we'll get there.

From "Cuttin' the Body Loose: Dancing in Defiance of Death in New Orleans" (pp. 78–79), by K. Salaam, 1991, September-October, Utne Reader.

Sometimes they draped black crepe over pictures, mirrors, and other places throughout the house (Habenstein and Lamers, 1962, pp. 389–444).

By the time of the funeral, family members had swathed themselves in black mourning garb that symbolized their intimacy of relationship to the deceased and their depth of grief. After the funeral, custom encouraged the continued expression of grief,

During the latter part of the nineteenth century, mourners were expected to dress the part. In this picture, pallbearers wear mourning clothes, black sashes, badges, and dark hats.

as widows were expected to spend a year in "deep" mourning and a year in "second" mourning. For the first year, a bereaved woman wore dull black clothes, matched by appropriately somber accessories. In the second year, she gradually lightened her appearance by using a variety of materials in somewhat varying colors. Widowers and children were supposed to follow a similar regimen, but, in practice, women bore the burden of the nineteenth century mourning requirements. Social contact and correspondence followed similar rules, with widened social participation or narrowed black borders on stationery as indicators of different stages of mourning.

By the twentieth century, the somber nature of the nineteenth century mourning rituals had evolved along with attitudes toward death, which was now celebrated as a passage to eternal life, not a moment of judgment. Death was defined as "a normal change in an eternal process of growth" (Abbott, 1913). Mourning rituals were no longer somber dramatic affairs, with elaborate displays of grief. Instead, restraint and passivity characterized twentieth century bereavement practices.

The funeral director used the culture of professionalism to become the authority on American death rituals. His middle-class clients, who feared death anyway, were all too happy to allow the funeral director to take control. This led to a scenario where the funeral director was the stage manager and the family was the audi responding to the drama in prescribed ways in the hopes of achieving a catharsis death. Instead of the expressive grief of the nineteenth century, family members wer expected to contain and control their emotions and to meet death stoically. At the turn of the century, some religious liberals saw grief as a lack of faith in the imminence of immortality. Others reacted to the central place of the mourners in the mournful Victorian funeral and charged that "over-much grief would seem mere selfishness" (Mayo, 1916, p. 6). Over and over again, writers proclaimed that "the deepest grief is the quiet kind" (Sargent, 1888, p. 51). An 1890s etiquette book suggested that "we can better show our affection to the dead by fulfilling our duties to the living, than by giving ourselves up to uncontrolled grief" (Pike & Armstrong, 1980, p. 125). Portraying grief as a selfish ploy to stop the ongoing business of life, mourners were persuaded to keep their grief controlled and private. In the long run, they predicted the modern practice of grief therapy in which grief is seen as a disorder (Mayo, 1916, p. 6; Pike & Armstrong, 1980, p. 125; Sargent, 1888, p. 51).

With this new ideal, Americans reduced their symbolic expressions of grief such as mourning wear, because such expressions were thought to offend those who preferred to focus on life, not death. In concealing the uncouth and discordant expressions of grief, Americans reversed the nineteenth century tradition of "good mourning," and isolated mourners were forced to discover their own private mourning rituals (Hillerman, 1980; Oxley, 1887).

The changes brought about by funeral directors early in the twentieth century are still evident in mourning behaviors today. Private expression of grief in grief counseling sessions is preferred to public expression of grief. More recently, funeral services have become even briefer, often without the eulogy, which might be too upsetting. Psychologists are concerned about this decline in the expression of grief. Although many people do seek grief counseling, others do not, repressing their feelings of grief and becoming depressed or drowning their sorrow in alcohol. The decision to seek grief counseling is clearly gender related, with women being much more likely to receive professional therapy and participate in self-help groups for the bereaved.

Gender Differences in Mourning Behaviors

The Rosenblatt et al. (1976) study mentioned earlier noted that, in the 78 cultures studied, there are significant gender differences in emotions during bereavement, with women tending to cry and self-mutilate more than men. The men tend to direct anger and aggression away from self. One of the traditional theories to explain gender differences in emotional expression was that it may be easier to socialize women than men to be overtly nonaggressive; thus, crying may represent a female expression of aggression. Another theory is that women will be more affected by the loss from a death because of their stronger attachments through their role as mothers. On the other hand, women may not experience death more strongly, but they may simply be used as the persons symbolizing publicly, in burdensome or self-injuring ways, the loss that all experience. The analysis of data from these 78

hat the kind of data needed to "explain" the emotional differing.

y of emotions is found among the Kapauku Papuans of west
1963). Their ritual at death requires female relatives of the
al expression of their grief as soon as the soul leaves the
es, cut off their fingers, tear their garments and net carry-
faces and bodies with mud, ashes, or yellow clay. A loud
ws.

vomen among the Cheyenne Indians cut off their long
as the blood flows (Hoebel, 1960). If the deceased in-
enemies, they slash their legs until caked with dried blood.
the women their own masochistic outlet. Cheyenne men, on the
hand, simply let down their hair in mourning and do not bother to lacerate
themselves. In traditional African funerals (Barrett, 1992) women tend to wail,
whereas men sing and dance; men are not to cry in front of women because they
would appear weak before the very group that they are to protect.

The Dinka women of the Sudan (Deng, 1972) cut their leather skirts and cover
their bodies with dirt and ashes for as long as a year. Widows among the Swazi in
Africa (Kuper, 1963) shave their heads and remain "in darkness" for 3 years before
given the duty of continuing the lineage for the deceased through the levirate—
required marriage to her dead husband's brother. Mourning imposed on the Swazi
husband is less conspicuous and shorter than that imposed on the wife.

Somewhat less dramatic than the previous examples, the Huicholes of Mexico
(Weigand & Weigand, 1991) display a "great deal of crying and wailing" at the actual moment of death. This subsides during the preparation of the corpse but resumes
again at the funeral site. However, Huichol women who have suffered the death of a
mother or a child often express their loss by suicidal gestures—the stated goal of such
behavior is to accompany the deceased. In reality, however, these gestures seldom result in suicide, according to C. G. Weigand and P. C. Weigand (1991).

In conclusion, cross-cultural studies of the ritual of mourning at death reveal
that a double standard prevails among some cultures—different "scripts" for different genders. Women in many societies, including the United States, are expected
to display more of their emotions than are men. Mourning rituals also tend to last
for a set period of time in many societies, and women, in general, are expected to
mourn longer than are men. One indication of a different duration of mourning by
gender is that, when compared to women, men seem to remarry sooner without social sanction or ridicule and women are more inclined to remain single after being
widowed.

CUSTOMS AT DEATH

As will be discussed in chapter 11, in the United States a professional is called upon
to prepare the body for final **disposition** because we are a very specialized society

with a high division of labor. The funeral director takes the body away and returns it later for viewing. The kin and friends in the United States normally play no significant role in handling the corpse. Compared with most societies, we are unique in the level of professional specialization relative to the preparation of the corpse and the actual disposition of the body. Most of the cultures discussed in this section encourage families and friends to become very involved in preparing the corpse for its final disposition.

There are two perspectives for viewing customs at death. From one viewpoint, it is possible to conclude that each society creates different practices unique to the culture in which they occur. An alternative point of view is to recognize that there are common human needs (e.g., to dispose of the body), and each society creates practices, rites, meanings, and rituals, which are functional equivalents to those found in other societies. The social anthropological perspective employs the latter interpretation.

Norms Prior to Death

Some societies have specific norms just before death. It is important in many societies, including the United States, that one be with the dying person at the time of death. So often it is said, "If only I had gotten there a few minutes earlier . . ." A visit before death allows one to say goodbye. Among the Dunsun of northern Borneo (Williams, 1965), relatives come to witness the death. The dying person is propped up and held from behind. When the body grows cold, the social fact of death is recognized by announcing "he exists no more" or "someone has gone far away."

The Salish Indians of the northwestern United States (Habenstein & Lamers, 1974) leave the dying person alone with an aged man who neither receives pay nor is expected to have any special qualifications for the task. One who is near death must confess his or her misdeeds to this man. The confession is to prevent the ghost from roaming the places that were frequented in life by the dying person.

The Ik of Uganda (Turnbull, 1972) place the dying person in the fetal position because death for them represents a "celestial rebirth." The Magars of Nepal (Hitchcock, 1966) purify a dying person by giving him or her water that has been touched with gold.

The Maori of New Zealand (Mead, 1991) have a ceremony just before the person dies to send the spirit of the dying person away while the person is still alive and conscious of what is happening. After the spirit is gone, the individual may not be medically dead, but, in a Maori sense, is dead. Now that the person is a corpse, ceremonies begin and a space is set aside in the meeting house for the corpse to lie in an open coffin.

Having a cultural framework prescribing proper behavior at the time of death provides an established order and perhaps gives comfort to the bereaved. These behavioral norms give survivors something to do during the dying process and immediately thereafter, thus facilitating the coping abilities of the bereaved.

Preparing the Corpse

It seems significant in most societies that the body be cleaned and steps be taken to assure a tolerable odor before final disposition. A reverence for the body also seems prevalent. As in the United States, it is important in many other societies that the deceased person "look good" to the mourners.

Cleaning, Decorating, and Clothing the Body

Ritual preparation of the body for final disposition, including cleaning and wrapping the body (**mummification**), is common in many societies, especially those that do not practice embalming. The material used for wrapping is often white cloth, which is considered natural and pure, but may also consist of other natural materials. For example, the Tiwi of Australia (Hart & Pilling, 1960) wrap the body in bark. The Qemant of Ethiopia (Gamst, 1969) wrap the body in a piece of white cloth and cover it with a mat of woven grass. After cleaning the body, the Ulithi in Micronesia (Lessa, 1966) cover it with a tuberous plant and decorate the head and hands with flower garlands.

Anthropologist Linda Connor (1995) notes that corpse washing in Bali is a pivotal point in the process of grieving for relatives. For the Bornu of Nigeria (Cohen, 1967), family members are required to wash the body, wrap it in a white cloth, place it onto a **bier,** and take it to the burial ground. Similarly, the Semai in Malaya (Dentan, 1968) have the housemates bathe the corpse and sprinkle it with perfume or sweet-smelling herbs to mask the odor of decay. They then wrap the body in swaddling clothes.

The Mapuche Indians of Chile (Faron, 1968) sometimes smoke the body, then wash and dress it in the person's best clothes, lay it out on a bier in the house, and place it into a pine coffin. Among some groups in southern Thailand (Fraser, 1966), the body is held and bathed with water specifically purified with herbs and clay. The corpse is then rinsed and dried, and all orifices plugged with cotton.

In the French West Indies (Horowitz, 1967) the neighbors wash the body with rum and force a liter or more of strong rum down the throat as a temporary preservative before dressing the body and placing it onto a bed. Rather than use a strong drink like rum, the Zinacantecos of Mexico (Vogt, 1970) pour water into the mouth of the deceased about every half hour "to relieve thirst" while the grave is being dug.

The Tewa Indians in Arizona (Dozier, 1966) bury a woman in her wedding outfit and wrap a man in a blanket for burial. The Hopi Indians wrap a man in buckskin and a woman in her wedding blanket, bury them in the clothes that they were wearing at the time of death, and do not wash or prepare the body other than to wash and tie back the hair (Cox & Fundis, 1992). The Navajo in the southwestern United States (Cox & Fundis, 1992) bathe the corpse, then dress it in fine clothes and put the right moccasin onto the left foot and the left moccasin onto the right foot.

In Oregon, there is a Russian immigrant group whose body preparation practices date back to the 17th century (Morris, 1991). Death is a "village" affair, and all turn their attention to it, when it happens. They rather rapidly do what they consider necessary to lay the body to rest, rarely taking more than 24 hours. When death occurs, the body is washed and dressed in a white, loosely based cloth. The body is then placed inside the casket with the arms crossed on the chest and the hands formed into the sign of the cross in the old style: the first two fingers are extended and the thumb is joined with the third and fourth fingers. The service occurs in the living room of the home, where the casket has been placed.

Body Preparation by Specialists.   Most of the practices cited have used family or friends to prepare the corpse for burial. However, this is not always the case. In many societies, there are designated individuals who perform body preparation rituals. For example, Muslims bring their dead to the mosque directly after death, where there are special rooms for washing the body. During this preparation, prayers and passages from the Qur'an will be recited by a *hoca*—a lay holy man, not a priest—and care will be taken that the body always faces Mecca.

The Huicholes of Mexico (Weigand & Weigand, 1991) handle the corpse with a great deal of respect and care. A singer/curer who is usually closely related to the deceased supervises the preparation of the corpse for burial or placement into a cave. Clothing is changed, personal items are arranged to accompany the body, and the individual's hands and face are washed with water. Most adults aid in the preparation unless "emotional collapse prevents such activity." However, most people are rather calm during these preparations.

Anglo-Canadians in and around Toronto, Canada (Ramsden, 1991) prefer to remove the corpse from the company of the living as soon as possible. The corpse is disposed of immediately after death, never to be seen again. This rapid removal of the corpse precludes any opportunity to note changes in the individual after death, thus reinforcing the perception of the Anglo-Canadians that death is a static state. They wish to convey the idea that the dead are physically and socially gone. From the moment of death, the physical condition of the body is of no consequence, with the focus on permanently removing the body from the world of the living and banishing the dead person from social relations. The event of death is perceived as an instantaneous occurrence and marks a distinct boundary between a state of being alive and a state of being dead.

Embalming

In contemporary American culture, the process of final disposition of the body should be studied within its cultural and historical context. As we have demonstrated earlier, it is wrong to believe that the funeral is either unique to Western culture or has been invented or created by it. To bury the dead is a common social practice. The methods to accomplish burial, and the meanings associated with it, are culturally determined.

This burial process often requires the involvement of a functionary, who may be a professional, tradesperson, religious leader, servant, or even a member of the family. The functionary in each society is closely associated with the folkways and mores of the society and its philosophical approach to life and death.

The Hebrew scripture reveals in Genesis 50:2 that physicians embalmed the body of Jacob, the father of Joseph. This description is followed by a detailed description of the funeral and burial. The historian Herodotus records embalming preparation as early as circa 484 B.C. These two sources and archaeological discoveries of earlier cultures give evidence of the disposition of the dead.

With the discovery of the circulatory system of the body, circa 1600, and the possibility of diffusion of preserving chemicals through that system, more sophisticated methods of embalming the body were developed. Dr. Hunter of England and Dr. Gannal of France, independently of each other, furthered the process in the early 1600s. In France, this work coincided with the advent of the bubonic plague. During this period of time, extensive attempts were made to preserve the bodies of the dead to protect the health of those who survived the plague.

Whereas the ancient Egyptian process of embalming required 70 days to perform, body preparation today is completed within a few hours and is more effective and acceptable. Body preparation may be as simple as bathing the body, closing the eyes and mouth, and dressing the body for final disposition. This procedure, though infrequently selected, is utilized by families who wish direct disposition, which we will discuss later in this chapter as a procedure that may be an acceptable and logical choice.

In the United States, it is estimated that four out of five bodies are embalmed before final disposition. **Embalming,** by definition, is the replacement of normal body fluids with preserving chemicals. This process is accomplished by using the vascular system of the body to both remove the body fluids and to suffuse the body with preserving chemicals. The arterial system is used to introduce the chemicals into the body, and the venous system is used to remove the body fluids. This intravascular exchange is accomplished by using an embalming machine. The machine

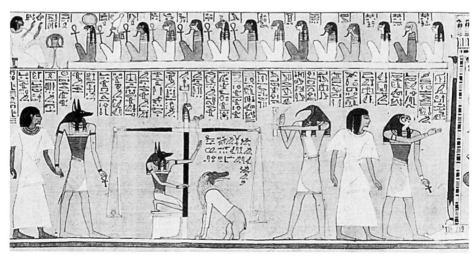

The most elaborate and famous burial customs were those of the Egyptians.

DEATH ACROSS CULTURES
Death in Ancient Egypt

The earliest burials known in Egypt date to a period well before 3000 B.C. and display evidence, through funerary gifts in the graves, of a belief of continued existence after death. During the middle and late Predynastic Period (before 3100 B.C.), the practice of wrapping the body in animal skins gradually gave way to other forms of protection, particularly the use of basket trays upon which the body was laid out.

Wrapped bodies of the first three dynasties were not truly mummified, since no treatment other than the use of linen bandages and resin was used. By the Fourth Dynasty, however, evidence was found of deliberate attempts to inhibit decomposition by removal of the soft internal organs from the body—accomplished by means of an incision in the side of the abdomen. Removal of the liver, intestines, and stomach improved the chances of securing good preservation because the emptied body cavity could be dried more rapidly. The removed organs were deposited in a safe place in the tomb in order for the body of the deceased to be complete once more in the netherworld.

Bandages soaked in resin were carefully molded to the shape of the body to reproduce the features, particularly in the face and the genital organs. As the resin dried, it consolidated the linen wrapping in position, preserving the appearance of the body for as long as it remained undisturbed. The corpse itself decomposed very rapidly within this linen shell, leaving the innermost wrappings in close contact with the skeleton.

An important factor in the development of the Egyptian tomb was the necessity to provide storage space for the items of funerary equipment considered essential for continued use by the deceased in the hereafter. A significant part of the material provided for the dead took the form of actual offerings of food and drink—required for the "very survival" of the deceased before enjoying all the other possessions in the tomb. Because of the need to provide offerings of food at the tomb, the tomb had to combine the function of a burial place with that of a mortuary chapel in which the priests could officiate.

From Death in Ancient Egypt, *by A. J. Spencer, 1982, New York: Penguin Books.*

can best be described as an artificial heart outside of the body that produces the pressure necessary to accomplish the exchange of fluids. This, together with the filling of the chest and abdominal cavities with embalming fluid, constitutes the embalming procedure.

In addition to embalming and thorough bathing of the body, cosmetic procedures are used to restore a more normal color to the face and hands. When death occurs, the pigments of the skin, which give the body its normal tone and color, no longer function. Creams, liquids, and/or sprays are used to give the appearance of normal skin coloration.

The question "But why embalm or cosmeticize the dead body?" is based on the assumption that one of the needs of the family is the reality of death; thus, the body should be left in its most deathlike appearance. Those who have seen a person die (especially if the dying process was painful, prolonged, and emaciating) know that the condition of the body at the time of death can be very repulsive. Many people cannot accept this condition. This is why contemporary funeral directors embalm and cosmeticize the body.

Another reason for embalming is the mobility of the American population. Viewing, which is practiced in over 75 percent of the funerals today that involve earth burial and 22 percent of cremation services (Dawson, Santos, & Burdick, 1990), often requires more than a bathing and dressing of the body. Embalming is necessary to accomplish a temporary preservation of the body to allow time for the family to gather—as many as 2 or 3 days may be needed. If the body were to remain unembalmed for this length of time, the distasteful effects of decomposition would create a significant problem for grievers.

Though arguments favoring embalming have been presented, it may not always be necessary or desired. Embalming is not required in all states. In many states, for example, if the body is disposed of within 72 hours, if it is not transported on a common carrier or across state lines, and/or if the person did not die of a contagious disease, embalming is not required. If a body is to be cremated and no public viewing is held, embalming would not be necessary. Many consumers just assume that embalming should or must occur.

A frequently asked question is, "If a body is embalmed, how long will it last?" There is no simple answer to this question. This is why funeral directors talk in terms of temporary preservation. Most families are interested in a preservation that will permit them to view the body, have a visitation, and allow the body to be present for the funeral. Beyond that, they are not concerned with the lasting effects of embalming.

Final Disposition

Rituals and rites related to final disposition serve the functions of maintaining positive relationships with ancestral spirits, reaffirming social solidarity, and restoring group structures dismembered by death.

In many cultures and religions, individuals are considered to be composed of several elements, each of which may have a different fate after death (Palgi & Abramovitch, 1984). Thus, the actual destruction of the corpse, whether through cremation, burial, or decomposition, is thought to separate the elements—the various bodies and souls. For example, in India, when a son breaks the skull of his father on the funeral **pyre** (a combustible pile for burning a corpse), he is demonstrating that the body no longer has any value—because it is worn out—but the soul lives on (Tully, 1994).

As Gerry Cox and Ronald Fundis (1992) state in their discussion of Native American burial practices, "If nothing else is known, it is that tribal groups did not abandon their dead. They provided them with ceremony and disposal." Indeed, a lot of ceremony and a way of disposing of the dead tend to go hand in hand with death among different cultures. These disposition rites not only reaffirm group structure, but also enhance social cohesiveness.

There are two primary forms of final disposition, earth burial and **cremation.** As we will discuss, the dominant religions that practice earth burial are Judaism, Christianity, and Islam, while Buddhism and Hinduism practice cremation. Another practice, which used to be common among Plains Indians in the United States, is **above-ground burial** on scaffolds or in trees. This type of burial is often the first part of **secondary burial** rites, which allows for open-air decomposition of the body, followed by earth burial, and possibly subsequent reburials.

Earth Burial

Earth burial as a method of final disposition is by far the most widely used in the United States. It is used in approximately 75 percent of the 2.3 million American deaths annually. Almost without exception, earth burial takes place within established cemeteries. In some instances, earth burial can take place outside of a cemetery if the landowner where the interment is to be made and the health officer of jurisdiction grant their permission. In 1997 Bill Cosby acquired permission from local authorities to bury his son Ennis on the grounds of his estate. By law, cemeteries have the right to establish reasonable rules and regulations to be observed by those arranging for burial in them. A person does not purchase property within a cemetery, but rather purchases the "right to interment" in a specific location within that cemetery. Most cemeteries require that the casket be placed into some kind of outer receptacle or burial vault. The cemetery will also control how the grave can be marked with monuments or grave markers.

Entombment, occurring in less than 5 percent of all final dispositions, might be considered as a special form of earth burial. It consists of placing the body (contained within a casket) into a building designed for this purpose. Cemeteries offer large buildings (**mausoleums**) as an alternative to earth burial or cremation. In some instances, families may purchase the right to interment in a cemetery and on the designated space build a private or family mausoleum that will hold as few as one or two bodies or as many as 12 to 16. Both types of mausoleums must be designed and constructed to provide lasting disposition for the body. Most states and/or cemeteries regulate the specifications and construction of the mausoleum.

Some societies keep the corpses in or near the home of the deceased. The Yoruba of Nigeria (Bascom, 1969), for example, dig the grave in the room of the deceased. In Uganda, a Lugbara male (Middleton, 1965) is buried in the center of the floor of his first wife's hut. The Swazi of Africa (Kuper, 1963) bury a woman on the outskirts of her husband's home.

To assure that the spirit will be reborn, some societies go to great lengths. For example, the Dunsun of northern Borneo (Williams, 1965) kill animals to accompany the deceased on the trip to the land of the dead. The Ulithi in Micronesia (Lessa, 1966) place a loincloth and a gingerlike plant in the right arm of the deceased so that gifts can be presented to the custodian at the entrance of the other world. For the Zinacantecos of Mexico (Vogt, 1970), a chicken head is put into a bowl of broth beside the head of the corpse. The chicken allegedly leads the "inner soul" of the deceased. A black dog then carries the "soul" across the river.

The Russian Orthodox community in Oregon, mentioned earlier (Morris, 1991), shares this concern for the rebirth of the spirit, and constructs the grave site to accommodate that rebirth. All the graves are lined up facing east, with an Orthodox cross at the foot of each grave. At the Second Coming, the dead will rise from the grave and stand next to the cross, facing East, toward Christ.

As with the Zinacantecos of Mexico, a chicken is used in the burial rites of the Yoruba in southwestern Nigeria (Bascom, 1969). A man with a live chicken precedes the carrier of the corpse, plucking out feathers and leaving them along the trail for the soul of the deceased to follow back to town. Upon reaching the town gate, the chicken is killed by striking its head against the ground. The blood and feathers are then placed into the grave so that others will not die. A second chicken is killed, and its blood put into the grave so that the soul of the deceased will not bother the surviving relatives.

In the old days of the Wild West in the United States, one was buried "6 feet under with his boots on." It is true that graves were 6 feet deep in earlier periods of United States history, but efficiency (and perhaps agnosticism regarding the belief in ghosts) dictates that today graves are less than 6 feet deep—typically around $4\frac{1}{2}$ feet with 18 inches of dirt above the top of the casket/vault. With the sealed, heavier caskets of today—often placed within a steel or concrete vault—it is not necessary to place the body so deep in the ground, as was the case with the unsealed pine box of an earlier period.

The Kalingas of the Philippines (Dozier, 1967) bury adults in graves 6 feet deep and 3 feet wide. The Mardudjara aborigines of Australia (Tonkinson, 1978) dig a rectangular hole about 3 feet deep, line the bottom with leafy bushes and small logs, and then place the body inside. Similarly, the Semai of Malaya (Dentan, 1968) dig the grave 2 to 3 feet deep.

Cremation

Cremation is the other method of final disposition. As mentioned before, this is the preferred method among Buddhists and Hindus. The practice of cremation among both of these religious groups will be discussed in detail later in the chapter. However, here we will discuss the growing practice of cremation in the United States.

Although earth burial is the dominant practice in the United States, the number of cremations has increased dramatically in the last two decades. In 1989, 16 percent of deaths in the United States involved cremation. In 2000, 25 percent of deaths in the United States involved cremation. Nevada leads the nation with 65 cremations per 100 deaths, followed by Hawaii, Washington, Oregon, and Alaska (in descending order, from 62 percent in Hawaii to 57 percent in Alaska). States with the fewest cremations are Mississippi, Alabama, and West Virginia (6 percent each) and Tennessee and Kentucky (8 percent each). In Canada 42 percent of all deaths involve cremation (Cremation Association of North America, 2000).

A 1992 study of 1,000 Americans conducted by the Funeral and Memorial In-formation Council (1992) found that 37 percent of respondents older than age 30 said that they would choose cremation for themselves, whereas 32 percent said they would choose it for a loved one. The Cremation Association of North America pre-dicts that in the year 2010, 40 percent of all deaths in the United States will involve cremation and that in the state of Alaska cremation will be selected 86 percent of the time (Cremation Association of North America, 1997).

Asked why they were likely to choose cremation for themselves or a loved one, the respondents in the study conducted by the Funeral and Memorial Information Council (1992) gave the following explanations:

19 percent said it saves land

18 percent said it saves money

13 percent said they did not like to think of the body in the earth

8 percent said it is "convenient"

Until recently, the **crematory** was generally located within the cemetery. With the increase of cremation as an option for final disposition, however, many funeral homes have now installed crematories. There is a trend in the funeral industry to change the name of the establishment from "funeral home" to "funeral and crema-tion services." In 1995 there were 1,155 crematories in the United States perform-ing nearly a half million cremations annually. For Canada, the respective numbers are 132 and 76,000 (Cremation Association of North America, 1997).

Crematories generally require containment of the body in an appropriate casket or other acceptably rigid container. The containerized body is not removed or disturbed after it arrives at the crematory and is placed into a furnace or retort. Cremation is ac-complished by the use of either extreme heat or direct flame. In either instance, reduc-ing the casket (or alternative container) and the body to "ashes" takes 2 to 3 hours. Cremated remains do not have the appearance or chemical properties of ashes; they are primarily bone fragments. Some crematories process cremated remains to reduce the overall volume; others do not. Depending on the size of the body, cremation results in 3 to 9 pounds of remains (National Funeral Directors Association, 1997).

After the cremation, the remains are collected, put into an **urn** or box, and then disposed of according to the wishes of the family. The **cremains** may be buried in a family plot or cemetery, placed into a **niche** in a **columbarium** (a special room in a cemetery), or kept in another place of personal significance, such as the home or church crypt. Subject to some restrictions, cremated remains can be scattered by air, over the ground, or over water. Some cemeteries provide areas for scattering and may provide a space where families can place a commemorative plaque or other memorial.

Many people choose to memorialize the site of disposition because they find consolation in knowing that there is a specific place to visit when they wish to remember and feel close to the person they have lost, regardless of whether the de-ceased person's remains are actually located at that place. Families should always choose an option that best fits their emotional needs.

The columbarium provides families with an additional option for body disposition. For approximately 25 percent of the 2.3 million annual deaths in the United States, cremation is used as the method of final disposition.

One might assume that cremation would be the least expensive form of final disposition because the typical cost of a simple "no-frills" cremation is approximately 40 percent of the cost of the traditional funeral service with burial (Lino, 1990). However, a cremation service may be as simple or as elaborate as family members wish. Some people are surprised to learn that cremation does not preclude a funeral with all of the traditional aspects of the ceremony. Visitation and viewing with a funeral ceremony and church or memorial services are options to be considered. In some states, funeral homes are permitted to rent caskets for viewing and services (National Funeral Directors Association, 1997). It is entirely possible to spend more money on a funeral involving cremation if plans include a "traditional funeral" that includes viewing a body in a casket and placing the cremated remains into an urn in a niche in a columbarium.

Case Study: Burial and Mortuary Practices of Native Americans Living on the Plains (adapted from an article written by Gerry Cox, and used with his permission)

There is probably more generalized knowledge of Plains tribes than of most Native Americans, since they are the ones most often portrayed in movies and television. Unlike the more sedentary tribes of the Southwest, the Plains tribes were mobile on a large scale. With the great temperature variations of the Great Plains, the inhabitants needed to adapt to all kinds of climatic changes. They were primarily dependent upon the bison as a source of food, clothing, and shelter. Other animals and plant life were also major sources of food, but the bison offered the most dramatic picture of the life of the Plains tribes. There were sedentary tribes, but the various groups that spoke the Siouan language were the hunters and nomads of film and television. The coming of the horse, following the arrival of the Europeans, added greatly to their prowess as warriors and as hunters. A war-like tribe, the Sioux have made their name in history by fighting against the European-Americans in famous battles such as "Little Big Horn" and "Wounded Knee." The former was recently portrayed in Kevin Costner's film *Dances with Wolves*, and earlier in the film *Little Big Man*, which starred Dustin Hoffman.

The eight (seven main tribes and one, the Assiniboin, that was outside of the loose confederation) Sioux tribes had relatively similar burial and mortuary practices. Like many other Western tribes, the Plains tribes believed that everything in the world around them was filled with spirits and powers that affected their lives whether from the sun, the mountains, the buffalo, or the eagle (Capps, 1973).

The Sioux feared the dead and would burn the dwelling of the deceased, forbid using the individual's name, and bury personal goods with the corpse to keep the ghost of the deceased from coming along to live with friends and relatives (LaFarge, 1956). An individual was expected to pursue honored roles as he or she progressed through life and to learn how to play his or her roles well. The roles included a spirit of generosity, which meant giving to others from birth to death (Malan, 1958). This spirit meant that the Dakota cared for their dead with comradeship. Yet, death in old age was not feared nor were the ghosts of dead persons, who were often thought to remain for a time after their death (Spencer & Jennings, 1965). The Lakota practiced "Ghost Keeping" ceremonies to try to keep the soul of the deceased on earth to purify it and to ensure that it would return to its creator (Hirschfelder & Molin, 1992). To keep a ghost requires that a family endures a great sacrifice and will ultimately give away all of their personal possessions to the needy in memory of the ghost (Powers, 1977).

The Sioux took the position that death will occur to all regardless of one's achievements, fame, wisdom, bravery, or whatever and that the mortuary practices allowed the living a way of showing their reverent respect for the dead (Hassrick, 1964). The Dakota (Lakota or Sioux) tribes would prepare a tipi to honor the deceased. In front of the tipi, they would place a rack upon which robes and articles of clothing would be displayed, while inside the tipi mourners would prepare themselves for their bereavement (Hassrick, 1964). If the deceased was a young person, particularly a child, the mourners would gash their arms and legs and engage in ritual crying (Spencer & Jennings, 1965). When death occurred in the home, the burial would be delayed for a day and a half in hopes that the deceased might revive (Hassrick, 1964). The body would be dressed in the finest available clothes, provided by a relative, if the deceased had none. The corpse would be wrapped tightly in robes, with the weapons, tools, medicines, and pipe included with the corpse. Then the bundle would be placed on a scaffold for air burial, with food and drink placed beneath the scaffold (Spencer & Jennings, 1965).

Some Dakota or Sioux groups used earth burial. There is evidence that in earlier times they used mound burial (Spencer & Jennings, 1965). During the winter, when scaffolds could not be built, trees were often used for burial (Hassrick, 1964). After the body was prepared and properly wrapped, the adult members of the family began wacekiyapi or worship rite for the deceased in which men might run pegs through their arms and legs, women might slash their limbs and cut off their little fingers at the first joint, and both men and women might cut their hair and express their grief by singing, wailing, or weeping (Hassrick, 1964). The favorite horse of the deceased would be killed beneath the scaffold of its owner and its tail tied to the scaffold, and the mourning would continue for as long as a year (Powers, 1977).

By placing the corpse in a tree or scaffold, the Dakotas and other similar tribes believed the soul would then be free to rise into the sky if the person died of natural causes. If the person died in battle, the Dakotas would often leave the person on the plains where he was slain to rise into the sky (Capps, 1973). For the Dakota or Sioux, the spirits of the dead are not gone and lost to humankind, but rather continue to exist here and can be reached by the living for support and aid (DeMallie & Parks, 1987). The Lakota also have a memorial feast, which is held around the time of the anniversary of the death. This ceremony ends the mourner's duties, and the final giveaway will occur at this time (Hirschfelder & Molin, 1992). The ghosts of the dead appear at their will and communicate with the living (Powers, 1977).

There is evidence that Plains tribes used all known methods of disposal of the dead including burial (both ground and air), cremation, and mummification. It is also probable that the cause of death, where the death occurred, and the age, sex, and social status of the deceased person had an impact on the mortuary and burial practices of the tribe, but information about how such factors influence burial practices is not conclusive.

Evidence suggests a general pattern that the tribes exhibit a fear of the dead. It is also likely that climate, weather, availability of materials to dispose of the body, and religious beliefs were major determinants in how bodies of the dead were disposed. Burial practices also seemed to remain stable for a remarkably long period of time among the tribes (Voegelin, 1944). Almost universally, tribes provide provisions for a spirit journey whether for a single or for a group burial (Atkinson, 1935). If nothing else is known, it is clear that tribal groups did not abandon their dead. They provided them with ceremony and disposal.

Influence of Social Position on Death-Related Behaviors and Rituals

A deceased person's social position and gender will often determine the way the body is prepared for burial and the final disposition of the corpse. For example, a

DEATH ACROSS CULTURES
Burial Rites Among the Kapauku Papuans

Among the Kapauku Papuans of West New Guinea, burial rites are determined by the deceased's status and cause of the death. The simplest burial is given to a drowned man, whose body is laid flat on the bank of the river and protected by a fence erected around it. The body is then abandoned to the elements. Very young children and individuals not particularly liked and considered unimportant are completely interred. Children, women, and elderly persons who were unimportant but loved are tied with vines into a squatting position and semi-interred with the head above ground. A dome-shaped structure of branches and soil is then constructed to protect the head.

A respected and loved adult male among the Kapauku Papuans receives a tree burial. Tied in a squatting position, the corpse is placed in a tree house with a small window in front. Corpses of important individuals, who are feared by their relatives, and of women who died in childbirth require a special type of burial. Their bodies are placed in the squatting position on a special raised scaffold constructed in the house where death occurred. The house is then sealed and abandoned.

The most elaborate burial among the Kapauku Papuans is given to a rich headman. A special hut is built on high stilts, the body is tied in a squatting position, and a pointed pole is driven through the rectum, abdomen, chest cavity and neck with its pointed end supporting the base of the skull. The body is then placed in the dead house with the face appearing in the front window of the structure. The body is pierced several times with arrows to allow the body fluids to drain away. Years later, the skull of the respected man may be cleaned and awarded a second honor of being placed on a pole driven into the ground near the house of the surviving relatives.

From The Kapauku Papuans of West New Guinea, *by L. Pospisil, 1963, New York: Holt, Rinehart and Winston.*

deceased Buddhist common person in Thailand (Leming & Premchit, 1992) will be cleaned, dressed, and placed into a casket. A high-status Buddhist individual, however, will be embalmed and then bathed and dressed with new clothes. The face will be covered with gold leaves before the body is placed into the casket.

In Greece (Brabant, 1994) the majority of individuals cannot afford "permanent" burial; thus, they rent a grave for 3 years, and then the remains are **exhumed** and the bones are placed into the "bone room." More prosperous individuals in Greece can afford to pay to have their remains stay buried permanently. The "Death Across Cultures" box reading about the Kapauku Papuans gives details about position in the community determining how and where burial or final disposition will occur.

Gender is also a factor. Among the Abkhasians near the Black Sea (Benet, 1974), women are buried 10 centimeters deeper than men. The Barabaig of Tanzania (Klima, 1970) place the bodies of women and children out into the surrounding bush, where they are consumed by hyenas. Only certain male and female elders will receive a burial.

In examining men's and women's eighteenth- and nineteenth-century gravestone epitaphs in cemeteries in the northeastern United States, Tarah Somers (1995) revealed major differences in the social expectations of men and women. Women were described in more passive and private terms ("meek and affectionate" and "joyfully departed life"), whereas men were often memorialized with active and public terms ("skillful and valiant in truth" and "triumphant at the approach of death").

Thus, in death as in life, one's social status and gender determine how one is treated. Whether one is buried or left to the elements is often determined by gender, age, standing in the community, and cause of death. If one is buried in the ground, even the depth of burial may vary by one's social position.

DEATH RITUALS OF MAJOR RELIGIOUS GROUPS

Whatever their differences, religious rituals for the dead are always communal events. Feeding the survivors and the telling of stories cuts across all traditions.

Jewish Customs

Jewish mourning rituals focus more on the bereaved than on the body. By custom, Jews try to bury their dead within 24 hours—if possible, without embalming—in a plain, wooden coffin. Traditional Jews do not put the body on view or have it cremated.

According to P. S. Knobel (1987), traditional Jewish burial customs require that the body be cleansed by members of the Jewish burial society (hevra'qaddisha', "holy society") in a washing process called tahorah or "purification." Custom forbids embalming, cremation, and autopsy unless local laws require these procedures. The body is then dressed in plain linen shrouds (takhrikhim); men are usually buried with their prayer shawls (tallit). The body is then placed into a plain wooden casket and buried before sunset on the day of death, if at all possible. Reform Judaism allows for cremation and entombment, but burial is the most frequent form of body disposition. Throughout this process, it is considered inappropriate to use the funeral as a means for displaying one's social position and wealth.

For the bereaved, the Jewish mourning ritual begins by the rending (tearing) of garments—survivors cut their clothing with a razor—on the left for a parent; on the right for a spouse, child, or sibling to symbolize the tear in life that death has produced. For some, the ripping of a black ribbon, which is then attached to the clothing, has symbolically replaced the process of rending garments.

After a ritual healing meal, shiva begins. For the first week, men don't shave, survivors are not supposed to wash their whole bodies, and the entire family receives visitors while sitting on the floor or on low chairs. From the death until burial, mourners are exempt from normal religious obligations (e.g., morning prayers) and must not engage in the following activities: drinking wine, eating meat, attending parties, and engaging in sexual intercourse (Knobel, 1987).

The liturgy for the funeral will consist of the recitation of psalms, a eulogy, and the following *El Male' Rahamin* memorial prayer (Knobel, 1987, p. 396):

> O God full of compassion, You who dwell on high! Grant perfect rest beneath the sheltering wings of Your presence, among the holy and pure who shine as the brightness of the heavens, unto the soul of [name of the deceased] who has entered eternity and in whose memory charity is offered. May his/her repose be in the Garden of Eden. May the Lord of Mercy bring him/her under the cover of His wings forever and may his/her soul be bound up in the bond of eternal life. May the Lord be his/her possession and may he/she rest in peace. Amen.

During the interment service the body is lowered into the grave and covered with earth. The interment service consists of an acclamation of God's justice, a memorial prayer, and the recitation of *Qaddish*—a doxology reaffirming the mourner's faith in God despite the fact of death. After the burial service, the people in attendance form two lines between which the primary mourners pass. Those present comfort the mourners as they pass, saying, "May God comfort you among the rest of the mourners of Zion and Jerusalem" (Knobel, 1987, p. 397).

After the *shiva*, mourners continue to avoid social gatherings for 30 days after the death. When one is mourning the death of a parent, the restrictions are observed for 1 year. After 1 year, all ritual expressions of grief cease with the exception of the *Yahrzeit*—the yearly commemoration of the person's death. *Yahrzeit* is observed by lighting a memorial light, performing memorial acts of charity, and attending religious services to recite the *Qaddish* prayer (Carse, 1981; Knobel, 1987).

Christianity

The Christian funeral service is primarily a worship service (or Mass of Christian burial for the Roman Catholic) that reflects the twin themes of victory and loss (described in chapter 4). During the service, hymns are sung and scriptural passages are read to emphasize the resurrection of the dead and to provide consolation for the bereaved. It is common for a eulogy or biographical statement concerning the deceased to be read.

In the United States, Christian teaching does not discourage the process of embalming, nor does it prohibit the autopsy, cremation, or any other form of final disposition (as is the case among traditional Jews)—provided that such practices do not indicate a rejection of a belief in the resurrection of the body. American Christian funerals are conducted by members of the clergy in a church, funeral chapel, and/or cemetery. Memorial services—religious services in which the dead body is not present—are becoming increasingly popular in many Protestant churches. At the occasion of death, the family of the deceased is expected to disengage from most normal social

PRACTICAL MATTERS:
Jewish Group Buries Its Own

When a Jewish congregation here first began the practice of offering simple, inexpensive burials for its dead, some members were upset. But they now increasingly condone it. Rabbi Arnold M. Goodman, spiritual leader of Adath Jeshurun Congregation, says volunteers of its society to honor the dead—*Chevra Kevod Hamet*—now handle about half the funerals of members. The *Chevra* was formed in 1976 after Goodman, in a sermon, dealt with the impact of American values upon the funeral practices of Jewry. He suggested a committee study the requirements of the *Halacha*, or Jewish law, for responding to death.

Months of study convinced committee members that a simple wooden coffin should be used, the body should be washed in a ritual process called *tahara* and, because dust is to return to dust as quickly as possible, there should be no formaldehyde in the veins, no nails on the coffin. The society decided to offer traditional funerals free to Adath Jeshurun members. The congregation provided seed money. Memorial donations and voluntary contributions from the bereaved are accepted.

Here's how the *Chevra* functions: When death occurs, *chaverim* (friends) call on the family, aid in writing the obituary, explain death benefits, aid in other ways, and remain available for help. *Chevra Kadisha* (sacred society), people of the same sex as the deceased and usually five in number, wash the body at the mortuary while saying prayers. The body is dressed in a shroud sewn by *Chevra* members and placed in a wooden coffin.

Shomrin (guards) watch over the body, in blocks of 2 hours, until burial. The coffin with rope handles is light enough to be borne by pallbearers, including women. Spurning mechanical contrivances, the pallbearers lower the coffin into the grave. *Chaverim*, the rabbi and cantor shovel in dirt. Family members may participate.

Judaism historically insists that the greatest commandment is to take personal involvement in burying the dead, Goodman says, but affluence enables people to pay surrogates to do it. Goodman says a funeral costs the *Chevra* less than one third the normal price charged by a local funeral home director.

From Rocky Mountain News,
June 25, 1982.

functioning until after the funeral. Funeral arrangements are typically made with the professional assistance of a funeral director and/or member of the clergy. For most Roman Catholics and many Protestants, on the day before the funeral, a wake or visitation service will be held at the funeral home. During this time (approximately 5 hours in duration), friends may view the body (if on view) and visit with the family of the deceased. Roman Catholic families may also have a rosary service and/or prayer service during the wake.

The funeral is typically held 2 to 4 days after the death. Occasionally, the funeral is delayed if family members are unable to make travel arrangements on such short notice (this would not be the case for a traditional Jewish funeral). If final disposition involves

LISTENING TO THE VOICES
The Ritual Solution

I was introduced to death early in life. In the Roman Catholic grade schools of my youth, funerals were part of the informal curriculum. When a classmate's parent died, we all assembled for the funeral mass, passing by the (usually) open casket and sharing—as best we could—the sorrow of the grieving family. Occasionally, it was a fellow pupil lying in the casket, snatched from life by an accident or—like my closest fifth-grade friend—from an illness, in his case a fatal epileptic seizure. What brought us together, young and old, was sacred ritual.

The liturgy that gathered and directed our emotions was familiar and color-coded. For funerals, the priest wore black vestments symbolizing death, just as he wore red for feast days of the martyrs and—on ordinary Sundays—green for the hope that all Christians have of life eternal. The music varied, too. Long before I learned a modern language, I knew by heart the somber Latin funeral hymn "Dies Irae" ("Day of Wrath"). I knew, too, that the clouds of incense billowing around the body were meant to honor flesh that was soon to turn to dust. By such appeals to the senses we children were inducted into the abstract mystery of death—our own as well as that of others. Sad? Yes. But never morbid. Death is real, the liturgy instructed us. But so is the promise of Resurrection.

From "The Ritual Solution" (P. 62),
by Kenneth L. Woodward
with Anne Underwood, 1997,
September 18, Newsweek.

B.C.   by johnny hart

cremation, the cremation can take place either after the funeral or before the memorial service. Burial and entombment dispositions are usually accompanied by rites of committal. On those occasions when there is no graveside service, words of committal will be read at the conclusion of the funeral service. At the conclusion of the funeral process, family members and others who have attended the services are often invited to share a meal together. This becomes a community rite of reincorporation.

Hinduism

When a death takes place in Hindu society, the body is prepared for viewing by laying it out with the hands across the chest, closing the eyelids, anointing the body with oil, and placing flower garlands around it. This is done by individuals of the same gender as the deceased, and this process will be presided over in the home of the deceased by the dead person's successor and heir (Habenstein & Lamers, 1974).

Because Hindus believe that cremation is an act of sacrifice, whereby one's body is offered to God through the funeral **pyre**, cremation is the preferred method of body disposition. In preparation for cremation, family members will construct a bier, consisting of a mat of woven coconut fronds stretched between two poles and supported by pieces of bamboo. The uncasketed body of the deceased will be borne on the bier from the deceased's home to the place of cremation by close relatives. This funeral procession will be led by the chief mourner—usually the eldest son—and will include musicians, drum players, and other mourners. The wife of the deceased will always remain behind in the home (Habenstein & Lamers, 1974).

According to R. W. Habenstein and L. M. Lamers, when the body reaches the place of cremation, usually a platform (*ghat*) located on the banks of a sacred river, the body will be removed from the bier and immersed in the holy waters of the river. During this process, a priest will perform a brief disposal ceremony. The body will then be smeared liberally with *ghi* (clarified butter) and placed on the pyre for burning. At this point, the chief mourner, who has brought burning coals from the house of the deceased, lights the pyre, and a priest recites an invocation similar to the following:

> Fire, you were lighted by him, so may he be lighted from you, that he may gain the regions of celestial bliss. May this offering prove auspicious (Habenstein & Lamers, 1974).

After the body has been consumed by the fire, and only fragments of bones remain, the mourners will ritually wash themselves in the river in a rite of purification. They will then make offerings to the ancestral spirits of the deceased. Upon completion of this duty they will recite passages from the sacred texts.

Three days after this ritual, a few relatives of the deceased will return to the cremation site in order to gather the bones. A priest will again read from the sacred texts and sprinkle water onto the *ghat*, while any remains of the deceased will be placed in a vase and given to the chief mourner. It is then the obligation of the chief mourner to cast these remains into the Ganges or another sacred river (Habenstein & Lamers, 1974).

Between 10 and 31 days after the cremation, a *Shraddha* (elaborate ritual feast) is prepared for all mourners and priests who have taken part in the funeral rituals. During the *Shraddha*, gifts are given to the *guru* (religious teacher), the *purohita* (officiating priest), and other Brahmins (religious functionaries). The social status of the family will determine how elaborate the *Shraddha* will be—for the poor this ritual will last 8 to 10 hours, while the wealthy may give a *Shraddha* lasting several days. At the close of the *Shraddha*, the mourning period officially ends, even though later *Shraddhas* may be given as memorial remembrances (Habenstein & Lamers, 1974). Although it

---

DEATH ACROSS CULTURES
The Sati

Though not the common practice, in 1987 an 18-year-old woman in India rested the head of her dead husband on her lap, sat on his funeral pyre, and was burned alive (Salamat, 1987). No one could stop her from committing **sati,** the most noble act of loyalty that a widow can perform for her husband, who, in the Hindu religion, is supposed to be considered a god by his wife. A widow cannot remarry and is treated as an evil omen and an economic liability. If she chooses to live, she has to remain barefoot, sleep on the floor, and can never go out of the house because she would be slandered if seen talking to a man. Some would argue that she is better off dead. Before this widow's act, however, sati had not been performed in that community since early in the twentieth century.

From "A Young Widow Burns in Her Bridal Clothes" (pp. 54–55), by A. Salamat, 1987, Far Eastern Economic Review, 138.

---

is believed that these ritual meals provide nourishment to the spirit of the deceased in its celestial abode, from a sociological perspective they serve as rites of family reincorporation and also differentiate the family regarding social status.

According to Habenstein and Lamers (1974), next to Hindu wedding ceremonies, funerals are the most important religious ceremonies. Although the funeral process is very costly to families, often resulting in their impoverishment, not providing the *Shraddha* creates more social problems for families than do the financial consequences of these rituals.

Buddhism

As the religious teachings of Buddha were disseminated throughout Asia, the beliefs and practices were adapted to indigenous cultural traditions dealing with death. Consequently, it is not possible to discuss Buddhist funeral rituals per se. Rather, there are Japanese, Korean, Chinese, and Thai funeral customs practiced within a Buddhist context.

In general, all Buddhist funeral ceremonies have some similarities. At the Buddhist temple, priests assist families as they engage in this important rite of passage. Prayers of the priests illustrate the "lesson of death"—that life is vanity. At funerals, Buddhist priests often read the following words from Buddha (cited by Habenstein & Lamers, 1974, p. 97):

> The body from which the soul has fled has no worth. Soon it will encumber the earth as a useless thing, like the trunk of a withered tree. Life lasts only for a moment. Birth and death follow one another in inescapable sequence. All that live must die. That man indeed is fortunate who achieves the nothingness of being.

All animal creation is dying, or dead, or merits to be dead. All of us are dying. We cannot escape death.

For most Buddhists, cremation is the preferred form of body disposition, but earth burial is also frequently practiced. In Buddhism, unlike Hinduism, there is no "soul"—both the body and the idea of a soul distract from the proper meditation and attainment of nirvana. Cremation serves the function of promoting the process of liberation of the individual from the illusion of the present world.

During and after the funeral, family members will make offerings through the priests to the spirit of the deceased. They will also give ritual feasts for the priests and other mourners. As in other religious traditions, all of these funeral activities will emphasize the importance of the religious worldview, promote community cohesiveness, and reincorporate chief mourners into routine patterns of social life.

Michael Leming has lived for significant periods of time in Northern Thailand. During his residence, he has had a number of opportunities to observe the Thai Buddhist funeral. As in rituals of any culture, there are always some variations, yet the following description is generally accurate in most situations.

When a typical Buddhist person dies, the body is cleaned, dressed, and placed within a casket. The casket is kept either within the home or at the *wat* (temple) for a period of 3 days. During this period, monks will come every evening to chant the Buddhist scriptures (*Abhidhamma*). Friends will attend these services and will offer gifts of floral tributes. On the fourth day, the body will be taken to the charnel-ground (cremation site), which is usually a good distance from the wat.

The procession is a significant rite in the Buddhist funeral. The casket is put on a carriage and taken from the *wat* compound to the cremation site. During the procession, a long white cotton cord is attached to the casket and eight (the number is flexible) monks together with lay devotees will carry the cord. In the urban areas the carriage will be motorized, while in traditional funerals and/or funerals in rural areas the carriage will be pulled by walking members of the procession.

After arriving at the site of cremation, the relatives pose for pictures by the casket. Walking around the casket three times (which symbolizes the traveling in the cycle of death and rebirth), the relatives then place the casket in front of the crematorium. Many times, young male family members will be ordained as monks for a period of a few days during the funeral ceremonies. In addition, the merit earned by these young men can be dedicated to benefit the dead.

As a final merit-making rite, 5 or 10 important persons will come forward (one by one) and place a set of yellow robes on the long white cotton cord, which is linked to the coffin. The most senior monk attending the service will collect the robes after "contemplating symbolically the dead." According to religious practice in Buddhism, the contemplation of the corpse by Buddhist monks will bring merit to those who provide opportunity for the monk to do so. The merit earned can be dedicated further to the dead as well.

The stage is now set for the actual cremation. Just prior to the lighting of the fire, the biography of the deceased is read while *dok mai chan* is distributed to all in attendance. The *dok mai chan* is a sandalwood flower with one incense-stick and

After arriving at the site of cremation, the relatives pose for pictures by the casket. Walking around the casket three times, they then place the casket in front of the crematorium. Many times, young male family members will be ordained as monks for a period of a few days during the funeral ceremonies. The merit earned by these young men can be dedicated to benefit the dead, as well.

two small candles attached. The mourners are all invited to come forward and deposit the *dok mai chan* before the casket. By so doing, mourners are deemed participants in the actual cremation.

The chairman of the ceremony then ignites the fire, and the casket is consumed. The next morning, the ashes (several pieces of bones) are gathered and made into a shape of a human being with the head facing east. Four monks attend this ritual, which culminates when the ashes are placed into a receptacle. Afterward, the ashes are enshrined in a reliquary built in the compound of the monastery.

Throughout the funeral ceremony sorrow or lamentation is not emphasized. Rather the focus is upon impermanence of all things. The funeral is primarily a social event that affirms community values and group cohesiveness. Furthermore, it is a time when people are expected to enjoy the fellowship surrounding the rituals. Typically there is entertainment (dancing and musical performances) associated with funeral rituals for sorrow and loneliness to dissipate. It is believed that this assists the bereaved to conceptualize a happy and pleasant paradise in which the deceased will reside.

Islam

Islamic tradition requires that the funeral service take place without unnecessary delay and that burial rites be simple and austere (Rahman, 1987). When it is time for the funeral, the body will be transported from the home or hospital to the mosque on the shoulders of the pallbearers.

In preparation for burial, Islamic family members will call into their home or the hospital a person of the same gender as the deceased who knows the prescribed ritual for washing and preparing the body. The eyes and mouth of the deceased will be closed, the arms will be straightened alongside the body, and the body will be washed and wrapped in a white seamless cloth (shroud) similar to that worn for the pilgrimage to Mecca (Eickelman, 1987).

Muslims mourn their dead in mosques, never at funeral parlors. The body is washed in a special room (men prepare males; women prepare females) while the family and friends recite *suras* from the Koran as blessings for the deceased. At the mosque the body will be placed onto a stone bier (*musalla*) in the outer courtyard. The funeral service will be a part of one of the five regular daily religious services (usually the noon service). Because Muslims consider burying the dead a good deed, when worshippers leave the mosque and see the coffin in the courtyard, they will participate in the procession to the cemetery even though they were not acquainted with the deceased (Habenstein & Lamers, 1974). Within hours of death, the body is buried.

The body is then transported from the mosque to the cemetery on the shoulders of those male mourners in the procession. In most Islamic traditions, all the male members of the community walk in a procession to the gravesite; women visit later during a 40-day mourning period (Habenstein and Lamers, 1974).

It is customary for every man in good health to carry the coffin on his shoulders for seven steps at least, and for passers-by to accompany the procession for at least seven steps. When a new bearer pushes under the coffin, another steps away so that 8 or 10 people are always under the load. These customs ensure that the remains will have an escort, even though the dead person may have no living relatives. At a prearranged spot, hearse and funeral cars await the procession. Where distances to the cemetery are short, the body will be borne to the grave totally on foot.

At the cemetery the body is placed into the grave, and mourners place handfuls of dirt on top of it. A close friend of the deceased climbs into the grave to read final instructions to the dead in preparation for his or her meeting with Allah. Then, according to Islamic belief, the angels who accompany every believer in life enter the grave to question the departed soul on matters of faith and life: "Who is your Lord? Who is your prophet? What book do you follow?" The sexton then fills the remainder of the grave using a shovel.

At the gravesite, rather than placing cut flowers, Muslim mourners plant flowers because they believe that every living plant utters the name of God. During this process, prayers are recited, and the service concludes with the preaching of a sermon (Habenstein & Lamers, 1974). After returning from the cemetery, all of the participants will partake of a meal that is served at the home of the deceased. Occasionally, some food from this meal is placed over the grave for the first 3 days after the death. Mourning continues for another 3 days, while family members receive social support and consolation from friends and members of the community.

Although Islamic women are allowed to openly express emotion in the bereavement process, Islamic men are encouraged to retain their composure as a sign that they are able to accept the will of Allah. A widow is required by the Qur'an to go into seclusion for 4 months and 10 days before she is allowed to remarry (Eickelman, 1987).

CONCLUSION

As discussed in chapter 2, dying is more than a biological process. One does not die in a vacuum but rather in a social milieu. The act of dying has an influence on others because it is a shared experience. The sharing mechanism is death-related meanings composed of symbols.

Because death meanings are socially constructed, patterns of "correct" or "incorrect" behavior related to dying and death will largely be determined within the social setting in which they occur. Death-related behavior of the dying person and of those relating to him or her is in response to meaning relative to the audience and the situation. As noted, death-related behavior is shared, symboled, and situated.

Because death generally disrupts established interaction networks, shared "scripts" aid in providing socially acceptable behavior for the bereaved. Such "scripts" are essential because they prevent societal breakdowns while providing social continuity. Burial rituals are important in assuring social cohesion at the time of family dismemberment through death. Because death is a family crisis (Dickinson & Fritz, 1981), appropriate networks for coping must be culturally well grounded.

Death-related meanings are socially created and transmitted. Through participant observation, small children learn from others how to respond to death. If children are sheltered from such situations, their socialization will be thwarted. As noted in this chapter, family involvement plays an important role in most societies as individuals prepare for death, as they prepare the corpse for final disposition, and as burial rituals that follow are performed. Therefore, whether death rituals involve killing a chicken, scraping the meat from the bones of the corpse, crying quietly, wailing loudly, mutilating one's own body, or burning or burying the corpse, all bereavement behavior has three interconnected characteristics—it is shared, symboled, and situated.

SUMMARY

1. Death occurs in all societies, yet it evokes an incredible variety of responses.

2. People in all societies are inclined to symbolize culturally defined feelings in conventional ways. Ritual behavior is an effective means of expressing or reinforcing these important sentiments.

3. Funeral or mortuary rites in many societies are critical because they ensure that the dead make the transition to the next stage of life or nonlife.

4. All rites of passage will have three subrites—rites of separation, rites of transition, and rites of reincorporation.

5. Societies encourage the expressions of both sadness and joy in their mourning rituals. In the former, participants are given the opportunity to collectively express their grief. In the latter societies individuals will engage in activities that will enable them to transform grief and anger over a death into joy.

6. Cross-cultural studies of the ritual of mourning at death reveal that a double standard prevails among some cultures—different "scripts" for different genders.

7. There are common human needs (e.g., to dispose of the body) and each society or religious group creates practices, rites, meanings, and rituals, which are functional equivalents to those found in other societies or groups.

8. Customs for caring for the dying before death assist both the dying individual and the survivors in coping with the impending death.

9. For some, death is viewed as the end, whereas for others it is viewed as a continuation of life in a different form.

10. Burial rituals serve the functions of appeasing the ancestral spirits and the soul of the deceased, bringing the kin together, reinforcing social status, and restoring the social structure.

11. In most societies the body of the deceased is cleaned and prepared for burial. Some groups even keep parts of the body for ornamental or special purposes.

12. Anthropologists are interested in the meanings that different events, such as death, have for different cultures and religious groups.

13. Being familiar with cross-cultural death customs should help one to better understand the American concept of death.

DISCUSSION QUESTIONS

1. Why is it important to learn about bereavement patterns in other cultures? What are some of the manifest and latent functions of burial rites discussed in this chapter?

2. Drawing upon your own knowledge, discuss any U.S. behavior patterns for the dying just before death. How do these customs compare with those cited in this chapter?

3. Describe mourning rituals commonly found in the United States. Mourning rituals may differ by region of the country or ethnicity. Discuss these differences.

4. It is suggested that an explanation for crying when someone dies may be rather complex. Discuss the reasons for crying over a death.

5. Discuss the importance of the family being involved in some way related to the process of the final disposition of the deceased.

6. In death as in life, gender discrimination occurs. Discuss with a local funeral director the differences in funerals for males and females.

7. When you visit cemeteries in the United States, do you notice differences in graves (such as size of headstones and length of epitaphs) between males and females? How does social class or status make a difference in the treatment of the dead within U.S. cemeteries?

8. What are some of the functions of burial rites discussed in this chapter?

GLOSSARY

Above-Ground Burial: The practice of placing the body on a scaffold or in trees for the purpose of allowing the body to decay or decompose. This type of burial is often the first step in the final disposition of the body. In the second stage the body is buried, which is referred to as **secondary burial.**

Bier: A framework upon which the corpse and/or casket is placed for viewing and/or carrying.

Columbarium: A building or wall for above-ground accommodation of cremated remains.

Cremains: That which is left after cremation.

Cremation: The reduction of a human body by means of heat or direct flame. The cremated remains are called cremains or ashes and weigh between 3 and 9 pounds. "Ashes" is a very poor word to describe the cremated remains because they are actually processed bone fragments and calcium residue that have the appearance of crushed rock or pumice.

Crematory: An establishment in which cremation takes place.

Disposition: Final placement or disposal of a dead person.

Embalming: A process that temporarily preserves a deceased person by means of displacing body fluids with preserving chemicals.

Entombment: Opening and closing of a crypt, including placing and sealing of a casket within.

Ethnography: The systematic description of a culture based on firsthand observation.

Exhume: To remove a corpse from its place of burial.

Latent Function: Consequences of behavior that were not intended (for example, a funeral brings the family together, usually in an amiable way).

Manifest Function: Consequences of behavior that are intended and overt (such as going to a funeral to pay respects to the deceased person).

Mausoleum: A building or wall for above-ground accommodation of a casket.

Mummification: The process of wrapping the body with cloth before its final disposition.

Niche: A chamber in a columbarium in which an urn is placed.

Pyre: A combustible pile (usually of wood) for burning a corpse at a funeral rite.

Rites of Passage: Ceremonies centering around transitions in life from one status to another (including baptism, the marriage ceremony, and the funeral).

Ritual: The symbolic affirmation of values by means of culturally standardized utterances and actions.

Sati: A controversial practice rarely exercised in some areas of India in which the widow throws herself onto the funeral pyre of her deceased husband. This "most noble act of loyalty" makes her a goddess who is worshiped at her cremation site.

Urn: A container for cremated remains.

SUGGESTED READINGS

Barber, P. (1988). *Vampires, burial, and death: Folklore and fealty.* New Haven: Yale University Press. How people in preindustrial cultures look at the processes and phenomena associated with death and the dissolution of the body.

Bendann, E. (1930). *Death customs: An analytical study of burial rites.* New York: Alfred A. Knopf. A thorough anthropological analysis of death customs, including an analysis of burial rites, causes of death, attitudes toward the corpse, mourning, and beliefs in the afterlife.

Bloch, M., & Parry, J. (1982). *Death and the regeneration of life.* Cambridge: Cambridge University Press. Focuses on the significance of symbols of fertility and rebirth in funeral rituals from China, India, New Guinea, Latin America, and Africa.

Bond, G. C., Kreniske, J., Susser, I., & Vincent, J. (1997). *AIDS in Africa and the Caribbean.* Boulder, CO: Westview. Detailed ethnographic studies to explain AIDS in a global and comparative Third World context.

Counts, D. R., & Counts, D. A. (Eds.). (1991). *Coping with the final tragedy: Cultural variation in dying and grieving.* Amityville, NY: Baywood Publishing. An anthology that discusses death and grieving in various parts of the world.

Crissman, J. K. (1994). *Death and dying in central Appalachia: Changing attitudes and practices.* Urbana, IL: University of Illinois Press. An exploration of cultural traits related to dying and death in Appalachia, showing how they have changed since the 1600s.

Dettwyler, K. A. (1994). *Dancing skeletons.* Prospect Heights, IL: Waveland Press. Reports from a biocultural anthropologist about her fieldwork on living and dying in west Africa.

Farrell J. (1980). *Inventing the American way of death, 1830–1920.* Philadelphia: Temple University Press. Examines the transformation from the Puritan way of death to the American way of death.

Habenstein, R. W., & Lamers, W. M. (1974). *Funeral customs the world over* (Rev. ed.). Milwaukee: Bulfin Printers. A review of the cultures of the world especially significant in cross-cultural study of the various practices of funerals in all cultures.

Huntington, R., & Metcalf, P. (1992). *Celebrations of death: The anthropology of mortuary ritual* (2nd ed.). Cambridge, MA: Cambridge University Press. An anthropological analysis of dying and death, including universals and culture, death as transition, and the royal corpse and the body politic.

Irish, D. P., Lundquist, K. F., & Nelson, V. J. (1993). *Ethnic variations in dying, death and grief.* Washington, DC: Taylor and Francis. An anthology that discusses dying, death, and grief in the following groups: African Americans, Mexican Americans, Hmong, Native Americans, Jews, Buddhists, Muslims, Quakers, and Unitarians.

McGuire, R. H. (1992). *Death, society, and ideology in a Hohokam community.* Boulder, CO: Westview Press. Explorations of the nature of Hohokam social organization in the prehistory of southern Arizona by an archaeologist.

Narasimhan, S. (1990). *Sati: Widow burning in India.* New York: Doubleday Anchor Books. Reasons why women choose to become, or are forced to become, sati and what this reveals about the society as a whole as outlined by a noted Indian journalist.

Parry, J. K., & Ryan, A. S. (1995). *A cross-cultural look at death, dying, and religion.* Chicago: Nelson-Hall Publishers. This book is about dying, death, and religion and how these subjects are integrated into the belief systems of various cultures: African Americans, Buddhists, Roman Catholics and other Christians, Chinese, Dominicans, Filipinos, Muslims, Jews, Koreans, and Mexican Americans.

Spiro, H. M., Curnen, M. G. M., & Wandel, L. P. (Eds.). (1996). *Facing death: Where culture, religion, and medicine meet.* New Haven: Yale University Press. Discussion of Christian, Judaic, Islamic, Hindu, and Chinese perspectives on death and rituals of mourning.

Stannard, D. E. (1977). *The Puritan way of death.* New York: Oxford University Press. A portrayal of death in the Western tradition, including discussions of death, childhood, and burial.

Strocchia, S. T. (1992). *Death and ritual in Renaissance Florence.* Baltimore: Johns Hopkins University Press. Historical account of how death rites in Renaissance Florence reflected Florence's quick rise to commercial wealth in the 14th century and steady progress toward power in the 15th and 16th centuries.

Subedi, J., & Gallagher, E. B. (1996). *Society, health, and disease: Transcultural perspectives.* Upper Saddle River, NJ: Prentice-Hall. Discusses the sociocultural context of health and disease, sociopolitical constraints in health and health care, the psychology of health and well-being, the threat of AIDS, and emerging areas in international health.

THE BUSINESS OF DYING

One could imagine a more pleasant means of livelihood, but almost any trade is bearable if the customers are sure.

> —Clarence Darrow, discussing an ancestor reported to have been an undertaker

There is a major misconception about how much money funeral directors make. When I was at mortuary school, a teacher asked how many of us thought funeral directors made a lot of money. Several raised their hands. Then he told us how much funeral directors actually made. Two weeks later, about half those people were gone.

> —John Everly (From H. Cox, *Insight*)

THE BUSINESS OF PREPARING THE DEAD

A Minneapolis television station had a special "Dimension Report" on a new Twin Cities (Minneapolis and St. Paul, Minnesota) business called "Doggie Do Do Pick-Up." For a fee, these people would come to your home in the spring of the year (after the snow had melted) and remove a winter's supply of pet "waste." Michael Leming and his son, who share an entrepreneurial spirit, concluded that there will always be money to be made by providing necessary services that no one wants to do. Furthermore, as the service becomes more necessary and the task more onerous, the price for the services rendered will escalate.

The Changing American Funeral

Funerals in the 21st century are obviously not the same as they were in colonial times in the United States. Indeed, "the times they are a-changing," as singer Bob Dylan wailed in the 1960s. Like other aspects of our society, we have gone from being generalists, where a lot of people know how to do a lot of things, to a society of specialization and division of labor. The birth of the funeral industry in the late 19th century is a good example of this specialization.

The Puritan Funeral

For the Puritans, the tolling of the town bell announced the onset of a funeral, and participants gathered at the home for prayer and the procession. Death was feared by the Puritans, who prayed not for the soul of the deceased but rather for the comfort and instruction of the living. They believed that judgment occurs at death and that the dead are beyond human aid, so they prayed to reaffirm their faith and to glorify their God. In addition to prayers, Puritans might also read an **elegy,** which generally depicted the dead person as a saint freed from the world and entering eternal bliss. Such elegies confirmed the new, separate status of the deceased, helping to bring grief under control, and provided a good example for structuring life after bereavement. Sometimes a copy of the elegy was pinned to the coffin or hearse for the funeral procession (Geddes, 1981).

The mourners walked from the home to the graveyard, where the men of the family, or the **sexton,** had opened a grave. They carried the coffin on a **bier,** covered with a black cloth "pall"—both of which were the property of the town instead of the church, as was the custom in England. During the procession, mourners were supposed to "apply themselves to meditations and conferences suitable to the occasion." At the grave site, the pallbearers lowered the coffin into the earth, and the grave was refilled (Geddes, 1981, p. 111).

After the burial, the mourners returned home, where they shared food, drink, prayers, and comforting words. The family members thanked the pallbearers and participants and sometimes gave additional presents. They might also ask the minister to deliver a funeral sermon, which would occur not on the day of burial, but rather at the next regular meeting of the congregation. Sometimes families had such sermons and/or the elegies published and distributed to friends as *memento mori.* At a later date, the family might also erect a marker over the grave to proclaim the imminence of death or God's promise of salvation. Such markers near the much-frequented meeting house were another way of maintaining the vitality of death in early American culture (Geddes, 1981; Ludwig, 1966).

The Puritan funeral was the primary social institution for channeling the grief of survivors. It provided for the disposition of the body and the acknowledgment of the absence of the deceased. It drew the community members together for mutual comfort, and it allowed mourners to honor the dead, to express their sorrow at separation, and to demonstrate their acceptance of God's will. After the funeral, Puritans expected mourners to return to their calling and to resume their life's work.

DEATH ACROSS CULTURES
Funeral Rituals in a Polynesian Community

In a Polynesian village in Micronesia in the Pacific Ocean, anthropologist Michael D. Lieber observed funeral rituals. Funeral preparations tend to vary little by age or sex, except for more elaboration with age. The female head of the deceased's descent group assigns initial tasks for cooking food, preparing the body for viewing (washing, dressing, and closing orifices), and preparing the house for the wake. No weeping or other emotional expressions are allowed during this phase.

At the wake the body is viewed, women give forth a high-pitched wail, and weeping is permitted. Bodies of older people are buried under floors of the ancestral houses called "grave houses." Others are weighted with stones and buried in the lagoon. Valuables are buried with the body. After the body is disposed of, food is served to all guests. Chanting begins after people have eaten and continues until the next morning—the mood is more festive and entertaining.

Following these activities, all of the closest kin are given new names. The names are derived from chant lyrics and are usually names that ancestors used. Old names are "thrown away" to symbolize the rupture of a relationship, rather than simply as a sign of loss. To continue to use the old name would be a painful reminder of one's personal loss of an important relationship.

From "Cutting Your Losses: Death and Grieving in a Polynesian Community" (pp. 169–189), by M. D. Lieber, 1991. In D. R. Counts and D. A. Counts (Eds.), Coping with the Final Tragedy: Cultural Variation in Dying and Grieving, *Amityville, NY: Baywood Publishing.*

They did not approve elaborate or extended mourning of the sort that became customary in the 19th century because it undermined the cheerful resignation to God's will that was essential to the Puritan experience.

The Victorian Funeral

Victorian times brought a focus on aesthetics and soothing the mourner. The attitude was one of beautifying the whole funeral process and making the funeral more tolerable for bereaved individuals. To "get on" with the funeral and move on with life seemed to be emphasized. This reflected the "dying of death movement" discussed in chapter 2.

The Focus on Aesthetics.   Between 1850 and 1920 funeral service changed significantly in the United States. Because "the growing wealth and prosperity of our country has caused people to demand something more in accordance with their surroundings" (Benjamin, 1882, p. 3) and because funeral directors cultivated a "steadily advancing appreciation of the aesthetics of society" ("Funeral Directors," 1883), the new Victorian funeral would be a work of art that would attempt to

---

WORDS OF WISDOM
Display of the Dead

Truly, we need to do away with some of the false ideas of Death which are shown in so many gruesome ways at funerals, and strive to give the young a different and truer idea of what the passing away of a soul means. The awfulness of some funerals is nothing short of criminal, especially as it affects the minds of the young. If there is work cut out for the minister of today, it is the enlightenment of his people on the subject of death and the funeral. But the minister must, first of all, imbibe a wholesome lesson of self-restraint for himself, and abolish the fulsome and tiresome eulogy which is the bane of so many funerals. He must learn for himself, and teach to his people, the beauty and solemnity of the brief service as prescribed by his church and attempt nothing more, and he must also relentlessly oppose the tendency which exists to turn the modern funeral, especially in the country, into a picnic. The present outpouring of a heterogeneous mass of folk from every point of the countryside is a farce that cannot be too soon abolished. A funeral is essentially a time for the meeting of the family and relatives and the closest friends, and the fewer the number of outsiders present the better. Nor is there anything quite so barbarous as the present custom at so many funerals of "viewing the remains" by a motley collection of folk, many of whom never even knew the dead in life, or, if they did, never thought enough of him to come and see him. The vulgar curiosity that prompts "a last look" at a loved one cannot be too severely denounced. Only second to it is the pretentious line of vehicles that "escort the remains to the grave" and the mental caliber of a community that bases the popularity of a man on the number of carriages that follow him to the grave! If there is a crying need of the gospel of simplicity it is in connection with funerals. It seems inconceivable that Death should be made the occasion for display, and yet this is true of scores of funerals. The flowers, including those fearful conceptions of the ignorant florist, such as "Gates Ajar"; the quality of casket and even of the raiment of the dead, the "crowd" at the "obsequies," the number of carriages in the "cortege"— oh, oh, "what fools these mortals be," to say naught of the wicked and wanton waste of much-needed money. It is difficult to conceive that a national love of display should have become so deep-rooted as to lead to the very edge of the grave!

Ladies Home Journal, *September 1903.*

---

restore order to the middle classes while hiding as much as possible the "unpleasantries" of death.

The demand for a new funeral service came from the American middle class, but it was created and supplied by casket manufacturers and funeral directors. The National Funeral Directors Association (NFDA) was founded in 1882 in Rochester, New York, the home of the Stein Casket Manufacturing Company. The association's official journal was *The Casket*, founded and funded for several years by the Stein Company. As this name suggests, the first widespread innovation in funeral service was the casket, a stylish container for the corpse. Before 1850, most Americans were laid to rest in a coffin, a six-sided box that was constructed to order by the local cabinetmaker. By 1927 "the old wedge-shaped coffin (was) obsolete. A great variety of styles and grades of caskets (were) available in the trade, ranging from a cheap, cloth-covered pine box to the expensive cast-bronze sarcophagus" (Gebhart, 1927, p. 8). The rectangular shape of the new caskets complemented the artwork in concealing the uncouth corpse. In applying for a casket patent in 1849, A. C. Barstow (cited in Habenstein & Lamers, 1962, p. 270) explained:

> The burial cases formerly used were adapted in shape nearly to the form of the human body, that is they tapered from the shoulders to the head, and from the shoulders to the feet. Presently, in order to obviate in some degree the disagreeable sensations produced by a coffin on many minds, the casket, or square form has been adopted.

The adoption of the word *casket* also accelerated the "dying of death," as the word had previously denoted a container for something precious, like jewels. Accepting the associated idea of the preciousness of the body, Americans decided that a dead-looking corpse looked out of place in an elaborate silk-lined casket. Rather than remove the casket, they decided to stylize the body. Originally a way of preserving bodies for shipment home from Civil War battlefields or Western cities, embalming soon became a way of preserving appearances. Responding to the germ theory of disease and the public health movement, funeral directors attempted to gain professional status by emphasizing the disinfectant qualities of embalming. Most funeral directors, however, wanted simply "to retain and improve the complexion" so that the corpse would look "as natural as though it were alive" (Hohenschuh, 1921, p. 82). To do this, they began to cosmeticize the corpse and to clothe and position the body naturally. They replaced the traditional shroud with street clothes, and they tried "to lay out the body so that there will be as little suggestion of death as possible." By 1920, they succeeded so well that a Boston undertaker supposedly advertised (Dowd, 1921, p. 53):

For composing the features $1

For giving the features a look of quiet resignation $2

For giving the features the appearance of Christian hope and contentment $5

Bereavement practices were affected by the change from coffin to casket and by the "restorative art" of the embalmer; they were also affected by the movement of the funeral from the domestic parlor to the funeral parlor. As people began banishing death from their homes to hospitals, they started moving the funeral from the family parlor to a specialized funeral parlor. After the Civil War, middle-class Americans began to exclude the formal parlor from their homes and to replace it with a "living room." At the same time, funeral directors wanted full control of the corpse and the funeral. The ease and efficiency of directing funerals in a funeral home made them more profitable. Despite these benefits, the transition to the funeral parlor was slow, extending well into the 20th century (Farrell, 1980).

The practice of laying the deceased on a bier for viewing in the home persisted through the end of the 19th century.

The Focus on Soothing the Mourner.   Both in the domestic parlor and in the funeral parlor, the procedure of the turn-of-the-century funeral changed. In conjunction with the reform forces of religious liberalism, funeral directors began to redirect funerals to be shorter, more secular, and more soothing. They shortened the service by trying to revise the long sermon with its exhortations of repentance and renewal. Although some clerics resisted, funeral directors wanted the sermon redirected from theology to psychology, from preaching to grief therapy, and from the state of survivors' souls to the state of their emotions. The funeral director took care of all of the details of the funeral and performed as much as a stage manager as a mortician. "Really it is much the same," wrote one director, "I work for effect—for consoling and soothing effect" ("The Man Nobody Envies," 1914).

After 1880, funeral directors used their arts and the culture of professionalism to effect a massive change in the American way of bereavement. Professionalism was part of the middle-class strategy of specialization. It required education in an area of expertise and an ethic of service, and it provided autonomy and income for its practitioners. The American undertaker sought professional status because it would help to become "enough of an authority to convince his clients, without offense, that there are better methods than are prescribed by custom" (Hohenschuh, 1921, p. 9). Etiquette books reinforced this culture of professionalism by advising readers that "the arrangements for the funeral are usually left to the undertaker, who best knows how to proceed" (Wells, 1887, p. 303). To the middle-class people who feared death anyway, this established a situation in which the public passively accepted changes in funeral service suggested by funeral directors (Hohenschuh, 1921; Wells, 1887).

The modern bereavement practices described proceeded from a simple desire to make death as painless for survivors as for the deceased. It came from a widespread cultural attempt "not to mention trouble or grief or sickness or sin, but to treat them as if they do not exist, and speak only of the sweet and pleasant things of life" ("The Ideas of a Plain Country Woman," 1913). This "dying of death" came from the desire of the middle class for control—of self, society, and the environment. It ended exactly where de Tocqueville (1835/1945, pp. 2, 4) predicted:

> As they perceive that they succeed in resolving without resistance all the little difficulties their practical life presents, [the Americans] readily conclude that everything in the world can be explained, and that nothing in it transcends the limits of the understanding. Thus they fall to denying what they cannot comprehend.

Denying and disguising death, middle-class Americans achieved, on the surface at least, the dying of death.

The Contemporary American Funeral: Meeting the Needs of the Bereaved

The modern thanatology movement, which was influenced by Jessica Mitford's *The American Way of Death* (1963) and Elisabeth Kübler-Ross' *On Dying and Death* (1969), effectively resurrected death in the consciousness of American life and thereby transformed the contemporary American funeral. Along with public education that attempted to provide some relevance for the American learner, funerals too attempted to be relevant to grievers and transform romanticism and sentimentality into an opportunity to honestly feel one's emotions and loss.

Mourners were encouraged to feel and experience the reality of death and then attempt to adapt to the environment in which the deceased is gone. As part of this process of combating the tendency to deny death, the modern funeral eschewed euphemisms and allowed mourners to personalize funeral rituals and work with professional funeral directors to create meaningful rituals that both reflected and celebrated the unique life of the person who had died. The process of personalization led family members to deliver their eulogies and create memorial tables (consisting of pictures, artifacts, and mementos of the deceased) as attempts to combat depersonalization and overprofessionalization that had characterized funerals of an earlier era.

Paul Irion (1956) has described the following needs of the bereaved: reality, expression of grief, social support, and meaningful context for the death. For Irion, the funeral is an experience of significant personal value insofar as it meets the religious, social, and psychological needs of the mourners. Each of these must be met for bereaved individuals to return to everyday living and, in the process, resolve their grief.

The psychological focus of the funeral is based on the fact that grief is an emotion. Edgar Jackson (1963) has indicated that grief is the other side of the coin of love. He contends that if a person has never loved the deceased—never had an emotional investment of some type and degree—he or she will not grieve upon death. As discussed in the opening pages of chapter 2, evidence of this can easily be demonstrated by the number of deaths that we see, hear, or read about daily that do not have an impact on us unless we have some kind of emotional involvement with those deceased persons. We can read of 78 deaths in a plane crash and not grieve over any of them unless we personally knew the individuals killed. Exceptions to the preceding might include the death of a celebrity or other public figure, when people experience a sense of grief even though there has never been any personal contact.

In his original work on the symptomatology of grief, Erich Lindemann (1944) stressed this concept of grief and its importance as a step in the resolution of grief. He defines how the emotion of grief must support the reality and finality of death. As long as the finality of death is avoided, Lindemann believes, grief resolution is impeded. For this reason, he strongly recommends that the bereaved persons view the dead. When the living confront the dead, all of the intellectualization and avoidance techniques break down. When we can say, "He or she is dead, I am alone, and from this day forward my life will be forever different," we have broken through the devices of denial and avoidance and have accepted the reality of death. It is only at this point that we can begin to withdraw the emotional capital that we have invested in the deceased and seek to create new relationships with the living.

On the other hand, viewing the corpse can be very traumatic for some. Most people are not accustomed to seeing a cold body and a significant other stretched out with eyes closed. Indeed, for some this scene may remain in their memories for a lifetime. Thus, they remember the cold corpse, not the warm, responsive person. Whether or not to view the body is not a cut-and-dried decision. Many factors should be taken into account when this decision is made.

Grief resolution is especially important for family members, but others are affected also—the neighbors, the business community in some instances, the religious community in most instances, the health-care community, and the circle of friends and associates (many of whom may be unknown to the family). All of these groups will grieve to some extent over the death of their relationship with the deceased. Thus, many people are affected by the death. These affected persons will seek not only a means of expressing their grief over the death, but also a network of support to help cope with their grief.

Sociologically, the funeral is a social event that brings the chief mourners and the members of society into a confrontation with death. The funeral becomes a vehicle to bring persons of all walks of life and degrees of relationship to the deceased together for expression and support. It is for this reason that in our contemporary culture the funeral becomes an occasion to which no one is invited but all may come. This was not always the case, and some cultures make the funeral ceremony an "invitation only" experience. It is perhaps for this reason that private funerals (restricted to the family or a special list of persons) have all but disappeared in our culture. (The possible exception to this statement is a funeral for a celebrity—where participation by the public may be limited to media coverage.)

At a time when emotions are strong, it is important that human interaction and social support become high priorities. A funeral can provide this atmosphere. To grieve alone can be devastating because it becomes necessary for that lone person to absorb all of the feelings into himself or herself. It has often been said that "joy shared is joy increased"; surely grief shared is grief diminished. People need each other at times when they have intense emotional experiences.

A funeral is in essence a one-time kind of "support group" to undergird and support those grieving persons. A funeral provides a conducive social environment for mourning. We may go to the funeral home either to visit with the bereaved family or to work through our own grief. Most of us have had the experience of finding it difficult to discuss a death with a member of the family. We seek the proper atmosphere, time, and place. It is during the funeral, the wake, the shivah, or the visitation with the bereaved family that we have the opportunity to express our condolences and sympathy comfortably.

Anger and guilt are often deeply felt at the time of death and will surface in words and actions. They are permitted within the funeral atmosphere as honest and candid expressions of grief, whereas at other times they might bring criticism and reprimand. The funeral atmosphere says in essence, "You are okay, I am okay; we have some strong feelings, and now is the time to express and share them for the benefit of all." Silence, talking, feeling, touching, and all means of sharing can be expressed without the fear of their being inappropriate.

Another function of the funeral is to provide a theological or philosophical perspective to facilitate grieving and to provide a context of meaning in which to place one of life's most significant experiences. For the majority of Americans, the funeral is a religious rite or ceremony (Pine, 1971). Those grievers who do not possess a religious creed or orientation will define or express death in the context of the values that the deceased and the grievers find important. Theologically or philosophically, the funeral functions as an attempt to bring meaning to the death and life of the deceased individual. For the religiously oriented person, the belief system will perhaps bring an understanding of the afterlife. Others may see only the end of biological life and the beginning of symbolic immortality created by the effects of one's life on the lives of others. The funeral should be planned to give meaning to whichever value context is significant for the bereaved.

"Why?" is one of the most often asked questions upon the moment of death or upon being told that someone we know has died. Though the funeral cannot provide the final answer to this question, it can place death within a context of meaning that is significant to those who mourn. If it is religious in context, the theology, creed, and articles of faith confessed by the mourners will give them comfort and assurance as to the meaning of death. Others who have developed a personally meaningful philosophy of life and death will seek to place the death in that philosophical context.

Cultural expectations typically require that we dispose of the dead with ceremony and dignity. The funeral can also ascribe importance to the remains of the dead. In keeping with the specialization found in most aspects of American life (e.g., the rise of professions), the funeral industry is doing for Americans that necessary task they no longer choose to do for themselves.

THE AMERICAN PRACTICE OF FUNERAL SERVICE

Education and Licensure

With the evolution of the funeral, there likewise has been an evolution of the funeral functionary. Contemporary Americans refer to that functionary as a "funeral director," yet a hundred years ago this functionary was a "layer out of the dead," often a member of the family who physically and emotionally could perform the necessary tasks of bathing the body, closing the eyes and mouth, and dressing the body. It was not unusual for a midwife or other person who provided nursing-like services in the community to be called on to assist the family. Early advertisements indicate that nurses did offer such services.

As early as the 1890s, various states began to enact legislation to protect the public's health by licensing embalmers. Early licensing agencies directed their attention to the **embalming** process, and it was not until the 1920s and 1930s that licensing agencies began to regulate other aspects of the funeral and the operation of funeral homes.

The rationale for this licensure was based upon the need to protect the public, primarily in the financial area. However, regulations also addressed how a funeral should be conducted when the cause of death was a contagious disease. Another issue of public health protection was the transportation of the dead from the place of death to the location of final disposition. Public health authorities claim that regulating the treatment of the dead has significantly contributed to the advanced standard of health of our country.

Based on these concepts, the licensing agencies most often charged with responsibility to regulate the funeral industry have been the various state boards of health. In some states, special boards have been established to implement regulation and enforcement procedures.

Licensure has been reserved for the individual states and includes three basic licenses. A license as an embalmer permits a person to legally remove the dead from the place of death and prepare the body through the process of embalming for viewing and holding the funeral. All states (with the exception of Colorado) require persons who perform these functions to be licensed. A second license, to practice as a funeral director, permits the holder to arrange the legal details of the funeral, including preparing the death certificate and counseling with the family to arrange and implement the kind of funeral desired for the deceased. A third license is one that

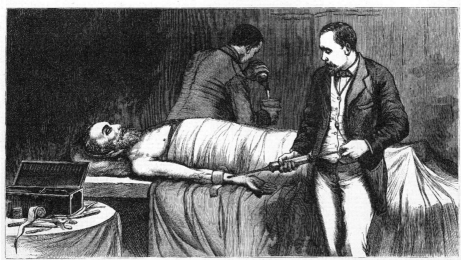

EMBALMING THE BODY OF THE DECEASED, ON THE MORNING OF SEPTEMBER 20TH.

DEATH OF PRESIDENT GARFIELD.
THE NATION'S GREAT AFFLICTION.

An artist's sketch of the embalming of the body of President Garfield on the morning of September 20, 1881, in preparation for the train ride from Elberon, New Jersey, to Washington, DC. Though originally a way of preserving bodies for shipment home from Civil War battlefields, embalming soon became a way of preserving appearances.

permits the holder to practice mortuary science—an all-inclusive specialty that covers the practices of both embalming and funeral directing.

A few states issue a funeral director license, which may be held by only one person in each firm (usually the owner or manager) that serves to give the licensing agency control over all of the practitioners within that firm. A greater number of states have created a funeral home license or permit that is required for each funeral home and permits the state to close the funeral home by withdrawing the license without taking action against the licensees employed by that firm. In addition to requiring embalmers to be licensed, with the exception of Colorado, all states and the District of Columbia require embalming practitioners to also be licensed as funeral directors. The exact number of states requiring funeral home licenses or permits is difficult to determine because some are required by law, some by regulation, and some by local ordinance. Approximately 25 states have some requirement that governs the operation of the funeral homes located in them.

The qualifications for licensure chiefly concern age, citizenship, and specific education. As of 1995 (with the exception of Colorado), all states require a high school education and some postsecondary education in mortuary science. Fourteen states require 2 years of college, including the 1-year major in funeral service or mortuary science, 16 states require 3 years of college, including the 1-year funeral service program, and one state requires a bachelor's degree for licensure in funeral service. The typical funeral service student graduates with an associate degree in funeral service or mortuary science.

Following academic preparation and in some cases before, all states require applicants to serve an apprenticeship that varies from 1 to 3 years, depending on individual state regulations. Upon the completion of academic and internship or apprenticeship requirements, applicants for licensure are required by all states to pass a qualifying examination before the issuance of the license to practice (Bigelow, 1997).

Currently, 30 states mandate continuing education for renewal of the license to practice. The Academy of Professional Funeral Service Practice certifies and approves all continuing education programs and courses. The average number of continuing education hours per year required by states for relicensure is 8, but it varies from 3 hours every third year (required by West Virginia) to 24 hours (required every 2 years by Iowa) (NFDA, 1999).

According to the National Director of Morticians/The Red Book (NFDA, 1999), there are 22,107 funeral homes in the United States with 35,000 licensed personnel and 80,000 funeral service and crematory personnel. Approximately 5 percent of the licensees in the United States are women (Goldman, 1993). However, in the last decade, the number of women entering colleges of funeral service education and becoming licensed has greatly increased. In 1996, 32 percent of the 2,168 graduates from the 48 accredited schools of mortuary science in the United States were women. In 1976 only 8 percent of graduates were women (Bigelow, 1997).

In a recent survey of funeral homes by the Federated Funeral Directors of America (1997), it was determined that in 1995 the average salaries for full-time, licensed personnel ranged from $33,200 per year in communities of 10,000 to 30,000 population to $35,000 in communities with more than 100,000 population.

The Role of the Funeral Director

Rabbi Earl Grollman (1972) describes the role of the funeral director as that of a caretaker, caregiver, and gatekeeper. He indicates that the etymology of the word *undertaker* is based upon the activities of the early undertaker who "undertook" to do for people at the time of death those things that were crucial in meeting their bereavement needs. The funeral director, from the perspective of the community, was viewed as a secular gatekeeper between the living and the dead.

John Brantner (1973), elaborating upon the caregiver role, emphasizes that the funeral director is a crisis intervenor. This idea can be documented in the vast amount of literature on the counseling role of crisis intervenors, who are not clinical practitioners by training but to whom the public turns in crisis.

The funeral director serves families by determining their needs and responding to them (Raether & Slater, 1974). This service includes, but is not limited to, the funeral (or its alternative) that the director and the family plan and implement

PRACTICAL MATTERS
Should Funeral Directors Professionalize?

Raymond DeVries

Like many other occupations, funeral directors would like to be thought of as professionals. Should we as consumers support the attempts of funeral directors to become more professional?

Our first response to this question is, "Yes, of course." But let us not be so hasty. We must first consider what it is that makes an ordinary job a "profession." How do we distinguish a profession from a regular job? One way is to list the features or traits of occupations commonly accepted as professions. If asked which occupations are professions, most of us would answer: physician or lawyer. What sets these occupations apart? They are characterized by the following:

1. A specialized body of knowledge
2. A long period of training
3. An orientation toward service rather than profit
4. A commonly accepted code of ethics
5. Legal recognition (most often through licensure)
6. A professional association

Implicit in this definition is the assumption that professionals have the best interest of the public in mind. After all, they submit to a long period of training, look forward to serving others, abide by a code of ethics, and police themselves through their professional associations. All occupations should become professions!

But this is not the only way to define a profession. Others look more cynically on the role of professions in society. George Bernard Shaw said, "Professions are a conspiracy against the laity." What did he mean? Shaw's comment hints at an alternative definition of the professions, a definition that there is just one distinguishing characteristic of the professions: power. Professions are those occupations that have accumulated enough power to control the definition and substance of their work. For example, this view contends that physicians are professionals by virtue of their complete control of matters of health. Through their associations they control the number and training of doctors, they limit the practices of competitors (e.g., chiropractors or nurse practitioners), and they set rates of reimbursement for health care. Adherents to this view point out that professionals in fact incapacitate us: they limit our choices, make us feel unable to help ourselves, and encourage dependency.

Should funeral directors become more professional? Not all would agree. Funeral directors subscribe to the first definition and assert that professionals can better attend to the needs of the public. Followers of the second definition conclude that the move toward professionalization would limit competition, drive prices up, and, by promoting dependency, make us less able to deal with death.

Professor of Sociology, St. Olaf College, Northfield, MN

BY PERMISSION OF JOHNNY HART AND CREATORS SYNDICATE, INC.

together. As a licensee of the state, the funeral director handles such details as recording the death properly as required by law and files permits for transportation and final disposition of the body. The funeral director serves as a liaison with other professionals, working with the family's medical personnel, clergy, lawyers, cemetery personnel, and, when necessary, law enforcement officials.

According to the NFDA (1999), 87 percent of funeral homes are family-owned and average 69.5 years in operation in the same community. A 1999 survey by the NFDA indicates that the average funeral home arranges 182 funerals a year and has 1.6 locations; 40 percent of these funeral homes are located in small towns or rural areas, and 26 percent are in large or moderate-sized cities (NFDA, 1999).

Funeral Expenses

Charges made by a funeral home ordinarily are for the services of the professional staff, the use of the funeral home facilities and equipment, transportation, and the casket or other container. In addition, most funeral homes provide burial vaults or other types of outer enclosures for the casket and ancillary items that may be purchased from the funeral director—clothing, register books, acknowledgment cards, and stars of David, crucifixes, or crosses.

The other major cost of the funeral is the cemetery charge—either for the purchase of cemetery property for the right to interment therein, for **mausoleum** space, or for an **urn** for the **cremated** remains (in some instances, there will be a charge for providing a space in a **columbarium** for the urn). Most families will also select, in one form or another, a monument or marker to identify the grave or other place of final disposition.

A final category of expenses incurred by the family is money that the funeral home sometimes advances, at the request of the family, to other people involved in the final disposition. Such cash advances typically pay for the following: charges for opening and closing the grave, **crematory** costs, honoraria for clergy and musicians, **obituary** notices, flowers, and transportation costs in addition to the transportation ordinarily furnished by the funeral home.

In a survey by the Federated Funeral Directors of America (1999) of more than 1,468,280 funerals conducted in 1999, it was determined that the average cost for a funeral for an adult in the United States was $5,023. This cost did not include grave or cremation expenses, the cost of a vault, clothing, or extra service requirements, such as the clergy, florist, and musicians. The cost is broken down below:

Personnel (owner salary, employee salaries, retirement plan, payroll taxes, insurance, professional services) $1,561

Cost of facilities (rent allowance/rent, depreciation, heat, water, electricity, insurance/general maintenance and repairs, taxes, telephone, leased music, interest expense) 753

Automotive equipment (depreciation and insurance, automotive expenses, livery expense, leased autos) 191

Promotion (advertising, business contributions, convention/meeting, organization dues, subscriptions, business promotion, meals/entertainment) 171

Supplies (preparation room supplies, miscellaneous funeral supplies) 96

Business services (legal services, postage, office supplies, consultants, accounting) 171

Cost of sales (casket cost, other merchandise sold to families) 1,431

After-sale expenses (bad debts and discounts) 76

Miscellaneous and sundry (tips and gratuities, laundry and cleaning, freight, travel) 76

Profit 497

Total cost of average adult funeral $5,023

A 1999 survey of NFDA members determined that nationally, for every dollar taken in by affiliated funeral homes, money was distributed in the following manner:

32 cents for salaries and benefits

15 cents for other operating expenses

28 cents for merchandise (caskets, vaults, etc.)

15 cents for facilities

10 cents for before-tax profits

The 1999 NFDA financial surveys provide us with a number of other important findings. First, the data suggest that funeral service is a competitive industry. When the costs of average funerals are compared, the difference between the highest and the lowest prices charged varies less than $400 regardless of demographic category (region of the country, number of families served, number of facilities operated, or metro/rural location) (NFDA, 1999).

Second, approximately 64 percent of firms operate only one funeral home facility, and 22 percent operate two facilities (NFDA, 1999). Finally, the average funeral service firm's actual return on net worth is 9.9 percent (before taxes). This relatively low return on equity makes funeral service (like other capital-intensive industries) a very difficult one to enter as a new entrepreneur. For the entrant, the price of land and buildings would result in higher operating expenses to perform the same service than those of a more established funeral home (NFDA, 1999).

Pricing Systems

Until 1984 a funeral home charged for its services and merchandise in basically three ways. The first was the unit method of pricing. In this method, all of the costs involved with the funeral home (including the staff, the facilities, the automotive equipment, and the casket) were included in a single charge. In making a selection under this method, the family looked at a "bottom line" figure to which only other charges paid by the funeral home (e.g., a vault and additional burial merchandise) might be added.

The second method was the biunit or triunit pricing system. The biunit system made separate charges for professional services and the casket, and the triunit system made separate charges for professional services, the use of facilities, and the casket selected. This method enabled families to understand the charges for the three basic components making up funeral costs.

A third method of presenting costs was referred to as either functional, multiunit, or **itemization.** In this method, each and every item of service, facility, and transportation was shown as a separate item together with its cost. In this method usually a minimum of 8 to 10 items was listed, and the family decided in each instance whether or not that item would become a part of the funeral service. In 1984 the Federal Trade Commission (FTC) mandated that all funeral homes in the United States itemize their fees and that consumers can have access to pricing information over the phone.

A major advantage of itemization is that family members have greater flexibility in arranging a funeral and controlling costs. The family members should have the freedom to decline those items that they do not want, and a proper allowance should be made for the items that are not used. For example, one may ask to see "the pine box"—usually a cloth-covered casket of wood or pressed wood. These caskets are relatively inexpensive and may or may not be in the display room. If the body is to be transported a great distance to the grave site or crematorium, perhaps the funeral director's van or station wagon could be used rather than the expensive hearse. Similarly, the consumer should have the right to choose among types of **vault liners.** The greatest advantage of itemization is that one can look carefully at the itemized services and obtain the most adequate services at the best price.

There are two major disadvantages of the FTC policy requiring funeral homes to give prices over the phone and to itemize funeral expenses. The first is that when consumers receive prices over the phone it is difficult for them to make accurate price comparisons. One firm can say that it sells an oak casket for $1,800, and

THE WIZARD OF ID   **Brant parker and Johnny hart**

BY PERMISSION OF JOHNNY HART AND CREATORS SYNDICATE, INC.

another will price a different oak casket at $3,800. It is entirely possible that the price/quality ratio is better for the more expensive casket. A helpful analogy might be to imagine calling two import automobile dealerships, Volkswagen and Mercedes, to ask each, "How much do cars cost in your dealership?" The only way that one can make accurate price comparisons is to personally inspect both products. The same is true for the funeral industry. Families are better served when they can make careful comparisons between the costs of services provided by funeral firms before the death of a significant other. Most people do not wait until the car breaks down before they shop for a new one. Furthermore, if they did find themselves in this situation, they would not purchase a car over the telephone.

The disadvantage of the FTC ruling requiring itemization of funeral expenses is that itemization did not uniformly result in decreased expenses to families. Before the 1984 ruling, many funeral firms included some merchandise and services as part of the standard funeral. With itemization, funeral directors could provide justification to raise the cost for funeral services. The analogy of restaurant pricing may be appropriate at this point—it is often less expensive to order a complete meal than to order food a la carte.

ALTERNATIVES TO THE FUNERAL

People often ask if there are alternatives to the traditional funeral. There are three alternatives: immediate disposition of the body of the deceased, the bequest of the body by him or her to a medical institution for anatomical study and research, and the memorial service.

Immediate **disposition** occurs when the deceased is removed from the place of death to the place of cremation or earth burial without any ceremony; proper certificates are filed and permits received in the interim. In these instances, the family is not present, usually does not view the deceased after death, and is not concerned with any further type of memorialization. Disposition is immediate in that it is accomplished as quickly after death as is possible. In this situation the body probably will not be embalmed, and the only preparation will consist of bathing and washing. The 1999 survey of funeral directors found that 21 percent of all deaths involved immediate disposition. The frequency of this alternative differs by region of the country—it is performed most in the northwest part of the United States (approximately 55 percent of cases) and least in southern states (less than 6 percent of cases) (NFDA, 1999).

Body-bequest programs have become better known in the last four decades and permit the deceased (before death) or the family (after the death) to give the body to a medical institution. A compendium of body donation information (NFDA, 1981) indicates that, when the family desires, 75 percent of the donee institutions permit a funeral to be held before the delivery of the body to the institutions for study or research. Some medical schools will pay the cost of transporting the body to the medical schools; others will not. This is the least expensive means of final disposition, especially if a memorial service is conducted without the body present. The compendium also indicates that in almost every instance the family may request that either the residue of the body or the cremated remains be returned when they are of no further benefit to the donee. In those instances when the family does not desire to have the body or the cremated remains returned, the donee institution will arrange for cremation and/or earth burial—oftentimes with an appropriate ceremony. People who are considering donating their bodies should be aware of the fact that at the time of death the donee institution may not have the need of a body. If this does happen, the family will have to find another institution or make other arrangements for the disposition of the body. The 1985 NFDA survey found that gifts for anatomical research are made in less than 1 percent of all deaths (NFDA, 1986).

To some, body donation may not appear to be an alternative to the funeral (especially when a ceremony is held before delivery of the bequested body), but inasmuch as the procedure is different from the most common methods of disposition, it may be considered as an alternative. According to the NFDA (1999), approximately 7,000 such donations are made out of 2,300,000 deaths each year.

The memorial service is defined as a service without the body present. It is true that every funeral is a memorial service—inasmuch as it is in memory of someone—but a memorial service, by our definition, is an alternative to the typical funeral. It may be conducted on the day of the death, within 2 or 3 days of the death, or sometimes as much as weeks or months after the death. Those who wish to have a viewing can do so on the evening before the day of the memorial service. The memorial service typically places little or no emphasis on the death. Instead, it usually is a service of acclamation of philosophical concepts. Religious or nonreligious in content, such a service can meet the needs of the bereaved individuals.

Organizations called **memorial societies** exist for consumers. An example is found in Ithaca, New York. The bylaws of this particular nonprofit and nonsectarian organization establish the following as purposes of this society:

1. To promote the dignity, simplicity, and spiritual values of funeral rites and memorial services

2. To facilitate simple disposal of deceased persons at reasonable costs, but with adequate allowances to funeral directors for high-quality services

3. To increase the opportunity for each person to determine the type of funeral or memorial service that he or she desires

4. To aid its members and promote their interests in achieving the foregoing

Thus, such a memorial society would help educate consumers regarding death before the actual death of a significant other and present options for final disposition of the body. Likewise, many funeral directors today serve as valuable resource persons by sharing information about death with various community groups.

NEW TRENDS IN FUNERAL SERVICE

The newest trend in funeral service is to cultivate new business by courting the living with special services. Included in these services are working with the public in preneed funeral planning, providing aftercare for survivors, and assisting community and religious organizations in providing death education for the public. All of these efforts can be thought of as an extension of past efforts to market and advertise funeral services, but the funeral industry has made concerted efforts to make these efforts effective in a very competitive industry where, in the past 15 years, the number of no-frills cremation services have increased significantly and corporate profits have declined by more than one fourth—as a percent of sales, from 13.2 percent in 1981 to 9.9 percent in 1999 (Federated Funeral Directors of America, 1999).

Preneed Funerals: A "New" Trend in Planning Funerals

One of the more controversial issues today in funeral service is the trend for **preneed funerals. Prearranging** is the process of arranging funerals in advance of need. This process can include selecting merchandise, planning the service, determining the method of viewing and final disposition, and selecting persons to be involved in the funeral. In addition to prearranging, **prefunding,** or making the legal commitment of money to pay for the funeral service, is also common. This is usually accomplished through insurance or a trust. *Preneed* is a generic term that refers to both processes of prearranging and prefunding (Hocker, 1987).

According to a 1983 national survey conducted by the National Research and Information Center (Will, 1988), 9.2 percent of Americans have made prearrangements for their funerals, and 62 percent feel that they should make funeral prearrangements with a funeral director. Presently, approximately one million people a year prearrange their funerals, compared with 22,000 in 1960 (Anderson, 1997). A 1995 survey of its members conducted by the American Association of Retired Persons (AARP) determined that seven million people had prearranged their funerals for a total cost exceeding 15 billion dollars (American Association of Retired Persons, cited by National Funeral Directors Association, 1997).

According to James Will (1988, p. 367) and William Hocker (1987, pp. 4–11), consumers prearrange and/or prefund funerals for these reasons:

1. To provide a forum for death-related discussions that is not profoundly affected by the grief that naturally accompanies death

2. To make one's funeral preferences known to one's survivors, thus assuring that survivors will not select a type of funeral that differs from the one desired

3. To provide an opportunity to personalize the funeral

4. To provide the dying with peace of mind—planning one's funeral can be one of the last pieces of unfinished business that one can accomplish for one's survivors

5. To give individuals an opportunity to get the most for their money through comparison shopping at a time when they are not faced with urgent need or overwhelmed by grief

6. To unburden loved ones of the obligation of planning and paying for a funeral

7. To protect an estate from funeral expenses in the future

8. To assure that funds are available in the future for the type of funeral that is desired

The disadvantage of preneed funerals from the perspective of consumers and funeral service professionals is the potential for consumer fraud. Some consumers have paid unethical funeral salespersons (some licensed and others not) for funerals and later found at the time of death that either the firm was no longer in business, the money had not been put into a trust account, and/or the deceased had moved to a location not served by the firm. For these reasons Thomas D. Bischoff (cited by Kelly, 1987, p. 32), senior vice president of the Prearranged Funeral Division of Service Corporation International (the largest chain of funeral homes in the world), has made the following recommendation:

> We, as funeral service professionals, must encourage that everything be done to minimize these eventualities. We would again suggest that an insurance-funded prearranged funeral program is the best safeguard that the funeral director and the consumer have. Insurance companies are very tightly and closely regulated and are required to maintain sufficient funds on hand to meet requirements. An insurance-funded prearranged funeral program has very little potential for mismanagement or fraud and as such should become the standard for the industry.

Ultimately, consumers must protect themselves against fraudulent entrepreneurs and dishonest "get-rich-quick" salespersons. Insurance-funded prearranged funeral programs are important potential safeguards, but the words of Robert W. Ninker (cited by Kelly, 1987, p. 32), executive director of the Illinois Funeral Directors Association, should also be heeded:

> The best protection a buyer can have is to purchase from a funeral director with a long history of success, after asking and understanding that this person is licensed and, in fact, trusts his funeral funds. . . . The buyer should also check references with acquaintances. That is about as much certainty as there is in life. Of course, if the buyer responds to door-to-door sellers or boiler room operations, he is his own victim! No one can protect someone from his own fool-like actions.

Aftercare: The Future Business Is to Be Found in Today's Customers

The funeral industry has always known that 80 percent of its business is with families served in the past. However, aftercare is one method to increase the likelihood of getting repeat business. The newest trend in funeral service is to provide extensive aftercare services and products for widows and widowers. Among these services and products are grief therapy, bereavement support groups, video tributes, and even greeting cards sent to survivors to mark the anniversary of death or the deceased's birthday (Scott & Dolan, 1991).

In 1984 Accord Aftercare Services began providing bereavement support materials to funeral homes, public and private organizations, hospices, hospitals, support groups, counselors, and individuals. Through its professional development seminars and grief programs, Accord offers training in the areas of grief counseling, communications skills, and aftercare for funeral service personnel and the families they serve. Accord designs materials, such as a self-study grief workbook, a quarterly magazine, the videotape *The Positive Power of Grief*, and brochures for "forgotten grievers," that can be redistributed by local funeral homes to maintain contact with families after the funeral service. Accord serves 50,000 bereaved families annually through its products, programs, and seminars.

An example of such aftercare efforts can be found at Bradshaw Funeral and Cremation Services in the Twin Cities (Minnesota) and in the work of Paul Johnson. Johnson taught classes in sociology and dying and death at Bethel College (St. Paul) for 10 years until he resigned in 1985 to become the bereavement services director at the Twin Cities chain of seven funeral homes. Johnson's work requires that he be involved in service plans of all deaths for which problematic grieving is likely—the death of a child, accidental death, homicide, suicide, and death from AIDS. Johnson also contacts the surviving spouse or other family members a few weeks after the funeral to "hear their concerns about their own particular situation." Johnson recommends grief support groups and provides families with information about the grieving process. Johnson also conducts bereavement workshops for the public, sponsored by the funeral home. Recently, Johnson created a special self-help support group of widowers that meets monthly with six to eight men in attendance.

For many years Johnson has edited *Caregivers Quarterly*—a newsletter that is produced by Bradshaw Funeral and Cremation Services and distributed without charge to 1,200 clergy and caregivers in the Twin Cities area. In the near future Johnson plans to distribute a set of four pamphlets to newly bereaved persons at prescribed time intervals after the death anniversary (3 months, 6 months, 9 months, and 1 year). The purpose of these pamphlets is to empower bereaved persons in their grief work.

Many other funeral homes in the United States are now offering grief counseling and support groups for their clients. In the 1970s the Carbon Funeral Home in Windsor Locks, Connecticut, created a support group called Begin Again. Begin Again meets weekly and sponsors seminars on a wide range of topics, including financial management, car maintenance, and occupational reentry. Begin Again even has meetings the primary function of which is entertainment—featuring magicians and ventriloquists (Anderson, 1997).

Other funeral establishments are sponsoring tree-planting ceremonies and annual memorial observances as a service to the families of the deceased. In Chicago, the Blake-Lamb Funeral Home provides free limousine services for weddings in families of the deceased (Scott & Dolan, 1991). An optional service sponsored by the Fitzgerald and Son Funeral Home of Rockford, Illinois, is the annual Walk to Remember for families mourning stillbirths, miscarriages, and early infant death. During the walk, participants plant a tree in a public park, write their babies' names on paper, and put them next to the tree (Anderson, 1997).

In addition, a new professional association is emerging for funeral service personnel who are involved in bereavement counseling and aftercare. With 150 members and a 15-member board of directors, the National Association of Bereavement Support Providers in Funeral Service has held annual meetings with the National Funeral Directors Association and the Association for Death Education and Counseling in order to create an organization to support the funeral industry in providing aftercare as a vital part of the funeral services that it offers.

Funeral Services on the Internet

Innovation and change are constants in any business and death-related businesses are no exception. When the FTC in 1984 mandated that all funeral homes in the United States itemize their fees and that consumers have access to pricing information over the phone and then when e-commerce through the World Wide Web became accessible for all Americans at the end of the 20th century, the stage was set for death-related business decisions to be affected by high-technology communications. Today, many funeral homes and cemeteries provide consumers with price information and preneed and at-need funeral planning via the Internet. One can purchase funerals, caskets, vaults, cemetery plots, urns, flowers, obituary notices, cremation services, and scattering services all online and pay for these services with credit cards without ever meeting face to face or talking with funeral personnel. Because the FTC has determined that the cost of funerals must be itemized and that consumers must only pay for those services they order from funeral establishments, it is possible to purchase caskets, vaults, and urns online from e-business vendors and have them delivered to funeral directors who are making the arrangements for the final disposition. Today's Internet consumer will find cost-comparison shopping possible when making funeral preparations.

Presently there are three major e-commerce full-service cooperative death-related businesses on the Internet—HeavenlyDoor.com, Funeralnet.com, and Funeral.com. These Web sites include direct links to web sites of individual funeral homes, online obituaries, virtual cemetery visits, professional grief and religious counselors, a bereavement chat room, and an e-commerce component that allows the visitor to shop for and order services and products online. Consumers also have online access for sympathy gifts and cards, for making charitable contributions in a loved one's name, and for last-minute travel arrangements. Internet e-funeral Web sites are free to consumers because of fees paid by the funeral-related industries who make sales through the site.

THE BUSINESS OF BURYING THE DEAD

Disposing of the dead in American society meant burying the deceased in such a manner that respect was given to the body, the life that had been lived was memorialized, and the values of the living were reinforced; yet, the survivors were allowed to detach themselves from the deceased and return to their everyday lives. Like funeral functionaries, cemetery sextons and other grave-related workers assisted families as they dealt with the practical problems of disposing of the dead and returning to the task of living.

The Changing American Cemetery

Like funeral service, contemporary American cemeteries have evolved from colonial landscapes that looked much different. Before the Industrial Revolution, cemeteries provided for the community a "*memento mori.*" For our 18th century ancestors it was important to "remember death": the fact of our death should motivate us to work hard and do good works as a preparation for a final judgment by a righteous God.

The Puritan Cemetery

Before the development of the new cemeteries in the 19th century, Americans interred their dead in various types of places (Sloane, 1991). The earliest burials by pioneers were in unorganized, isolated places. The graves were typically unmarked because the grave would not be maintained. The European influence was present: the European custom was that the corpse was not highly regarded and dead bodies were relegated to trenches or places for the bones of the dead. In some cases a burial place on a family farm might be used.

Clusters of graves soon grew as the pioneers settled into small villages. These pioneers were breaking from European tradition in not burying their dead next to churchyards. Domestic burial grounds were found in all of the colonies, but were less popular in New England where the Puritans were more likely to follow English custom and bury their dead in a central burial ground beside the new churches, notes Sloane. Occasionally, a church member could be **entombed** beneath the church in public or in a private vault. African-American slave burial places tended to be in community graveyards rather than in family plots. Religious graveyards, whether Protestant, Catholic, or Jewish, were remarkably similar in their layout, monument, and management, states Sloane.

Graves in these churchyards were designated with wood and stone markers, primarily using slate and sandstone. Carvings on gravestones appeared in the 1700s. Earlier stones had simply noted the deceased person's initials, age, and birth and death years. The appearance on gravestones of the cross bones and skull and the soul faces symbolized the struggle between mortality and immortality of 17th century Puritan New England's "almost obsessive concern with the oppressive inevitability of death and the necessity of reminding the living of the uncertain fate of the soul" (Combs, 1986, p. 18). Yet, the appearance of the soul figure embodies the Christian's confidence in a resurrection and offers a reassurance of salvation.

Most towns had a potter's field in which to bury the destitute or those "not accepted" into other burial places. Potter's fields were transient places and were often abandoned after a severe epidemic.

Segregation by race occurred in life as well as in death. Both in the South and in the North, public and private graveyards were racially segregated, observed Sloane. Sometimes separate graveyards were established for African Americans and whites.

The City Garden Cemetery

Nineteenth century cemetery reform derived from religious liberalism, romantic naturalism, middle-class family sentimentality, and scientific concerns. In keeping with the middle-class philosophy of death with order, cemeteries were changed into small family gardens that hid the harsh reality of death, and attempted to create instead a "forgetfulness of the certainty of death" and sentimental feelings about "peaceful afterlife."

In the early 19th century, embroidered or painted mourning pictures flourished as a form of memorialization. These pictures often showed stylized graveyard scenes, including standard features such as the weeping willow, the gravestone and **epitaph** (gravestone inscription) of the deceased, the mourners, and a *memento mori*. These remained popular until the 1830s, when printed memorials with spaces for names and dates came into vogue. During the middle third of the century, posthumous mourning portraiture also flourished, depicting the deceased with conventional symbols of mortality like the broken shaft or roses held with blooms downward. These paintings were drawings of the corpse, and they expressed "the desire for the restoration of the dead through art." All of these forms of mourning art provided icons for the bereaved family to contemplate as a part of the extended mourning ritual (Lloyd, 1980).

LISTEN TO THE VOICES
Grave Remarks: I'll Write My Own Epitaph Before I Leave, Thank You

Epitaphs are footnotes chiseled on tombstones. They are parting shots taken at or by the deceased. They can be patriotic, poetic, profound, or pathetic. They can be wise, witty, or just weird. They can glorify, be grievous, or be gruesome. Few of them have summed up a person's attitude toward life as well as one found in a Georgia cemetery:

I told you I was sick!

Perhaps the most famous epitaph is one credited to W. C. Fields, written for himself:

On the whole I'd rather be in Philadelphia.

And in a Moultrie, Georgia, cemetery:

Here lies the father of twenty-nine
He would have had more but he didn't
have time.

On the tombstone of a hanged sheep-stealer from Bletchley, Bucks, England:

Here lies the body of Thomas Kemp
Who lived by wool and died by hemp.

In a Thurmont, Maryland, cemetery:

Here lies an atheist—All dressed up and
no place to go.

In a Uniontown, Pennsylvania, cemetery:

Here lies the body of Jonathan Blake;
Stepped on the gas instead of the brake.

Written by a widow on her adulterous husband's tombstone in an Atlanta, Georgia, cemetery:

Gone. But not forgiven.

Similarly, a Middlesex, England, widow put this on the gravestone of her wandering husband:

At last I know where he is at night!

An English epitaph over the grave of Sir John Strange, a lawyer:

Here lies an honest lawyer, and that is
Strange.

An 1890 epitaph of Arthur C. Hormans of Cleveland, Ohio, puts things into startlingly clear perspective:

Once I wasn't. Then I was.
Now I ain't again.

From The People's Almanac (pp.
1312–1320), by D. Wallechinsky and
I. Wallace, 1975, Garden City, NY:
Doubleday, and Lexington Herald,
Lexington, KY, September 21,
1979, p. D5.

In 1895 a writer in *American Gardening* wrote that "the modern garden cemetery, like the modern religious impulse, seeks to assuage the cheerlessness and the sternness of life and to substitute the free and gracious charity of One who came to rob death of its hideousness" ("Extracts," 1895). By the 1830s many physicians had begun to worry about the possible health hazards of city graveyards. By the same time, commercial development had raised the price of land on which graveyards were located, and Romantic ideas of landscape architecture had begun to affect the aesthetically oriented members of the middle class. Also, space limitations of city graveyards prevented the possibility of family plots, and many of the burial places had become overcrowded, unkempt, and unsightly.

The Rural Cemetery

The solution to these problems was the rural cemetery—a landscaped garden in a suburban setting. The rural cemetery movement sought to simplify the cemetery and provide an aesthetic environment in which grave markers did not detract from the beauty of the natural surroundings of grass, trees, flowers, and rolling hills. With this sense of romantic naturalism, cemeteries were a place to go and reflect upon the peace and beauty of the new world where the dearly departed had gone to reside.

In 1831 Mount Auburn Cemetery was founded 4 miles west of Boston, and its success stimulated the spread of such cemeteries all over the country. In rural cemeteries, family plots averaging 300 square feet were nestled among trees and shrubs upon the slopes of soft hills or on the shores of little lakes. Paths curled throughout the grounds, passing lots enclosed by stone coping or wrought iron fences and surmounted by a monument of some sort. Such cemeteries were established to bury bodies; to ease the grief of survivors; to bring people into communion with God, with nature, and with deceased family and friends; to teach them important lessons of life; to surround bereavement with beauty and to divert the attention of survivors from death to the setting of burial; to display "taste" and "refinement"; and to reinforce the class stratification of the status quo (Bender, 1973; French, 1975; Rotundo, 1973).

The founders of Mount Auburn were liberal Unitarian reformers—progressive professionals and businesspeople who considered the cemetery in the context of social developments of their day. They equated the family with the garden and imagined both as a counterpoint to a society of accumulation. In the same way that upper-middle-class families moved to suburbs where curved roads and greenery contrasted with the grid-block plan of the cities and offered space to raise a family, they moved from the "cities of the dead" to rural cemeteries that offered space to "plant" a family.

This was also the period of "the discovery of the asylum," when social deviants were located in restorative rural settings that would rehabilitate people away from the contaminating influence of urban life. Many reformers saw the cemetery, like the insane asylum, the orphanage, and the penitentiary, as an "asylum" from urban ills. In the cemetery, "the weary and worn citizen" was also rehabilitated. Cleaveland (1847, p. 13) noted:

> Ever since he entered these greenwood shades, he has sensibly been getting farther and farther from strife, from business, and care. . . . A short half-hour ago, he was in the midst of a discordant Babel; he was one of the hurrying, jostling crowd; he was encompassed by the whirl and fever of artificial life. Now he stands alone in Nature's inner court—in her silent, solemn sanctuary. Her holiest influences are all around him.

Mount Auburn Cemetery in Massachusetts, founded in 1831, was the first rural, landscaped cemetery in the United States. By moving cemeteries away from the cities, they were "out of sight," thus "out of mind" (fitting of the "dying of death" period from 1830–1945).

The rehabilitation of the rural cemetery paralleled the philosophy of education found in the common school. Like the new compulsory schools, rural cemeteries responded to middle-class fears of mobilization of the masses in the Age of Jackson. Educational reformers such as Horace Mann both reflected and affected cemetery proponents in their belief that "sentiment is the great conservative principle of society" and that "instincts of patriotism, local attachment, family affection, human sympathy, reverence for truth, age, valor, and wisdom. . . constitute the latent force of civil society" (Tuckerman, 1856, p. 338).

Like these other reforms of antebellum society, rural cemeteries "took the public mind by storm." Cities throughout the United States established rural cemeteries, and people flocked to visit them. In New York, Baltimore, and Philadelphia, Andrew Jackson Downing estimated that more than 30,000 people a year toured the rural cemeteries. Consequently, he wondered whether they might not also visit landscaped gardens without graves. In articles such as "Public Cemeteries and Public Gardens" and "The New York Park," Downing (1921) advanced the idea that would result in New York's Central Park and a new direction for cemetery development.

The Lawn Park Cemetery

The new direction was the lawn park cemetery, emphasizing a new aesthetic—efficiency—and the absence of death. By the end of the century, members of the Association of American Cemetery Superintendents routinely wrote in journals like *Park and Cemetery* that "a cemetery should be a beautiful park. While there are still some who say 'a cemetery should be a cemetery,' . . . the great majority have come to believe in the idea of beauty" (Simonds, 1919, p. 59). This new aesthetic emphasized the open meadows of the beautiful style over the irregular hill-and-dale outcroppings of the picturesque style and the irregular outcroppings of obelisks and monuments in the unregulated rural cemetery. This aesthetic coincided with considerations of efficiency as the uncluttered landscape required less upkeep than the enclosures and elaborate monuments of the rural cemetery. Finally, this aesthetic buried death beneath the beauty of the design. Andrew Jackson Downing (1921, p. 59) had said that "the development of the beautiful is the end and aim of all other fine arts. . . . And we attain it by the removal or concealment of everything uncouth or discordant." The cemetery superintendents practiced what Downing preached. "Today cemetery making is an art," said one superintendent in 1910, "and gradually all things that suggest death, sorrow, or pain are being eliminated" (Hare, 1910, p. 41).

Cemetery superintendents eliminated suggestions of death by banning lot enclosures and grave mounds and by encouraging fewer gravestones and fewer inscriptions. They banned lot enclosures (fencing or stone coping) because these broke up the unified landscape, blocked the path of the lawn mower, and signified a "selfish and exclusive," possessive individualism. Whereas grave mounds obstructed the view and the lawn mower, they also reminded people of death. Without them, a cemetery lot would evoke "none of the gruesomeness which is invariably associated with cemetery lots. . . . *No grave mounds are used, so save the headstones, there is nothing to suggest the presence of Death*" (Smith, 1910, p. 539).

Some superintendents did not want to save the headstones either, proposing instead "a cemetery where there is no monument, only landscape" (Association of American Cemetery Superintendents, 1889, p. 59). Most superintendents favored sunken stones at the site of the grave. Howard Evarts Weed (1912, p. 94), author of the influential publication *Modern Park Cemeteries*, argued that "with the headstones showing above the surface we have the old graveyard scene, but buried in the ground they do not appear in the landscape picture and we then have a park-like effect." Some superintendents zoned the cemetery to permit monuments only on large "monument lots." This allowed the superintendent, like a real estate agent, to charge premium prices for such lots and for prime locations for corner lots or hillside or lakeside property. This **social stratification** of the cemetery allowed for the social mobility of the dead as ambitious dead people moved to better "neighborhoods" as their survivors saw fit. Where superintendents could not ban or limit the number of monuments, they tried to make them as unobtrusive as possible, preferring horizontal monuments to the earlier upright markers and preferring inexpressive epitaphs to the poetic epitaphs of earlier times. They wanted to replace the *memento mori* of earlier stones with "forgetfulness" as they buried death with the dead (Farrell, 1980).

Cemetery Services to Soothe the Mourners

Superintendents tried to structure cemetery services "to mitigate the harshness and cruelty of death and its attendant details and ceremonies" (Seavoy, 1906, p. 488) and to provide a sort of grief therapy for bereaved individuals. They encouraged private, family funerals, and they tried to remove or conceal the uncouth and discordant aspects of interment. They carted the dirt away from the grave, or they hid it beneath cloth, flowers, or evergreens. They lined the grave with cloth to make it look like a little room. They suggested changes in religious services, and they escorted mourners away from the grave before filling it in order to shield them from the finality of death. In all of these services, "everything that tends to remove the gloomy thoughts is done. . . . The friends cannot but leave the sacred spot with better, nobler thoughts, freed from the gloom and terror that otherwise would possess them" (Hay, 1900, p. 46).

CEMETERIES TODAY

As one can see, there have been many changes in American cemeteries. Cemeteries constructed in the 20th and 21st centuries, where there are still grave sites available, often have all for diversity of memorialization. There may be some sections with large above ground tombstones and even grave enclosures with head and foot stones to provide a *memento mori*. There might also be sections that only allow flat markers and often have a single above-ground memorial centerpiece to commemorate all those buried in these more simple surroundings.

Although the appearance of the modern cemetery will vary to some degree, in general, there are three basic types of cemeteries found in the United States: government cemeteries, nonprofit cemetery associations, and for-profit cemeteries.

The Government Cemetery

The federal government, through the Veterans Administration, has established 118 national cemeteries in 39 states (and Puerto Rico) to honor individuals who have given their lives in the military service of their country. The first and probably most famous of these national cemeteries is Arlington Cemetery, located in Arlington, Virginia, where we find more than 250,000 graves of not only soldiers but also of political leaders (e.g., John and Robert Kennedy), 3,800 former slaves who were released as a result of the Civil War, and many war correspondent journalists who died in combat. Annually, four million visitors pay their respects at this 612-acre cemetery, which was formerly part of a plantation owned by Mary Anne Randolph Custis, wife of Robert E. Lee. Along with 14 other cemeteries, Arlington was set aside by President Abraham Lincoln in 1862 as a national cemetery. Today there are more than 2.5 million national cemetery grave sites including full-casket grave sites, in-ground grave sites for cremated remains, and columbarium **niches**. The largest of the 118 national cemeteries is Calverton National Cemetery (1045 acres), located in

New York, and the smallest is Hampton National Cemetery (0.03 acre), found in Virginia. Twenty-seven of the 118 national cemeteries are closed to new interments, but may still accommodate family members in already occupied graves. The two newest national cemeteries were dedicated in Dallas-Fort Worth, Texas, and in Rittman, Ohio, in 2000, and there are major expansion projects planned for six existing cemeteries, which will add another 167,000 grave sites to the system.

Any honorably discharged veteran and his or her spouse and/or dependent children are eligible for burial within the 118 national cemeteries. With this free burial (including the cemetery plot, grave liner or vault, and grave marker), families may be provided with a dignified military funeral ceremony including folding and presenting the United States burial flag, the playing of "Taps," and a memorial certificate of appreciation signed by the President of the United States. The Veterans Administration will also furnish, upon request, a free grave marker (made of bronze, marble, or granite), which can be installed in a private, nongovernment, cemetery. This latter benefit is not provided to spouses and eligible dependents who choose not to be interred in national cemeteries.

At the time of a death, viewing and funeral services are performed and then the deceased is transferred to the national cemetery for a committal service and/or burial. For reasons of safety, committal services are held at sites located away from the

Arlington Cemetery in Virginia is the most famous of U.S. national cemeteries. Note the precision with which the gravestones are lined up, very fitting for a military cemetery.

grave site after which the casket is placed in the vault and then later taken to the actual graveside for burial. Committal services last approximately 15 minutes and are performed with the assistance of two or more uniformed military persons with at least one a member of the veteran's parent service of the Armed Forces. Veterans' organizations may also assist in the provisions of military funeral honors. Although national cemeteries are opened 7 days each week, burials are not normally conducted on Saturdays or Sundays.

Cities and states also build cemeteries as a service to their residents and as a way of celebrating local history and honoring civil leaders. Government cemeteries are not designed as profit-making ventures; rather they are administered as services to the public and are paid for by fees charged to those receiving the benefits. A portion of these charges are set aside for endowments or trust funds, the interest from which is used to pay for perpetual care of the cemetery.

The Not-For-Profit Cemetery

The second type of cemetery is the not-for-profit cemetery, which is run by a cemetery association for the benefit of private members. Like a condominium association, a cemetery association consists of individuals and families who own lots in the cemetery. Typically meetings are called on a regular basis to employ workers (sextons, grounds crew, and grave diggers), set lot prices and fees, and plan for maintenance and improvements in the cemetery. For example, Michael Leming's town of Northfield, Minnesota (population 15,000), has three not-for-profit cemeteries—one was created by the city of Northfield in the late 19th century and deeded to the cemetery association after its creation, Mt. Calvary Cemetery is owned and run by the Roman Catholic Dioceses for the benefit of St. Dominic's Church members, and Oak Lawn Cemetery was established by private citizens as a nonprofit cemetery.

The normal way for nonprofit cemeteries to charge for their services is to establish a fee for the lot and burial (opening and closing) and then take a portion of these two charges (typically 10 to 15 percent) and place this money in an interest-bearing trust fund or endowment. The annual income from the endowment should be enough to pay for repairs, maintenance, and upkeep of the cemetery. When annual income is not enough to pay for annual costs, the prices for lots and burials are increased.

Unlike government cemeteries, lot owners and their families establish rules, covenants, and ordinances to regulate the behaviors of the owners. Typical rules would include the following: all bodies will be buried in an east-west direction, lots will be 4 feet by 8 feet in size, a vault or grave liner will be required for all earth burials to keep the ground from sinking, monuments or markers will be placed on the east edge of the grave lot, plastic flowers will be removed from all graves, the cemetery will be open to visitors at sunrise and close at sundown, and there will be no more than one burial for each full-casket grave site and no more than two in-ground grave site burials for cremated remains.

The following story is an example of local nonprofit cemetery politics. A few years ago all three nonprofit cemeteries in Northfield contracted grave digging to the same earth-moving firm. This firm provided a favorable price to the cemeteries, subject to the requirement that there would be no winter burials and that burials would only take place from April 1 through December 1. Because the cemetery associations were interested in keeping costs reasonable and their competition was following the same procedures, they were willing to make concessions to the gravedigger. A few years later, however, a prominent member of the community died in the dead of winter and his casket was to be kept in the cemetery crypt (actually the lawn mower maintenance building) until April 1. As a result of this "important" person's death and the holding of his body in the "maintenance building" until spring, a grassroots social movement took place soon thereafter to reform the cemetery's rules. Today, earth burials take place at any time of year in all three Northfield cemeteries.

The For-Profit Cemetery

The final type of cemetery is the for-profit cemetery. These are cemeteries whose goals are to provide a service for a fee to the consumer and a profit for the owners of the cemetery. The most famous of all for-profit cemeteries is Forest Lawn Memorial-Parks & Mortuaries of California. This chain of five cemeteries, located in southern California, has served more than 500,000 families in its 90-year history. Forest Lawn is a full-service facility providing funeral and mortuary services, cemetery plots, mausoleums and columbariums, florists, and the selling of death-related merchandise (monuments, vaults, caskets, clothing, and sundry items). Like nonprofit cemeteries, Forest Lawn cemeteries have a huge endowment of more than 200 million dollars that pays for perpetual maintenance, upkeep, repairs, and improvements. The principal of the endowment funds can never be touched (by law), only the income generated from the endowment.

Present minimal costs at Forest Lawn for cemetery plots and services are as follows:

Property for burial beginning at $700 to $1190 (depending on cemetery selected)

Contribution to the Endowment Fund 15 percent of lot price

Charges for full-size interment (Saturday $285 extra) $550

Charges for cremated remains in niche or ground $150

Vault $300

Marker $600

Forest Lawn also provides funeral services and sells funeral merchandise and flowers. Forest Lawn Memorial-Parks have more than a million visitors each year who pay respect to the memories of loved ones, participate in Forest Lawn special events (such as Easter Sunrise Services and Memorial Day ceremonies) and educational programs, and enjoy the magnificent works of art including exquisite stained glass windows, a collection of marble statuary, a rare combination of original works by modern masters and exact reproductions of Michelangelo's most timeless creations, a museum filled with original paintings and bronze statuary, and many historic artifacts. As a child growing up in Los Angeles, Michael Leming's school classes and YMCA day-camp programs would visit Forest Lawn Memorial-Parks, as they would the other cultural and entertainment centers of the area—Disneyland, Universal Studios, and Sea World.

There are many other for-profit cemeteries such as those run by Service Corporation International (SCI), the world's largest provider of death care services. With nearly 4,500 funeral service locations, cemeteries and crematoria in 20 countries, SCI has more than 40,000 employees who produce annual revenues of approximately $3.3 billion. Investors can purchase stock in this and other for-profit corporations in hopes of earning an income from corporate funeral homes, cemeteries, and floral businesses. If ever there was a business that specialized in providing a necessary service that no one wants to do, it is the death-related business of caring for and burying the dead.

Graveyard Symbols of Death

When one goes to a cemetery, the purpose is usually to attend a burial service. At such times the individual reflects on the life of the recently deceased person and pays little attention to the surrounding gravestones. Try visiting a cemetery, especially an old one, under less stressful circumstances to examine the gravestones.

If you have access to gravestones dating back to the early 18th century, you may see some markers with cross bones and a skull, pointing out the stark reality of death—after death, muscle and skin deteriorate, and the skeleton remains. Other markers may have a "soul face," usually a smiling, portrait-like face, often with wings. Unlike the cross bones and skull, the soul face with wings symbolizes the idea of an afterlife. Some markers from the early 1700s may have a bust of the deceased engraved on the stone. These busts are usually reserved for someone of prominence, such as a minister. Eighteenth-century markers are typically made of slate or sandstone.

Walking on through the cemetery, an individual might see various symbols of the passing of life and of death on the gravestones: an hourglass ("like sands in the hourglass, so are the days of our lives"), babies and women weeping (men weeping are rarely seen), Greek and Roman urns (the open urn is symbolic of the separation of the soul from the body), broken columns (not whole and complete but broken), weeping willow trees (not just any willow tree), fallen trees (thus, will die), broken limbs (severed from life), the eternal flame (life continues), the Grim Reaper ("cometh" in the night to "take one away"), and a lamb on children's markers (innocence and gentleness).

The idea that we do not die in the United States, but simply go to sleep, is evident in a cemetery. The word *cemetery* itself derives from the Greek and means "sleeping place." "R.I.P." ("rest in peace") is sometimes seen on a gravestone. Bed-shaped markers, complete with headboard, footboard, and sideboards surrounding the grave are symbolic of "resting." Double beds, single beds, and "cribs" for children can be found. One 19th century marker in Columbia, South Carolina, gives the birth date of the deceased, but rather than saying "died" says "went to sleep" on a certain date.

The stroll through an old cemetery might also reveal "tabletop" markers with four "legs" and a top. These markers served a function during the Civil War in that they were used as operating tables in some southern cemeteries. With a limited number of hospitals available and with a need for surgical facilities, these were available and about the right height. There are also false crypts that are "enclosed tables" with sides filled in between the "legs." False crypts are part of the English "table tombs," the preferred burial style of socially prominent people.

Epitaphs (inscriptions) on the markers were often rather lengthy in the 18th and 19th centuries and gave a brief history and personal traits of the persons buried there. Though worn by the passing of time, many of the epitaphs on these gravestones are still very legible. The viewer can sometimes note historically when various plagues or diseases occurred from the numerous members of a family who died within days of each other. An inscription of particular note is on a marker in Charleston, South Carolina, for a sea captain, aged 37, who went down with his ship in 1754. A reclining, smiling skeleton is displayed with its head resting on a winged hourglass. Above the smiling figure of death is the inscription: "yesterday for me, today for thee."

Greek and Roman columns, false crypts, and tabletop markers are prominent in Charleston, South Carolina, where the oldest gravestone dates back to the 1690s.

In contrast to the earlier U.S. cemeteries, modern cemeteries are more uniform and less personalized and have limited gravestone variation in size and shape. Contemporary markers, typically made of granite or marble, generally have the name of the deceased and the birth and death dates. Rarely does a marker today have an epitaph other than perhaps an occasional quotation.

Thus, a walk through a cemetery, especially an older one, could prove to be a peaceful and enlightening experience for the observer. In a place for the dead, the cemetery seems to have life with all the flowers and trees. The venture can be like walking through a history book. In a *memento mori* field that reminds the individual of mortality, one is surrounded by immortality.

LIFE INSURANCE

That innovative institution, life insurance, has tried to exorcise the anxiety and financial insecurity that would otherwise possess people contemplating death. Very early evidences of life insurance were seen in Barcelona in the 15th century, when the lives of slaves were insured, and in England in the late 17th century (Smith, 1941). The Aztecs in the 15th century and the ancient Babylonians in the Code of Hammurabi created elaborate systems of compensation for deaths (Greer, 1985). In the United States, as early as 1759, Presbyterian ministers established a company to provide a means whereby a minister of the Presbyterian faith could buy an annuity for his widow by making annual payments for a specified number of years (Presbyterian Ministers' Fund, 1938). By 1850 in the United States, 48 companies held policies valued at $97 million; by 1920, 335 companies held 65 million policies worth $40 billion. Life insurance contributed to "the practical disappearance of the thought of death as an influence bearing upon practical life," and thus, it contributed to the "dying of death" (U.S. Bureau of the Census, 1975, p. 1050).

The tremendous growth of the purchase of life insurance in the United States, after its beginnings in the mid 19th century, was due in part to our capitalistic economy (Clough, 1946). In earlier times society was organized on an essentially agricultural basis; many families were largely self-sufficient in an economic sense with the exchange of goods conducted mainly on a barter basis. The farm workers and the farmer had an equity in the land and, upon death, the children would then till the fields and carry on the tradition of the parents or continue as landholders. The farm was left to the heirs and the farm workers could typically continue working on the farm on which they grew up, if they so desired.

With the Industrial Revolution and the move to urban areas, however, the security of an agrarian society began to dissolve, observed Clough (1946). In a capitalistic economy, capital is a mobile surplus, which can be used for the accumulation of more surplus or profit, and the capitalist system is one in which the means of making a living are primarily based upon the employment of capital. In the evolution of capitalism, the means of production have in part passed out of the possession or direct control of the worker and the individual does not consume what he or she produces and must supply daily needs by going to the market. Livelihood is obtained either by money wages, by earnings on capital, or by the expenditure of surplus. Society has evolved away from economic self-sufficiency to economic interdependence, notes Clough. Such an evolution fits the description of French sociologist Emile Durkheim (1893/1946): We have moved from a society based on **mechanical solidarity** to one based on **organic solidarity**—from a fairly "simple," homogeneous society in which individuals were generalists and could more or less "take care of themselves" off the land to a more complex, diverse, heterogenous society in which we are more dependent on each other. Organic solidarity fits the analysis of society from a structural-functional perspective, as discussed in chapter 1.

Life insurance emerged from the same historical context as did funeral reform and rural cemeteries; it assumed the uniformity and continuity of death as natural occurrences. Life insurance depended on the science of statistics and on a species perspective of death and immortality that focused attention not on the life of the individual policyholder, but rather on the lives of beneficiaries. Over time, with experience in rates of mortality and the development of the laws of probability, life insurance on a scientific basis was possible. Yet, it was necessary to have enough people interested in sharing the risks of death so that there would be an average of expected losses before the scientific principles of insurance could operate correctly.

The value of life used by insurance agents, economists, legal experts, scientists, and agency administrators ranges from a few dollars to many millions of dollars, depending on the formulas used (Greer, 1985). One way to figure value is to break down the body into chemical elements—5 pounds of calcium, $1\frac{1}{2}$ pounds of phosphorus, 9 ounces of potassium, 6 ounces of sulfur, 6 ounces of sodium, a little more than 1 ounce of magnesium, and less than 1 ounce each of iron, copper, and iodine. On that basis a human life is worth $8.37. Another way to measure the monetary value of human life, notes Greer, is the going price of contract murder. Such a "kill," might pay several thousand dollars to the killer, depending on how badly someone wants the person killed and how much the person wants to carry out the assignment. Insurance companies, however, do not figure the value of a person based on chemical elements in the body or on the "price" of a hired killing. The insurance company, in figuring what a person's life is worth, determines this by what one would have earned had he or she lived—earning power over the course of a working life.

Like the rural cemetery, life insurance was praised for its educational benefits as it taught lessons of self-reliance, forethought, thrift, discipline, and (very) delayed gratification. For these reasons, clergymen such as Henry Ward Beecher endorsed the system of life insurance, responding to critics that, in effect, God helps those who help themselves. Also, life insurance accentuated the importance of the family, as did Romantic sentimentalism and the family plot of the rural cemetery. Finally, life insurance provided families with money to pay for elaborate funerals, a fact that affected both the development of funeral service and the history of bereavement.

Life Insurance as Protection

The middle-class American virtue of seizing control and planning ahead for the future is exemplified by the fact that the majority of Americans have life insurance. The purpose of life insurance is to protect an individual's dependents and to give the insured person a feeling of security in knowing that his or her dependents will have some coverage in the event of death.

The amount of life insurance needed is related to the number of dependents a person has and the financial security of each. For example, if the person is the primary provider in a family having three small children, he or she would have a greater need for life insurance than a single person with no dependents. If the primary provider were to die, these dependents would have some financial coverage, at least for the time being. At the time of death, the added burden of the loss of income is not needed by the survivors.

The age of the individual and that person's place in the life cycle are also important considerations. If the individual is aged and there are no dependents or the individual's spouse has enough retirement income to provide for his or her living expenses, the primary need will be to cover death-related expenses. However, if one's only purpose for life insurance is to provide for one's dependents, term life insurance is probably the best choice. **Term life insurance** tends to have the lowest premiums for the greatest amount of coverage. Group policies, such as those provided by one's place of employment, tend to have lower premiums than policies obtained outside of a group. Term life insurance is for a specific number of years, such as 5 or 10, with premiums going up as one ages. "My suggestion is to start with term, but if there is a permanent insurance need, convert to a good **universal life** policy," says Andrew Gross, a financial planner in Washington, DC (Smith, 1988, p. 154).

Life Insurance as an Investment

Today, however, life insurance for many individuals is more than security for their dependents. Some policies, universal life policies for example, are a form of tax-deferred investment from which one can also borrow. Thus, one can have coverage in the event of death and can have the opportunity to build cash assets at the same time. This type of insurance policy is called **whole life** from which, for a set premium, one receives life insurance and a savings fund. Universal life insurance is more flexible, has investment opportunities, and allows one to raise or reduce premiums and the amount of coverage on one's life. **Variable life** insurance does not allow for the premium or minimum coverage on one's life to change, but one can switch the savings from among money markets or various forms of stock.

Another advantage of life insurance is that it is a way to pass assets to relatives and other survivors without having any estate tax assessed by the government. Proceeds from life insurance policies do not go through **probate** and therefore the tax-favored treatment of life insurance allows one to exceed the deductible allowances for estate tax and pass on more resources to one's survivors.

It is advisable to shop around and compare the benefits of the different life insurance programs. Premiums vary considerably for the same coverage. Be cautious of high-powered salespersons trying to sell coverage not needed. Become informed about life insurance by talking with knowledgeable consumers or by reading consumer magazines before purchasing any life insurance. Comparison shopping should pay off.

CONCLUSION

In our discussion of the business of dying, we have described funerals and their alternatives within a cultural and historical perspective. In the United States an evolutionary, not a revolutionary, process has occurred. Americans did not invent the funeral or the funeral functionary. However, contemporary Americans have found an expression for their bereavement.

In this chapter we have also discussed the development of cemeteries, the building of funeral homes, and the establishing of life insurance companies, and demonstrated how certain needs of Americans have been fulfilled. Security comes from the sheer orderliness and structure of these "institutions." One should know what to "expect" from these services, and payment turns the responsibility over to the professionals. We Americans differ significantly from persons in nonliterate societies, in which such functions are completed within the kin network. However, paying someone else to perform a service fits middle-class Americans' ideas about specialization and division of labor. This historical perspective on changes in the cemetery, funeral, and life insurance industries helps us understand our experience with the contemporary American death customs as we know them today. The consumer can be well informed about burial practices and insurance in our society by staying abreast of various funeral home regulations and various life insurance offerings.

SUMMARY

1. Death was feared by the Puritans, who prayed not for the soul of the deceased but rather for the comfort and instruction of the living. The funeral was the main social institution for channeling the grief of Puritan survivors.

2. The Victorian funeral brought a focus on aesthetics and soothing the mourner, reflecting the "dying of death" movement.

3. The contemporary American funeral focuses on meeting the needs of bereaved individuals in terms of social support and grief resolution.

4. Licensure in the funeral industry covers embalming, practicing as a funeral director, and practicing mortuary science. Education requirements vary from state to state, but all states (except Colorado) require a 1- to 3-year apprenticeship.

5. The role of a funeral director is that of a caregiver, caretaker, and gatekeeper.

6. Funeral bills have been presented to customers using the following pricing systems: unit pricing, biunit or triunit pricing, and itemization. The latter is mandated in all states by the Federal Trade Commission. In 1999 the cost of the average adult funeral was $5,023.

8. Alternatives to funerals include immediate disposition, body donation, and memorial services.

9. New trends in funeral services include preneed funerals, aftercare services, and funeral services on the Internet.

10. The Puritans were buried in small, unadorned family plots, graveyards, and churchyards. The Victorian era ushered in the city garden cemetery, filled with elaborate headstones and epitaphs and designed to reflect the romantic naturalism and sentimentality of the period. Health concerns and crowding led to the development of the rural cemetery, designed to soothe the mourners by diverting attention from death to the rural surroundings. Whereas the rural cemeteries were characterized by elaborate monuments, the next phase of cemetery building, the lawn park cemetery, discouraged the use of gravestones, focusing on the aesthetics and efficiency of an uncluttered landscape of open meadows.

11. There are three types of cemeteries found in the United States: government cemeteries, nonprofit cemetery associations, and for-profit cemeteries.

12. Life insurance emerged from the same historical context as did funeral reform and rural cemeteries; it assumed the uniformity and continuity of death as natural occurrences. Today, most Americans have life insurance, which is seen as a way to protect dependents, as well as a tax-deferred investment from which one can borrow.

DISCUSSION QUESTIONS

1. Describe and discuss the Puritan view of death. Describe the procedures and atmosphere surrounding the typical Puritan funeral.

2. How did the funeral reflect the values of the Victorian era?

3. What are the influences of the following occupations upon the "dying of death": life insurance agents, cemetery superintendents, and funeral directors?

4. How do funeral directors address the needs of the bereaved?

5. Describe the changes that have taken place with regard to the role of the family in the funeral process.

6. What are the requirements for becoming a funeral director, and how have they changed over the years?

7. Discuss the factors affecting costs after death and the expenses related to funerals and final disposition.

8. What would be your choice of final disposition of your body? Why would you choose this method, and what effects might this choice have upon your survivors?

9. What are the advantages and disadvantages of preneed funeral arrangements?

10. Describe and explain the reforms that have taken place over the years in the construction and maintenance of the cemetery.

11. What are the advantages and disadvantages of the following: term life insurance and whole life insurance? What factors should be taken into consideration when one is in the process of purchasing life insurance?

GLOSSARY

Bier: A framework upon which the corpse and/or casket is placed for viewing and/or carrying.

Cremation: The reduction of a human body by means of heat or direct flame. The cremated remains are called *cremains* or *ashes* and weigh between 3 and 9 pounds. *Ashes* is a very poor word to describe the cremated remains because they are actually processed bone fragments and calcium residue that have the appearance of crushed rock or pumice.

Crematory: An establishment in which cremation takes place.

Columbarium: A building or wall for above-ground accommodation of cremated remains.

Disposition: Final placement or disposal of a dead person.

Elegy: A song or poem expressing sorrow, especially for one who is dead.

Embalming: A process that temporarily "preserves" a deceased person by means of displacing body fluids with preserving chemicals.

Entombment: Opening and closing of a crypt, including placing and sealing of a casket within.

Epitaph: An inscription, often on a gravestone, in memory of a deceased person.

Itemization: A method of pricing a funeral in which every item of service, facility, and transportation is listed with its related cost. This is a service mandated by the Federal Trade Commission.

Mausoleum: A building or wall for above-ground accommodation of a casket.

Mechanical solidarity: Emile Durkheim's concept to describe the form of social cohesion that exits in small-scale societies that have minimal division of labor. In such societies there is little specialization or individuation. The "conscience collective" embraces individual awareness.

Memento mori: Any reminder of death.

Memorial Society: A group of people joined to obtain dignity, simplicity, and economy in funeral arrangements through planning.

Niche: A chamber in a columbarium into which an urn is placed.

Obituary: Notice of a death, usually with a brief biography.

Organic solidarity: Emile Durkheim's concept to describe the form of social cohesion that exits in complex societies with a high division of labor. Such societies have an organic character because of the necessary interdependence of their specialized and highly individuated members.

Prearranging: Arranging funerals in advance of need. This process can include selecting merchandise, planning the service, determining method of viewing and final disposition, and selecting persons to be involved in the funeral.

Prefunding: Legally committing money to pay for the funeral service.

Preneed Funerals: A generic term that refers to both prearranged and prefunded funerals.

Probate: The judicial determination of the validity of a will.

Sexton: A church custodian charged with the upkeep of the church and parish buildings and grounds.

Social Stratification: A ranking of social status (position) in groups; upper, middle, and lower classes are basically distinguished in the U.S. social class system, for example, whereas India's stratification is a caste system.

Term Life Insurance: A type of insurance policy covering the insured for a fixed period of time (e.g., 10, 20, or so years). Premiums are usually lower for a greater amount of coverage than with other types of life insurance.

Universal Life Insurance: A type of insurance policy that is flexible and allows one to raise or reduce premiums and the amount of coverage on one's life.

Urn: A container for cremated remains.

Variable Life Insurance: A type of insurance policy that does not allow for the premium or minimum coverage on one's life to change but allows one to switch the savings from among money markets or various forms of stock.

Vault or Grave Liner: A concrete or metal container into which a casket or urn is placed for ground burial. Its function is to prevent the ground from settling.

Whole Life Insurance: A type of insurance policy with which, for a set annual premium, one receives life insurance coverage and, at the same time, invests one's money.

Suggested Readings

American Association of Retired Persons. (1996). *AARP Product Report: Funeral Goods and Services, 2*(3). American Association of Retired Persons. (1992). *AARP Product Report: Pre-Paying Your Funeral? 2*(2). Two reports by the American Association of Retired Persons for consumers on prepaying and preplanning conventional funerals, burials, and alternatives, including cremation, direct burial, and body donation.

Carlson, L. (1997). *Final act of love: Caring for your own dead.* Hinesburg, VT: Upper Access Book Publishers. Morgan, E. (1990). *Dealing creatively with death: A manual of death education and simple burial* (12th ed.). Bayside, NY: Barclay House Books. Two consumer resources that assist bereaved individuals in bypassing the funeral director and arranging everything themselves, including burial at home, where it is feasible and permitted.

Combs, D. W. (1986). *Early gravestone art in Georgia and South Carolina.* Athens, GA: University of Georgia Press. A look at 18th century gravestones in the South, carved primarily by New England carvers.

Farrell J. (1980). *Inventing the American way of death, 1830–1920.* Philadelphia: Temple University Press. Examines the transformation from the Puritan way of death to the American way of death.

Habenstein, R. W., & Lamers, W. M. (1962). *The history of American funeral directing.* Milwaukee: Bulfin. An excellent source for the study, review, and analysis of the history of funeral directing in the American culture from its introduction in colonial times to 1962.

Little, M. R. (1998). *Sticks and stones: Three centuries of North Carolina gravemarkers.* Chapel Hill: University of North Carolina Press. Book with more than 230 illustrations looking at gravestones in North Carolina.

Meyer, R. E. (Ed.). (1993). *Ethnicity and the American cemetery.* Bowling Green, OH: Bowling Green State University Popular Press. Anthology of eight essays exploring the manner in which representative ethnic groups in America have made their cemeteries. The book has an interdisciplinary focus from folklore, cultural history, historical archaeology, landscape architecture, and philosophy.

Sloane, D. C. (1991). *The last great necessity: Cemeteries in American history.* Baltimore: Johns Hopkins University Press. Explores changing attitudes about cemeteries as well as changing landscapes and describes many of America's most famous cemeteries.

CHAPTER 12

THE LEGAL ASPECTS OF DYING

Life's too short for worrying.
Yes, that's what worries me.

—Origin Unknown

Most of the law that regulates the disposition of a person's body and property upon death is state law, not federal law, and considerable variation exits between the laws of different states (Scheible, 1988). As a general rule, the law of the state where a person lives at the time of death controls disposition of the body and of property, although real estate (land and the structures thereon) is governed by the law of the state where the property is located. It is unlawful to dispose of a corpse until the cause of death has been satisfactorily ascertained (Mims, 1999). Several state laws, therefore, may have an impact upon a person's death, in addition to some federal laws. Though an individual does not have to have legal counsel to handle some of the legal issues involved in death, because lawyers are the experts on these topics, it might behoove one to consider legal counsel. Perhaps "doing it one's self" and being one's own attorney can be likened to trying to "fix the plumbing." After trying to "fix the plumbing," the amateur often has to bring in a plumber to undo what he or she has done, and usually at a much greater expense than would have originally been the case.

What does the law say regarding ownership of a dead body? It turns out that in English law, going back to 1614, no one owns a dead body (Mims, 1999). In other words, in England no one can be arrested for stealing a body, a fact that led

425

to difficulties in the prosecution of "body snatchers." In England in 1614, a Mr. Haynes was accused of stealing the shrouds (burial sheets) from a corpse. In the court ruling, it was stated that the corpse could not own the sheets, and this has been misinterpreted as meaning that the corpse itself could not be owned. Thus, in 1856 in England, a son removed his mother's remains from a graveyard, but because the corpse was not anyone's property, he could only be charged with trespassing in the graveyard. Exactly who owns a human body even today is not clear, under British law, notes Mims (1999). A corpse, it seems, unlike a urine sample or a lock of hair, is not an item of property and cannot be stolen!

However, the same "lack of ownership" is *not* the case in the United States. As discussed in chapter 8, various ethical questions revolve around "ownership" of one's body. Legally, however, the right to control the disposition of one's own body is more limited in the United States than the relatively broad power to dispose of property upon death. The decedent's body is not "property" in any conventional sense of the word and is not part of the decedent's estate (Scheible, 1988). Therefore, the right to control one's own body, like other legal rights, ends at death, and any remaining rights pass to other individuals. Thus, any preferences for disposition of one's body expressed during life (e.g., funeral arrangements) will be considered but are not legally binding. Instead, the deceased person's next of kin have the right to arrange such details, because in the United States, unlike England, the next of kin have property rights in the corpse (Mims, 1999). For example, if an individual had requested (either orally or in writing) that the funeral director cremate upon death and the next of kin prefer earth burial, the funeral director may be in an awkward position, yet the wishes of the next of kin are legally binding.

As previously discussed in chapter 2, social and biological deaths differ in various cultures. Our focus in this chapter is on biological deaths, legally determined.

ESTABLISHING THE CAUSE OF DEATH

If an individual is enrolled in a hospice program and dying of brain cancer, the cause of death is usually rather obvious. The process leading to death is probably related to the malignancy and can be tracked from there. The ultimate cause of death may be heart failure brought on by the weakened condition from the cancer. Yet, as discussed in chapter 1, determining the cause of death is not always a clear-cut case. For example, an 88-year-old woman suffering from chronic heart disease, diabetes, and high blood pressure, may fall and break her hip, forcing her to be bedridden and fitted with a catheter for urination purposes. She develops an infection from the catheter. Pneumonia sets in, and her lungs fill with fluid. She dies. What is the "cause" of her death? Such a question will not be easy to answer, yet the attending physician has to seek accuracy in filling out the death certificate.

Death Certificate

A **death certificate** is a permanent record of an individual's death that must be completed soon after death. The purpose of the death certificate is to obtain a simple description of the sequence or process leading to death rather than a record describing all medical conditions present at death (National Center for Health Statistics, 1997). Certified copies of a death certificate are needed to collect insurance proceeds and other death benefits. Copies can be obtained through the funeral home and added to the cost of the funeral bill (typically $4 to $15 each, depending on the state). Otherwise, copies can be obtained by writing to the vital statistics office or county health department in the county where the person died. If you wait for several weeks, copies can be obtained from the state's vital statistics office.

Cause-of-death data are important for surveillance, research, design of public health and medical interventions, and funding decisions for research and development. The death certificate is a legal document used for legal, family, and insurance purposes. It is also used to compile mortality statistics. Although these data are overseen by the National Center for Health Statistics, physicians shape the content of mortality statistics by virtue of what they write on the death certificate and whether they obtain consent for an autopsy.

Causes of death on the death certificate represent a medical opinion that might vary with individual physicians. For situations in which the cause of death is difficult to certify, the certifier should select the causes that are suspected to have been involved and use words such as "probably" or "presumed" to indicate that the description provided is not completely certain (Hazlick, 1994). If the certifier really is not sure, he or she should state that the etiology (cause or origin) is unknown, undetermined, or unspecified; thus, it is clear that the certifier did not have enough information to provide even a qualified opinion about the etiology.

When preparing the death certificate for an elderly person, causes should not include terms such as *old age* because they have little value for public health or medical research. If malnutrition is noted, the certifier should consider if other medical

---

DEATH ACROSS CULTURES
India's Living Dead Form Their Own Club

An association for dead people has been established in northern India by a group of men and women claiming that reports of their demise have been greatly exaggerated. The Union of Dead People was set up to represent the growing number of unfortunates who have been declared dead when they are actually very much alive.

The living dead are victims of a particularly nasty form of land-grabbing. They have been pronounced dead by relatives working in collusion with crooked bureaucrats to defraud them of their property. Once a death certificate has been issued and the person's land transferred into an-

other person's name, it seems little can be done to rectify the error.

Ten "dead" people recently staged a sit-in protest against their plight outside local government offices. The range in ages of the protestors was from the teens to age 85. The 85-year-old woman said she was swindled out of her land so long ago that nobody now believes her claim. Apparently there are "thousands of people" like this. Many individuals are afraid to complain in case they are really killed by their relatives.

From "India's Living Dead Form Their Own Club." (p. 17), by D. Orr, 1999, June 18, The London Times.

---

conditions could have led to malnutrition. If several conditions resulted in death, the certifier should choose a single sequence to describe the process leading to death and list the other conditions in another section. "Multiple system failure" could be included as an "other significant condition." Finally, if the certifier cannot determine a descriptive sequence of causes of death, death may be reported as "unspecified natural causes" (Hazlick, 1994).

With the death of an infant, maternal conditions may have initiated or affected the sequence contributing to the death. Such conditions should be reported in addition to the causes related to the infant. When sudden infant death syndrome (SIDS) is suspected (discussed in chapter 14), a complete investigation should be conducted. If the infant is younger than 1 year of age and no cause of death can be found, then the death can be reported as SIDS. If the investigation is not completed, the death may be reported as presumed to be SIDS.

Despite the importance of death certificates and the fact that physicians are the professionals who sign these documents, the majority of doctors in the United States have very little training in completing death certificates (Wass, Berardo, & Neimeyer, 1988). To address this problem, the death certificate has been revised to include instructions for completing the form, along with material for additional information. For example, the primary cause of death is defined as "the disease or injury that initiated the train of morbid events leading directly to death, or the circumstances or violence which produced the fatal injury" (Kircher, 1992, p. 1268). It is stated that, without the underlying cause (primary cause), death would not have occurred. The "immediate

---

WORDS OF WISDOM
"Value" of an Autopsy

"Doctor," complained the patient, "all of the other physicians called in on my case seem to disagree with your diagnosis."

"Yes, I know they do," said the doctor, "but the autopsy will prove that I'm right."

---

cause" of death is that which "directly precedes death" and is the ultimate consequence of the "underlying cause." The interval between the underlying and immediate causes of death may last for years or just seconds.

Though the importance of death certificates is pointed out above, death certificates are not universally used. Indeed, the World Health Organization, in reporting causes of death around the world, observes that death certificates are made out for only about 25 percent of all deaths (Mims, 1999). They are not used for deaths due to great famines, massacres, wars, or natural disasters.

Autopsy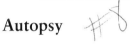

A dissection and examination of a just-deceased patient is an **autopsy.** The **pathologist** performing the autopsy assumes a mechanical, physical cause for the death. First, the pathologist gathers a history as to how the death occurred and often obtains the past medical history of the deceased. The pathologist then examines the body externally and internally, taking biopsies of tissues to identify any diseases or abnormalities that may have contributed to the death. During the course of the autopsy, various laboratory tests may be performed including x-rays, toxicological analysis of blood and urine, and cultures of body fluids and organs for evidence of infection.

Autopsies for cases of natural death or deaths occurring among patients under the care of a physician are usually performed at the hospital where the death occurred and with the permission of the next of kin. Local statutes may require an autopsy for cases of traumatic or sudden, unexpected deaths or for deaths due to external causes. Such an autopsy is requested by either a coroner or a medical examiner. A report summarizing the findings is prepared by the pathologist. Often, this medical specialist is then subpoenaed to testify in court about the findings.

A *complete autopsy* refers to an examination of the organs of the three major cavities of the body—the abdomen, chest, and head. The pathologist makes a Y-shaped incision extending from each armpit to the center of the lower abdomen. Various internal organs are removed and weighed, and blood, urine, and other fluids are sampled. Much can be determined from the size, color, and feel of various body parts. For example, the liver of an alcoholic patient may be pale and shriveled, and a diseased heart is likely to be flabby and grossly enlarged.

DEATH ACROSS CULTURES
Native Canadian Beliefs Regarding the Autopsy

An anthropological study of Native Canadian patients in a hospital setting found conflict between indigenous people and the medical establishment regarding postmortem care of the body. One case involved the death of a 10-month old female Ojibway child from a reserve community. The pediatric hospital contacted the parents for consent to perform an autopsy, because the cause of death was undetermined, though the child was being treated for a wide range of neurological problems. The parents refused to consent to the procedure, saying that it conflicted with their spiritual beliefs about the care of the body of the deceased. Though the parents continued to object, the medical examiner ordered the autopsy, nonetheless.

After this unpleasant experience, the medical staff of the pediatric hospital decided to hold an in-service training workshop dealing with the general issue of the cultural interpretation of autopsies among Native people. The meeting opened with a presentation of the biomedical perspective of the physicians, emphasizing the importance of an autopsy for improving the general level of medical knowledge. Native speakers argued that a primary value in Native culture was the corporeal integrity of the body after death, a value linked to spiritual understandings about the length of time required for the transition of a person's soul to the afterworld.

A pathologist from the hospital stated that Native clients' objections to autopsies must be balanced against the wider community interest in assuring that deaths do not occur without determination of negligence or malevolent cause. The medical examiner also stressed that the post-mortem examination was sometimes necessary to control contagious or environmentally caused diseases.

(continued)

Other than seeking the cause of death in an "unknown" situation with possible criminal activity involved, there are other reasons for performing autopsies, as noted by the National Association of Medical Examiners: (1) The autopsy serves as a check on the accuracy of the clinical diagnoses and historical data. Did the physicians correctly identify the patient's problem? Such information could be helpful in the event of a law suit against the hospital and staff regarding the death. (2) Likewise, the autopsy can be a check on the appropriateness of medical and surgical therapy which followed the diagnoses. Were the doctors treating the patient in appropriate ways? (3) An autopsy can help gather data on new and old diseases and surgical procedures. Such gathered information could help in treating future patients and thus benefit many individuals. (4) Information from the autopsy can also be useful to the deceased's family. The autopsy could identify inherited diseases that constitute a risk for other family members (e.g., heart disease, cancer, and certain kinds of kidney disease).

The autopsy has ancient roots and has produced countless medical advances over time (Clark & Springen, 1986). The cumulative data gathered through autopsies

(continued from previous page)

The director of the Native Services program was asked to provide an overview of the concerns of Native clients regarding current policies governing autopsies. She contrasted the analytical approach of the medical community with Ojibway beliefs about dying and post-mortem care, emphasizing that traditional beliefs focused on the importance of understanding the reasons for a person's death in terms of violation of traditional beliefs or moral transgressions. She stressed that traditional believers were also concerned with identification of cause of death but that the Ojibway fostered an integrative approach, not a narrow understanding of causation as in the biomedical culture. Funeral rituals provided mechanisms that transformed death into an integrative event for the community. The Ojibway beliefs emphasize any opposition to a procedure that would disturb corporal integrity, including amputation or tissue removed in biopsies. The Ojibway people believe that the spirit resided in the body for a defined period after death. To open the body cavity or to remove tissue would disrupt the departure of the spirit and provide the possibility for retribution.

The conference was followed up with continuing informal consultations between clinicians and Ojibway representatives. In some cases family acceptance of the autopsy resulted. In other situations the pathological examination was limited to minimally invasive procedures. Yet, in other instances the opposition to autopsy was pursued through legal channels. In the end, culturally sensitive support for dying Native patients and their families may require that individuals assert their prerogative to die in their homes, not in a hospital setting.

From "Cultural mediation of dying and grieving among Native Canadian patients in urban hospitals" (pp. 231–251), by J. M. Kaufert & J. D. O'Neil, 1991. In D. R. Counts & D. A. Counts (Eds.), Coping with the Final Tragedy: Cultural Variation in Dying and Grieving. Amityville, NY: Baywood Publishing.

have enabled physicians to treat or prevent a number of diseases. Autopsies confirmed the link between cigarette smoking and lung cancer. They helped in the discovery of abnormalities involved in congenital heart defects, multiple sclerosis, Alzheimer's disease, and viral infections in the brain that may cause dementia shown by some victims of AIDS. The purpose of an autopsy is summed up by the statement on the entrance to the autopsy room at the Medical University of South Carolina in Charleston: "This place is where death rejoices to come to the aid of life."

Despite the usefulness of autopsies, there has been a sharp decline in the number of autopsies performed in the United States (Brody, 2001). In the mid-1940s half of the patients who died in a hospital were dissected, whereas as of 1992 only 12 percent are subjected to a postmortem examination (Kircher, 1992). Autopsy findings are believed to increase the reliability of death certificate data and therefore to provide an essential check on the accuracy of diagnoses—although there may be legitimate disagreement between the attending physician and the pathologist as to precisely what killed the patient.

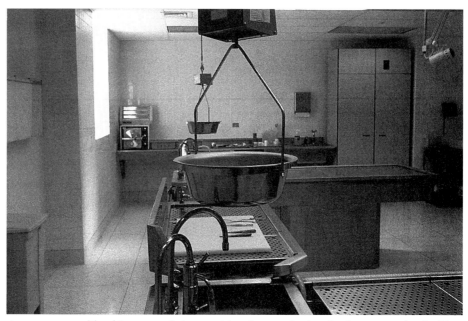

Autopsy rooms like this one are being used less frequently than in the past, and valuable medical information is going undiscovered.

Physicians, hospital administrators, and families of the deceased have shown reduced interest in autopsies. Physicians in particular fear that an autopsy will uncover an error and thus spur malpractice suits. High-tech methods are not necessary. Autopsies also cost hospitals $2,500 each (Brody, 2001) and must be subsidized by general charges because they are not directly covered by health insurance. In addition, physicians are reluctant to ask bereaved next of kin for permission to perform a procedure that many regard as upsetting and an invasion of privacy.

In the "Listening to the Voices" account, Elizabeth Maxwell, a former St. Olaf College student visiting a medical center during a health science internship, discusses her observation of an autopsy and her emotional reactions to it. Although it is difficult to generalize from one experience, it is possible that her first encounter with the autopsy is not unlike that of many first-year medical students in gross anatomy class.

Coroners and Medical Examiners

When a person dies of unnatural causes, by violence or accident, the body may be made available to the appropriate police authorities for evidentiary and record-keeping purposes (Wolf, 1995). The coroner or public medical examiner is authorized to view the body and conduct an autopsy. If criminal conduct or suicide is suspected or if public health may be affected, an autopsy may be performed over the objection of the deceased's family in extreme circumstances.

LISTENING TO THE VOICES
Reflections on an Observed Autopsy

After observing an autopsy, health intern Elizabeth Maxwell wrote about her experience.

We thanked the resident and autopsy specialist and left the room. It was nice to take off the gloves, masks, and aprons we had been wearing. Fresh air was welcome, too. But I began thinking about what I had seen that morning. I was upset by the whole situation: an innocent, older man suddenly dies, and with the loss of his life, he loses his identity and dignity. He was just a body for anatomical study. He was not a person, though he was still wearing his wedding ring.

As I contemplated this paradox, I thought again of the whole autopsy process, and was dismayed again. It had been so brutal. It was nothing like the delicate surgery I had expected. Since an autopsy must show every detail, huge incisions must be made. Nothing needs replacement, so little care is taken to keep the parts intact. This carelessness upset me.

I tried to sort everything out in my mind—the patient, the tragic accident, the brutal procedure, the stoic autopsy specialist, the helpful and concerned medical student, and the joking male medical students. All of this was too much for me to comprehend.

All that day and night, I kept smelling the autopsy odor in various places I went. I could not escape it! The next morning I wrote down my feelings in my journal concerning what I had seen and felt. I also talked with people about my frustrated emotions for several weeks. With the passing of time I can now talk about it, but sometimes I think about that old man.

Elizabeth Maxwell, former undergraduate student intern at St. Olaf College, Northfield, MN.

A **coroner** is a public official, appointed or elected, in a particular geographic jurisdiction, whose official duty is to make inquiry into deaths in certain categories (e.g., if the death is unexpected or unexplained or if there is the possibility of an injury or poisoning). The office of the coroner ("crowner") dates back to medieval times when the crowner was responsible for looking into deaths to be sure death duties were paid to the king (National Association of Medical Examiners, 1996). For example, it was the job of the coroner to maintain the private property of the British Crown. The coroner was responsible for investigating accidents such as shipwrecks to see what money could be obtained for the "royal purse" (Mims, 1999). Primary duties today, however, are to make inquiry into a death and complete the certificate of death. The coroner assigns a cause and manner of death and lists them on the certificate of death. The coroner also decides if a death occurred under natural circumstances or was due to accident, homicide, suicide, or undetermined means.

Coroners are called upon to decide if a death was due to foul play. Depending on the jurisdiction and the law, the coroner may or may not be trained in the

PRACTICAL MATTERS
Permit for Postmortem Examination (Autopsy)

Date _____

Name of deceased _____ Hosp. No. _____

Date and time of death _____

Month   Day   Year   Hour   P.M.   A.M.

1. I hereby authorize any qualified member of the Staff of Physicians of the Hospital to perform an autopsy upon the body of the above named at such place as the hospital or physician may designate. My relation to the deceased being that of _____.

2. I know of no survivor of the deceased who is closer kin than I to authorize this postmortem examination.

3. The post-mortem examination here authorized may be either complete or partial and such parts of the body may be removed as may be necessary for subsequent study in order to accomplish its purpose. If the nature and extent of this examination or the right to remove parts of the body are to be limited in any way, these limitations must be clearly stated below. In the absence of any stated limitations, it is to be understood that the physician by whom the operation is performed is hereby permitted to examine any portion of the body that, in his or her opinion, should be examined, including organs within the chest, abdomen, skull and extremities.

4. I further authorize him or her to have present at the post-mortem examination such persons as he or she may deem proper.

Limitations (if any) _____

Signed _____

_____ (Witness)

_____ (Witness)

medical sciences. The coroner in some states is a licensed physician, and in other states the coroner may have no more than a high school degree. The coroner often employs pathologists to perform autopsies when a questionable death occurs.

A **medical examiner** is a physician charged within a particular jurisdiction to investigate and examine persons dying sudden, unexpected, unexplained, or violent deaths. Unlike the coroner, the medical examiner is expected to bring medical expertise to the evaluation of the medical history and physical examination of the deceased. Though not required to have training in pathology, many part-time medical examiners are encouraged to take medical training to increase their ability to investigate death scenes.

ADVANCE DIRECTIVES

All states have legislation authorizing the use of legal documents such as durable power of attorney for health care (sometimes referred to as health-care proxy, surrogate, agent, or attorney-in-fact) and living wills, which together are known as **advance directives (ADs)**. However, only 20 percent of adult Americans have living wills and durable powers of attorney for health care (Chatzky, 2000). These documents allow individuals to maintain some control over medical treatments if they become incapacitated. To complete them requires thought and talk about disability and the circumstances of one's death, observe Kathryn Braun and Reiko Kayashima (1999).

In the absence of a legally binding advance directive, a court action may be necessary to decide whether treatment of an incompetent person or minor child may be terminated (Wolf, 1995). Such judicial determinations in the recent past have focused on ascertaining what the patient would most likely have chosen had she or he been competent and whether the judgment of a third party (e.g., parent, guardian, or physician) may be substituted for the patient's consent. In some cases, since 1976, a hospital ethics committee has made these decisions, rather than require the often prolonged process of litigation in court. By 1990, an estimated 60 percent of hospitals in the United States had established such bioethics committees (Wolf, 1995).

As noted in chapter 8, since the implementation of the Patient Self-Determination Act in December 1991, health maintenance organizations (HMOs) and other health-care institutions have been required to inform all adult patients about ADs, and hospitals have been required to ask all patients whether they have completed advance directives. A study in California (Gordon & Shade, 1999) of more than 5,000 persons 65 years of age or older regarding end-of-life care preferences revealed that less than one third in this HMO group had completed and filed an AD, even though an all-out effort had occurred to encourage participation——waiting-room posters, articles about advance directives in their health plan newsletter, and personal efforts on behalf of the health plan staff. Rates of completion were higher among the older persons who were in poorer health. A number of other educational strategies to increase advance directive completion rates have been tried, with varying amounts of success; however, most have been unsuccessful (Braun & Kayashima, 1999).

Researchers Braun and Kayashima (1999) decided to encourage advance directives through churches and synagogues. The outcome of their 12-month project in Hawaii, using interviews and focus groups and distributing handbooks developed by their religious leaders to the congregations, was that the respondents saw real value in having the church/temple serve as a venue for death education. They suggest that an AD project would probably succeed in churches and temples where ministers encourage these efforts.

Living Wills

As previously mentioned in chapter 8, a **living will** is a document that states that an individual does not want medical intervention if the technology or treatment that keeps her or him alive cannot offer a reasonable quality of life or hope for recovery. Living wills were first introduced in the 1970s as the public came to fear dying while being connected to feeding tubes in the intensive care unit even as they sought the cures hospitals and doctors offered (Plech, 2000). Despite a somewhat growing acceptance of living wills, the issue is still complicated by emotion and questions of when such a will should be invoked. Many persons are reluctant to take steps leading to a patient's death, and some are willing to withhold further medical treatment but do not want to disconnect existing life support equipment. Others cannot even distinguish living wills from "do not resuscitate" (DNR) orders (see chapter 8).

A. M. Cugliari and T. E. Miller (1994) determined that 29 percent of New York State hospitals would not honor a patient's request to withdraw or withhold treatment. In another study by G. Almgren (1993) in which 140 nursing home directors of nursing were interviewed, it was discovered that the competency of the patient and type of nursing home ownership were the primary factors influencing decisions to withdraw nutrition from terminally ill patients—the fact that a patient had signed a living will had little influence on these decisions. Furthermore, S. H. Miles and A. August (1990) have demonstrated that the courts have a gender bias in the forced implementation of living wills. Because men were perceived as more rational and women as more unreflective, emotional, and immature, the courts accepted the treatment preference of 75 percent of the men but only 14 percent of the women. This pattern of gender bias had been discovered in many states and remains unchanged when one controls for the patient's age, condition, and treatment modality (Miles & August, 1990). Consistent with the research findings just cited, Donald Dilworth (1996) conducted his own research in five hospitals and reached the following conclusion:

> Living wills do virtually nothing to reduce patient suffering, because doctors and hospitals ignore them. Some 80 percent of doctors either misunderstood or set aside the dying requests, and a program to help patients avoid painful life-prolonging treatments had no effect at any of five hospitals I studied. The cause of the problem is the medical culture that rejects death and promotes technology.

Another problem with living wills is that they often address specific medical scenarios, yet possible treatments sometimes happen in "gray" areas (Chatzky, 2000). For example, many patients' living wills rule out some treatments if they are terminally ill. However, a profound stroke or end-stage Parkinson's disease may not be defined as a terminal condition.

A solution that forces one to think about the end-of-life gray areas ahead of time is called **Five Wishes.** Five Wishes was designed by lawyer Jim Towey, legal counselor to Mother Teresa. This "friendly to use" living will document guides people in making essential decisions about the care they want at the end of their life (Chatzky, 2000). The idea of Five Wishes is to sit around the dinner table with your significant others and work through your wishes step by step. Examples of questions to be answered are: Who should make care decisions when I cannot? What kind of medical treatment do I want toward the end, and what do I not want? How comfortable do I want to be while dying (e.g., pain-relief medication

---

PRACTICAL MATTERS
A Living Will

To my family, my physician, my lawyer and all others whom it may concern:

Death is as much a reality as birth, growth, maturity, and old age. Death is the one certainty of life. If the time comes when I can no longer take part in decisions for my own future, let this statement stand as an expression of my wishes and directions, while I am still of sound mind.

If at such a time the situation should arise in which there is no reasonable expectation of my recovery from extreme physical or mental disability, I direct that I be allowed to die and not be kept alive by medications, artificial means or "heroic measures." I do, however, ask that medication be mercifully administered to me to alleviate suffering even though this may shorten my remaining life.

This statement is made after careful consideration and is in accordance with my strong convictions and beliefs. I want the wishes and directions here expressed carried out to the extent permitted by law. Insofar as they are not legally enforceable, I hope that those to whom this will is addressed will regard themselves as morally bound by these provisions.

Signed _____

Date _____

Witness _____

Witness _____

From Death: Confronting the Reality,
by W. E. Phipps, 1987, Atlanta:
John Knox Press.

---

even if it makes me drowsy)? How do I want to be treated, and do I want to die at home, if possible? What do I want my significant others to know about me and my feelings after I am dead? Five Wishes is currently valid in 34 states. About one million individuals have filled out the eight-page questionnaire in the past 2 years (Matousek, 2000). Five Wishes is easy to use and understand, not written in heavy legal or medical language. It allows people to discuss comfort, dignity, and forgiveness, not just medical concerns. Also, the price is right. A copy of Five Wishes can be obtained for $5.

If you complete a living will or Five Wishes, make several copies of it and share with physicians and other health-care providers, family members, and friends. A copy should also be given to your attorney to be kept in her or his office. This is not something that you should limit to a single copy, lock up in a safe deposit box, or keep as a big secret. Share the news!

Durable Power of Attorney for Health Care

A **durable power of attorney** for health-care decisions allows an individual to designate someone else as a legal proxy to make treatment decisions if the person becomes unable to make them (Braun & Kayashima, 1999). *Durable* means that the document will be recognized in a court of law. There has been some disenchantment with the rather anemic power of living wills. Living wills seem to have no teeth, concludes Margaret Battin in *The Least Worst Death* (1994). However, states began in the 1990s to modify existing living will statutes or to adopt new ones to provide a durable power of attorney for health care. States provided some statutory mechanism for appointing a surrogate, so that a person could select someone—a friend, relative, or anyone she or he trusted—to make medical decisions authorizing the withholding or withdrawing of life-sustaining treatment if the individual were no longer able to do so. Durable power of attorney documents (and living wills) do not necessarily reliably represent a patient's true choices over time, notes Battin (1994). Indeed, in thorough research of this topic, Markson and colleagues (1997) conclude that those who commonly participate in surrogate decisions for incapacitated patients—family members and physicians—may place a different value on medical interventions than the patients they are supposed to represent and may not predict patient preferences accurately.

"WE HAVE A STRONG PULSE, REGULAR HEARTBEAT AND ADVANCE DIRECTIVES."

The durable power of attorney for health care can be structured to give an agent as little or as much authority as the patient chooses (Bennett, 1999). However, in general, the following are authorized by the agent: (1) use or withholding life support and other medical care; (2) placing the patient in or taking her or him out of a health facility such as a hospital; and (3) decisions not specifically covered by a living will.

On the other hand, the durable power of attorney for health care does *not* (Bennett, 1999): (1) give any authority with respect to financial matters, administering involuntary psychiatric care, or sterilization; or (2) generally cover medical treatment that would provide comfort or relieve pain, which means that a health-care proxy cannot refuse pain relief on your behalf.

The durable power of attorney for health care generally only takes effect when two physicians, including the patient's physician of record, certify that the patient is not capable of understanding or communicating decisions about her or his health. Durable health-care power of attorney copies should be distributed to the primary physician, the pharmacist, the individual's agent, and the hospital of choice, suggests Jarratt Bennett (1999). An example of a durable power of attorney for health care from the state of Oregon appears in the "Practical Matters" box.

DISPOSITION OF PROPERTY

Upon death, a person's business affairs must be concluded (debts, taxes, and funeral expenses), and the property owned at death must be distributed to others. Real property, which includes land and what is attached to it, such as a house (as opposed to personal property such as clothes or jewelry), is transferred in three ways—through a deed, a will, or an inheritance. The latter two deal with property transfer at death. A will is important if one wishes to have any say in the transfer.

Intestate and Inheritance

If an individual dies without a will, that person dies **intestate.** In such a situation the property of the **decedent** (the deceased person) will pass through inheritance. To die intestate means that important decisions such as who gets your assets and raises your children may be made by a state judge in ways that you might not like. The court will appoint a person or institution (e.g., a bank), called an **administrator,** to administer the estate of a person who died intestate. Since only about 30 percent of Americans have a will (Tyson, 1997), dying intestate is the way most Americans die. Each of the 50 states has its own laws regarding wills, thus what is true of one state may not necessarily be true of another.

If one dies intestate, classes of heirs (also called laws of descent) are established by the various states to determine how the property passes. Typically, if one is married at the time of death and has no children, the spouse receives the entire inheritance. If one has children, they also take from the inheritance (if the children are minors, the

PRACTICAL MATTERS
Directive to Physicians

This directive is made this _____ day of _____, (month) _____ (year).

I, _____, being of sound mind, willfully and voluntarily make known my desire

_____ (a) That my life shall not be artificially prolonged and

_____ (b) That my life shall be ended with the aid of a physician under the circumstances set forth below, and I do hereby declare:

1. If at any time I should have a terminal condition or illness certified to be terminal by two physicians, and they determine that my death will occur within six months,

   _____ (a) I direct that life-sustaining procedures be withheld or withdrawn, and

   _____ (b) I direct that my physician administer aid-in-dying in a humane and dignified manner.

   (You must initial (a) or (b) or both.)

   _____ (c) I have attached Special Instructions on a separate page to the directive.

   (Initial if you have attached a separate page.)

   The action taken under this paragraph shall be at the time of my own choosing if I am competent.

2. In the absence of my ability to give directions regarding the termination of my life, it is my intention that this directive shall be honored by my family, agent (described in paragraph 4), and physician(s) as the final expression of my legal right to

   _____ (a) Refuse medical or surgical treatment, and

   _____ (b) To choose to die in a humane and dignified manner.

   (You must initial (a) or (b) or both, and you must initial one box below.)

   _____ If I am unable to give directions, I do not want my attorney-in-fact to request aid-in-dying.

   _____ If I am unable to give directions, I do want my attorney-in-fact to ask my physician for aid-in-dying.

3. I understand that a terminal condition is one in which I am not likely to live for more than six months.

4. (a) I, _____ do hereby designate and appoint _____ as my attorney-in-fact (agent) to make health care decisions for me if I am in a coma or otherwise unable to decide for myself as authorized in this document. For the purpose of this document, "health care decision" means consent, refusal of consent, or withdrawal of consent to any care, treatment, service, or procedure to maintain, diagnose, or treat an individual's physical or mental condition, or to administer aid-in-dying.

   (b) By this document I intend to create a Durable Power of Attorney for Health Care under the Oregon Death With Dignity Act and ORS Section 126.407. This power of attorney shall not be affected by my subsequent incapacity, except by revocation.

(continued)

(continued from previous page)

(c) Subject to any limitations in this document, I hereby grant to my agent full power and authority to make health care decisions for me to the same extent that I could make these decisions for myself if I had the capacity to do so. In exercising this authority, my agent shall make health care decisions that are consistent with my desires as stated in this document or otherwise made known to my agent, including, but not limited to, my desires concerning obtaining, refusing, or withdrawing life-prolonging care, treatment, services and procedures, and administration of aid-in-dying.

5. This directive shall have no force or effect seven years from the date filled in above, unless I am incompetent to act on my own behalf and then it shall remain valid until my competency is restored.

6. I recognize that a physician's judgment is not always certain, and that medical science continues to make progress in extending life, but in spite of these facts, I nevertheless wish aid-in-dying rather than letting my terminal condition take its natural course.

7. My family has been informed of my request to die, their opinions have been taken into consideration, but the final decision remains mine, so long as I am competent.

8. The exact time of my death will be determined by me and my physician with my desire or my attorney-in-fact's instructions paramount.

I have given full consideration and understand the full import of this directive, and I am emotionally and mentally competent to make this directive. I accept the moral and legal responsibility for receiving aid-in-dying.

This directive will not be valid unless it is signed by two qualified witnesses who are present when you sign or acknowledge your signature. The witnesses must not be related to you by blood, marriage, or adoption; he or she must not be entitled to any part of your estate; and he or she must not include a physician or other person responsible for, or employed by anyone responsible for, your health care. If you have attached any additional pages to this form, you must date and sign each of the additional pages at the same time you date and sign this power of attorney.

Signed _____

City, County, and State of Residence _____

(This document must be witnessed by two qualified adult witnesses. None of the following persons may be engaged as witnesses: (1) a health care provider who is involved in any way with the treatment of the declarant, (2) an employee of a health care provider who is involved in any way with the treatment of the declarant, (3) the operator of a community care facility where the declarant resides, (4) an employee of an operator of a community care facility who is involved in any way with the treatment of the declarant.)

From The Oregon Death With Dignity Act,
Oregon Revised Statutes, Chapter 97, 1990.

inheritance is to be held in trust until they reach a certain age). If one is married with children, the spouse takes half and the children take the other half. (For example, if there is one child, she or he takes the entire half or 50 percent; if there are two children, they split the 50 percent, thus receiving 25 percent each—one half of one half; if there are three children, they each receive approximately 17 percent—one third of one half or one sixth, etc.). If the decedent has children but no spouse, then the children take the entire inheritance. (If there are two children, they each receive 50 percent of the inheritance, for example). If one dies intestate, the classification of heirs typically follows this pattern:

> Spouse (no children), then spouse takes all.

> Spouse and children, then the spouse takes half and the children split the other half.

> Child or children and no spouse, then the child takes all or the children split equally.

> If no spouse or children, then parents take from the inheritance.

> If no spouse, children, or parents, typically siblings (brothers and sisters) take from the inheritance.

> If "none of the above," then usually grandparents take from the inheritance, followed by aunts and uncles. If one dies intestate and without heirs, one's property goes to the state (the legal term for this is **escheat**).

Wills

A **will** is essential if one does not want the state to determine the distribution of her or his property and who will raise the children (if children are involved). A well-executed will can provide orderly distribution of property, get the decedent's assets to the people to whom she or he wants them to go, reduce the expense of probate court, and allow the **testator** (the one making the will) to name an individual of her or his choosing to administer the estate and someone to be legally responsible for young children, rather than have the court appoint an administrator and place minor children (in the event of a single parent's death). In order to make out a will, the following criteria generally apply, though these may vary by states:

1. Must be at least 18 years of age or living in one of the few states that permit younger persons to make a will if they are married, in the military, or otherwise considered "emancipated."

2. Must be of sound mind. One does not have to prove mental state in most cases. The testator seemingly might mentally be "way out" and act "crazy," yet be of "sound mind" in a legal sense. A minimum mental capacity is imposed to ensure that persons writing wills have an appropriate appreciation of their actions, their property, and the parties to whom they wish their estates to pass.

3. The will must be in writing—typewritten or computer-printed.

4. The will must be signed by the testator.

5. The will must have two witnesses to the signing (Vermont requires three witnesses). The witnesses must indeed *witness* the testator signing the will. What one cannot do is sign the will, then rather casually ask two people to sign as "witnesses." They would not have "witnessed" the signing, their sole function. The witnesses do not have to read the will. They are simply confirming that they saw the testator sign her or his will. Often the attorney and one of her or his employees will sign. Someone who is to benefit from the will should not be a witness.

6. The will must be dated.

A will must state who the **executor** (or "personal representative") of the estate will be (one responsible for supervising the distribution of property after death and seeing that the debts and taxes are paid). The executor may also have the durable power of attorney, giving her or him the ability to handle finances (write checks and sell stock, for example), if the testator becomes incapacitated before death. The executor may perform in this capacity, though someone else could be chosen to have the durable power of attorney (Chatzky, 1998).

The will typically names one or more persons who will benefit from the will (called **beneficiaries**) and specifies the personal guardian for minor children. The will should be kept somewhere so that the executor can find it easily. Wills need to occasionally be updated, if one's situation changes (e.g., another child is born into the family). Generally, an individual is free to leave her or his property to whomever or whatever he or she desires.

If a testator does not wish to leave anything to a particular child, that child should be named and left something, say one dollar, so that she or he cannot protest the will later and claim that the parent simply "forgot" her or him. If the child is named in the will, then the parent did not "forget."

It is not necessary to have an attorney make out a will, though as noted earlier, to hire a professional may be a good idea. If one has good self-help materials, it is not difficult to make a will that names who receives your earthly possessions and who will be guardians of your minor children (if applicable). For rather simple wills, which most individuals have, an attorney has a standard form in the computer and has the secretary type in the information. Unless a will is complicated, having a lawyer prepare the will should cost around $500 (Chatzky, 1998). Attorneys usually charge by the hour, but some will draw up a simple will for a set fee.

A will can be changed by adding an amendment called a **codicil.** A codicil allows one to amend the will by adding new provisions without having to rewrite it entirely. For example, if additional property is acquired and the testator wishes to designate a certain person or organization to receive this, rather than write a completely new will, this information is simply "added on" as a codicil. Or, if a baby is born into the family or adopted, the testator will probably want to add her or him as a recipient from the estate.

There are several ways in which a will can be revoked (canceled or rescinded). Included among those ways are the following: the testator divorces or remarries, the testator writes another will (thus the importance of the date on the will), or the will is literally destroyed (tearing, burning, or destroying in any way). If the will was accidentally destroyed, its terms may be proved by other competent evidence, usually a copy of the original will. For this reason having two originally signed wills is recommended. The safest way to revoke a will is to execute a new will that expressly states that all previous wills are revoked (Scheible, 1988). A will can be broken (shown to be useless) in various ways. For example, if the testator signed under coercion (someone forced her or him to sign against her or his wishes), and this can be proven in court, the will is then not legally binding. If it can be proven in court that the testator, at the time of signing the will, was incompetent, then the will can be broken. In addition, if fraud (deception or trickery) can be proven, the will can be broken.

A parent or parents with small children might prefer to set up a trust fund for each child, rather than have the assets go directly to minors. The trust does not usually require witnesses, but it should be notarized. The will can be used to name a trustee who will handle any property the child receives until the child reaches the age of majority, which in most states is age 18. Until the child reaches such an age, the trustee can spend the money for things such as health care, education, and general support of the child. Unless stated otherwise, when the child reaches the specified age, the trustee ends the trust and gives the child whatever is left. This arrangement affords wide flexibility in the management of the estate. The children have someone handle the money until the day when they are old enough to handle it themselves. An advantage of such a trust is that the child does not suddenly come into a lot of money and perhaps simply blow it all in a short period of time. A responsible adult is "managing" the assets for her or him.

Holographic wills

Though legal in only about half of the 50 states (Wolf, 1995), **holographic wills** (written in one's own handwriting) differ from other wills. Such wills do not require witnesses. The will must be signed by the testator and should be dated. Even in the states where holographic wills are recognized, probate judges are often reluctant to accept them. Such wills are rather unreliable because proving their authenticity is often difficult. If, through handwriting comparisons, the holographic will can be authenticated, there is the additional problem of proving that the tesator was in full possession of mental faculties when she or he signed the will. Also, the requirements for a holographic will vary among those states that accept them. Words that are part of a printed form or that have been typed and thus are not written in the testator's own hand may render the entire instrument invalid—it is then not wholly "written in one's own handwriting."

George Dickinson recalls an attorney speaking to his death and dying class a few years ago in Kentucky, where holographic wills are legal. He said that an elderly

"And my newish Cadillac, with no equity and only
57 more payments, goes to my trusted banker."

woman came to him rather upset that her sister had recently died intestate. She said that she and her sister were very close. The sister's two adult children, however, paid no attention to their mother—never contacted her and did not send even a card on her birthday or Mother's Day. Yet, because the sister died without a will, from the classification of heirs, the two children would take 100 percent from her estate—50 percent each (the sister's husband had predeceased her). The woman, rather casually, mentioned to the lawyer that her sister had written to her a few months before she died and stated that she wanted her to have everything she owned when she died. The attorney asked if she still had the letter. She produced the letter, written in her sister's own handwriting, signed, and dated by her sister. This letter in essence was a holographic will, which the lawyer took and successfully executed in court. The sister took all from the estate, and the two nonattentive children received nothing. Her sister had written a will—a holographic will—and indeed did *not* die intestate.

DEATH ACROSS CULTURES
Keep the Dead Content or Suffer the Consequences

The bodies of Zimbabwean soldiers killed in the Congolese civil war are being dug up to prevent their spirits from returning to haunt their families. It is believed that if the corpses of the soldiers killed fighting for the Democratic Republic of Congo are left in their graves or abandoned to decompose on the battlefield, their spirits will trouble their relatives. The number of soldiers estimated to have been killed ranges from a few dozen to a few hundred over a 10-month period. The spirit of the dead will worry the family as a way of punishing them, since the person was "thrown away in a foreign country." The spirit will cause problems "for many generations with sickness and death and other kinds of misfortune." To appease the spirits, the bodies will be exhumed for reburial at home. If the body itself cannot be located, soil from the area where the soldier died could be returned home.

From "Zimbabwe to Appease Spirits by Exhuming Soldiers in Congo" (p. 14), by Jan Raath, 1999, May 31, The London Times.

Probate

Probate ("to prove") is a legal process that takes place after someone dies. Probate involves proving in court that a deceased person's will is valid, identifying and inventorying the deceased person's property, having the property appraised, paying debts and taxes, and distributing the remaining property as the will directs (Probate FAQ, 1999). Probate generally involves paperwork and court appearances by attorneys. Probate can take up to 1 year to complete. A will must be probated to be effective. In many states, it is a crime to withhold a will from probate (Scheible, 1988).

After a death, the executor of the estate or administrator appointed by a judge files papers in the local probate court. The executor proves the validity of the will, presents the court with lists of the deceased person's property, lists the debts, and names who is to benefit from the will. Creditors are notified of the death either directly or through an ad in the local newspaper, requesting anyone to come forward who is owed money by the deceased person's estate. This may take several months. The executor may have to make a decision as to whether or not to sell some liquid assets to pay off debts.

Not all property goes through probate court. Most states allow a certain amount of property to pass free of probate. In addition, money from life insurance policies goes directly to the person(s) named in the policy—the beneficiary.

If one wishes to avoid probate court to save time and legal expenses, there are several ways to do this (Randolph, 1999). Examples include setting up "payable-on-death accounts" so that one's liquid assets (like cash) go directly to a person(s) or

---

PRACTICAL MATTERS
Federal Estate Tax Exemption Changes

Year of Death	Exempt Amount
2001	$675,000
2002, 2003	$700,000
2004	$850,000
2005	$950,000
2006 and after	$1,000,000

From Plan Your Estate, *by D. Clifford &*
C. Jordan, 1999, Berkeley, CA: Nolo Press.

---

institution. Or one could have a "transfer-on-death" to beneficiaries for stocks and bonds and vehicles. Or one could have property in joint tenancy with right of survivorship. Under this legal arrangement, two or more parties are given equal rights to the property during their lifetimes. When one party dies, that person's ownership rights dissolve, leaving nothing to pass through probate. The co-owner(s) becomes the complete owner of the property. If there are several co-owners, the last co-owner to survive becomes the complete owner.

Estate and Inheritance Taxes

Death taxes are imposed in two general forms: estate tax and inheritance tax. Estate taxes are imposed on the decedent's estate and are paid from the estate before the remaining assets are distributed to beneficiaries and heirs. Inheritance taxes are levied against individuals who receive property through inheritance (Scheible, 1988).

The federal government currently imposes estate tax at death only if the decedent's property is worth more than $670,000. However, all property left to the spouse is exempt from the tax, as long as the spouse is a U.S. citizen. In addition, federal estate tax will not be assessed on any property left to a tax-exempt charity (Clifford & Jordan, 1999). Beginning in the 21st century, the amounts exempt from taxation are increasing, as noted in the "Practical Matters" box.

New rules apply to family-owned businesses and farms, which may receive a special $1.3 million exclusion from estate taxes. To qualify for the exemption, the business or farm must constitute more than 50 percent of the estate and must be located primarily in the United States. Interest in the business or farm must be left to family members or individuals who have been actively employed by the business for at least 10 years before the death of the owner (Clifford & Jordan, 1999).

Federal estate tax rates start at 37 percent. The maximum is 55 percent for property worth more than $3 million (Clifford & Jordan, 1999). There are ways to

> ### WORDS OF WISDOM
> #### Because I Could Not Stop for Death
>
> Because I could not stop for Death—
>   He kindly stopped for me—
>   The Carriage held but just Ourselves—
>   And Immortality.
>
> We slowly drove—He knew no haste
>   And I had put away
>   My labour and my leisure too,
>   For His Civility—
>
> We passed the School, where Children strove
>   At Recess—in the Ring—
>   We passed the Fields of Gazing Grain—
>   We passed the Setting Sun—
>
> Or rather—He passed Us—
>   The Dews drew quivering and chill—
>   For only Gossamer, my Gown—
>   My Tippet—only Tulle—
>
> We paused before a House that seemed
>   A Swelling of the Ground—
>   The Roof was scarcely visible—
>   The Cornice—in the Ground—
>
> Since then—'tis Centuries—and yet
>   Feels shorter than the Day
>   I first surmised the Horses' Heads
>   Were toward Eternity—
>
> *Emily Dickinson*

avoid or save on federal estate taxes. The most popular method is used by married couples with children and is called an **AB trust** (sometimes called credit shelter trust, marital life estate trust, or marital bypass trust). Property is put in the trust. When one spouse dies, her or his half of the property goes to the children. The surviving spouse obtains the right to use the property for life and is entitled to any income generated from it. When the second spouse dies, the property goes to the children outright. Such an arrangement keeps the second spouse's taxable estate half the size it would otherwise have been, thus helping to avoid estate tax. Another way to save on estate taxes is to make a sizable gift to a tax-exempt charity. Also, while still living, an individual can give away up to $10,000 per year to any person or noncharitable institution without having to pay estate tax. For example, an individual could give each of her or his three children $10,000 a year for 4 years, and the $120,000 over 4 years would be free of estate taxes. In addition, the recipients would not have to pay any taxes on the $10,000 each year.

A few *states* have death taxes—inheritance and estate (Clifford & Jordan, 1999). Inheritance taxes are paid by those who inherit (the recipient). The taxes vary by state and according to the relationship of the two parties (e.g., Nebraska imposes a 15 percent tax if you leave money to a friend, but only 1 percent if left to a child). Currently, 17 states have inheritance taxes. State estate taxes are similar to federal estate taxes, but only Mississippi, New York, North Carolina, Ohio, and Oklahoma currently have such taxes (New York is currently phasing out its tax). In the other 45 states, the state takes part of the money paid to the federal government, but it does not increase the tax bill to the individual.

CONCLUSION

From the alpha to the omega of the life cycle—birth to death—a "certificate" is needed to legitimize one's existence and then to legitimize one's "lack of existence." One can't "just" be born and "just" die, it all must be recognized within our legal system. Likewise, at death one's possessions go to other individuals, and this process is laid out within the laws of the various states. If one does not want the state to determine where the possessions go, a will is a must. Also, if an individual desires to have a say in end-of-life issues, she or he should have an advance directive (living will and legal power of attorney for health care). Americans sometimes joke about how many lawyers we have in our society, but, given all the legal requirements accompanying our various moves (birthing to dying), perhaps they are needed.

SUMMARY

1. Autopsies are sometimes performed after death to aid in determining the cause of death.

2. Though now legal throughout the United States, advance directives are not popular, perhaps suggesting a reluctance to face death.

3. Medical examiners differ from coroners in that the latter usually do not have medical degrees. Both deal with trying to determine the causes of death in situations in which the cause of death is not known.

4. The death certificate states the reasons for the death of a person.

5. If one wishes to have any say in the disposition of one's property upon death and who would be legally responsible for the placement of minor children, he or she should have a will.

6. To die intestate (without a will) means that the state, via inheritance, will determine who receives one's property and also how minor children will be placed.

7. A holographic will (written in one's own handwriting) is very simple to make out, though sometimes difficult to process in court and not legal in some states.

8. Probate is a legal process that takes place after someone dies and involves proving in court that a deceased person's will is valid.

9. The federal government imposes an estate tax at death if the decedent's property is worth more than $670,000. The amount exempt from federal estate tax is gradually going up over the next few years.

10. Individuals who want to make sure their wishes for medical treatment are honored should prepare an advance directive.

DISCUSSION QUESTIONS

1. Why are fewer autopsies occurring in the United States today?

2. What dangers are inherent in writing one's own will?

3. Would you be interested in being a coroner? Why or why not?

4. Discuss some of the problems with determining the causes of death.

5. Discuss the importance of a death certificate.

6. Discuss the advantages of an advance directive (living will and/or durable power of attorney for health care).

7. Discuss why most individuals do *not* have advance directives.

8. What are the advantages of having a will versus dying intestate?

9. What are the criteria for making out a will?

10. Discuss who needs to make out a will.

GLOSSARY

AB Trust: Half of a couple's property placed in a trust. When one spouse dies, that half of the property goes to the children, but the surviving spouse gets the right to use the property for life. When the second spouse dies, the property goes to the children outright, thus reducing estate taxes by half.

Administrator: A person or institution (e.g., a bank) appointed by the court either to administer the estate of a person who died intestate or to administer the estate of a person who died with a will that does not properly name an executor or that names an executor who fails or refuses to qualify.

Advance Directives (ADs):   Legal documents stating an individual's wishes regarding end-of-life decisions such as a living will and durable power of attorney for health-care decisions.

Autopsy: A pathologist's medical examination of the organs of the dead body to determine the cause of death.

Beneficiary: One receiving benefit or advantage from a will, insurance policy, or trust.

Codicil: A supplemental modification of an existing will.

Coroner: A public official, appointed or elected, in a particular geographic jurisdiction, whose official duty is to make inquiry into deaths in certain categories.

Death Certificate: A permanent record of an individual's death.

Decedent: The deceased person.

Durable Power of Attorney (for Health Care):   A document that ascribes to a surrogate decision-maker the legal authority to consent (or refuse to consent) to medical treatment should the patient lack competency to make such decisions.

Escheat: Property of a decedent goes to the state if not disposed of by a will and if the decedent has no heirs.

Executor: A person or institution (e.g., a bank) named in the will to administer an estate—to pay creditors and taxes and to distribute the estate to the beneficiaries. In some states the phrase *personal representative* is used.

Five Wishes: A living will in the form of an eight-page questionnaire that guides people in making essential decisions about their care at the end of life.

Holographic will: A will written wholly in one's own handwriting that generally requires no witnesses.

Intestate: Describes a person who dies without a will.

Living Will: A document stating that an individual does not want medical intervention if the technology or treatment that keeps one alive cannot offer a reasonable quality of life or hope of recovery.

Medical Examiner: A physician charged to investigate and examine persons dying a sudden, unexpected, or violent death.

Pathologist: A physician who deals with the nature of disease, especially with the structural and functional changes caused by disease. A pathologist conducts autopsies to try to determine the cause(s) of death.

Probate: A court proceeding resulting in a judgment that, when recorded, vests the title to property passing under a will to the beneficiaries the same way a deed does. The will then becomes effective. The procedure is the legal process of proving that a will is valid.

Testator: A person who makes a will.

Will: A legal document in which a person states how he or she wants property and possessions distributed after death.

SUGGESTED READINGS

Appel, J. C., & Gentry, F. B. (1990). *The complete will kit.* New York: John Wiley. A self-help guide to steps for writing or amending a will with state-specific information on legal requirements and estate taxes.

Bennett, J. G. (1999). *Maximize your inheritance for widows, widowers and heirs.* Chicago: Dearborn Financial Publishing. A helpful guide to estate planning and dealing with monetary issues *after* the funeral, written by a financial planner.

Cassel, C. K. (Ed.). (1999). *The practical guide to aging: What everyone needs to know.* New York: New York University Press. Gives some very practical advice for estate planning and dealing with legal issues, whether one is younger or older than age 65.

Clifford, D., & Jordan, C. (1999). *Plan your estate.* Berkeley, CA: Nolo Press. A comprehensive estate planning tool that covers everything from simple wills to sophisticated tax savings trusts.

Dacey, N. F. (1990). *How to avoid probate.* New York: Macmillan. An exhaustive review of probate and trusts—largely a self-help book that discusses how to avoid probate.

Norrgard, L. E., & DeMars, J. (1992). *Final choices: Making end-of-life decisions.* Santa Barbara, CA: ABC-CLIO. Has a good directory, organized by state, of information regarding legal aspects of death.

Strauss, P. J., Wolf, R., & Shilling, D. (1990). *Aging and the law.* Chicago: Commerce Clearing House. Pertinent federal and state statutes and regulations regarding dying and death, among other issues. This lengthy volume was written for attorneys who specialize in the practice of law relating to the elderly.

Wiegold, C. F. (Ed.). (1997). *The Wall Street Journal lifetime guide to money: Everything you need to know about managing your finances—For every stage of life.* New York: Hyperion. A good, easy-to-follow guide for individuals of all ages that provides information on investing, taxes, insurance purchasing, and retirement planning.

Wolf, S. S. (1995). "Legal perspectives on planning for death." In H. Wass and R. A. Neimeyer (Eds.). *Dying: Facing the facts* (pp. 163-184). Washington, DC: Taylor and Francis, pp. 163-184. A nice summary by Sheryl Wolf, an attorney, on the legal aspects of planning for death in a book edited by two well-respected thanatologists, Hannolore Wass and Robert Neimeyer.

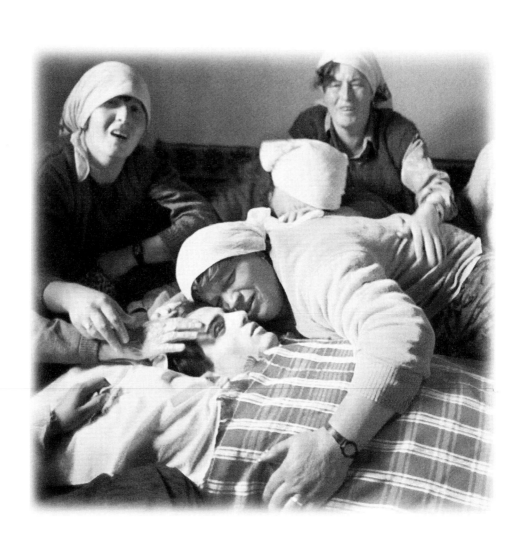

CHAPTER 13

COPING WITH LOSS

Only people who avoid love can avoid grief. The point is to learn from it and remain vulnerable to love.

—John Brantner (From J. William Worden,
Grief Counseling and Grief Therapy)

If at the conclusion of the funeral service grief work were finished, the process of reintegration of the bereaved into society would be completed. The funeral service and the final disposition of the dead, however, mark only the end of public mourning; private mourning continues for some time.

THE BEREAVEMENT ROLE

In earlier chapters we discussed bereavement behavior within historical and cross-cultural perspectives. We have given a general description of the norms and cultural patterns that prescribe proper conduct for the bereaved within American society. When we apply these bereavement norms to particular persons occupying statuses within a group or social situation, we are talking about the **bereavement role.**

As discussed in chapter 5, Talcott Parsons (1951) describes the sick role as being composed of rights and obligations. The first right of the sick person is to be exempted from "normal" social responsibilities. The extent to which one is exempted is contingent upon the nature and severity of the illness. The second right is to be taken care of and to become dependent on others as one attempts to return to normal social functioning. In exchange for these rights, the sick person must express a desire to "want to get better" and must seek technically competent help.

J. D. Robson (1977) suggests that behavior related to the death of a significant other (e.g., spouse or parent) is similar to illness behavior patterns. At the onset of death, bereaved persons are exempted from their normal social responsibilities. Depending on the nature and the degree of relationship with the deceased, the bereaved are awarded time away from employment in much the same way that they are given sick leave—spouses and children may be given a week, whereas close friends and relatives may be given time only to attend the funeral.

Bereaved individuals are also allowed to become dependent on others for social and emotional support and for assistance with tasks related to the requirements of normal, daily living. In offering this type of support, neighbors and friends call on survivors with gifts of food, flowers, and other expressions of sympathy. This custom led Robert Kavanaugh's younger brother to ask if "dead people ate meatloaf and chocolate cake" (Kavanaugh, 1972).

In exchange for these privileges created by the death of a loved one, those adopting the bereavement role not only are required to seek technically competent help from funeral directors and clergy members, but also are expected to return as soon as possible to normal social responsibilities. The bereavement role is considered a temporary one, and it is imperative that all role occupants do whatever is necessary to relinquish it within a reasonable period of time. Time extensions are usually granted to spouse and children, but there is a general American value judgment that normal grieving should be completed by the first anniversary of the death.

American folk wisdom would contend that "time heals"—with the intensity of the grief experienced diminishing over time. According to Paul Rosenblatt (1983), however, a more accurate notion of mourning is that there are fewer periods of intensity-felt grief with the passing of time. Furthermore, in analyzing 19th-century diaries, Rosenblatt discovered that it was quite common for a person to experience periods of mourning for losses that occurred many years earlier. What is abnormal behavior, from the perspective of the American bereavement role, is a person's being preoccupied with the death of the loved one and refusing to make attempts to return to normal social functioning. Examples of deviant behavior of this type would include the following:

1. Malingering in the bereavement role and memorializing the deceased by refusing to dispose of articles of clothing and personal effects and living as if one expected the dead person to reappear

2. Rejecting attempts from others who offer social and emotional support, refusing to seek professional counseling, and taking up permanent residence in "Pity City"

3. Rejecting public funeral rituals and requesting that the funeral functionaries merely pick up the body and dispose of it through cremation without any public acknowledgment of the death

4. Engaging in aberrant behavior such as heavy drug or alcohol usage

LISTENING TO THE VOICES
A Wednesday Afternoon

Jane Soli

The procession crept down Division Street. The neon sign at the Rock County Bank flashed 94 degrees and 3:06. Behind the plate glass of The River Inn, businessmen shook dice for their afternoon coffee. Lois Palmer came out of Hanson's Variety Store carrying a lamp shade.

Why did you die, Dan? Did you have to die when the children were so young? You were lean and fit, physically active, no apparent health problems. You were fine when you went to bed, but I heard your death rattle; a small cough, I thought. You were dead when I came to bed five minutes later.

Is there any money? Will Karen be able to finish her last year of college? How about Peter—he's only seventeen. And then Johanna and Tina, thirteen and eleven. At least Sig is out of school. Such problems. Such a loss. How will I ever manage?

Two things came back to me during those early days of worry, fear, and grief. I recalled twenty years of bridge games with Judy and Elmer and the late-night discussions of death and funerals as we ate brownies and drank coffee. We spoke of our own deaths and our personal wishes. Dan, always the most vocal, stated again and again, "Don't spend any money on me when I am dead," and "Give me a military funeral." There was not a man in the USA more proud to be a Marine than Dan Soli. To hear him tell it, the Marines fought World War II single-handedly. Though he was 33 and the others in boot camp were 18, he was a member of the United States Marine Corps and never mind if the recruits called him "Grandpa."

About one hundred yards into the cemetery the procession came to a stop. I saw the fake grass surrounding the open grave over by the fence. Close by stood the caretaker's tool shed with a barrel for refuse in front. Someone had lettered a crude sign above the barrel. It said, "RUBBAGE." The grave diggers stood behind the tool shed waiting to complete their job.

As the casket moved from the hearse to the waiting pallbearers, the honor guard, smart in their black uniforms, aligned themselves on either side of the men carrying the casket and accompanied them to the grave site.

The children and I left the car to follow. Tina cried, "I'm too young to be without a father," and clung to me. I had vowed to remain dry-eyed. My silly little purse held a lawn hanky—no Kleenex, no billfold, no lipstick. I would not open that purse.

(continued)

5. Rushing into major life changes such as hastily remarrying or moving to a new home

Behaviors such as these may be sanctioned by others through social avoidance, ostracism, and criticism. As a consequence, most people are not only encouraged but also forced to move through the grieving process.

(continued from previous page)

Friends and relatives surrounded the grave as Pastor Jensen gave the committal service. Somehow the next part took me by surprise. Two of the honor guards stepped forward and together removed the American flag from the casket. With ceremony they folded it in the traditional way and presented it to me—the widow.

The two uniformed men rejoined the honor guard and they all stepped back from the mourners. At the command of the captain they fired a twenty-one gun salute.

At that point Peter moved away from the family, raised his trumpet and blew Taps for his dad. Sweet, melancholy notes rose plaintively over the gathered crowd. From the other side of the hill came the faraway echo. My heart ached with the pain of that moment.

Later, as we were leaving, Alice Overbeck came over to me, reached for my hand and said, "Too bad your husband died, but you'll get over it."

She was wrong; I didn't get over it.

I cry when I hear Taps.

Jane Soli was the retired secretary to the academic dean of Saint Olaf College, Northfield, MN. The above account of her husband's death was written in 1989. Michael Leming read the above at Jane Soli's memorial service in 1999.

THE GRIEVING PROCESS

Grief is a very powerful emotion that is often triggered or stimulated by death. Thomas Attig makes an important distinction between grief and the grieving process. Although grief is an emotion that engenders feelings of helplessness and passivity, the process of grieving is a more complex coping process that presents challenges and opportunities for the griever and requires energy to be invested, tasks to be undertaken, and choices to be made (Attig, 1991).

Most people believe that grieving is a diseaselike and debilitating process that renders the individual passive and helpless. According to Attig (1991, p. 389):

> It is misleading and dangerous to mistake grief for the whole of the experience of the bereaved. It is misleading because the experience is far more complex, entailing diverse emotional, physical, intellectual, spiritual, and social impacts. It is dangerous because it is precisely this aspect of the experience of the bereaved that is potentially the most frustrating and debilitating.

Death ascribes to the griever a passive social position in the bereavement role. Grief is an emotion over which the individual has no control. However, understanding that grieving is an active coping process can restore to the griever a sense of autonomy in which the process is permeated with choice and there are many areas over which the griever does have some control. James Miller provides several options for grievers involved in active coping (see the "Practical Matters" box).

Normal and Abnormal Grief

Harvard psychiatrist Erich Lindemann (1944), over a half century ago, wrote an article in which he discussed "normal" and "abnormal" grief. Basically, he argued that normal (ordinary) grief will resolve on its own, whereas abnormal grief may require some help, either from a support group or a clinical psychologist or psychiatrist.

In a normal grieving process, bereaved individuals may have difficulty "letting go." They may have difficult and painful dreams about the deceased person. They may have delusions or assume habits of the individual. They may "see" the dead person or find themselves "talking" to her or him, turning to her or him while watching a television program and saying something, as they have done for perhaps decades. According to research by Glennys Howarth (2000), communication with a deceased person may occur during semiconsciousness or in dreams or may occur while one is fully conscious. Douglas Davies (1997, p. 154) recently reported that approximately 35 percent of the people contacted for his survey of 1,603 people in the United Kingdom "had gained some sense of the presence of the dead." Sensing the presence of a parent is the most common type of experience, followed by grandparents, then spouses. It is "difficult to let go" of something you have had for so much of your life.

It is also normal to try to avoid the pain of grief. Pain comes in waves. One may try to avoid reminders such as birthdays, religious holidays, or anniversaries. The survivor may avoid going to certain places where he or she often went before the individual died—a favorite restaurant, a walk on the beach, or a climb in the mountains. Tom Hanks in the movie *Sleepless in Seattle* moved from Chicago to Seattle to avoid those places he could not go since his wife had died. The opening scene of the movie has Hanks contemplating what to do. He decides to move— to avoid the pain of grief. He simply starts over in a new environment and thus will not be faced with trying to go to those places where he had memories of his deceased spouse.

PRACTICAL MATTERS
Grief Tips—Help for Those Who Mourn

Following are ideas to help people who are mourning a loved one's death. Treat this list as a gathering of assorted suggestions that various people have tried with success.

Talk regularly with a friend—Talking with another about what you think and feel is one of the best things you can do for yourself.

Walk—Go for walks outside every day if you can. If you like, walk with another.

Visit the grave—Not all people prefer to do this. But if it feels right to you, then do so. Don't let others convince you this is a morbid thing to do. Spend whatever time feels right there.

Plant something living as a memorial—Plant a flower, a bush, or a tree in memory of the one who died.

If you're alone, and if you like animals, get a pet—The attention and affection a pet provides may help you adapt to the loss of the attention and affection you're experiencing after this significant person has died.

Plan ahead for special days—Birthdays, anniversaries, holidays, and other special events can be difficult times, especially for the first year or two. Give thought beforehand to how you will handle those days. Do things a little differently than you used to, as a way of acknowledging this change in your life. But also be sure to invoke that person's presence and memory somehow during the day.

Allow yourself to laugh—Sometimes something funny will happen to you, just like it used to. You won't be desecrating your loved one's memory.

Allow yourself to cry—Crying goes naturally with grief. It may feel awkward to you, but this is not unusual for a person in your situation. A good rule of thumb is this: if you feel like crying, then cry.

Plan at least one thing you'll do each day—Even if your grief is very painful and your energy very low, plan to complete at least one thing each day, even if it's a small thing. Then follow through with your plan, day after day.

Keep a journal—Write out your thoughts and feelings. Do this whenever you feel the urge, but do it at least several times a week, if not several times a day.

Consider a support group—Spending time with a small group of people who have undergone a similar life experience can be very therapeutic. You can discover how natural your feelings are.

Vent your anger rather than hold it in—You may feel awkward being angry when you're grieving, but anger is a common reaction. Yell, if in an empty house. Cry. Hit something soft.

Give thanks every day—Whatever has happened to you, you still have things to be thankful for. Perhaps it's your memories, your remaining family, your support, your work, your own health—all sorts of things.

From The Caregivers Book:
Caring for Another, Caring for Yourself,
by James E. Miller, 1996, St. Paul, MN:
Augsburg Fortress Press.

WORDS OF WISDOM
The Brevity of Life

How short is life.
How soon comes death.

Epitaph on gravestone in Haworth, England.

A total absence of grief, on the other hand, is not normal. The survivors react as if nothing has happened. Life goes on. The television program *Benson* some years ago had an episode in which Benson's mother had come to visit him and died unexpectedly. Benson's reaction was to call the funeral home and have the body shipped back to her home for burial. Benson stayed there and went to work that day! His fellow employees did not know how to deal with him, because such behavior was "not normal." He was displaying a total absence of grief.

Though some degree of depression is normal, attempting suicide, or a prolonged inability to perform daily activities, is a cause for concern. Most individuals are able to return to work and get out in the community "within a few days." Though we may never "get over" the death of a significant other (like we "get over" the flu), we learn to live with the fact that the person is dead. We must accept the fact that that person will never again sit at her or his place at the breakfast table. "Life goes on," however, and we must move on ourselves.

Stages of Grief

The grieving process, like the dying process, is essentially a series of behaviors and attitudes related to coping with the stressful situation of a change in the status of a relationship. As discussed in chapter 5, many individuals have attempted to understand coping with dying as a series of universal, mutually exclusive, and linear stages. Not all people, however, will progress through the stages in the same manner. Seven behaviors and feelings that are part of the coping process are identified by Robert Kavanaugh (1972): shock and denial, disorganization, volatile emotions, guilt, loss and loneliness, relief, and reestablishment. It is not difficult to see similarities between these behaviors and Kübler-Ross's five stages (denial, anger, bargaining, depression, and acceptance) of the dying process. According to Kavanaugh (1972, p. 23), "these seven stages do not subscribe to the logic of the head as much as to the irrational tugs of the heart—the logic of need and permission."

Shock and Denial

Even when a significant other is expected to die, at the time of death there is often a sense in which the death is not real. For most of us our first response is, "No, this can't be true." With time, our experience of shock diminishes, but we find new ways to deny the reality of death.

Some believe that denial is dysfunctional behavior for those in bereavement. However, denial not only is a common experience among the newly bereaved but also serves positive functions in the process of adaptation. The main function of denial is to provide the bereaved with a "temporary safe place" from the ugly realities of a social world that offers only loneliness and pain.

With time, the meaning of loss tends to expand, and it may be impossible for one to deal with all of the social meanings of death at once. For example, if a man's wife dies, not only does he lose his spouse, but also his best friend, his sexual partner, the mother of his children, a source of income, and so on. Denial can protect an individual from some of the magnitude of this social loss, which may be unbearable at times. With denial, one can work through different aspects of loss over time.

Disorganization

Disorganization is the stage in the bereavement process in which one may feel totally out of touch with the reality of everyday life. Some go through the 2- to 3-day time period just before the funeral as if on "automatic pilot" or "in a daze." Nothing normal "makes sense," and they may feel that life has no inherent meaning. For some, death is perceived as preferable to life, which appears to be devoid of meaning.

This emotional response is also a normal experience for the newly bereaved. Confusion is normal for those whose social world has been disorganized through death. When Michael Leming's father died, his mother lost not only all of those things that one loses with a death of a spouse, but also her caregiving role—a social role and master status that had defined her identity in the 5 years that her husband lived with cancer. It is only natural to experience confusion and social disorganization when one's social identity has been destroyed.

Volatile Reactions

Whenever one's identity and social order face the possibility of destruction, there is a natural tendency to feel angry, frustrated, helpless, and/or hurt. The volatile reactions of terror, hatred, resentment, and jealousy are often experienced as emotional manifestations of these feelings. Grieving humans are sometimes more successful at masking their feelings in socially acceptable behaviors than other animals, whose instincts cause them to go into a fit of rage when their order is threatened by external forces. However apparently dissimilar, the internal emotional experience is similar.

In working with bereaved persons over the past 20 years, Michael Leming has observed that the following become objects of volatile grief reactions: God, medical personnel, funeral directors, other family members, in-laws, friends who have not experienced death in their families, and/or even the person who has died. Mild-mannered individuals may become raging and resentful persons when grieving. Some of these people have experienced physical symptoms such as migraine headaches, ulcers, neuropathy, and colitis as a result of living with these intense emotions.

WORDS OF WISDOM

It's funny the way most people love the dead. Once you are dead, you are made for life.

Jimi Hendrix, Rolling Stone, *December 2, 1976.*

The expression of anger seems more natural for men than expressing other feelings (Golden, 2000). Expressing anger requires taking a stand. This is quite different from the mechanics of sadness, where an open and vulnerable stance is more common. Men may find their grief through anger. Rage may suddenly become tears, as deep feelings trigger other deep feelings. This process is reversed with women, notes Golden. Many times a woman will be in tears, crying and crying, and state that she is angry.

As noted earlier, a person's anger during grief can range from being angry with the person who died to being angry with God, and all points in between. Golden's mentor, Father William Wendt, shared the story of his visits with a widow and his working with her on her grief. He noticed that many times when he arrived she was driving her car up and down the driveway. One day he asked her what she was doing. She proceeded to tell him that she had a ritual she used in dealing with her grief. She would come home, go to the living room, and get her recently deceased husband's ashes out of the urn on the mantle. She would take a very small amount and place them on the driveway. She then said, "It helps me to run over the son of a bitch every day." He concluded the story by saying, "Now that is good grief." It was "good" grief because it was this woman's way of connecting to and expressing the anger component of her grief.

Guilt

Guilt is similar to the emotional reactions discussed earlier. Guilt is anger and resentment turned in on oneself and often results in self-deprecation and depression. It typically manifests itself in statements like "If only I had . . .," "I should have . . .," "I could have done it differently . . .," and "Maybe I did the wrong thing." Guilt is a normal part of the bereavement process.

From a sociological perspective, guilt can become a social mechanism to resolve the **dissonance** that people feel when unable to explain why someone else's loved one has died. Rather than view death as something that can happen at any time to any one, people can **blame the victim** of bereavement and believe that the victim of bereavement was in some way responsible for the death—"If the individual had been a better parent, the child might not have been hit by the car," or "If I had been married to that person, I might also have committed suicide," or "No wonder that individual died of a heart attack, the spouse's cooking would give anyone high cholesterol." Therefore, bereaved persons are sometimes encouraged to feel guilt because they are subtly sanctioned by others' reactions.

WORDS OF WISDOM
My Wars Are Laid Away in Books

Emily Dickinson

My wars are laid away in Books—
   I have one Battle more—
   A Foe whom I have never seen
   But oft has scanned me o'er—
   And hesitated me between
   And others at his side,

But chose the best—Neglecting me—till
   All the rest have died—
   How sweet if I am not forgot

By Chums that passed away—
Since Playmates at threescore and ten
Are such a scarcity—

From The Poems of Emily Dickinson
(Reading Ed.), edited by R. W. Franklin,
1999, Cambridge, MA: The Belknap
Press of Harvard University Press.

Loss and Loneliness

Feelings of loss and loneliness creep in as denial subsides. The full experience of the loss does not hit all at once. It becomes more evident as bereaved individuals resume a social life without their loved one. They realize how much they needed and depended upon their significant other. Social situations in which we expected them always to be present seem different now that they are gone. Holiday celebrations are also diminished by their absence. In fact, for some, most of life takes on a "something's missing" feeling. This feeling was captured in the 1960s love song "End of the World."

> Why does the world go on turning?
> Why must the sea rush to shore?
> Don't they know it's the end of the world
> Cause you don't love me anymore?

Loss and loneliness are often transformed into depression and sadness, fed by feelings of self-pity. According to Kavanaugh (1972, p. 118), this effect is magnified by the fact that the dead loved one grows out of focus in memory—"an elf becomes a giant, a sinner becomes a saint because the grieving heart needs giants and saints to fill an expanding void." Even a formerly undesirable spouse, such as an alcoholic, is missed in a way that few can understand unless their own hearts are involved. This is a time in the grieving process when anybody is better than nobody, and being alone only adds to the curse of loss and loneliness (Kavanaugh, 1972).

Those who try to escape this experience will either turn to denial in an attempt to reject their feelings of loss or try to find surrogates—new friends at a bar, a quick remarriage, or a new pet. This escape can never be permanent, however, because loss and loneliness are a necessary part of the bereavement experience. According to Kavanaugh (1972, p. 119), the "ultimate goal in conquering loneliness" is to build a new independence or to find a new and equally viable relationship.

Relief

The experience of relief in the midst of the bereavement process may seem odd for some and add to their feelings of guilt. Michael Leming observed a friend's relief 6 months after her husband died. This older friend was the wife of a minister, and her whole life before he died was his ministry. With time, as she built a new world of social involvements and relationships of which he was not a part, she discovered a new independent person in herself whom she perceived was a better person than she had ever before been.

Relief can give rise to feelings of guilt. However, according to Kavanaugh (1972, p. 121): "The feeling of relief does not imply any criticism for the love we lost. Instead, it is a reflection of our need for ever deeper love, our quest for someone or something always better, our search for the infinite, that best and perfect love religious people name as God."

Reestablishment

As one moves toward reestablishment of a life without the deceased, it is obvious that the process involves extensive adjustment and time, especially if the relationship was meaningful. It is likely that one may have feelings of loneliness, guilt, and disorganization at the same time and that just when one may experience a sense of relief, something will happen to trigger a denial of the death. What facilitates bereavement and adjustment is fully experiencing each of these feelings as normal and realizing that it is hope (holding the grieving person together in fantasy at first) that will provide the promise of a new life filled with order, purpose, and meaning.

Reestablishment occurs gradually, and often we realize it has been achieved long after it has occurred. In some ways it is similar to Dorothy's realization at the end of *The Wizard of Oz*—she had always possessed the magic that could return her to Kansas. And, like Dorothy, we have to experience our loss before we really appreciate the joy of investing our lives again in new relationships.

Any time we experience a loss, especially a loss as significant as the death of someone close, it changes us. In most cases, we do not have control over the loss we have experienced. We do, however, have some control over the ways in which we respond to that loss. Some people grow through a grief experience while others seem to get stuck. How are such opposite results possible?

The answer to this question is closely related to how individuals grieve following their loss. Individuals who shut themselves off from the grief process also shut themselves off from the transformative experience which can result (Johnson, 1997). The transformative nature of the grief focuses not only on the process of "getting through" a time of sadness and loneliness immediately following the loss, but also on reconstructing one's life following the loss.

The transformative process which grief encourages includes three components that can best be understood by responding to the following three questions (Johnson, 1997): What have I lost? What do I have left? and What may be possible for me? Honestly answering these questions in response to one's situation may help facilitate a grieving process that shifts from limits to opportunities.

What have I lost? This question seeks to discover how extensive the loss is, that is, to clearly identify what has been lost. The grief process cannot begin or progress to completion until the loss, and secondary losses which accompany it, have been identified. Once the losses have been identified, and we believe we can go on, the healing process of grief has begun.

What do I have left? In responding to this question we allow the meaningful aspects of what remains in our lives to be recognized, remembered, and valued. Initially, and often depending upon our loss, it is possible to wonder if enough remains to even make our life worthwhile. No matter how small what is left may seem, it is enough to build upon. It is important to remember to be appreciative of what one has and go forward from there, not ungrateful for what one does not have or has lost.

What may be possible for me? Once we have determined what is left, we are able to move on and begin to determine what is possible, in spite of the significant loss that has occurred. Success in responding to this question is based on our perspective. Rather than looking at our limitations, our opportunities are viewed instead. By focusing on what is possible, we allow ourselves to discover new ways to relate, understand, create, and commit to an on-going process of renewal and discovery. The process of transformation takes time, and each question mentioned above deserves attention. Honestly answering each of them is a significant step toward a healthy grief experience.

Disenfranchised Grief

Like other aspects of death-related behavior, grief is socially constructed—people grieve only when they feel it is appropriate. Social scripts are provided for grievers, and social support is given to those who are recognized as having experienced loss and who act in accordance with the norms. However, not all loss is openly acknowledged, socially sanctioned, and publicly shared. Kenneth Doka (1989) uses the term **disenfranchised grief** when referring to phenomena of this type.

According to Doka (1989) four types of situations lead to disenfranchised grief. The first situation is one in which the relationship to the deceased is not socially recognized. Examples of this type of disenfranchised grief include nontraditional relationships—such as extramarital affairs, heterosexual cohabitations, and homosexual relationships (Doka, 1987). If outsiders are unaware that a relationship exists and the bereaved individuals are unable to publicly acknowledge their loss, they will not receive social support for their grief, and their bereavement will be problematic.

A second type of disenfranchised grief occurs when the loss is not acknowledged by others as being a genuine loss. An abortion or miscarriage is often deemed to be of lesser significance because the mother never had the opportunity to develop a face-to-face relationship with the child (Thornton, Robertson, & Mlecko, 1991). In the case of induced abortion, it is assumed that because the pregnancy was unwanted, grieving is unnecessary. According to Idell Kesselman (1990, p. 241), whatever one's position on abortion, "we must acknowledge that at least one death

The grave of a pet indicates the meaningful relationship that existed between the pet and its owner.

occurs—in addition to the fetus, there is often the death of youth, of innocence, of dreams, and of illusion." Kesselman maintains that women who have had an abortion need to express "unresolved feelings of loss" and to deal with "issues of death, loss, and separation." Kesselman concludes that grief therapy is a necessary part of abortion counseling.

Other examples of unacknowledged losses are the death of a pet companion and the death of a former spouse. According to Avery Weisman (1990–1991), the loss of a pet companion is often accompanied by intense grief and mourning but is seldom recognized by others as an important and authentic occasion for bereavement. Likewise, the death of a former spouse is rarely thought of as a legitimate loss because most people believe that grief work should be completed shortly after the divorce.

Related to the unacknowledged loss is the third type of disenfranchised grief, in which the grievers are unrecognized. The grief over the death of an adolescent peer or friend is rarely openly acknowledged and socially sanctioned. One of the reasons why wakes and visitation services often attract larger audiences than do funerals is that employers are increasingly unwilling to provide employees with time from work

person who is not a family member (Sklar & Hartley, 1990).
 evers are young children, mentally incompetent and/or re-
 erly adults. In each of these cases the bereavement needs of
 n ignored by most social audiences (Sklar & Hartley, 1990;
 9).

 senfranchised grief occurs when the death is not socially
 e of a death occurring in the act of a crime or when death
 utoerotic asphyxia (Thornton, Wittemore, & Robertson,
 90; Murphy & Perry, 1988). When people feel ambivalent,
 ortable about the cause of the death, they are generally un-
 de the social support needed by the bereaved.

 989), whenever disenfranchised grief occurs, the experi-
 and the normal sources of social support are lacking. Dis-
enfranchised grievers are usually barred from contact with the deceased during the dying process. They are also frequently excluded from funeral rituals as well as from care and support systems that may assist them in their bereavement. Finally, they may often experience many practical and legal difficulties after the death of their loved one. All of these circumstances intensify the problematic nature of bereavement for disenfranchised grievers.

Four Tasks of Mourning

In 1982 J. William Worden published *Grief Counseling and Grief Therapy*, which summarized the research conclusions of a National Institutes of Health study called the Omega Project (occasionally referred to as the Harvard Bereavement Study). Two of the more significant findings of this research, displaying the active nature of the grieving process, are that mourning is necessary for all persons who have experienced a loss through death and that four tasks of mourning must be accomplished before mourning can be completed and reestablishment can take place.

According to Worden (1982), unfinished grief tasks can impair further growth and development of the individual. Furthermore, the necessity of these tasks suggests that those in bereavement must attend to "grief work" because successful grief resolution is not automatic, as Kavanaugh's (1972) stages might imply. Each bereaved person must accomplish four necessary tasks: (1) accept the reality of the loss, (2) experience the pain of grief, (3) adjust to an environment in which the deceased person is missing, and (4) withdraw emotional energy and reinvest it in another relationship (Worden, 1982).

Accept the Reality of the Loss
Especially in situations when death is unexpected and/or the deceased lived far away, it is difficult to conceptualize the reality of the loss. The first task of mourning is to overcome the natural denial response and realize that the person is dead and will not return.

LISTENING TO THE VOICES
Don't Reduce My Grief to Something Logical and Universal

Grief discriminates against no one. It kills. Maims. And cripples. It is the ashes from which the phoenix rises, and the mettle of rebirth. It returns life to the living dead. It teaches that there is nothing absolutely true, or untrue. It assures the living that we know nothing for certain. It humbles. It shrouds. It blackens. It enlightens.

Grief will make a new person out of you, if it doesn't kill you in the making. People say "How are you?" and they have that look in their faces again. And I look at them like they're crazy. I think, how do you think I am? I'm not strong enough for talk. A woman said to me, in a comforting tone of conspiracy, "My mother had breast cancer and lost her right breast. . . ." "Oh," I say, "that's awful . . ." and wonder if that is supposed to be some sort of comparison. A tit, a husband— same thing? Strange how people want to join you in your intensity, your epiphany, by trying to relate—grabbing at straws of a so-called "like" experience while simultaneously they are repulsed by the agony. They see you on the street, driving by and pretending they don't see you. We all have our stuff to deal with. I suppose it's hard enough dealing with your own.

I can't tell them to shut up, not because I'm not rude by nature (it takes too much energy to be polite) but because I am afraid they'll go away. One more abandonment. "I lost my father two years ago . . ." and I think, schmuck, that's supposed to happen, it's a natural part of becoming adult—children are supposed to outlive their parents. Did you lose the only other person in the world who would love your child the way you do? Did you lose the person you held all night, who slept next to you, warmed your bed so much you didn't need an extra blanket in the winter? Do you know how many blankets it takes to replace a husband? Did you lose the person who would worry about bills with you? Screw in a light bulb when you were busy with the baby? Don't reduce this experience to something logical, universal. Even if it is, I walk alone amongst the dead, it's my death, my pain. Don't pretend you know it, like you know batting averages. Don't sacrilege all over my crucifixion.

From "The Agony of Grief"
(pp. 75–78), by S. Ericsson, 1991,
September–October, Utne Reader.

Bereaved persons can facilitate the actualization of death in many ways. The traditional ways are to view the body, attend the funeral and committal services, and visit the place of final disposition. The following is a partial list of additional activities that can assist in making death real for grieving persons.

1. View the body at the place of death before preparation by the funeral director.

2. Talk about the deceased person and the circumstances surrounding the death.

3. View photographs and personal effects of the deceased person.

4. Distribute the possessions of the deceased person among relatives and friends.

Experience the Pain of Grief

Part of coming to grips with the reality of death is experiencing the emotional and physical pain caused by the loss. Many people in the denial stage of grieving attempt to avoid pain by choosing to reject the emotions and feelings that they are experiencing. As discussed by Erich Lindemann (1944), some do this by avoiding places and circumstances that remind them of the deceased. Michael Leming knows one widow who quit playing golf and quit eating at a particular restaurant because these were activities that she had enjoyed with her husband. Another widow found it extremely painful to be with her dead husband's twin, even though he and her sister-in-law were her most supportive friends.

Worden (1982, pp. 13–14) cites the following case study to illustrate the performance of this task of mourning:

> One young woman minimized her loss by believing her brother was out of his dark place and into a better place after his suicide. This might not have been true, but it kept her from feeling her intense anger at him for leaving her. In treatment, when she first allowed herself to feel anger, she said, "I'm angry with his behavior and not him!" Finally she was able to acknowledge this anger directly.

The problem with the avoidance strategy is that people cannot escape the pain associated with mourning. According to Bowlby (cited by Worden, 1982, p. 14), "Sooner or later, some of those who avoid all conscious grieving, break down—usually with some form of depression." Tears can afford cleansing for wounds created by loss, and fully experiencing the pain ultimately provides wonderful relief to those who suffer while eliminating long-term chronic grief.

Assume New Social Roles

The third task, practical in nature, requires the griever to take on some of the social roles performed by the deceased person or to find others who will. According to Worden (1982), to abort this task is to become helpless by refusing to develop the skills necessary in daily living and by ultimately withdrawing from life.

An acquaintance of Michael Leming's refused to adjust to the social environment in which she found herself after the death of her husband. He was her business partner, as well as her best and only friend. After 30 years of marriage, they had no children, and she had no close relatives. She had never learned to drive a car. Her entire social world had been controlled by her former husband. Three weeks after his funeral she went into the basement and committed suicide.

The alternative to withdrawing is assuming new social roles by taking on additional responsibilities. Extended families who always gathered at Grandma's house for Thanksgiving will be tempted to have a number of small Thanksgiving dinners at different places after her death. The members of this family may believe that "no one can take Grandma's place." Although this may be true, members of the extended family will grieve better if someone else is willing to do Grandma's work, enabling the entire family to come together for Thanksgiving. Not to do so will cause double pain—the family will not gather, and Grandma will still be missed.

Reinvest in New Relationships

The final task of mourning is a difficult one for many because they feel disloyal or unfaithful in withdrawing emotional energy from their dead loved one. One of Michael Leming's family members once said that she could never love another man after her husband died. His twice-widowed aunt responded, "I once felt like that, but I now consider myself to be fortunate to have been married to two of the best men in the world."

Other people find themselves unable to reinvest in new relationships because they are unwilling to experience again the pain caused by loss. The quotation from John Brantner at the beginning of this chapter provides perspective on this problem: "Only people who avoid love can avoid grief. The point is to learn from it and remain vulnerable to love."

However, those who are able to withdraw emotional energy and reinvest it in other relationships find the possibility of a newly established social life. Kavanaugh (1972, pp. 122–123) depicts this situation well with the following description:

> At this point fantasies fade into constructive efforts to reach out and build anew. The phone is answered more quickly, the door as well, and meetings seem important, invitations are treasured and any social gathering becomes an opportunity rather than a curse. Mementos of the past are put away for occasional family gatherings. New clothes and new places promise dreams instead of only fears. Old friends are important for encouragement and permission to rebuild one's life. New friends can offer realistic opportunities for coming out from under the grieving mantle. With newly acquired friends, one is not a widow, widower, or survivor—just a person. Life begins again at the point of new friendships. All the rest is of yesterday, buried, unimportant to the now and tomorrow.

ASSISTING THE BEREAVED

In his book *Bereavement: Studies of Grief in Adult Life*, Colin Parkes (1972) notes that the funeral often precedes the "peak of the pangs" of grief that tends to be reached in the second week of bereavement. The face put on for the funeral can no longer be maintained, and the bereaved individuals have a need to be freed to grieve. The most valued person at this time is the one making few demands on the bereaved person, quietly completing household tasks, and accepting the bereaved person's vented anguish and anger—some of which may be directed against the helper.

It is important to recognize that the bereaved person has painful and difficult tasks to perform that cannot be avoided or rushed. The best assistance that one can offer to those in grieving is to encourage them to attend to the four tasks of mourning described earlier.

One can help bereaved individuals come to grips with the reality of the death by talking about the deceased person and encouraging them to conceptualize the loss. Parkes observes that it is often reassuring to the bereaved person when others show that they are not afraid to express feelings of sadness. Such expressions make the bereaved person feel understood, help reduce a sense of isolation, and help them to express their own feelings of sadness.

It is not uncommon for one approaching a newly bereaved person to be unsure how to react. Parkes suggests that although a conventional expression of sympathy can probably not be avoided, pity is the last thing that the bereaved person wants. Pity makes one into an object—the bereaved person somehow becomes pitiful. Pity puts the bereaved person at a distance from, and in an inferior position to, the intended comforter. Parkes maintains that it is best to get conventional verbal expressions of sympathy over as soon as possible and to speak from the heart or not at all. There is not a proper thing to say at this time; a trite formula serves only to widen the gap between the two persons.

The encounter between the bereaved person and the helper may not seem satisfactory because the helper cannot bring back the deceased individual and the bereaved person cannot gratify the helper by seeming to be helped. Bereaved people do, however, appreciate the visits and expressions of sympathy paid by others. These tributes to the dead loved one confirm to the bereaved individual the belief that the deceased person is worth all of the pain. Bereaved individuals are also reassured that they are not alone and feel less insecure.

The Hallmark Card Company sells its products with the slogan, "When you care enough to send the very best." Although sending a card is one way to validate the loss of the bereaved individuals and to assist them in their grieving, if you really care enough to send the very best, a personal visit or a handwritten letter is better. On the other hand, if you find it impossible to communicate a personal expression of sympathy, sending a card is preferable to doing nothing. In the "Practical Matters" box entitled "The Art of Condolence," seven suggestions are given for expressing condolence.

Although many bereaved people are frightened and surprised by the intensity of their emotions, reassurance that they are not going mad and that this is a perfectly natural behavior can be an important contribution of the helper (Parkes, 1972). On the other hand, absence of grief in a situation where expected, excessive guilt feelings or anger, or lasting physical symptoms should be taken as signs that not all is going well. These persons may require special help, and the caregiver should not hesitate to advise the bereaved person to get additional help if the caregiver is uncertain about the course of events.

Support groups empower persons to cope with social crisis and loss in a dealienating environment. According to Louis LaGrand (1991, p. 212), the power of support groups lies in the "introduction of new ways of looking at one's problems and the development of belief systems that enhance the twin attributes of self-determination and interdependence, all of this accruing in a socially secure setting."

The grieving process often socially isolates the bereaved by marginalizing them from normal patterns of social interaction. Bereavement self-help groups, by their very

PRACTICAL MATTERS
The Art of Condolence

Based upon their study of thousands of condolence letters and analysis of their structure, Leonard and Hillary Zunin (1991) share seven components that provide the writer with a practical, simple, and clear outline for sharing his or her thoughts. Though all seven components need not be included in every letter of condolence, keeping them in mind will provide an effective guide for sharing thoughts in such a letter. The components include the following:

1. **Acknowledge the loss.** Mention the deceased by name and indicate how you learned about the death. It is very acceptable to relate personal shock and dismay at hearing such news, and such an acknowledgment sets the tone for your letter.

2. **Express your sympathy.** Share your sorrow in an honest and sincere fashion. In so doing you are showing that you care and relate in some way to the difficult situation they are facing. Do not hesitate to use the words *died* or *death* in your comments.

3. **Note special qualities of the deceased.** As you reflect upon the individual who has died, think about those characteristics you valued most in that person and share them in your letter. These may be specific attributes, personality characteristics, or other qualities. Sharing these with the bereaved help them realize that their loved one was appreciated by others.

4. **Recount a memory about the deceased.** At the time of a death, memories we have of the deceased person are a most valued possession and something which

can never be in too great a supply. Because the bereaved often have difficulty keeping those memories in the forefront of their thoughts, your sharing of memories will be very gratefully received. Feel free to recall humorous incidents as they can be very beneficial at this time.

5. **Note special qualities of the bereaved.** Grieving people also need to be reminded of their personal strengths and other positive qualities—those characteristics which will help them through this difficult time. By reminding them of the qualities you have observed in them, you will be encouraging them to use these qualities to their advantage at this time.

6. **Offer assistance.** Offering help need not be part of a condolence letter, but if help is offered, it should be for a specific thing. An open-ended offer of help places the burden for determining what that help will be on the bereaved who have enough burdens already. Making an offer to do something specific and then doing it is a most welcome extension of yourself.

7. **Close with a thoughtful word or phrase.** The finals words in your letter of condolence are especially important and should reflect your true feelings. Flowery or elaborate phrases do not help. Honest expressions of your thoughts and feelings communicate best.

From "The Art of Condolence"
(pp. 3–4), by P. V. Johnson, 1991,
Fall, Caregivers Quarterly, *6(3).*

nature, challenge the assumption that grievers are unique by confronting them with caring people who understand their experience of loss and are willing to practice patterns of coping that reduce feelings of despair and depression (LaGrand, 1991).

Some national support groups for grievers are Compassionate Friends, Candlelighters, Empty Arms, Begin Again, Widow to Widow Program, the Omega Project, Parents Without Partners, and various hospice programs and/or hospital grief therapy groups. Another resource is the Internet or World Wide Web. There are chat rooms for grievers, and one can do grief work by creating memorial tributes to the person who has died.

COPING WITH VIOLENT DEATH

Deaths due to **acute diseases** (e.g., pneumonia), accidents, disasters, murder, war, and suicide provide special problems as well as advantages for survivors. For all quick deaths there is the problem of being unprepared for the death. The survivors did not have a support group in place and did not have the advantage of having experienced anticipatory grief. Some of the grieving that has preceded a death caused by a chronic disease cannot be expressed in deaths of this type. Consequently, grief is usually more intense when the dying takes place in a short period of time. Survivors may also experience more intense guilt: "If only I had done something, she wouldn't have died." Suicide creates special problems for survivors because they can become stigmatized by having a relative commit suicide: "They drove him to it." Finally, when people die without warning, survivors often are troubled because they did not have a chance to mend a broken relationship or say goodbye.

Michael Leming recently gave a eulogy for a man who had died from a drug overdose. The man's wife wanted the cause of death to be mentioned, but the mother and brother of the man protested. The brother became so angry that he threatened Leming with bodily harm if he mentioned that the death was drug related. The "compromise" was to have a second service for friends, when most of the family members would have gone, at which time Leming would deliver the eulogy.

HI AND LOIS REPRINTED WITH SPECIAL PERMISSION OF KING FEATURES SYNDICATE, INC.

The irate brother actually stayed in the wings of the church for the second service and listened. After the eulogy, he came up to Leming and apologized for his anger and inappropriate behavior. Such volatile emotions are a normal part of the grieving process, as Leming told him. Studies (Walter, 1999) of bereavement clearly indicate that grieving people do not act rationally. Like falling in love and giving birth, grief temporarily makes people say and do nutty things.

Though various problems evolve from sudden deaths, survivors of sudden deaths are spared the following problems associated with **chronic diseases:**

1. Dying persons may not be willing to accept death, and when learning of their fates, may act in unacceptable ways.

2. Families may be unwilling to accept the death.

3. The dying process may be a long and painful process, not only for the dying patient, but also for the family.

4. The cost of dying from a chronic disease can be, and usually is, very expensive. The entire assets of a family can be wiped out by the medical bills of a chronically ill patient.

Because we have limited choices in the matter of how we die (unless we intervene), we will simply have to take whatever comes along. Who knows? Perhaps death will come when we are very old, at home, unexpectedly, in our own beds, while sleeping—and with our full mental and physical capabilities. But we may not die "the all-American way" at home, as described in the previous sentence or slowly in an institution. Death may be sudden and violent from an accident, a disaster, murder, or war.

Accidents

Accidents occur in disproportionately higher rates among younger individuals and are the leading cause of death among all persons from birth to age 24 years (Fingerhut & Warner, 1997). As Therese Rando (1993) observes, a high correlation exists between accidental death and violence, mutilation, and destruction. Such events leave survivors with losses that are often more shocking and difficult to assimilate. In addition, traumatic memories of the event or imagined scenes of what happened produce problems for the bereaved individuals. Guilt may be greater than usual as a result of the strong need to assign blame and responsibility in accidental deaths. For example, "*If only* I had not asked her to go to the store for the milk, this never would have happened," a parent may say. Such guilt can get in the way of getting on with bereavement, suggests Rando. There is often a need on the part of survivors to determine how much the individual suffered at the end; thus, they have a desire to talk to others who witnessed the event. The survivors want some reassurance that she or he did not suffer at the end.

The issue of unfairness and injustice often must be addressed with accidental deaths. The fact that the individual died because she or he was in the wrong place at the wrong time is difficult to accept. Such an "injustice" makes getting on with her or his own life somewhat difficult for the survivor and often grates on her or him day and night. Anger may be a reaction for survivors *if* the deceased person contributed in any way to her of his death (e.g., drinking and driving). The survivor may be angry because the deceased person caused the losses in part and contributed to a most painful experience for family and friends (i.e., "How could you do this to us?" kind of reaction). Bereavement in accidental deaths can also be complicated by the media continually replaying the accident. This makes it difficult for bereaved individuals to "get away from" the accident, notes Rando.

What part do the media play in violent deaths? Friedrich Nietzsche, writing over a century ago in 1882, had this to say about the media and violent deaths (cited in Guignon, 1993, p. 290):

> Our age is an agitated one, and precisely for that reason not an age of passion; it heats itself up constantly because it feels that it is not warm—basically it is freezing. . . . In our times it is merely by means of an echo that events acquire their "greatness"—the echo of a newspaper.

William Merrin (1999) suggests that Nietzsche provides a powerful image of a cold public looking to its media to fill an emptiness—an image of an era trying to warm up by overheating its events in order to feel alive, if only temporarily. Merrin looks at the automobile accident that took the life of Princess Diana and how the media responded. The media are not a mirror reflecting the world but are themselves inseparable from the event. Today, the event and its broadcast have become a single phenomenon—*the media event.*

The media may cover violent deaths for several days; then the coverage begins to wane. Such withdrawal does not mean that those grieving have completed their task, but the media's part in the situation is over. The media may return on the anniversary of certain tragic events, but the response may or may not be there. For example, on the first anniversary of the death of Princess Diana a "Remembrance Walk" was planned and was expected to attract 15,000 people—300 came (Merrin, 1999). The accident that took Princess Diana's life attracted media attention to the extent that it did because of who she was. However, the extensive media coverage then attracted millions of viewers and readers. The media do contribute to highlighting violent deaths and to some extent play them up beyond what they are. British theologian Douglas Davies (1999) observes that the "Diana event" was, primarily, a media event, yet with a difference. The difference consisted of two elements—the active and practical participation of millions of people, on the one hand, and the idea that the media were, maybe, responsible for the death on the other hand. The media were reporting on something for which many thought them responsible.

In trying to make some sense out of the "Diana event," Davies notes the loss of a highly charged symbol of social meaning—Diana was a condensed symbol of love, divorce, fragmented families, stardom and failure, beauty and eating disorders, and also of the marginalized groups of AIDS, leprosy, and land mine victims. Diana possessed the charisma that only certain royals have—a regality that sets them apart, along with the ability to connect with ordinary people. Through her the ordinary person could connect symbolically to society itself. Those who actually met and touched her could connect physically.

Disasters

Some common threads cut across deaths caused by disasters, according to B. Raphael (1986). Identification of the dead bodies may cause problems contributing to complications for the bereaved. Dismemberment may make it difficult to put remains into a human form. When remains are not distinguishable (e.g., charred or mutilated), survivors are left with fear about the suffering that must have accompanied the death. Though one may feel traumatized at the thought of viewing the body, Raphael cites numerous studies which indicate that bereaved individuals who view the body do not regret it. Yet, family members are often advised not to see the body.

If unable to see the body of the deceased person, family members must deal with uncertainty. "Is that really her or him who was identified?" they may ask. Perhaps the individual is still alive. Thus, an element of hope may appear through the uncertainty. Such may have been the case, at least for a few days, for family members and friends with the sinking of the *Titanic* in April 1912. Nonetheless, a memorial service for the victims of the sinking took place at St. Paul's Cathedral, London, on April 19, 1912, with many thousands unable to get into the church (Davies, 1999). There were soldiers and sailors who had heard the playing of the "Dead March" for dead comrades, yet they too stood erect, with tears streaming down their faces, not ashamed of tears. The outpouring of grief was overwhelming, according to London papers.

If multiple deaths occurred in the disaster, not only is a survivor suffering from several deaths, but the number of significant others who could have offered social support is reduced. Thus, the whole issue is compounded. Also, families might be separated because of the disaster, adding to uncertainty. Problems resulting from food and water shortages or contamination and unsanitary conditions and poor medical care may result. Mobility may be impaired due to torn up streets and highways, thus simply getting around may be difficult and add to a survivor's dilemma.

Murder

When the murder of a child is discussed, the child is described as *the victim*. Yet, the parents are also victims (Conrad, 1998). They are not dead, but they have been injured—permanent changes have been brought about in their lives. Permanent voids are created in their lives. They think of the future *with dread*. They may lose interest in activities and events that once gave purpose to their lives. They may lose their faith in the "goodness of mankind" and their trust in the world—they feel betrayed.

LISTENING TO THE VOICES
The Aftermath of a Storm

After Hurricane Hugo, with sustained winds of 130 miles per hour and gusts up to 165 miles per hour, whipped through Charleston, South Carolina, in September of 1989, George Dickinson recalls the aftermath of the storm.

The eerie sound of silence (no birds chirping, no cars moving on the streets) leaves one with a funny sensation. With the devastation all around (houses destroyed, fallen trees everywhere, cars caved in) and the silence that follows the storm, one stands in awe of nature. There is concern for survivors of those killed and for those injured. Funerals were put "on hold" until the city had some semblance of order again. Thus, the process of grief is delayed while individuals spend time and energy stopping their leaking roof (if they still had one) and, in many cases, seeking water and food. Survivors spent the next several days and weeks simply getting their own house back in a livable fashion before they could really concentrate on "burying the dead." "Life goes on," and an individual must take care of immediate needs *before* disposing of the dead.

During the storm, emergency personnel had the frustration of determining which distress call was most serious. In one case the fire department went out during the eye of the storm to rescue an elderly man whose house had fallen in, leaving him trapped under the debris. Fire department personnel worked feverishly to remove him before the eye had passed and the storm resumed. The chief had to make a decision. Time was running out, and they were not going to have the man out before the eye had passed. They could hear the man's fee-

ble calls for help. Should the chief continue and risk the lives of some of his crew or stop the rescue efforts? He opted to withdraw his crew. When the storm was finally over, they returned to the scene to find the man dead. Such dilemmas are not everyday occurrences for rescue personnel. One's decision in such a situation will probably stay with him or her throughout a lifetime—aftereffects of a disaster.

In the aftermath of the hurricane, individuals were generally cooperative, caring, and generous. Neighbors who did not previously know each other helped clean debris from and within houses. They shared what was in their freezers and refrigerators (with no electricity for 2 weeks, it would soon be ruined anyway). Then as day turned into night, absolute calm would set in—no lights anywhere that the eye could see, only light from the stars and moon above—the eerie calm of nightfall.

A natural disaster such as a hurricane leaves scars on individuals for a long time. Months and years later, when the wind picks up, perhaps even on a perfectly cloudless day, it sends reminders of the hurricane. The grief of those who lost significant others was indeed eased by the love and concern shown by others during the crisis. A feeling of social solidarity and unity prevailed. Life would never be quite the same for the survivors. Not only did many have a death in the family, but for them and others, personal treasures were also lost (e.g., pictures, family furniture, pets), never to be replaced. Thus, grieving after a disaster encompasses more than for lost life—lost personal treasures also leave a void in an individual's future.

A natural disaster such as a hurricane destroys human lives, physical property, and personal possessions. Thus, there is grieving because of lost life but also because of lost treasures.

When a child is murdered, the initial problem is to tell the parents, a job not cherished by police, yet it is their job. Joan Schmidt (1986) suggests that one giving such "bad news" be short and to the point: "I have bad news; your child has been hurt; your child is dead." Because the recipient of the news will probably be in shock and will disbelieve immediately after receiving the news, then is the time to add the simplest of details: the child was murdered, where, perhaps briefly how, and where the child is now.

If the murderer is known, the family may have divided alliances, conflict, and anger (Olson, 1997). Guilt is evident in most cases of family violence that ends in homicide. If the killer is unknown, relatives may be in fear for their own lives. Coping for these survivors begins when they find out about the loss. They may need to spend time with the body. If it is impossible for them to see the body, just being in the same building or area where the body was may be helpful. These survivors need information about legal issues, such as an autopsy, and when results will be known.

As survivors of a murdered individual, there are certain enemies, observes Therese Rando (1986). Aside from the murderer(s), there are lawyers and courts, the media, and well-meaning "imbeciles." Swift justice does not often exist, and an attorney can many times find a loophole or a judge can grant a stay or overturn a conviction or give a delay. Friends may comment, "The trial is still going on?"

WORDS OF WISDOM

Circumstances rule men. Men do not rule circumstances.

Herodotus.

Life after the homicide of a significant other involves numerous practical and concrete changes developing out of the survivors' fear (Rando, 1993). Changes may be abstract (e.g., paralyzing fear) or concrete (e.g., keeping a gun for protection). Also, death imagery becomes intense and frightening for survivors who did not witness the trauma directly or view the body. Survivors are compelled to work through an internalized fantasy of grotesque dying that increases fears and complicates mourning by presenting the task of assimilating the violence and transgression implicit in this type of death, notes Rando.

The bereaved person's perception of the death as preventable is a high-risk factor for complicated mourning no matter the circumstance. If the death is construed to have been preventable, anger becomes a central effect of the perception of this preventability, states Rando (1993). It is easy to assign blame and thus punish one's self or others. The outrage and indignation at the ultimate arrogance of someone who assumes the right to exterminate another human being contributes to a bereaved person's frustrations.

Assuming the murder is solved and the convicted murderer goes to prison (if not given the death penalty), the stress for the family is not over. "Life in prison" does not always literally mean that. For example, the convicted murderer may come up for parole within a few years. Such a situation gives the murderer of the family's child a chance to be free and produces fear that the murderer may strike again and cause grief to yet another family. A recent case in South Carolina has the murderer, though serving seven life sentences plus 355 years in prison, up for parole on an annual basis after serving only 17 of those years. The 78-year old mother and family of the murdered woman every year gather 8,000 signatures and take them to the parole board to encourage them to keep the convicted murderer behind bars. Thus, on an annual basis, the murder/rape of this elderly woman's daughter will surface. There are many victims of a single murder—the family continues to grieve and suffer.

War

In war, an individual faces huge numbers of casualties, a threat to self and country, and the violence of modern weapons—all leading to a massive threat of death and the destruction of large numbers of individuals (Raphael, 1983). War deaths typically have all the risk factors and issues found in deaths caused by accidents, disasters, and homicides. Other factors include (Rando, 1993): the service person's and survivor's philosophical agreement with war in general, the type of death and

In memory of all who
died or fought in the
BURMA CAMPAIGN 1941·45

When you go home
Tell them of us. and say
For your tomorrow
We gave our today

ERECTED BY THE MEMBERS OF THE DERBY & DISTRICT BRANCH 1979

As noted from this plaque on a church in Derby, England, we need to be mindful of those who gave their lives so that we could live in a free society.

circumstances, the stress of separation experienced by the survivor while the service person was deployed, and the extent of social support available.

For combat soldiers, external group demands and strong group cohesion keep them going. They interpret combat as a sequence of demands to be responded to by precise military performances (Bourne, 1970). Yet, some soldiers display long-lasting effects from combat (posttraumatic stress disorders) and may later have feelings of anger and guilt. A friend of George Dickinson's, a helicopter paramedic in Vietnam, was in a helicopter crash. He was the lone survivor out of a crew of eight. He felt guilty and often asked himself, "Why me? Why should I survive when everyone else died?"

Some bereaved individuals may experience postdeath reactions similar to those seen after the loss of a family member from a lingering illness, depending on the length and nature of the stress of the separation and the degree to which the service person's death is anticipated (Rando, 1993). Nonetheless, such an encounter with

---

DEATH ACROSS CULTURES
The English Way of Grief

Noted British sociologist Tony Walter writes about the emotional reserve and the English way of grief. He says that in England the bereaved are not supposed to weep and hug each other, especially men—except on the soccer field. In England the "style of mourning" is really a matter of individual choice. However, when given total freedom, most individuals do not know how to express grief and look around anxiously to see how others are reacting.

The British feel most comfortable grieving in private. However, it is important that others know you are grieving, showing that you have cared for the deceased. Thus, one must provide clues of deep-seated feelings without actually breaking down in public and embarrassing others—not an easy task. The norm of private grief makes public appearances problematic; thus the funeral can be dreaded.

Over the past two to three decades, absent and private grief have been criticized by expressivists because it short-circuits the process of working through a range of feelings. However, the British have not immediately taken to this new way of grief, notes Walter. The English are now encouraged to cry, yet are aware that weeping spontaneously in the supermarket or outside the school gates will disturb others. Also, at the funeral they are encouraged to express their emotions today, yet uncontrollable emotion at funerals is frowned upon. Walter concludes that to talk of an English way of grief is not to claim that it is unchanging or unproblematic.

From "Emotional Reserve and the English Way of Grief," by Tony Walter, 1997. In K. Charmaz, G. Howarth, & A. Kellehear (Eds.), The Unknown Country: Death in Australia, Britain, and the USA, *London: Macmillan Press.*

---

death, violence, and mutilation often has profound and long-lasting effects: sights, sounds, and smells may linger for a lifetime. Without doubt, the death of a family member or friend in war is an unusual mixture of both sudden and anticipated loss.

CONCLUSION

In this chapter we have discussed the grieving process, bereavement roles, normal adaptations to experiences of loss, the four tasks of mourning, and coping with violent deaths. We have attempted to demonstrate that bereaved individuals must attend to grief work because successful grief resolution is not automatic. This grief work refers to Worden's (1982) four necessary tasks of bereavement: accepting the reality of the loss, experiencing the pain of grief, adjusting to an environment in which the deceased is missing, and withdrawing emotional energy and reinvesting it in other relationships.

Grievers need support and assistance in the bereavement process. One can help grievers come to grips with the reality of the death by talking about the deceased person and encouraging them to conceptualize the loss that they are experiencing.

It is often reassuring to the bereaved individual when others show that they are not afraid to express feelings of sadness. Such expressions make bereaved persons feel understood and reduce their sense of isolation. Many well-meaning friends find it difficult to find proper words to say. Unfortunately, their discomfort often causes them to do nothing. Even trite words are better than no words at all. A Hallmark card company slogan says, "When you care enough to send the very best" (send one of our cards). When comforting bereaved friends, however, a better personal slogan might be, "When you care enough to send the very best, send yourself." At this time, the very best that we have to offer is our own caring presence.

SUMMARY

1. The bereavement role is a temporary role that gives one the right to be exempted from normal social responsibilities and to become dependent upon others.

2. Abnormal bereavement behavior includes preoccupation with the death of the loved one and refusal to attempt to return to normal social functioning.

3. Disenfranchised grief involves loss that cannot be openly acknowledged, socially sanctioned, and publicly shared. There are four circumstances that lead to disenfranchised grief: (1) when the relationship to the deceased is not socially recognized; (2) when the loss is not acknowledged by others as being genuine loss; (3) when the grievers are unrecognized; and (4) when the death is not socially sanctioned.

4. The grieving process is similar to the dying process in that it is a series of behaviors and attitudes related to coping with the stressful situation of a change in the status of a relationship. The grieving process need not be passive because bereaved individuals can make a number of decisions as they attempt to cope with their loss.

5. It is important in grieving to let feelings emerge into consciousness and not be afraid to express sadness.

6. It is not uncommon to be unsure of how to act around a newly bereaved person.

7. Rather than suggesting that "time heals," an accurate description of the mourning process suggests that the time intervals between intense experiences of grief increase with the passing of time.

8. Seven behaviors and feelings are part of the normal bereavement process: shock and denial, disorganization, volatile emotions, guilt, loss and loneliness, relief, and reestablishment.

9. All persons who have experienced a loss through death will need to attend to the four necessary tasks of grief work before mourning can be completed and reestablishment can take place. These tasks involve accepting the reality of the loss, experiencing the pain of grief, adjusting to an environment in which the deceased person is missing, and withdrawing emotional energy and reinvesting it in other relationships.

10. Signs of "abnormal grief" suggest that an individual might need to seek professional help or at least affiliate with a support group.

11. Deaths from violent means create special bereavement problems for survivors.

DISCUSSION QUESTIONS

1. How can one avoid "deviant" or "abnormal" behavior regarding the bereavement role? What are some functions of defining bereavement roles as "deviant" or "abnormal"?

2. What is the relation between time and the feelings of grief experienced within the bereavement process?

3. Discuss how the seven stages of grieving over a death can also be applied to losses suffered from going through a divorce, moving from one place to another, or having an arm or a leg amputated.

4. If grief is a feeling that is imposed upon the individual through loss, how can grieving be conceived of as being an active process?

5. What are the unique problems faced by those whose grief is disenfranchised? What are the different types of disenfranchised grief?

6. Describe the four necessary tasks of mourning. What are some of the practical steps that one can take to accomplish each of these tasks?

7. What does Parkes mean by "The funeral often precedes the 'peak of the pangs'"? How can one assist friends in bereavement?

8. What could you do to assist people experiencing abnormal grief symptoms?

9. Distinguish between "normal" and "abnormal" grief.

10. How does grief from violent deaths differ from other deaths?

GLOSSARY

Acute Disease: A communicable disease caused by a number of microorganisms including viruses, fungi, and bacteria. Acute illnesses last for a relatively short period of time and result either in recovery or death. Examples of acute illnesses include smallpox, malaria, cholera, influenza, and pneumonia.

Bereavement Role: Behavioral expectations for the bereaved that are structured around specific rights and duties.

Blaming the Victim: A strategy asserted by individuals to relieve the dissonance experienced when innocent people suffer loss.

Chronic Disease: A noncommunicable self-limiting disease from which the individual rarely recovers, even though the symptoms of the disease can often be alleviated. Chronic illnesses usually result in deterioration of organs and tissues that makes the individual vulnerable to other diseases, often leading to serious impairment and even death. Examples of chronic illnesses include cancer, heart disease, arthritis, emphysema, and asthma.

Disenfranchised Grief: Grief that cannot be openly acknowledged, socially sanctioned, and publicly shared.

Dissonance: An inconsistency in beliefs and values—relative to a particular social situation—that causes personal discomfort or tension for the individuals involved.

SUGGESTED READINGS

Cahill, K. (Ed.). (1999). *A framework for survival: Health, human rights and human assistance in conflicts and disasters.* New York: Routledge. Focuses on mass death, destruction of the social order, and economic crisis.

Conrad, B. H. (1998). *When a child has been murdered: Ways you can help the grieving parents.* Amityville, NY: Baywood Publishing. This book was written to describe the painful experience of a child being murdered and to make individuals aware of the ways they can help the parents to survive their grief.

Doka, K. J. (1989). *Disenfranchised grief.* Lexington, MA: Lexington Books. A very helpful book for identifying "hidden" grievers and for expanding one's understanding of the grieving process.

Johns, A. (Ed.). (1999). *Dreadful visitations: Confronting natural catastrophe in the age of enlightenment.* New York: Routledge. A study of 18th century European disasters and responses to them.

Lattanzi-Licht, M. E., Kirschling, J. M., & Flemming, S. (1990). *Bereavement care: A new look at hospice and community based services.* Binghamton, NY: Haworth Press. Addresses the importance of delivery of bereavement care and services in a hospice setting. The book examines the grieving process and distinguishes between grief and clinical depression. The book is helpful for mental health professionals, social workers, chaplains, nursing personnel, and volunteers who work with the bereaved.

Pearson, C. & Stubbs, M. L. (1999). *Parting company: Understanding the loss of a loved one.* Toronto: Seal Press. Provides practical and emotional support for people confronting the death of a significant other through 14 first-hand accounts from caregivers reflecting on a wide range of experiences.

Rando, T. A. (1993). *Treatment of complicated mourning.* Champaign, IL: Research Press. A comprehensive coverage of loss and grief written by a noted clinical psychologist.

Schneider, J. (1994). *Finding my way: Healing and transformation through loss and grief.* Traverse City, MI: Seasons Press. An empowering book for grievers that helps to answer the following questions: "What have I lost?" "What do I have left?" and "What might be possible for me?"

Waszak, E. L. (1997). *Grief: Difficult times—simple steps.* Bristol, PA: Taylor & Francis. Practical information about what to do and what not to do for a person suffering from the death of a significant other.

Worden, J. W. (1991). *Grief counseling and grief therapy: A handbook for the mental health practitioner* (2nd ed.). New York: Springer Publishing. A very practical reference guide for the professional and lay grief counselor.

Zunin, L. M., & Zunin, H. S. (1991). *The art of condolence.* New York: HarperCollins. A book that will provide the reader with concrete skills in assisting the bereaved.

CHAPTER 14

GRIEVING THROUGHOUT THE LIFE CYCLE

Now I lay me down to sleep;
I pray the Lord my soul to keep.
If I should die before I wake,
I pray the Lord my soul to take.

—*New England Primer,* 1781

Perhaps no two individuals grieve in the exact same way. The way one grieves may depend on various situations, including past experiences with death, religious beliefs, personality, the particular person who died, the circumstances surrounding the death, and the age of the one grieving. It is the latter situation that we will address in this chapter—grieving throughout the life cycle.

GRIEVING PARENTS AND THE LOSS OF A CHILD

Adapting to the death of a family member is always difficult, but the death of a child is typically regarded as the most difficult of deaths. As discussed in chapter 3, within the life cycle, an individual expects grandparents to die first, followed by parents, then self, then offspring. That is the "logical" way life is supposed to work. Unfortunately, that which is "supposed" to happen does not always occur. This is the case with children and death. Parents are not "supposed" to have to deal with a dying child, let alone even think of having to attend their child's funeral. Yet, in the world of reality, not the ideal world, the life cycle is sometimes not played out as it is supposed to be, and parents must grieve over the loss of a child. Additionally, the death of a child symbolically threatens the family's hope for a future.

Research shows (Oliver, 1999) that spouses often grieve the loss of a child differently. The most consistent difference appears to be that, as a group, mothers tend to report more intense, long-lasting, and diverse grief reactions than do fathers. Gender differences have also been found in bereaved parents' coping styles. For example, bereaved mothers use more coping strategies as a whole and use more emotionally and verbally expressive forms of coping than men do. Mothers are more likely to feel a strong need to talk about the death and express their feelings, whereas fathers tend to try to resume regular activities and "keep busy," reports L. E. Oliver.

The Loss of a Fetus or an Infant

In order to conceptualize the nature of loss involved in the death of a child, we should remember that the relationship of a parent to a child begins long before birth (Raphael, 1983). For each parent, from the time of conception, the child becomes a source of fantasy—the imagined child whom he or she will become. These hoped-for extensions of self are very common among expectant parents. As the pregnancy progresses, the fantasy relationship with this imagined child intensifies with the selection of the name for the baby, the rehearsal for parenting, the fantasies shared with others, and finally the birth of the child. With the death of a newborn or a stillbirth, the bubble of one's fantasy world is suddenly burst.

Mourning the death of a fetus or **neonate** (infant in the first month of life) differs from mourning the death of another loved one (Furman, 1978). Mourning is a process of detachment from the loved one that is moderated by identification. The bereaved parents take into themselves aspects of the deceased, but because a fetus or newborn has not lived long enough as a separate person, the parents have little of their baby to take into themselves. Thus, they suffer detachment without identification. Parents must readjust their self-image with the knowledge that the baby will never again be part of them. In addition, bereaved parents are less likely to receive adequate validation for their loss—miscarriages, stillbirths, and infant deaths are often "discounted, minimized or negated," as are the deaths of adult children (Oliver, 1999).

In one of the first studies of **perinatal** loss (death occurring during the period closely surrounding the time of birth), J. H. Kennell, H. Slyter, and M. H. Klaus (1970) observed reactions of mothers who had lost a newborn infant and explored the strength of affectional ties between mothers and infants after first physical contact. The conclusion was that more intense mourning was associated with the previous loss of a baby either through miscarriage or death of a liveborn infant. Other studies (Hunfield, Wladimiroff, Verhage, & Passchier, 1995) have also demonstrated more intense grief in mothers who had a history of prior perinatal loss. Perinatal mortality rates vary by race. For example, perinatal mortality data for African

<table>
<tr><td>

WORDS OF WISDOM
Birth and Death

Every time an earth mother smiles over the birth of a child, a spirit mother weeps over the loss of a child.

Ashanti saying.

</td></tr>
</table>

Americans demonstrate that the incidence of perinatal loss is greater than twice that of whites (Guyer et al., 1997).

The death of an infant places severe strains on parents and members of the family—a sense of loss that will probably persist over a number of years. Studies show (Nicol, Tompkins, Campbell, & Syme, 1986) that up to one third of mothers experience a marked deterioration in their health and well-being after the loss of an infant. The death of a child also has dysfunctional consequences for family systems, like separation and divorce.

Longitudinal studies of grieving parents demonstrate that the experience of grief does not go away with time; rather its focus changes. According to S. G. McClowry and colleagues (1987), even after 9 years, parents are still experiencing significant pain and loss—which the researchers refer to as "an **empty space**." This "empty space" feeling takes on three recurring patterns of grieving. The first is "getting over it." In this pattern the grievers do what they can to get back to life as usual and accept death as a matter of fact.

The second pattern of dealing with the empty space is to "fill the emptiness" by keeping busy. Some parents attend grief groups, others immerse themselves in work, become more religious, increase their food or alcohol intake, or become involved as volunteers in organizations such as Candlelighters (a support group for parents of children with cancer) or Compassionate Friends or Empty Arms (two support groups for bereaved parents). Other parents attempt to replace the dead child with another by getting pregnant or by adopting.

The third pattern of dealing with the empty space is described as "keeping the connection." In this pattern of grieving, parents attempt to integrate this pain and loss into their lives (McClowry et al., 1987):

> Although most of the grievers who were "keeping the connection" expressed satisfaction with their present lives, they continue to reserve a small part of themselves for the loss of a special relationship which they view as irreplaceable.

Fetal Death

Fetal deaths include abortion. The abortion may be an **induced abortion** whereby the pregnancy is intentionally ended. This aspect of abortion continues to be very controversial in the United States, although it has been legal since the U.S. Supreme

Court decision of *Roe v. Wade* in 1973. Abortion may also be in the form of a **spontaneous abortion** or miscarriage. Such an abortion is not intentional and usually occurs during the early part of the pregnancy.

Induced Abortion.   Grief after an induced abortion, particularly for an adolescent, may be especially difficult, since many teenagers do not want to talk about the experience or their feelings, notes psychologist William Worden (1982). Since 1980 the number of abortions per 1,000 women aged 15 to 44 has declined about 20 percent, falling from 25 per 1,000 women to 20 per 1,000 women in 1995 (Vobejda, 1997). Since 1973, the Supreme Court has consistently reaffirmed the basic principles underlying *Roe v. Wade* and has refused to overturn that decision. Public opinion polls show that the majority of Americans believe that abortion should be legal under most circumstances—only 15 to 20 percent favor making abortion completely illegal (Stolberg, 1997).

In an induced abortion, a choice is made regarding life and death. Though the decision was made to abort the fetus, the individual(s) involved probably have very mixed emotions about this action. If they do not feel that the fetus is a human being, then the abortion is not killing but merely a medical procedure. The woman's right to terminate an unwanted pregnancy and have an abortion can be viewed as a service to women. The individual(s) involved may occasionally second-guess the decision and wonder what the fetus would have developed into and periodically note how old that child would be at a given time. For antiabortionists, to intentionally destroy a fetus is murder. Thus, the voices of those opposing abortion may echo in the ears of a woman who has had an induced abortion.

If the induced abortion is related to economic issues or is performed because the woman is unmarried, that is a very different situation from that of abortions because of rape or abortions because of evidence that the fetus has a serious illness. In the former cases feelings of grief may indeed be strong—the decision to abort may have been especially difficult and feelings of guilt may occur. The grief process may depend on the individual's views about abortion in general. In the latter cases (rape and illness of the fetus) the decision to abort would probably be more clear-cut. Though still a very difficult decision for most, no matter the reason, an individual's grieving could vary with the reasons for the choice.

Spontaneous Abortion.   When a fetus does not reach full term and life is terminated through a miscarriage, spontaneous abortion, the loss for the parents is particularly difficult. They had anticipated the forthcoming birth of their child, and this is now not going to happen. Perhaps, they had been preparing the child's room and were eagerly making other plans—all to no avail. The child will not be coming home. The parents had also been looking forward to bonding with their child and this also will not happen.

A miscarriage fits the description of disenfranchised grief. In the past, the survivors of miscarriage were viewed as "illegitimate mourners" (Nichols, 1984).

Though today their right to grieve is more acknowledged, such grief is still not socially sanctioned universally as "legitimate mourning." Not knowing what to say to the parents, some individuals may coldly say, "Don't worry, you can always have another baby!" But the parents wanted *this* baby and had anticipated such. It is like telling someone when their pet dies that they can always "get another pet." True, but that is not a very consoling statement to make. The bereaved parents probably feel cheated when a spontaneous abortion occurs and perhaps guilty—"Was it something from my sperm contribution?" the father may ask, or the mother may question her own carrying of the fertilized egg (**zygote**) and its becoming a fetus. "My fault, perhaps?" she may wonder.

In the 21st century, a current movement is on to mourn pregnancy losses, especially miscarriage. It seems that stillbirths have always been mourned. The movement is driven in part by the growing numbers of older women who, having put off childbearing, find themselves facing great difficulty trying to conceive healthy babies and carry them to term (Fein, 1998).

Perinatal Death

As noted earlier, death occurring during the period closely surrounding the time of birth produces grief for the parent(s). There is grief for that which did not occur—a bonding with the newborn. Such grief results from stillbirths, premature births, and deaths during the first month of life not related to sudden infant death syndrome.

Stillbirth.   The stillbirth experience is unique within the context of perinatal bereavement, as death is anticipated before birth, observes S. H. Hutchins (1986). After months of feeling movement, the mother's first sign may be unexpected stillness. The death of a baby *in utero* during labor or shortly before birth more nearly parallels a sudden death with surviving parents experiencing shock and fear. Parents experiencing stillbirth or other perinatal death are usually young and inexperienced in dealing with death. Subsequent attempts to have a child can be extremely stressful for them.

The shock of the death experience of a stillbirth has the family going from the anticipation of joy to sudden heartbreak. Such grief especially was not anticipated, because the pregnancy was seemingly going along smoothly and the day of birth was fast approaching. An intensive period of grief will follow such an unanticipated outcome. Holding the baby for as long as the parents like can be very therapeutic. Giving the baby a name helps to establish the identity of the offspring and makes it a "person" with whom the family can better identify. Having a memorial service may assist the family in coping with their loss.

For the people of southeast Tanzania in Africa miscarriages and stillbirths are considered types of disease (Wehbah-Rashid, 1996). Yet these events are causes for bereavement to the couple and their family. Although there is no elaborate mourning, the couple receive assistance and consolation from their relatives and friends. They are encouraged to forget the loss and look ahead with courage and enthusiasm because of the potential of having other babies.

With a stillbirth the parents do not have the opportunity of bonding. That baby which they so anticipated suddenly is not going to be. Nonetheless, for a brief moment, they had their child, though born dead, and must grieve over its loss.

Premmies.   Babies born before full term (9 months) are referred to as being born premature (**premmies**). Typically a baby with a birth weight of less than 5 pounds is considered to be premature. Such a birth may result in numerous health problems for the baby such as organs not being fully developed. Because the baby is born alive, the family has an opportunity to bond with him or her. Thus, they name and identify with their child. A premmie typically stays in the hospital for several weeks or months while the body continues to develop. Thus, the parents leave the hospital *without* their child. This separation in itself is cause for grief. The parents return home to the baby's room, minus the baby.

With the premmie still in the hospital, the parents spend much time going to and from to visit and determine the medical progress or lack thereof. The child is probably taped and wired to various pieces of medical equipment to enhance development and to keep the baby alive. If the prognosis is not good, the parents may begin to experience anticipatory grief whereby they begin grieving for the child before he or she dies. Their "letting go" process perhaps begins with the initial negative prognosis.

The parents may be faced with the decision at some point to have various life-support equipment removed, if the outcome of death is inevitable. Such a situation presents a very stressful dilemma for the parents. To decide about passive euthanasia is a decision that the parents had probably not previously faced. Thus, their grief over the upcoming death of their child is made even more difficult because of the decision to allow the child to die, if indeed such a situation occurs.

When the baby dies, preparing for and attending the service for final disposition can help the family move on with their grief. The support of others can be most comforting. Just to know that "others care" can indeed help one with the grieving process.

Neonatal Death. A death of a newborn baby during the first month of life can be especially devastating to the parents. Bonding is occurring, life is going on as expected, and quickly "all is not well." As with premmies, the parents may be faced with tough decisions regarding the baby's health.

George Dickinson recalls one situation when the prognosis was not good, and the newborn's condition was rapidly deteriorating. The decision was to let the baby starve to death, which is indeed the way many people die in Third World countries. Thus, the process of passive euthanasia was begun. Various hospice volunteer workers held the baby around the clock for several days until she finally died. Such an ordeal is exhausting both physically and mentally for the family, hospice workers, and volunteers. Many individuals grieve in such a situation.

For parents to make their adjustment to perinatal deaths, whether stillbirths, premmies, or neonatal deaths, they need support from those around them. In a study of 130 parents who had experienced a perinatal death, Judith Murray and Victor Callan (1988) discovered that a consistent predictor of better adjustment was the parents' level of satisfaction with the comfort and support provided by physicians, nurses, and other hospital staff. Parents who were pleased with the level of support that they received from medical personnel were also less depressed and had higher levels of self-esteem and psychological well-being. Furthermore, emotional support from the other parent, family, and hospital staff was linked to fewer grief reactions and better overall adjustment. Therefore, support from others can go a long way to facilitate parental bereavement at the death of a child.

Hospital-based intervention is especially necessary and helpful to parents who are experiencing a perinatal death, and medical personnel are in a unique position to assist parents in their grief (Davis, Stewart, & Harmon, 1988). However, although

PRACTICAL MATTERS
Behaviors of Professionals Identified as Helpful by Grieving Parents

Helpful Behaviors	Primary Responsibility for This Behavior
1. Informs the parents immediately of the condition of the baby	Doctor
2. Expresses feeling over the baby's death with consoling words to parents	Doctor/Nurse
3. Provides as much factual information surrounding the baby's death as is available	Doctor/Nurse
4. Describes the appearance of the baby in factual and tender terms before bringing the baby to them	Doctor/Nurse
5. Encourages parents to see and hold the baby and stays with them while they initially examine the baby	Doctor/Nurse
6. Touches parents affectionately and appropriately; words are not always necessary	Doctor/Nurse
7. Encourages parents to grieve openly	Doctor/Nurse
8. Acknowledges the baby's death at first contact with parents and daily thereafter; does not act like death has not occurred	Doctor/Nurse
9. Spends extra time with the parents to provide time to review the events surrounding the baby's death	Doctor/Nurse

From "Perinatal Death: Grief Support for Families" (p. 19), by P. Estok & A. Lehman, 1983, March, Birth, *10.*

the majority (70 percent) of bereaved parents wish to discuss their concerns with their physician (Clyman et al., 1979), 50 percent of parents received no physician follow-up, and many others had no contact with their physicians until a regularly scheduled postpartum check (Rowe et al., 1978). Thus, although parents seem to desire medical personnel to acknowledge their feelings of shock, guilt, and grief, such desires do not seem to be fulfilled in many instances.

Part of the emotional support needed by parents involves encouraging them to accept the reality of death and to express their feelings of loss and then validating these feelings. Communication and understanding by the medical staff will be great sources of support and comfort. By being available, medical personnel can help to reduce the isolation that parents often feel at this difficult time. Because grief is a necessary process, whenever bereavement is facilitated, the parents become subject to a lower risk for psychological and physical disturbances.

The "Practical Matters" box identifies various ways in which professionals can contribute to the adjustment of grieving parents.

Procedures Following a Perinatal Death.   The stillbirth or the death of a newborn infant is stressful not only for parents and siblings, but also for all people who are involved with the child. Increasingly, parents are being included in the direct care of critically ill children until the time of death and after. Familiarity with the types of procedures and decisions that parents may face at the death of their child is essential. Most bereaved parents will learn about hospital procedures and how to share in decision making only when they are faced with such a situation.

The nursing procedures are fairly clear, and nurses are instructed to respond in ways that can help in socioemotional adjustments. Nurses notify a nursing-social worker counselor about the infant's trauma at the time of admission. Baptism is even offered by some hospitals, where it may be done by anyone in the absence of a chaplain. The pastoral care department of a hospital is notified, as is the communications department, so that accurate information is available to others.

In the case of stillbirths and infant deaths, the parents may be given the option to see and hold their infant, to learn the infant's gender, and to decide on an autopsy and funeral arrangements. The medical staff explains what the parents can expect the infant to look like. Believing that the age of siblings is the major factor in determining their involvement in these settings, E. Furman (1978, p. 217) notes:

> Adolescents should decide for themselves. Elementary grade children are helped by attending a service but not helped by seeing a malformed dead body. Children under school age are particularly not helped by seeing their dead brother or sister, but they are sometimes helped by being in the company of the parents at the time of the funeral.

The extent of individual involvement depends upon the preference of individual family members.

The nurses' responsibilities include attaching identification bands to wrist and ankle, measuring the weight, length, and head circumference, taking footprints and possibly a handprint, and completing standard forms. These forms might include a fetal death certificate, an authorization for autopsy, and a record of the death for the receptionist. Full front and back view photographs as well as close-up shots of any abnormalities may be taken by the medical photographer. These are used by physicians to describe the infant's medical condition. Nurses are also encouraged to take nonmedical photographs that include the infant unclothed in a blanket, a close-up of the face, and the parents holding the child if they so desire. These photos are given to the counselor and later to the parents.

The dead infant is wrapped in a blanket, labeled, and taken to the morgue. In the case of a fetus, the body is sent to the pathology department with a surgical pathology laboratory slip. The physician may request that the placenta be included in the case of spontaneous abortions or fetal deaths. Genetic studies may also be requested: samples of cord blood, placenta, fresh tissue such as gonadal tissue or connective tissue around the kidneys, and skin may be sent to the state laboratory for a complete genetic study. A nurse then completes a checklist for assisting parents who are experiencing perinatal deaths, and she or he may, if procedure calls for it, place an identifier by the name tag at the entrance to the mother's room.

WORDS OF WISDOM
Epitaph Upon a Child That Died

Here she lies, a pretty bud,
   Lately made of flesh and blood:
   Who as soon fell fast asleep
   As her little eyes did peep.
   Give her strewings, but not stir
   The earth that lightly covers her.

Robert Herrick, "Epitaph Upon a Child
That Died." From Washington, P. (Ed.)
(1998) Poems of Mourning.
New York: Alfred A. Knopf.

Hospitals with a special program to help the survivors of perinatal deaths provide nursing, medical, social work, and/or pastoral counselors who may assist the surviving parents and siblings. Time is taken to explain and help in the following matters: (1) autopsies and hospital procedures after death; (2) funeral or cremation options; (3) the nature and expression of grief and mourning; (4) coping with the reactions of friends and relatives; (5) helping siblings to deal with the death of their brother or sister; and (6) decisions regarding another pregnancy. Monthly meetings of bereaved parents can be established to provide a setting for sharing and learning about grief. These experiences of sharing with other parents allow for reality-based comparisons and for active support of other parents whose loss is also great.

This list of hospital procedures is by no means complete, and the attitudes and responses of physicians, nurses, and others may vary greatly. At times, for example, the helpers are in need of socioemotional support along with parents and siblings. Many hospitals, on the other hand, have not dealt with the special needs of families experiencing perinatal deaths. Although these practices are becoming more common around the country, one should not be surprised if a nurse or physician appears stunned at the request of parents to spend some time with the body of their dead child. We can only hope that the human values of medical care will prevail over bureaucratic values as the welfare of the whole person is taken into account in the medical arena. This can be accomplished most effectively if parents are provided with accurate information, encouraged to ask questions, given plenty of time to make decisions, and given opportunities to share their experience in parental bereavement with others.

Sudden Infant Death Syndrome

There has been a continuing decrease in the rate of infant mortality in the United States, and the number of deaths attributed to the mechanism known as **sudden infant death syndrome (SIDS)** has been decreasing as well. In the Western world, SIDS is the most common cause of death of infants between 1 month and 1 year of age and accounts for approximately 50 percent of deaths of infants between 2 and 4 months of age (Klonoff-Cohen et al., 1995). Much of the decrease in SIDS is attributed to an emphasis by the American Pediatric Association and other western world pediatric groups on **Back to Sleep,** a program encouraging that the baby be placed on its back or side to sleep, not on its stomach.

Sudden, unexpected infant death is a major cause of death for infants between the ages of 1 week and 1 year in the United States—an estimated 5,000 to 6,000 deaths per year (Walling, 1996). According to a study by the Foundation for the Study of Infant Deaths, these babies do not cry out as if in pain, but simply die quietly in their sleep, after becoming unconscious. Sudden infant death is usually defined as the sudden, unexplained death of an infant younger than 1 year where no cause is found through a postmortem examination (Willinger, 1995).

Historically, unexpected, unexplainable deaths of infants were routinely attributed to the mother's lying on them because mothers often slept with infants. If a mother awoke and found her baby dead, she assumed that she had lain on the child, smothering or crushing her or him. Perhaps the earliest such death recorded is in the Bible in 1 Kings 3:19: "And this woman's child died in the night, because she overlaid it." It is believed that SIDS was occurring long before it was recognized and accepted as a diagnostic label (Beckwith, 1978).

The majority of parents whose infants have died of SIDS see it as the most devastating crisis that they have ever experienced. These parents show the strongest reactions among parents who experience death of their infants (Boyle, Wladimiroff, Verhage, & Passchier, 1996). Approximately one fourth of parents bereaved by SIDS move from their homes to other communities in an effort to escape the pain of the baby's death (DeFrain, Jakub, & Mendoza, 1992). A study of 34 pairs of parents bereaved by SIDS (Carroll & Schafer, 1994) reveals that these parents sought support from within the family most often and from outside resources least often. Parents generally tend to "recover" from the death in approximately 3 years or more. However, a recent Norwegian study (Dyregrov & Dyregrov, 1999) of parents bereaved by SIDS reported that more mothers than fathers claimed to be affected by their grief 12 to 15 years after the death.

Some parents "turn to God," whereas others "turn away from God" as a response to their loss (DeFrain, Taylor, & Ernst, 1982). Their reaction may be reflected in church attendance. From 1985 through 1988 a study (Thearle et al., 1995) in Australia examining the emotional health of parents after SIDS, neonatal death, or stillbirth concluded that the bereaved parents who attend church regularly have less anxiety and depression compared with irregular church attenders and nonchurch attenders. Similarly, in interviews with parents who had lost children to SIDS, researchers Wortman and Silver (1992) found that parents' religious devotion and participation in religious activities were positively related to coping. Among these parents a most important feeling was that they would someday see their children in heaven.

When an infant dies suddenly and unexpectedly, the sense of loss and grief may be overwhelming. When the sudden death is due to a known cause, the concrete character of the event can be incorporated into the normal rationalization of mourning. However, when death is due to an unknown mechanism, as in SIDS, feelings of inadequacy in caring for the child are reinforced for the medical staff and the parents (Mandell, McClain, & Reece, 1987).

Because there are still many unanswered questions about SIDS, parents have a tremendous guilt feeling and shoulder the responsibility for the infant's death. Marital conflict, difficulties with surviving children, and anxiety about future children becoming victims of SIDS are often experienced. With the cause of death in a SIDS case being questionable, the parents are likely to undergo a police interrogation, in addition to the stress of having lost their infant. In an era of child abuse, the parents may be suspected of smothering the child. Being questioned by the police is difficult enough, but the perceived inability of others, such as friends and relatives, to understand the depth of their despair is also very frustrating to these parents (DeFrain, Jakub, & Mendoza, 1992).

The Centers for Disease Control and Prevention ("Guidelines to Help Discern SIDS from Homicide," 1996) recently issued investigation guidelines to help coroners and police distinguish between SIDS and homicide in infants. The guidelines suggest noting aspects such as the position of the infant's body, any suspected injuries, and any evidence of drug use in the home. The death-scene investigation, the child's medical history, and an autopsy are necessary for a thorough investigation.

Young parents whose baby dies of SIDS have probably not experienced the death of a close relative; thus, they are not familiar with the social and emotional aspects of grief. The death of one's baby is traumatic under any conditions, but the sense of not knowing how to mourn adds to the difficulties of socially adjusting to the loss. The parents and siblings experience "**anomic grief**"—a grief without the traditional supports of family, church, and community.

A survey of newly trained SIDS counselors in North Carolina (Kotch & Cohen, 1985) reported that sharing the autopsy report with bereaved parents was a valuable part of the counseling process and removed some of the mystery surrounding the diagnosis of SIDS. The autopsy report, by documenting that the child died a natural death, may relieve some families of the feeling that they were somehow responsible for the death.

George Dickinson's own discussions with parents who have lost a child through SIDS suggest that friends may turn on them as if they are criminals. Parents tend to "blame" each other for the death—"If only you had. . . ." Parents become "victims" because SIDS is both personally traumatic and complicated by problems of social interaction. The uncertainty of the cause of death is frustrating to the parents and medical staff and clouds the whole issue from a societal perspective. A death due to SIDS must be one of the more traumatic experiences that parents can have in a lifetime.

The beginning of healing for parents of an infant who has died of SIDS is the closure of the relationship with the infant, advises Melodie Olson (1997). After the child has been pronounced dead, it is important for the parents to hold the child, perhaps several times between the time of death and the funeral. Parents' privacy needs to be respected. They should be allowed to take something from the child, such as a lock of hair or the baby blanket.

PRACTICAL MATTERS
Suggestions for Survivors of a Death in the Family

1. Don't judge the way people grieve—those who do not cry can be just as devastated as those who cannot stop crying.

2. Don't assume the death was for the best—"He was my dad, even if he was old."

3. Don't assume that because there are other children, the pain is any less—an amputated arm is still missed.

4. Don't say, "I know how you feel"—You are not that person.

5. Don't say, "Don't worry, you'll get married again" or "You'll have another baby" or "It's God's will."

From Don't Ask for the Dead Man's Golf Clubs: Advice for Friends When Someone Dies, *by Lynn Kelly, 1998, New York: Kelly Communications.*

The Loss of a Child

C. M. Sanders (1980) notes that the death of a child is one of the most grievous of losses, significantly greater, on average, than that of a parent. Based on interviews with 155 families suffering the loss of a child ranging in age from 1 to 28 years, Ronald Knapp (1986) observed that the death of a child represents in a symbolic way the death of the self. Symbolically, a parent will die along with the child, only to survive in a damaged state with little or no desire to live today or to plan for tomorrow. In addition to losing part of themselves, parents lose some of their hope for the future, concludes Reiko Schwab (1992) from a study of 20 couples who had lost children.

Because children are "not supposed to die," especially before their parents do, the death of a child seems more tragic and traumatic than the death of an older person. We often take our children for granted. Although one can *imagine* the loss of a child, how often does one have such thoughts? Even so, such thoughts bear little resemblance to reality. For parents who have lost a child through death, however, the reality of the situation lingers forever. A friend whose son had died told George Dickinson that the stark reality hit him in the face every morning when he woke up and realized that this was not just a bad dream—his son was really dead.

Knapp (1986) notes that parents "live" the child's death over and over in their imaginations as the end is near. This imaginary scenario takes them from the moment of death through the funeral of the child. Parents may wish to keep the child at home so that he or she will not die in strange surroundings. When parents are able to make the decision to terminate all further treatment and let the disease take its course, a sense of tranquillity results, and parents are ready to release their hold. Sometimes parents must take on the painfully hard task of telling the child that it is all right to let go. It is all right to stop fighting. It is all right to die! Giving permission to die is

Though he only lived a few weeks, David Crawley indeed is missed by his family. He was "loved so much" but "is waiting" for his mother "just over the hill."

difficult. Sometimes it takes only gentle encouragement from parents—gentle persuasion that all has been done and that nothing more remains.

As mentioned earlier, available literature (Schwab, 1992), based on research and clinical observations, indicates that the death of a child strains the parents' marital relationship, sometimes resulting in separation and divorce. One study (Klass, 1987)

DEATH ACROSS CULTURES
Coping with Death in the Philippines

The Philippines are some 600 miles southeast of the coast of mainland Asia. The people in the Philippines are predominately of Malay stock, and the religion of the majority is Roman Catholic. Death is viewed as a destiny or fate and part of the cycle of beginning and end—a natural occurrence. Dying of old age is seen as a positive occurrence. Death is viewed as more than sorrow and loss—happiness is also a part of death. Death is seen as the last cure for the incurably ill.

Though Filipinos are known culturally as shy people, they very openly discuss death. They believe that the more one shares of the situation, the easier the grieving will be. Socioeconomic background of the family may dictate how open they are when death occurs. Rich families, for example, are usually rather private, whereas those of lesser means will depend more on the community for support.

The death of a child is the most difficult death to face. Severe guilt is experienced by the parents. Children are perceived as "angels," innocent and "without sin"; thus they go directly to heaven after they die. They are expected to wait and meet their parents in heaven when their turn comes.

From "The Filipino Perspective on Death and Dying" (pp. 215–220), by Ihande Weber, 1995. In J. K. Parry & A. S. Ryan (Eds.), A Cross-Cultural Look at Death, Dying, and Religion, Chicago: Nelson-Hall Publishers.

concluded, however, that marriages did not end because of a child's death. Instead, parents in rocky marriages decided that their struggle with marital problems was no longer worth the effort. Nonetheless, even strong marriages often suffer after the death of a child.

Sometimes, the difficulties in the loss of a child are the result of gender differences in coping with grief, as noted earlier in a study by L. E. Oliver (1999). In another study composed of 145 bereaved parents, J. A. Cook (1983) reported that fathers find it difficult to grieve openly and thus keep their grief to themselves. Mothers, on the other hand, are more comfortable grieving openly and want to talk about their loss. They find their husbands' lack of expression of grief a barrier to communication. Cook (1988) noted that men are given little comfort and support in bereavement and are expected to be strong and to provide a source of support for others, yet their nonexpressiveness comes into conflict with their wives' needs for expression. Schwab (1992) concluded in a study of 20 couples who had lost children that husbands and wives appeared generally irritable and less tolerant of their spouse. On the other hand, Schwab noted that husbands and wives with a good marital relationship prior to a child's death appear to have come closer together through the tragedy that shattered their lives. Thus, a weak marriage may collapse with the death of a child, and a strong marriage may indeed be strengthened by the death. Whatever the result, the strain of the death of a child is certainly a test of the marital relationship.

Words of Wisdom
Oh, Danny Boy

. . . for you must go
and I must die.
But come ye back
when summer's in the meadow,
or when the valley's hushed and
white with snow.

For I'll be here
in sunshine and in shadow.

Oh Danny boy. . .
I love you so.

Irish Folksong.

The Loss of an Adult Child

A description of the bereavement of parents who lose an adult child is provided by British anthropologist Goeffrey Gorer (1965), who notes that it is literally true that parents never get over it. As noted earlier, since it is "against the order of nature" for a child to die before her or his parents, the parents seem to interpret this as a "punishment for their own shortcomings," observes Gorer. Their self-image seems to be destroyed. For them, a reliance on the "orderliness of the universe" has been undermined.

Research reported by Therese Rando (1986) concludes that the death of an adult child is difficult for both fathers and mothers in that they see the adult child's responsibilities left unattended with death—fatherless/motherless children and unfinished tasks in life. The children probably feel "shortchanged" in that they will grow up without that parent. The surviving parents of the deceased adult child note this void in the lives of their grandchildren, observes Rando. They may see life as seeming unfair to their grandchildren.

The loss of an adult child leaves the parents with unfulfilled dreams for their offspring. Life may seem "incomplete" to them in that the chain in their generational lines has been broken—a loss of continuity in the life cycle occurs.

GRIEVING CHILDREN AND ADOLESCENTS

A developmental approach to grieving the loss of a parent from very young children through adolescence reveals different manifestations of grief, concludes Grace Christ (2000). Children aged from 3 to 5 years tend to sleep with the surviving parent, suck their thumb, wet the bed, display clinging behavior, have bad dreams, and display various physical symptoms such as a stomachache. Children aged from 6 to 8 years talk freely about the deceased parent, feel the parent's presence, and talk to the parent. Older school-aged children from 9 to 11 years seem to be overwhelmed by their grief and do not like to talk about the deceased individual. Slightly older

Children often express their grief through crying. As parents, we should be open to children's crying and not discourage such display of emotions.

children, aged 12 to 14 years, avoid feelings and information about the illness and grieve alone. In addition, they grieve for the loss of the deceased's specific characteristics and special functions in the family. The grief of adolescents between the ages of 15 and 17 differs markedly from that of their younger peers in that they can become overwhelmed by their grief and not be able to control it.

Storytelling is a good way to keep memories alive for children and adolescents (Harvey, 1996). Writing down one's thoughts about the deceased person can be very insightful. Stories can both teach and heal. In telling stories about the deceased, even years later, one can help the young child—now grown—to better "know" her or his deceased parent. A close friend of George Dickinson's shared how he told the grown children of a mutual boyhood friend stories about their

---

WORDS OF WISDOM
Music

Percy Bysshe Shelley

Music, when soft voices die,
  Vibrates in the memory—
  Odours, when sweet violets sicken,
  Live within the sense they quicken.
  Rose leaves, when the rose is dead,
  Are heaped for the beloved's bed;

And so thy thoughts, when thou art
  gone,
  Love itself shall slumber on.

From Washington, P. (Ed.) (1998). Poems
of Mourning. *New York: Alfred A. Knopf.*

---

now-deceased dad when he was growing up. This helped the children, whose dad committed suicide when they were small children, to better know what a great guy their dad had been.

Drawing is another way for children to express themselves. Pictures are a bridge to expression (Cox, 1998). The child can draw a picture, then "explain" what is meant by the picture. Drawing is an excellent way to bring a child's feelings out, especially a very young child whose verbal gifts may not be well developed.

In addition, music is a form of expression that can be used without oral fluency. Music is a therapeutic tool that can be used to impact in a positive way on one's mood. Music can also be used to promote spiritual development (Lowis & Hughes, 1997).

Grieving children and adolescents need to know that it is okay to laugh during a period of bereavement. George Dickinson, as noted earlier, has had college students tell him that when they were young and a grandparent had died it bothered them that people gathered at their home could be laughing "at a time like this." Humor, however, can be very healing to a grieving person. David Spiegel (1998) has found that those grieving and dying lived an average of 18 months longer if they were happy—happiness is important to good health. Laughter gives the child and the adult something else to think about, observes Gerry Cox (1998). Given a reason to laugh, the individuals involved can forget their pain for a few minutes. Laughter promotes confidence and hope. Let us now take a closer look at grieving children in the loss of a parent, sibling, and grandparent.

Loss of a Parent

How do children's grief reactions differ from those of other individuals? ask John Baker and Mary Sedney (1996). Children's grief reactions appear to last longer than those of adults. Their initial reactions appear to be less intense. Children have different ways of

"You couldn't put on a tie?"

coping than adults—they are more likely to distract themselves and use fantasy to cope with the suddenness of the loss. They will tend to identify with the parent who has died, whereas adults are more likely to separate themselves from the characteristics of the person who died.

Parent death during these early years of attachment and dependency threatens the child deeply (Hatter, 1996). Basic survival needs may be met by another, but the unique qualities of the parental bond can never be replaced. The more positive and frequent the contact has been with the deceased parent, the more acutely a young child will be aware of the parent's absence (Norris-Shortle, Young, & Williams, 1993). Long-term implications of the death of a parent arise because children must grow up with the loss.

According to their self-report, 57 percent of children who lost a parent (Silverman, Nickman, & Worden, 1992) "spoke" to the deceased parent in some way, 55 percent dreamed of the parent who had died, and 81 percent felt the parent was watching them. Such an attachment to thoughts and memories of a dead parent may be a sign of positive adaptation rather than a sign of a pathological problem, note Silverman, Nickman, and Worden.

Preschool children up to 5 years of age most often show manifestations of anxiety or of aggressive behavior when a parent dies (Baker & Sedney, 1996). School-aged children between 6 and 10 years of age may appear to deny that the death has occurred and strive to maintain an appearance of emotional control.

Adolescents show some trends similar to those in younger children, although their emotional reactions are often kept private at this age in an effort to appear "normal." In addition, adolescents are more likely than younger children either to become depressed or to try to escape from their emotions through acting-out behavior (e.g., running away, excessive drug use, and risk taking). Adolescents may also be preoccupied with the "unfairness" of death, note Baker and Sedney.

Loss of a Sibling

Through the death of a sibling, bereaved brothers and sisters learn at an early age difficult lessons about the preciousness of human life, the importance of close personal relationships, the vulnerability of such relationships, the multiple impacts of loss on themselves and other family members, and the significance of the legacy left to them by the sibling who died (Stalman, 1996). Thus, the death of a sibling during childhood has immediate as well as long-term consequences for surviving siblings. In today's smaller modern family the surviving sibling may become the "only child" in the family.

In a study of 65 children (between the ages of 4 and 16) who were the siblings of deceased children, D. E. McCown and C. Pratt (1985) confirmed previous studies indicating that 30 to 50 percent of surviving children demonstrate increased behavior problems after the death of a sibling. Their studies showed that children in the middle age group, 6 to 11 years, developed more behavior problems than those in other age groups. Reasons cited for more problems in this age group are that the loss of a sibling at this phase may lead to feelings of vulnerability and inferiority and that for the child in this age group who is making the transition to concrete thought, the event and cause of sibling death may evoke confusion. The increased behavior problems may be a reflection of that confusion.

David Adams and Eleanor Deveau (1987) observe that, after a child's death, many siblings fear minor physical symptoms and worry about death occurring at the same age. Siblings resent parents for their preoccupation with the dead child and blame them for their inability to protect these siblings during illness. Parents are often so consumed by their own grief that they have little energy left to help surviving children. Problems also develop when parents expect surviving siblings to surpass or equal the achievement of a deceased child or to replace the deceased child.

Due to limited resources and energies in the home, peer support and special attention to the siblings of dying children are needed. Teachers, in particular, need to be alert to the special concerns and needs of these children. The personal worth of surviving siblings needs to be reinforced. As with siblings of a newborn coming into the family, siblings of a dying child have special needs. They, too, wish to be noticed and given some love and care.

Loss of a Grandparent

As noted in chapter 3, a child's first experience with death many times is that of a grandparent. Grandparents often are important figures in the child's life, giving

Siblings of seriously-ill children need attention also. Often-times, the siblings may feel neglected.

unconditional love and caregiving. Often a grandchild can do no wrong in the eyes of the grandparents. The positive reinforcement given to children by grandparents will no longer be a source of feedback for the bereaved grandchild.

On the other hand, in our mobile society grandchildren may not have bonded with their grandparents, due to their living in a different geographical place; thus the loss of a grandparent may not be so traumatic. Seeing their own parents upset and perhaps crying will probably bother the children, but their own personal loss may not seem severe. The child may feel that he or she is expected to grieve, but in fact may not feel like grieving.

If this is the child's first experience with death, she or he may ask a lot of questions regarding death, because the event is new. Parents need to be receptive to questions in order to help the child cope with death. Indeed, for many children the death of a grandparent may be very traumatic or it may have a limited impact, depending on their degree of relationship with the deceased.

GRIEVING ADULTS

For grieving adults, death is a reminder of our own mortality. The death of another adult highlights the fragility of life. As with any death, a void is left in one's life, especially if that person had been a part of the life *forever*, such as a parent, or for much of our adult life, such as a spouse or even a pet.

Loss of a Spouse

The death of a spouse, not unlike the death of a parent or child, is always unique and seems devastating. Two individuals have chosen to live together, and now that bond is broken. Life will be different, usually significantly so. Melodie Olson (1997) notes factors associated with the death of a spouse that may contribute to an especially prolonged adjustment: unexpected death of the spouse, high dependency on the deceased, and poor health of the surviving spouse before the death.

The dependency established in a marriage will definitely leave voids in life for the surviving spouse. The division of labor found in a familial relationship will suffer when one member is taken away. Spouses are often best friends. Even in marriages far from perfect, important bonds are formed nonetheless.

According to research reported by Catherine Sanders (1999), various situational variables affect bereavement in the loss of a spouse. If the surviving spouse has a good network of friends and family, such a support system helps tremendously in "making it," especially during the days immediately after the death. Eating alone can be a painful experience for many bereaved spouses. In fact, loneliness is a major problem of bereaved spouses. Sensing the presence of the deceased partner remains with the surviving spouse. Religion is a mainstay, providing hope to many when discouragement is almost overbearing.

Loss of a Parent

The younger the adult child, the stronger is the attachment to parents, because childhood memories of dependency are fresher in the minds of young adults than in those of older more mature adults, observes Sanders (1999). In addition, the younger adult child likely has fewer significant others to draw from for support during grief.

The death of a parent can have a profound effect on the ways that adult siblings deal with each other, notes Lawrence Kutner (1990). Adult siblings may experience

---

WORDS OF WISDOM
Remember

Christina Rossetti

Remember me when I am gone away,
  Gone far away into the silent land;
  When you can no more hold me by the
  hand,

Nor I half turn to go yet turning stay.

Remember me when no more day by day
  You tell me of our future that you
  planned:
  Only remember me; you understand

It will be late to counsel then or pray.

Yet if you should forget me for a while
  And afterwards remember, do not grieve:
  For if the darkness and corruption leave
  A vestige of the thoughts that once I had,

Better by far you should forget and smile
  Than that you should remember and
  be sad.

From Washington, P. (Ed.) (1998)
Poems of Mourning. *New York:*
Alfred A. Knopf.

---

intense emotions as they reevaluate the meaning of family and their roles within it. The death of a parent, especially the second parent, often accentuates a pattern already existing—as siblings grow older, good relationships become better, and rotten relationships become worse. The death of a second parent often changes the focal point of family rituals such as organizing Thanksgiving or other significant holiday celebrations, observes Kutner. While the parents are alive, the relationships between siblings tend to be balanced compared with the unbalanced parent-child relationship; yet when parents die, one sibling tends to break out of this balanced relationship (Kutner, 1990).

Joan Douglas (1991), in studying reactions of middle-aged persons to the death of their parents, concluded that for many the integration of the loss of a parent involves confronting the loss of parental power and the reality of one's own mortality as part of the developmental process. The tension created by the need to both sever and maintain the parent-child bond and to face one's own death without giving way to despair forced many to move to a new perspective—a new level of integration.

Loss of a Pet

Examining the death of pets clarifies sociological issues, such as the process by which human traits are attributed to animals. For many individuals, a pet is a significant member of the family. People talk to pets and care for them as if they were their children. Pets often live with a family as many years as children live at home before leaving for college or emancipation. Pets can make people feel needed, can relieve loneliness, and can serve as friends and companions. Such a death is a traumatic experience for family members. As would occur for any other member of the family, the resulting dismemberment leaves a big void.

Unlike the person who loses a friend or relative and receives outpourings of sympathy and support, one who loses a pet often receives ridicule for overreacting or for being foolishly emotional. This leads to the problem of disenfranchised grief as discussed in chapter 13.

Discussion of pet bereavement enables individuals to share their feelings of grief, fear, and loss. A. D. Weisman's work (1990–1991) with bereaved pet owners suggests that there were many similarities between grieving for a pet and grieving for a human—preoccupation was common, people reported mistaking shadows and sounds for their dead pet, guilt and ambivalence were experienced, and there were also corresponding feelings of loneliness and emptiness in the grieving process.

Like other professionals dealing with issues related to dying and death, veterinarians are concerned about discussing euthanasia and giving bad news to pet owners (Edney, 1988). When having a pet "put to sleep" is being contemplated, Herbert Nieburg and Arlene Fischer (1982) suggest asking the owner whether the pet can do the things that it once enjoyed, whether there is more pain or more pleasure in its life, whether the pet has become bad-tempered and snappish as a result of old age or illness, whether it has lost control of its bodily functions, and whether one can afford the expense and time involved in keeping a sick pet alive. Whatever the final decision, this is not an easy choice for an individual to make.

For those who do make the decision to euthanize their pet, at first there is a feeling of regret for having given permission for euthanasia, despite illness or invalidism. According to Weisman (1990-1991), this regret should not be interpreted as an indictment of euthanasia but rather as an expression of how guilty owners feel about invoking their power of life and death over their beloved companion.

DYING, DEATH, AND BEREAVEMENT IN THE 21ST CENTURY: A CHALLENGE

As we are now into the 21st century and living in an era when we grow to "very ripe old ages," the complexion of grief may take on a slightly different tone. Baby Boomers are reaching retirement, yet many still have one or two parents for whom they are caring. At the same time, Baby Boomers may fall into the "sandwich generation" and are taking care of their own young-adult offspring. In addition, with diseases such as Alzheimer's on the rise, these aged parents may be especially difficult to care for, complicating matters even more for the Baby Boomers.

These individuals in the "middle years" are being courted by funeral homes with the message that they are willing to provide whatever funeral a client might desire. Traditional funerals are not appealing to many today; thus, funeral homes seem receptive to change. The use of cremation is rapidly rising in the United States as land space becomes an issue in some areas and the whole idea of preservation of the body seems bizarre to many.

The "resurrection of death" period of bereavement, though beginning in 1945 with the dropping of the atomic bombs in Japan, seems to remain "alive and well" today. As Tony Walter (1999) notes, the signs of death are "becoming more visible" with more public forms of mourning coming back into fashion—wearing a ribbon to support those with AIDS or having a moment of silence at the precise minute, 1 year or 5 years later, to recall an earlier disaster. The topic of death and grief is frequently seen in the media, compared with a few decades ago. End-of-life issues are addressed on television, in newspapers, in popular magazines, in best-selling books, on the radio, and on the Internet. With better mass communications today, a disaster anywhere in the world can be reported "around the world" practically as it is happening. We can now see wars being fought "live" on television from the comfort of our homes. Research money is available for topics related to dying and grief, rarely heard of until quite recently. College courses on dying, death, and bereavement are now almost ubiquitous on college campuses, yet were unknown in the 1960s. Euthanasia issues in medicine, many unknown a few years ago, are now daily news items. Individuals are more knowledgeable about end-of-life issues and thus are more demanding about their rights, whether regarding euthanasia or funeral practices.

The overwhelming focus in traditional medicine in the United States has been to "prolong life." Yet, in the 21st century palliative care is slowly creeping into American medicine. The focus previously has been on *curing* diseases, whereas now *caring* for those with diseases is gaining momentum. As best-selling author Sherwin Nuland of Yale Medical School observes (Carlson, 2000), "Doctors today are filled with a kind of arrogance about their ability to defeat death." He notes that young physicians feel that if you cannot cure, you have failed. Nuland advocates a more humanistic bent to medicine. If the patient cannot be cured, then the doctor should try to make the patient comfortable—palliative care. Indeed, Jack Kevorkian alerted the medical world to the need of patients for more care in controlling pain and in symptom control in general. The palliative care trend is there and probably will continue. Such an emphasis should enhance dying for the individual.

With chronic illnesses being prevalent in the United States today, not infectious diseases as in the past, an individual may simply live on and on—living with dying. Is this limited quality of life that a very elderly person often has really worth the $4,000 to $5,000 per month that is being paid to keep her or him alive in a nursing home? Will the soaring costs of drugs subside in the 21st century? How long will the family savings last to pay the bills? Will the aging parent "outlive" the coverage for living expenses during the last few years of life? When family money is gone, will the government have enough money available from Medicaid and Medicare, and be willing to spend it, to cover the expenses? This issue of government responsibility for medical care cost for elderly persons is already a political football. As the population ages into the 21st century, will the problem only get worse before it gets better?

As a society, in an age when new technologies can miraculously prolong life, we duck questions about how to spend our health dollars on those for whom death is a close inevitability, says John Lantos (2000). As individuals, we also need to think about how we want to live as we are dying. Perhaps Daniel Callahan (1987) was correct in his book *Setting Limits: Medical Goals in an Aging Society* regarding governmental responsibility for the elderly—maybe we should "draw the line" regarding life-prolonging decisions with elderly persons. But then when "the elderly 'is' us," we might have a change of opinion! It seems that death can indeed be a relief in many situations. Should an individual have the right to say when and where she or he wishes to die? Given all that has happened in the latter part of the 20th century, dying, death, and bereavement in the 21st century should be a challenge. Let's hope we are prepared.

CONCLUSION

No matter where in the life cycle a death occurs, it is not easy for the survivors to make the adjustment. The realization that the deceased individual will never again be in our presence is a shock that comes to us all throughout the life cycle. Such a loss is something that we do not simply "get over," but we learn to live with the fact that the individual is gone, never to return. Life will not be the same again because of the void left in our lives by the death of that significant other. Elisabeth Kübler-Ross stated in the title of one of her books (though referring to the one dying) that death is the final stage of growth. Experiencing the death of someone else is indeed a "growth experience" for the survivors, because we must adapt to life without that person—a major challenge indeed. With the numerous support systems in society today, however, we *can* learn to live without the deceased person. "Life goes on," and we must go on with it.

SUMMARY

1. Parents often grieve the loss of a child differently.

2. The death of an infant places severe strains on parents and other family members.

3. Grief does not go away with time, rather one adjusts to the absence of the deceased person.

4. A perinatal loss leaves the parents with an anticipation that will never happen.

5. Medical staff should be prepared to deal with the grief of parents after a perinatal death.

6. Sudden infant death syndrome is especially difficult for parents. Not only is their child dead, but they are initially "suspected" by law enforcement in the death.

7. Humor, art, storytelling, and music are good ways to help children cope with death.

8. In today's modern family, the surviving sibling may become the only child in the family.

9. The death of pets can be devastating for family members.

10. Younger adult children tend to have stronger attachments to parents than their older siblings.

11. Grief in the 21st century should be especially challenging with today's somewhat different perspectives toward death.

DISCUSSION QUESTIONS

1. Discuss particular stresses placed upon parents in the death of a child.

2. How does grief involving parental reactions to perinatal deaths differ from reactions to other deaths?

3. How can health-care professionals give support to parents in the death of a child?

4. What nursing procedures follow perinatal deaths?

5. Why are deaths from SIDS especially difficult for parents?

6. How does the loss of older children and adolescents differ from other deaths?

7. How can storytelling, music, art, and humor serve as therapy for a grieving individual?

8. How do children manifest their feelings in the death of a parent?

9. What impact does the death of a sibling have on the surviving sibling(s)?

10. What might cause children to react differently to the death of a grandparent?

11. Why is grief for the loss of a pet so significant for families?

12. What "losses" are especially difficult in the death of a spouse?

13. How do you think that grief in the 21st century might differ from that in the 20th century?

GLOSSARY

Anomic grief: A grief without the traditional support of others and no norms to help the bereaved deal with grief.

Back to Sleep: A program which encourages placing a baby on its back or side to sleep, not on its stomach.

Empty space: Pain and loss still experienced years after the death of a child.

Induced abortion: Intentional aborting of the fetus.

Neonatal: Refers to an infant less than 1 month old.

Perinatal: Refers to the period closely surrounding the time of birth and includes miscarriages, stillbirths, and neonatal deaths.

Premmie: A baby born prematurely (pregnancy did not last the full term of 9 months). The baby's birth weight is usually less than 5 pounds.

Spontaneous abortion: Also called miscarriage—expulsion of a fetus from the womb before it is sufficiently developed to survive.

Sudden Infant Death Syndrome (SIDS):   The sudden, unexplained death of an infant when no cause is found in post-mortem examination.

Zygote: Fertilized egg that formed by the union of the sperm and egg.

SUGGESTED READINGS

Abrams, R. (1999). *When parents die.* London: Routledge. Discusses the death of a parent of young adults.

Blank, J. B. (1998). *The death of an adult child: a book for and about bereaved parents.* Amityville, NY: Baywood Publishers. A book about the personal experiences of the author, not a clinical orientation.

Cecil, R. (Ed.). (1996). *The anthropology of pregnancy loss: Comparative studies in miscarriage, stillbirth and neonatal death.* Oxford: Berg. Contemporary, historical and oral-history accounts from regions as diverse as rural India, urban America, South Africa and Northern Ireland providing insight into the experience and management of miscarriage across a number of different cultures.

Christ, G. H. (2000). *Healing children's grief: Surviving a parent's death from cancer.* New York: Oxford University Press. The author relates the stories of 88 families through the terminal illness and death of one of the parents from cancer.

Corr, C. A. & Corr, D. M. (Eds.). (1996). *Handbook of childhood death and bereavement.* New York: Springer Publishing. An anthology about death, bereavement, and intervention.

Doka, K. J. (Ed.). (1995). *Children mourning, mourning children.* Bristol, PA: Taylor and Francis. Put out by the Hospice Foundation of America, this book was written to assist children with death *and* to aid adults in coping with the death of children.

Felber, M. (1997). *Grief expressed: When a mate dies.* New York: Life Words. Helpful information to the bereaved in the death of a spouse from Marta Felber, a therapist whose husband died of cancer.

Harpham, W. S. (1997). *When a parent has cancer: a guide to caring for your children.* New York: HarperCollins Publishers. A physician gives advice about helping children cope with their dying parent.

Harvey, J. H. (1996). *Embracing their memory: Loss and social psychology of story telling.* Boston: Allyn and Bacon. A social psychologist writes about the importance of story telling in keeping the memory of a deceased significant other alive.

Heinlein, S., Brunett, G., & Tibbals, J. (Eds.). (1997). *When a lifemate dies: Stories of love, loss, and healing.* Minneapolis: Fairview Press. Accounts of the loss of their husband, wife, or domestic partner and tell how it has affected their lives from more than 50 contributors.

Kelly, L. (1998). *Don't ask for the dead man's golf clubs: Advice for friends when someone dies.* New York: Kelly Communications. A 34-year-old widowed mother of three gives practical advice for survivors of a death in the family.

Lopota, H. L. (1996). *Current widowhood: Myths and realities.* Thousand Oaks, CA: Sage Publications. Exploration of a range of subjects and issues surrounding widowhood including a historical and comparative perspective on the situation of widows in various parts of the world. The author is a pioneer in research on widowhood in the United States.

Rando, T. A. (Ed.). (1986). *Parental loss of a child.* Campaign, IL: Research Press Company. An excellent source for information on various aspects of a parent's death for children.

Riches, G., & Dawson, P. (2000). *An intimate loneliness: Supporting bereaved parents and siblings.* Buckingham, UK: Open University Press. Through a sociological approach, an understanding that grief is not an individual matter, because individuals live in family networks that are part of larger communities.

Rosen, E. J. (1998). *Families facing death.* San Francisco: Jossey-Bass Publishers. A guide for health-care professionals and volunteers in working with families suffering the loss of a family member.

Rosenblatt, P. C. (2000). *Parent grief: Narratives of loss and relationship.* Philadelphia: Taylor and Francis. Explores the grief of a parent upon the death of a child and the aftereffects of that death on the lives of parents as a couple.

Sanders, C. M. (1999). *Grief the mourning after: Dealing with adult bereavement.* New York: John Wiley & Sons. Presents theoretical considerations of bereavement, phases of bereavement, types of bereavement, the multidimensionality of grief, and rituals of loss.

Walter, T. (Ed.). (1999). *The mourning for Diana.* Oxford: Berg. A sociological perspective on the various aspects of the aftermath of the death of Princess Diana.

Washington, P. (Ed.). (1998). *Poems of mourning.* New York: Alfred A. Knopf. A comprehensive work with major verse forms of mourning including epitaphs and laments.

REFERENCES

Abbott, L. (1913, August 30). There are no dead. *Outlook, 104*, 979–908.

Achte, K. (1988, Spring). Suicidal tendencies in the elderly. *Suicide and Life-Threatening Behavior, 18*, 55–65.

Adamek, M. E., & Kaplan, M. S. (1996). The growing use of firearms by suicidal older women, 1979–1992: A research note. *Suicide and Life-Threatening Behavior, 26*, 71–78.

Adams, C. (1988). *More of the straight dope*. New York: Ballantine Books.

Adams, D. W., & Deveau, E. J. (1987). When a brother or sister is dying of cancer: The vulnerability of the adolescent sibling. *Death Studies, 11, 279–295*.

Aiken, L. R. (1991). *Dying, death, and bereavement* (2nd ed.). Boston: Allyn and Bacon.

Albery, N., Elliott, G., & Elliott, J. (Eds.). (1993). *The natural death handbook*. London: Virgin.

Albom, M. (1997). *Tuesdays with Morrie*. New York: Doubleday.

Almgren, G. (1993). Living will legislation, nursing home care, and the rejection of artificial nutrition and hydration: An analysis of bedside decision-making in three states. *Journal of Health and Social Policy, 4*(3), 43–63.

Altman, L. K. (1989, November 14). Physicians endorse more humanities for premed students. *New York Times*, p. 22.

Alvarez, A. (1971). *The savage god: A study of suicide*. London: Weidenfeld and Nicolson.

American Geriatrics Society Public Policy Committee. Voluntary active euthanasia. (1994). *Journal of the American Geriatrics Society, 39, 826*.

American Medical Association Council on Ethical and Judicial Affairs. (1992). *Current opinions*. Chicago: American Medical Association.

American Psychological Association. (1992). Ethical principles of psychologists and code of conduct. *American Psychologist, 47*, 1597–1611.

Anderson, J. (1997, May 27). Funeral industry, seeking new business, courts the living with special services. *New York Times*, p. B1.

Anderson, R. (1996). *Magic, science, and health*. Fort Worth, TX: Harcourt Brace College Publishers.

Angell, M. (1997). The Supreme Court and physician-assisted suicide: The ultimate right. *New England Journal of Medicine, 337*, 50–53.

APACVY (American Psychological Association Commission on Violence and Youth). (1993). *Violence and youth: Psychology's response* (Vol. 1). Washington, DC: American Psychological Association

Archives of the *Philadelphia Gazette*. (1751–1752). Folio II, Item 1841 (May 25) [CD-ROM edition].

Aristotle. (1941). Nicomachean ethics III. In R. Mckeon (Ed. and Trans.), *The basic works of Aristotle*. New York: Random House.

Asarnow, J. R., & Carlson, G. (1988, Summer). Suicide attempts in preadolescent child inpatients. *Suicide and Life-Threatening Behavior, 18*, 129–136.

Ashley-Cameron, S., & Dickinson, G. E. (1979, February). *Nurses' attitudes toward working with dying patients*. Unpublished paper presented at the Alpha Kappa Delta Research Symposium, Richmond, VA.

Association of American Cemetery Superintendents. (1889). *3, 59*.

Atchley, R. C. (1994). *Social forces and aging* (7th ed.). Belmont, CA: Wadsworth.

Atkinson, M. J. (1935). *Indians of the Southwest*. San Antonio, TX: Naylor.

Attig, T. (1991). The importance of conceiving of grief as an active process. *Death Studies, 15*, 385–393.

Ayer, M. (1964). *Made in Thailand*. New York: Knopf.

Baby born without brain dies, but legal struggle will continue. (1992, March 31). *New York Times*, p. A8.

Baker, J. E., & Sedney, M. A. (1996). How bereaved children cope with loss: an overview. In C. A. Corr & D. M. Corr (Eds.), *Adolescence and death* (pp. 109–129). New York: Springer Publishing.

Baring-Gould, W. S., & Baring-Gould, C. (1967). *The annotated Mother Goose*. New York: World Publishing.

Barrett, R. K. (1992). Psychocultural influences on African American attitudes toward death, dying and funeral rites. In J. Morgan (Ed.), *Personal care in an impersonal world*. Amityville, NY: Baywood Publishing.

Barrett, R. K. (1996). Young people as victims of violence. In R. G. Stevenson & E. P. Stevenson (Eds.), *Teaching students about death* (pp. 63–75). Philadelphia: The Charles Press.

Barrett, W. (1958). *Irrational man: A study in existential philosophy*. New York: Doubleday.

Bascom, W. (1969). *The Yoruba of southwestern Nigeria*. New York: Holt, Rinehart and Winston.

Battin, M. P. (1982). *Ethical issues in suicide*. Englewood Cliffs, NJ: Prentice-Hall.

Battin, M. P. (1994). *The least worst death: Essays in bioethics on the end of life*. New York: Oxford University Press.

Becker, E. (1973). *The denial of death*. New York: The Free Press.

Becker, H. (1964, March). Personal change in adult life. *Sociometry, 27*, 40–53.

Becker, H. S. (1963). *Outsiders: Studies in the sociology of deviance*. Glencoe, IL: Free Press.

Beckwith, J. B. (1978). *The sudden infant death syndrome* (DHEW Publication No. HSA 75–5137). Washington, DC: U.S. Government Printing Office.

Bee, H. L. (1992). *The journey of adulthood*. New York: Macmillan.

Beecher, H. K. (1968, August). A definition of irreversible coma. *Journal of the American Medical Association, 205*, 85–88.

Bell, D. (1977). Sex and chronicity as variables affecting attitudes of undergraduates towards peers with suicidal behaviors [microfilm]. *Dissertation Abstracts International, 38*, 3380B.

Bell, M. D. (1996, December). Magic time: Observations of a cancer casualty. *The Atlantic Monthly, 278*, 40–43.

Beloff, J. (1989). Do we have a right to die? In A. Berger, P. Badham, A. H. Kutscher, J. Berger, M. Perry, & J. Beloff (Eds.), *Perspectives on death and dying: Cross-cultural and multi-disciplinary views* (pp. 163–172). Philadelphia: Charles Press.

Bendann, E. (1930). *Death customs: An analytical study of burial rites*. New York: Alfred A. Knopf.

Bender, T. (1973, June). The "rural cemetery" movement. *New England Quarterly, 47*, 196–211.

Benet, S. (1974). *Abkhasians: The long-living people of Caucasus*. New York: Holt, Rinehart and Winston.

Benjamin, C. L. (1882, February). Essay. *The Casket, 7*, 2.

Bennett, J. G. (1999). *Maximize your inheritance for widows, widowers & heirs*. Chicago: Dearborn Financial Publishing.

Berger, P. L. (1969). *Sacred canopy: Elements of a sociological theory of religion*. New York: Doubleday.

Berman, A. L. (1986). Helping suicidal adolescents: Needs and responses. In C. A. Corr & J. N. McNeil (Eds.), *Adolescence and death* (pp. 151–166). New York: Springer Publishing.

Biblarz, A., Biblarz, D. N., Pilgrim, M., & Baldree, B. F. (1991, Winter). Media influence on attitudes toward suicide. *Suicide and Life-Threatening Behavior, 21*, 374–384.

Bigelow, G. (1997, January 29). Letter from Dr. Gordon Bigelow, executive director of the American Board of Funeral Service Education, Brunswick, ME.

Bingham, C. R., Bennion, L. D., Openshaw, D. K., & Adams, G. R. (1994). An analysis of age, gender and racial differences in recent national trends of youth suicide. *Journal of Adolescence, 17*, 53–71.

Black, D., Hardoff, D., & Nelki, J. (1989). Educating medical students about death and dying. *Archives of Disease in Childhood, 64*, 750–753.

Blackhall, L. J., Murphy, S. T., Frank, G., Michel, V., & Azen, S. (1995). Ethnicity and attitudes toward patient autonomy. *Journal of the American Medical Association, 274*, 820–825.

Blau, P. M. (1964). *Exchange and power in social life*. New York: Wiley.

Blauner, R. (1966). Death and social structure. *Psychiatry, 29*, 378–394.

Blendon, R. J., Szalay, U. S., & Knox, R .A. (1992). Should physicians aid their patients in dying? The public perspective. *Journal of the American Medical Association, 267*, 2658–2662.

Bloch, M., & Parry, J. (1982). *Death and the regeneration of life*. Cambridge, MA: Cambridge University Press.

Bluebond-Langner, M. (1978). *The private worlds of dying children*. Princeton, NJ: Princeton University Press.

Bluebond-Langner, M. (1989). Worlds of dying children and their well siblings. *Death Studies, 13*, 1–16.

Bomanji, J. B., Britton, K. E., & Clarke, S. E. M. (1995). *Oncology*. London: British Nuclear Medicine Society.

Bordewich, F. M. (1988, February). Mortal fears: Courses in "death education" get mixed reviews. *Atlantic Monthly, 261*, 30–34.

Bouma, H., Diekema, D., Langerak, E., Rottman, T., & Verhey, A. (1989) *Christian faith, health, and medical practice*. Grand Rapids, MI: Eerdmans.

Bourne, P. (1970). *Men, stress, and Vietnam*. Boston: Little, Brown.

Bowlby, J. (1980). *Attachment and loss* (Vol. III). New York: Basic Books.

Boyd, M. (1997, November–December). Move forward with life. *Modern Maturity*, p. 71.

Boyle, F. M., Vance, J. C., Najman, J. M., & Thearle, M. J. (1996). The mental health impact of still birth, neonatal death or SIDS: Prevalence and patterns of distress among mothers. *Social Science and Medicine, 43*, 1273–1282.

Brabant, M. (1994, September). The high price of everlasting peace. In *Worldwide report: Death*. London: BBC Worldwide.

Bradshaw, A. (1996). The spiritual dimension of hospice: the secularization of an ideal. *Social Science and Medicine, 43*, 409–419.

Brandt, R. B. (1986). The morality and rationality of suicide. In R. F. Weir (Ed.), *Ethical issues in death and dying* (2nd ed., pp. 330–344). New York: Columbia University Press.

Brantner, J. P. (1973, January). *Crisis intervenor*. Paper presented at the Ninth Annual Funeral Service Management Seminar, National Funeral Directors Association, Scottsdale, AZ.

Braun, K. L., & Kayashima, R. (1999). Death education in churches and temples: Engaging religious leaders in the development of educational strategies. In B. de Vries (Ed.), *End of life issues: Interdisciplinary and multidimensional perspectives* (pp. 319–335). New York: Springer Publishing.

Brody, J. E. (2001, January 9). A price to pay as autopsies lose favor. *The New York Times*, p. D8.

Brown, A. S. (1996). *The social processes of aging and old age*. Upper Saddle River, NJ: Prentice Hall.

Brown, J. H., Henteleff, P., Barakat, S., & Rowe, C. J. (1986, February). Is it normal for terminally ill patients to desire death? *American Journal of Psychiatry, 143*, 208–211.

Brown, R. (1996, February). A free market in human organs. *The Future of Freedom Foundation.*

Brozan, N. (1986, January 13). Adolescent suicides: The grim statistics. New York Times, p. 16.

Buckman, R. (1992). *How to break bad news: A guide for health care professionals.* Baltimore: Johns Hopkins University Press.

Bugen, L. A. (1980). Coping: Effects of death education. *Omega, 11*, 175–183.

Butler, R. N., Lewis, M. I., & Sunderland, T. (1991). *Aging and mental health: Positive psychosocial and biomedical approaches* (4th ed.). New York: Macmillan.

Byock, I. (1997). *Dying well: The prospect for growth at the end of life.* New York: Riverhead Books.

Calhoun, L. G., & Allen, B. G. (1991). Social reactions to the survivor of a suicide in the family: A review of the literature. *Omega, 23*, 95–107.

Califano, J. A., Jr. (1986). *America's health care revolution: Who lives? Who dies? Who pays?* New York: Random House.

Callahan, D. (1987). *Setting limits: Medical goals in an aging society.* New York: Simon & Schuster.

Callanan, M., & Kelley, P. (1992). *Final gifts: Understanding the special awareness, needs, and communications of the dying.* New York: Poseidon Press.

Callender, C. O. (1987). Organ donation in blacks: A community approach. *Transplantation Proceedings, 19*, 1551–1554.

Campbell, T. W., Abernethy, V., & Waterhouse, G. J. (1984). Do death attitudes of nurses and physicians differ? *Omega, 14*, 43–49.

CancerNet (2000, February). *Fever, chills, and sweats.* Bethesda, MD: National Cancer Institute. Available: http://www.cancernet.nci.nih.gov

Canetto, S. S. (1992, Spring). Gender and suicide in the elderly. *Suicide and Life-Threatening Behavior, 22*, 80–97.

Canning, R. R. (1965). Mormon return-from-the-dead stories: Fact or folklore? *Utah Academy Proceedings, 42*, I.

Capps, B. (1973). *The Indians.* New York: Time-Life Books.

Carlson, E. (2000, September). Taking back the end of life. *AARP Bulletin,* 18–20.

Carroll, R. M., & Schafer, S. S. (1994, May). Similarities and differences in spouses coping with SIDS. *Omega, 28*, 273–284.

Carse, J. (1981). Death. In K. Crim, R. A. Bullard, & L. D. Shinn (Eds.), *Abingdon dictionary of living religions.* Nashville: Abingdon Press.

Cassell, E. (1975). From the moral to the technical order: Dying in a technological society. In P. Steinfels & R. Veatch (Eds.), *Death inside out.* New York: Harper & Row.

Cassell, E. J. (1991). *The nature of suffering and the goals of medicine.* New York: Oxford University Press.

CDC (Centers for Disease Control and Prevention). (1999, November). *HIV/AIDS surveillance report* (Vol. 11). Washington, DC: U.S. Department of Health and Human Services.

Chambliss, D. F. (1996). *Beyond caring: Hospitals, nurses, and the social organization of ethics.* Chicago: University of Chicago Press.

Charmaz, K. (1975). "The announcement of death by the coroner's deputy." *Urban Life, 4*(3), 296–316.

Charmaz, K. (1980). *The social reality of death.* Reading, MA: Addison-Wesley.

Chatzky, J. S. (1998). Where there's not a will. *Money,* p. 118.

Chatzky, J. S. (2000, September). The last word: Making your health-care wishes clear—and comprehensive. *Money,* p. 172.

Chng, C. L., & Ramsey, M. K. (1984–1985). Volunteers and the care of the terminal patient. *Omega, 15*, 237–244.

Chochinov, H. M., Wilson, K. G., & Ennis, M. (1995). Desire for death in the terminally ill. *American Journal of Psychiatry, 152*, 1185–1191.

Choron, J. (1972). *Suicide.* New York: Charles Scribner's Sons.

Christ, G. H. (2000). *Healing children's grief: Surviving a parent's death from cancer.* New York: Oxford University Press.

Christakis, N. A. (1999). *Death foretold: Prophecy and prognosis in medical care.* Chicago: University of Chicago Press.

Clark, B. (1981). *Whose life is it anyway?* Chicago: Dramatic Publishing.

Clark, D., Dickinson, G., Lancaster, C., Ahmedzai, S., Noble, T., & Philip, I. (2000, March 27–29). *UK geriatric medicine physicians' attitudes to care at the end of life.* Unpublished paper presented at the Palliative Care Congress 2000, Coventry, UK.

Clark, M., & Springen, K. (1986, November 17). The demise of autopsies. *Newsweek,* p. 61.

Clark, N. (1992, February 26). The high costs of dying. *The Wall Street Journal.*

Cleaveland, N. (1847). *Green-wood illustrated.* New York: R. Martin.

Clifford, D., & Jordan, C. (1999). *Plan your estate.* Berkeley, CA: Nolo Press.

Clough, S. B. (1946). *A century of American life insurance.* New York: Columbia University Press.

Clyman, R. I., Green, C., Mikkelsen, C., Rowe, J., & Ataide, L. (1979). Do parents utilize physician follow-up after the death of their newborn? *Pediatrics, 64*, 655–667.

Cockerham, W. C. (1991). *The aging society.* Englewood Cliffs, NJ: Prentice-Hall.

Cockerham, W. C. (1998). *Medical sociology.* Upper Saddle River, NJ: Prentice Hall.

Cockerham, W. C. (2001). *Medical sociology* (8th ed.). Upper Saddle River, NJ: Prentice-Hall.

Codex juris canonici. (1918). Rome, Italy: Polyglot Press.

Coe, R. M. (1970). *Sociology of medicine.* New York: McGraw-Hill.

Cohen, J. S., Fihn, S. D., Boyko, E. J., Jonsen, A. R., & Wood, R. W. (1994). Attitudes toward assisted suicide and euthanasia among physicians in Washington state. *New England Journal of Medicine, 331,* 89–94.

Cohen, R. (1967). *The Kanuri of Bornu.* New York: Holt, Rinehart and Winston.

Cohn, F., Harrold, J., & Lynn, J. (1997, May 30). Medical education must deal with end-of-life care. *Chronicle of Higher Education,* A56.

Coleman, L. (1987). *Suicide clusters.* Boston: Faber and Faber.

Combs, D. W. (1986). *Early gravestone art in Georgia and South Carolina.* Athens, GA: University of Georgia Press.

Conley, H. N. (1979). *Living and dying gracefully.* New York: Paulist Press.

Connor, L. H. (1995, September). The action of the body on society: Washing a corpse in Bali. *Journal of the Royal Anthropological Institute, 1,* 537–560.

Conrad, B. H. (1998). *When a child has been murdered: Ways you can help the grieving parents.* Amityville, NY: Baywood Publishing.

Conrad, N. (1992). Stress and knowledge of suicidal others as factors in suicidal behavior of high school adolescents. *Issues in Mental Health Nursing, 13,* 95–104.

Cook, J. A. (1983). A death in the family: Parental bereavement in the first year. *Suicide and Life-Threatening Behavior, 13,* 42–61.

Cook, J. A. (1988). Dad's double binds: Rethinking fathers' bereavement from a men's studies perspective. *Journal of Contemporary Ethnography, 17,* 285–308.

Cooney, W. (1998). The death poetry of Emily Dickinson. *Omega, 37,* 241–249.

Cope, L. (1978, June 22). Is death, like pregnancy, an all-or-nothing thing? *Minneapolis Tribune,* pp. 1, 6A.

Copp, G. (1998). A review of current theories of death and dying. *Journal of Advanced Nursing, 28,* 382–390.

Copp, G., Richardson, A., McDaid, P., & Marshall-Searson, D. A. (1998). A telephone survey of the provision of palliative day care services. *Palliative Medicine, 12,* 161–170.

Cotton, C. R., & Range, L. M. (1990). Children's death concepts: Relationship to cognitive functioning, age, experience with death, fear of death, and hopelessness. *Journal of Clinical Child Psychology, 19,* 123–127.

Council on Ethical and Judicial Affairs (1991, July). Report 32: Decisions to forgo life-sustaining treatment for incompetent patients. In *Code of Medical Ethics.* Chicago: American Medical Association.

Counts, D. A., & Counts, D. R. (1991). Loss and anger: Death and the expression of grief in Kaliai. In D. R. Counts & D. A. Counts (Eds.), *Coping with the final tragedy: Cultural variation in dying and grieving* (pp. 191 212). Amityville, NY: Baywood Publishing.

Cox, G. R. (1998, May). *Using humor, art, and music with dying and bereaved children.* Unpublished paper presented at the International Conference on Death & Bereavement at King's College, London, Ontario, Canada.

Cox, G. R., & Fundis, R. J. (1992). Native American burial practices. In J. Morgan (Ed.), *Personal care in an impersonal world.* Amityville, NY: Baywood Publishing.

Cox, H. G. (1996a, December 30). While the Grim Reaper toils, corporations reap big profits. *Insight,* 42–43.

Cox, H. G. (1996b). *Later life: The realities of aging* (4th ed.). Upper Saddle River, NJ: Prentice-Hall.

Crandall, R. C. (1991). *Gerontology: A behavioral approach* (2nd ed.). New York: McGraw-Hill.

Crase, D. (1994, Spring). Important consumer issues surrounding death. *Thanatos, 19,* 22–26.

Cremation Association of North America. (1997). *1997 fact sheet.* Milwaukee: Author.

Cremation Association of North America. (2000). *2000 fact sheet.* Milwaukee: Author.

Crissman, J. K. (1994). *Death and dying in central Appalachia: Changing attitudes and practices.* Urbana, IL: University of Illinois Press.

Crowther, C. E. (1980). The stalled hospice movement. *The New Physician,* 26–28.

Cugliari, A. M., & Miller, T. E. (1994, April). Moral and religious objections by hospitals to withholding and withdrawing life-sustaining treatment. *Journal of Community Health, 19*(2), 87–100.

Culver, C. M. (1990). *Ethics at the bedside.* Hanover, NH: University Press of New England.

Cummins, R., Oranto, J. P., & Thies, W. H. (1991). Improving survival from sudden cardiac arrest: The 'chain of survival' concept. *Circulation, 83,* 1832–1847.

Cummins, V. A. (1978, September). *Death education in four-year colleges and universities in the U.S.* Paper presented at the First National Conference on the Forum for Death Education and Counseling.

Curran, D. K. (1987). *Adolescent suicidal behavior.* Washington, DC: Hemisphere Publishing.

Curzer, H. J. (1994, Winter). Withdrawal of life-support: Four problems in medical ethics. *Journal of Medical Humanities, 15*(4), 233–241.

Cuzzort, R. P., & King, E. W. (1995). *Twentieth-century social thought.* Fort Worth, TX: Harcourt Brace College Publishers.

Dalden, J. J. M. van (1999, February). Slippery slopes in flat countries—A response. *Journal of Medical Ethics, 25,* 22–24.

Dattel, A. R., & Neimeyer, R. A. (1990). Sex differences in death anxiety: Testing the emotional expressiveness hypothesis. *Death Studies, 14,* 1–11.

Davies, D. (1999). The week of mourning. In T. Walter (Ed.), *The mourning for Diana* (pp. 3–18). Oxford: Berg.

Davies, D. J. (1997). *Death, ritual and belief: the rhetoric of funerary rites*. London: Cassell.

Davis, D. L., Stewart, M., & Harmon, R. J. (1988, December). Perinatal loss: Providing emotional support for bereaved parents. *Birth, 14*, 242–246.

Dawson, G. D., Santos, J. F., & Burdick, D. C. (1990). Differences in final arrangements between burial and cremation as the method of body disposition. *Omega, 21*, 129-146.

Day, J. (1999, January). Alleviating bone pain using strontium therapy. *Professional Nurse, 14*(4), 1–4.

de Beauvoir, S. (1965). *A very easy death* (P. O'Brien, Trans.). New York: G. P. Putnam.

De Leo, D., Conforti, D., & Carollo, G. (1997, Fall). A century of suicide in Italy: A comparison between the old and the young. *Suicide and Life-Threatening Behavior, 27*(3), 239–249.

de Tocqueville, A. (1945). *Democracy in America*. New York: Vintage Books. (Original work published 1835)

DeFrain, J. D., Jakub, D. K., & Mendoza, B. L. (1992). The psychological effects of sudden infant death on grandmothers and grandfathers. *Omega, 24*, 165–182.

DeFrain, J. D., Taylor, J., & Ernst, L. (1982). *Coping with sudden infant death*. Lexington, MA: Lexington Books.

Demallie, R. J., & Parks, D. R. (1987). *Sioux Indian religion: Tradition and innovation*. Norman, OK: University of Oklahoma Press.

Deng, F. M. (1972). *The Dinka of the Sudan*. New York: Holt, Rinehart and Winston.

Dentan, R. K. (1968). *The Semai: A non-violent people of Malaya*. New York: Holt, Rinehart and Winston.

Derry, S. (1997). Dying for palliative care. *European Journal of Palliative Care, 4*(3), 66–71.

Devons, C. A. J. (1996, March). Suicide in the elderly: How to identify and treat patients at risk. *Geriatrics, 51*, 67–71.

DeVries, R., & Subedi, J. (1998). *Of bioethics and society: Constructing the ethical enterprise*. Upper Saddle River, NJ: Prentice Hall.

Dickinson, C. W. (1996, October 19). Personal correspondence to the author.

Dickinson, G. E. (1985, December). Changes in death education in U.S. medical schools during 1975–1985. *Journal of Medical Education, 60*, 942–943.

Dickinson, G. E. (1986, April 18–20). *Effects of death education on college students' death anxiety*. Unpublished paper presented at the Forum for Death Education and Counseling, Atlanta, GA.

Dickinson, G. E. (1988, January). Death education for physicians. *Journal of Medical Education, 63*, 412.

Dickinson, G. E. (1992). First childhood death experiences. *Omega, 25*, 169–182.

Dickinson, G. E., & Ashley-Cameron, S. (1986, April 4–6). *Sex role socialization versus occupational role socialization: A comparison of female physicians' and female nurses' attitudes toward dying patients*. Unpublished paper presented at the Eastern Sociological Society's Annual Meeting, New York City.

Dickinson, G. E., & Fritz, J. L. (1981, September 2). Death in the family. *Journal of Family Issues*, 379–384.

Dickinson, G. E., & Mermann, A. C. (1996, December). Death education in U.S. medical schools, 1975–1995. *Academic Medicine, 71*, 1348–1349.

Dickinson, G. E., & Pearson, A. A. (1979a.) Differences in attitudes toward terminal patients among selected medical specialties of physicians. *Medical Care, 17*, 682–685.

Dickinson, G. E., & Pearson, A. A. (1979b.) Sex differences of physicians in relating to dying patients. *Journal of the American Medical Women's Association, 34*, 45–47.

Dickinson, G. E., & Pearson, A. A. (1980–1981). Death education and physicians' attitudes toward dying patients. *Omega, 11*, 167–174.

Dickinson, G. E., & Tournier, R. E. (1993, January-February). A longitudinal study of sex differences in how physicians relate to dying patients. *Journal of the American Medical Women's Association, 48*, 19–22.

Dickinson, G. E., & Tournier, R. E. (1994). A decade beyond medical school: A longitudinal study of physicians' attitudes toward death and terminally-ill patients. *Social Science and Medicine, 38*, 1397–1400.

Dickinson, G. E., Lancaster, C. J., Summer, E. D., & Cohen, J. S. (1997). Attitudes toward assisted suicide and euthanasia among physicians in South Carolina and Washington. *Omega, 36*, 201–218.

Dickinson, G. E., Lancaster, C. J., Winfield, I. C., Reece, E. F., & Colthorpe, C. A. (1997). Detached concern and death anxiety of first-year medical students: Before and after the gross anatomy course. *Clinical Anatomy, 10*, 201–207.

Dickinson, G. E., Sumner, E. D., & Frederick, L. M. (1992, May–June). Death education in selected health professions. *Death Studies, 16*, 281–289.

Dickinson, G. E., Tournier, R. E., & Still, B. J. (1999). Twenty years beyond medical school: Physicians' attitudes toward death and terminally ill patients. *Archives of Internal Medicine, 159*, 1741–1744.

Dillard, J. (1983). *Multicultural counseling*. Chicago: Nelson-Hall.

Dilworth, D. C. (1996, February). Dying wishes are ignored by hospitals and doctors. *Trial, 32*(2), 79–81.

Doctors "kill" patient to save her life. (1988, September 22). Charleston, SC, *News and Courier*, p. 1A.

Doka, K. J. (1987). Silent sorrow: Grief and the loss of significant others. *Death Studies, 11*, 441–449.

Doka, K. J. (1989). *Disenfranchised grief*. Lexington, MA: Lexington Books.

Doka, K. J. (1993). *Living with life-threatening illness: A guide for patients, their families, and caregivers.* New York: Lexington Books.

Domino, G., Domino, V., & Berry, T. (1987). Children's attitudes toward suicide. *Omega, 17,* 279–287.

Donne, J. (1930). *Biathanatos.* New York: Facsimile Text Society. (Original work published 1644)

Donnison, D., & Bryson, C. (1995). Matters of life and death: Attitudes to euthanasia. In R. Jowell, J. Curtice, A. Park, L. Brook, & D. Thomson (Eds.). *British Social Attitudes: The 13th Report* (pp. 161–184). Aldershot, UK: Dartmouth Publishing.

Douglas, J. D. (1967). *The social meanings of suicide.* Princeton, NJ: Princeton University Press.

Douglas, J. D. (1991). Patterns of change following parent death in midlife adults. *Omega, 22,* 123–137.

Dowd, Q. L. (1921). *Funeral management and costs: A world survey of burial and cremation.* Chicago: University of Chicago Press.

Downey, M. (1997, July/August). Developing palliative care services for patients with AIDS. *European Journal of Palliative Care, 4*(4), 105–109.

Downing, A. J. (1921). *Landscape gardening* (10th ed.). New York: Wiley.

Dozier, E. P. (1966). *Hano: A Tewa Indian community in Arizona.* New York: Holt, Rinehart and Winston.

Dozier, E. P. (1967). *The Kalinga of Northern Luzon, Philippines.* New York: Holt, Rinehart and Winston.

Duff, R. W., & Hong, L. K. (1995). Age density, religiosity and death anxiety in retirement communities. *Review of Religious Research, 37,* 19–32.

Dumont, R. G., & Foss, D. C. (1972). *The American view of death: Acceptance or denial?* Cambridge, MA: Schenkman.

Dunn, R. G., & Morrish-Vidners, D. (1987). The psychological and social experience of suicide survivors. *Omega, 18,* 175–186.

Durand, R. P., Dickinson, G. E., Sumner, E. D., & Lancaster, C. J. (1990, Spring–Summer). Family physicians' attitudes toward death and the terminally-ill patient. *Family Practice Research Journal, 9,* 123–129.

Durkheim, E. (1915). *The elementary forms of religious life.* New York: George Allen and Unwin.

Durkheim, E. (1947). *The elementary forms of religious life.* Glencoe, IL: Free Press. (Original work published 1915)

Durkheim, E. (1954). *Elementary forms of the religious life.* New York: The Free Press. (Original work published 1915)

Durkheim, E. (1951). *Suicide: A study in sociology.* New York: The Free Press. (Original work published 1897)

Durkheim, E. (1961). *Moral education.* Glencoe, IL: Free Press.

Durkheim, E. (1964). *The division of labor in society* (G. Simpson, Trans.). New York: Free Press. (Original work published 1893)

Durkheim, E. (1964). *The rules of the sociological method.* New York: The Free Press.

Dyregrov, A., & Kyregrov, K. (1999). Long-term impact of sudden infant death: a 12- to 15-year follow-up. *Death Studies, 23,* 635–661.

Eagen, T. (1992, January 8). Creating a pleasant stop on the journey to death. *New York Times,* A10.

Earl, W. L., Martindale, C. J., & Cohn, D. (1992). Adjustment: Denial in the styles of coping with HIV infection. *Omega, 24,* 35–47.

Eddy, J. M., & Alles, W. F. (1983). *Death education.* St. Louis: C. V. Mosby.

Edney, A. T. B. (1988). Breaking the news: the problems and some answers. In W. J. Kay et al. (Eds.), *Euthanasia of the companion animal.* Philadelphia: The Charles Press.

Edwards, G., & Mazzuca, J. (1999, March 24). Three quarters of Canadians support doctor-assisted suicide. Gallup News Service, Toronto, Ontario, Canada.

Eickelman, D. F. (1987). Rites of passage: Muslim rites. In M. Eliade (Ed.), *The encyclopedia of religion* (12th ed.). New York: Macmillan.

Ellis, J. B., & Range, L. M. (1989). Characteristics of suicidal individuals: A review. *Death Studies, 13,* 485–500.

Emanuel, E., & Emanuel L. (1998). The promise of a good death. *Lancet, 351,* 21–29.

Emmons, N. (1842). Death without order. In J. Ede (Ed.), *The works of Nathaniel Emmons* (Vol. 3). Boston: Crocker and Brewster.

Enright, L. (1994, September). Keeping the corpse company with a whiskey. In *World-wide report: Death.* London: BBC World-wide.

Erikson, E. (1959). Identity and the life cycle: Selected papers. *Psychological Issues, 1,* 1–171.

Erikson, E. (1963). *Childhood and society.* New York: Norton.

Euthanasia: What is the good death? (1991, July 20). *The Economist,* 21–24.

Evans, E. J., & Fitzgibbon, G. H. (1992). The dissecting room: Reactions of first year medical students. *Clinical Anatomy, 5,* 311–320.

Ewing, C. P. (1990). *Kids who kill.* Lexington, MA: Lexington Press.

EXIT. (1980). *A guide to self-deliverance.* London: Author.

Extracts. (1895, August). *Park and Cemetery, 5,* 108.

Faber-Langendoen, K., & Bartels, D. M. (1992, May). Process of forgoing life-sustaining treatment in a university hospital: An empirical study. *Critical Care Medicine, 20*(5), 570–577.

Farberow, N. L. (1975). *Suicide in different cultures.* Baltimore: University Park Press.

Farnham, P. G. (1994, May). Defining and measuring the costs of the HIV epidemic to business firms. *109 Public Health Reports,* 311–318.

Faron, L. C. (1968). *The Mapuche Indians of Chile.* New York: Holt, Rinehart and Winston.

Farrell, J. J. (1980). *Inventing the American way of death, 1830–1920.* Philadelphia: Temple University Press.

Faulkner, J., & DeJong, G. F. (1966). Religiosity in 5-D: An empirical analysis. *Social Forces,* 45, 246–254.

Federated Funeral Directors of America. (1997). *1995 facts on funeral costs.* Springfield, IL: Author.

Federated Funeral Directors of America (1999). *Survey of funeral homes.* Springfield, IL: Author.

Feifel, H. (1959). *The meaning of death.* New York: McGraw-Hill.

Fein, E. B. (1998, January 25). For lost pregnancies, new rites of mourning. *New York Times,* p. 30.

Feldman, D. A. (1990). *Culture and AIDS.* New York: Praeger.

Fergusson, D. M., Woodward, L. J., & Horwood, L. J. (2000, January). Risk factors and life processes associated with the onset of suicidal behavior during adolescence and early adulthood. *Psychological Medicine,* 30, 23–39.

Field, D., & Howells, K. (1986). Medical students' self-reported worries about aspects of death and dying. *Death Studies,* 10, 147–154.

Fingarette, H. (1996) *Death: Philosophical soundings.* Peru, IL: Open Court Publishing.

Fingerhut, L. A., & Warner, M. (1997). *Injury chartbook. Health, United States, 1996–97.* Hyattsville, MD: National Center for Health Statistics.

Fletcher, J. (1966). *Situational ethics: The new morality.* Philadelphia: Westminster.

Flynn, C. P. (1986). *After the beyond: Human transformation and the near-death experience.* Englewood Cliffs, NJ: Prentice-Hall.

Foderaro, L. W. (1994, April 7). Death no longer a taboo subject. *Palm Beach Post.*

Fortner, B. V., & Neimeyer, R. A. (1999). Death anxiety in older adults: A quantitative review. *Death Studies,* 23, 387–411.

Frank, A. W. (1991). *At the will of the body: Reflections on illness.* Boston: Houghton Mifflin.

Fraser, T. M., Jr. (1966). *Fishermen of south Thailand: The Malay villagers.* New York: Holt, Rinehart and Winston.

Freidson, E. (1972). *Profession of medicine.* New York: Dodd Mead.

French, S. (1975). The establishment of Mount Auburn and the "rural cemetery" movement. In *Death in America.* Philadelphia: University of Pennsylvania Press.

Freund, P. E. S., & McGuire, M. B. (1995). *Health, illness, and the social body: A critical sociology* (2nd ed.). Englewood Cliffs, NJ: Prentice-Hall.

Fulton, G. B., & Metress, E .K. (1995). *Perspectives on death and dying.* Boston: Jones and Bartlett Publishers.

Fulton, R., & Owen, G. (1988). Death and society in twentieth century America. *Omega,* 18, 379–394.

Funeral and Memorial Information Council (1992, March). The Funeral and Memorial Information Council conduct study. *Canadian Funeral Director,* 8.

Funeral directors. (1883, June). *The Casket,* 8.

Fung, K. K. (1993). Wealth transfer through voluntary death. *Journal of Health and Social Policy,* 5(2), 77–86.

Furman, E. (1978) The death of a newborn: Care of the parents. *Birth Family Journal,* 5, 214.

Gabriel, T. (1991, December 8). A fight to the death. *New York Times Magazine,* 46–48.

Gallup, G. H. (1994). *Religion in America: 1994 (Suppl.).* Princeton, NJ: Gallup Poll.

Gallup Poll. (1996). Princeton, NJ.

Gallup Survey. (1993, April). Physicians need to detect suicide warning signs. *Geriatrics,* 48, 16.

Gamarekian, B. (1987, June 25). A support network for dying children. *New York Times,* pp. 19–20.

Gamst, F. C. (1969). *The Qemant: A pagan-Hebraic peasantry of Ethiopia.* New York: Holt, Rinehart and Winston.

Garfield, C. A. (1978). *Psychosocial care of the dying patient.* New York: McGraw-Hill.

Garfinkel, B. D., Froese, A., & Hood, J. (1982, October). Suicide attempts in children and adolescents. *American Journal of Psychiatry,* 139, 1257–1261.

Gaspard, N. J. 1970. The family of the patient with long-term illness, *Nursing Clinics of North America,* 5, 77–84.

Gasperson, K. R. (1996). *Delivering bad news in the clinical context: Current recommendations and student perspectives.* Unpublished master's thesis, University of Kentucky, Lexington.

Gebhart, J. C. (1927). *The reasons for present-day funeral costs.* Unpublished article.

Geddes, G. (1981). *Welcome joy: Death in Puritan New England.* Ann Arbor, MI: U.M.I. Research Press.

Gentile, M., & Fello, M. (1990, November). Hospice care for the 1990s: A concept coming of age. *The Journal of Home Health Care Practice,* 1–15.

Gervais, K. G. (1987). *Redefining death.* New Haven: Yale University Press.

Gibbs, J. T. (1997, Spring). African-American suicide: A cultural paradox. *Suicide and Life-Threatening Behavior,* 27(1), 68–79.

Gibson, J. A., Range, L. M., & Anderson, H. N. (1987). Adolescents' attitudes toward suicide: Does knowledge that the parents are divorced make a difference? *Journal of Divorce,* 163–167.

Gill, S. D. (1981). *Sacred words: A study of Navajo religion and prayer.* Westport, CT: Greenwood.

Gillick, M. (2000). *Lifelines: living longer, growing frail, taking heart*. New York: W. W. Norton & Company.

Gillon, R. (1990). Editorial: Death. *Journal of Medical Ethics, 16,* 3–4.

Gilman, C. P. (1935). *The living of Charlotte Perkins Gilman*. Madison: University of Wisconsin Press.

Ginsberg, E. (1991). Access to health care for Hispanics. *Journal of the American Medical Association, 165,* 238–241.

Ginsberg, H., & Opper, S. (1979). *Piaget's theory of intellectual development* (2nd ed.). Englewood Cliffs, NJ: Prentice-Hall.

Glaser, B., & Strauss, A. (1965). *Awareness of dying*. Chicago: Aldine.

Glock, C., & Stark, R. (1966). *Christian beliefs and anti-Semitism*. New York: Harper and Row.

Goffman, E. (1959). *The presentation of self in everyday life*. New York: Doubleday.

Goffman, E. (1963). *Stigma*. Englewood Cliffs, NJ: Prentice-Hall.

Golden, T. R. (2000). *Swallowed by a snake: The gift of the masculine side of healing*. New York: McDonald and Woodward Company.

Goldenberg, S. (2000, April 11). Many in America are resigned to pain. *New York Times*, p. D8.

Goldman, A. L. (1993, February 15). Increasingly, funeral business gets female touch. *New York Times*, p. A8.

Goldman, I. (1979). *The Cubeo: Indians of the northwest Amazon*. Urbana, IL: University of Illinois Press.

Goleman, D. (1989, December 5). Fear of death intensifies moral code, scientists find. *New York Times,* p. 19.

Goodwin, D. M., Higginson, I. J., Myers, K., Douglas, H. R., & Normand, C. E. (2000). *Methodological issue in evaluating palliative day care: A multi-centre study*. Paper presented at the Palliative Care Congress, Conventry, England, March 27–29.

Gordon, A. (1986). The tattered cloak of immortality. In C. A. Corr & J. N. McNeil (Eds.), *Adolescence and death* (pp. 16–29). New York: Springer Publishing.

Gordon, A. K., & Klass, D. (1979). *The need to know: How to teach children about death*. Englewood Cliffs, NJ: Prentice-Hall.

Gordon, N. P., & Shade, S. B. (1999, April 12). Advance directives are more likely among seniors asked about end-of-life care preferences. *Archives of Internal Medicine, 159,* 701–704.

Gorer, G. (1955, October). The pornography of death. *Encounter, 5,* 49–53.

Gorer, G. (1965). *Death, grief, and mourning*. Garden City, NY: Doubleday.

Gottlieb, B. (1993). *The family in the Western world: From the Black Death to the Industrial Age*. Oxford, UK: Oxford University Press.

Gould, M. S., & Shaffer, D. (1986, September 11). The impact of suicide in television movies. *New England Journal of Medicine, 315,* 690–694.

Gould, M. S., Wallenstein, S., & Davidson, L. (1989, Spring). Suicide clusters: A critical review. *Suicide and Life-Threatening Behavior, 19,* 17–29.

Grady, D. (2000a, May 30). Charting a course of comfort and treatment at the end of life. *The New York Times*, p. D-7.

Grady, D. (2000b, May 29). At life's end, many patients are denied peaceful passing. *The New York Times*, p. A-13.

Grassi, L., Magnani, K., & Ercolani, M. (1999). Attitudes toward euthanasia and physician-assisted suicide among Italian primary care physicians. *Journal of Pain and Symptom Management, 17,* 188–196.

Gray-Toft, P. A., & Anderson, J. G. (1986–1987). Sources of stress in nursing terminal patients in a hospice. *Omega, 17,* 27–38.

Greer, W. R. (1985, June 30). Putting a price on human life. Louisville, KY, *Courier-Journal*, p. 1D.

Grollman, E. (1967). *Explaining death to children*. Boston: Beacon Press.

Grollman, E. A. (1972, May). *Commencement address*. Department of Mortuary Science, University of Minnesota, Minneapolis, MN.

Gubrium, J. B. (1975). *Living and dying at Murray Manor*. New York: St. Martin's Press.

Guidelines to help discern SIDS from homicide. (1996, June 21). Charleston, SC, *Post and Courier*, p. 4A.

Guignon, C. (1993). *The Cambridge companion to Heidegger*. Cambridge: Cambridge University Press.

Guillon, C., & LeBonniec, Y. (1982). *Suicide, its use, history, technique and current interest*. Paris: Editions Alain Moreais.

Gustafson, E. (1972). Dying: The career of the nursing home patient. *Journal of Health and Social Behavior, 13,* 226–235.

Guterman, L. (2000, June 2). The dope on medical marijuana. *Chronicle of Higher Education, 46,* A21–A22.

Gutierrez, P., King, C. A., & Ghaziuddin, N. (1996). Adolescent attitudes about death in relation to suicidality. *Suicide and Life-Threatening Behavior, 26,* 8–18.

Guyer, B., Marin, J. A., & MacDorman, M. F., Anderson, R. N., & Strobino, D. M. (1997). Annual summary of vital statistics—1996. *Pediatrics, 100,* 905–918.

Guyer, R. L. (1998, February 6–8). When decisions are life-and-death. *USA Weekend*, 26.

Habenstein, R. W., & Lamers, W. M. (1962). The pattern of late 19th century funerals. In *The history of American funeral directing*. Milwaukee: Bulfin.

Habenstein, R. W., & Lamers, W. M. (1974). *Funeral customs the world over* (Rev. ed.). Milwaukee: Bulfin Printers.

Hafferty, F. W. (1991). *Into the valley: Death and the socialization of medical students*. New Haven: Yale University Press.

Haight, B. K. (1992). Long-term effects of a structured life review process. *Journal of Gerontology: Psychological Sciences, 47,* 312–315.

Hale, N. G. (1971). *The origin and foundations of the psychoanalytic movement in the United States, 1876–1918*. New York: Oxford University Press.

Hale, R. (1971, September). Some lessons on dying. *Christian Century, 88*, 1076–1079.

Hall, G. S. (1922). *Youth: Its education, regimen and hygiene*. New York: D. Appleton.

Hammond, P., & Mosley, M. (1999). *Trust me (I'm a doctor)*. London, Metro Books.

Hammond, W. R., & Yung, B. (1993). Psychology's role in the public health response to assaultive violence among young Afro-American men. *American Psychologist, 48*, 143–154.

Hanks, G. W., deConno, F., Ripamonti, C., Hanna, M., McQuay, H. J., Mercadante, S., Meynadier, J., Poulain, P., & Casas, J. R. (1996). Morphine in cancer pain: modes of administration. *British Medical Journal, 312*, 823–826.

Hanson, L. C., Henderson, M., & Rodgman, E. (1999). Where will we die? A national study of nursing home death. *Journal of General Internal Medicine, 14*, 101.

Hardin, G. (1985, September-October). Crisis on the commons. *The Sciences*, 21–25.

Hare, S. J. (1910). The cemetery beautiful. *Association of American Cemetery Superintendents, 24*, 41.

Hart, C. W. M., & Pilling, A. R. (1960). *The Tiwi of North Australia*. New York: Holt, Rinehart and Winston.

Harvey, J. H. (1996). *Embracing their memory: Loss and the social psychology of storytelling*. Needham Heights, MA: Allyn and Bacon.

Hassan, R. (1996). The euthanasia debate. *Medical Journal of Australia, 165*, 164–165.

Hassrick, R. B. (1964). *The Sioux: Life and customs of a warrior society*. Norman, OK: University of Oklahoma.

Hatter, B. S. (1996). Children and the death of a parent or grandparent. In C. A. Corr & D. M. Corr (Eds.), *Handbook of childhood death and bereavement* (pp. 131–148). New York: Springer Publishing.

Haviland, W. A. (1991). *Anthropology* (6th ed.). Fort Worth, TX: Holt, Rinehart and Winston.

Hawkins, A. H. (1990). Constructing death: Three pathographies about dying. *Omega, 22*, 301–317.

Hay, E. E. (1900). Influence of our surroundings. *Association of American Cemetery Superintendents, 14*, 46.

Hays, J. C., Gold, D. T., Flint, E. P., & Winer, E. P. (1999). Patient preference for place of death: A qualitative approach. In B. de Vries (Ed.), *End of life issues* (pp. 3–21). New York: Springer Publishing.

Hays, J. C., Gold, D. T., Flint, E. P., & Winer, E. P. (1999). Patient preference for place of death: A qualitative approach. In B. de Vries (Ed.), *End of life issues: Interdisciplinary and multidimensional perspectives* (pp. 3–21). New York: Springer Publishing.

Hazlick, R. (Ed.). (1994). *The Medical Cause of Death Manual*. Northfield, IL: College of American Pathologists.

Heidegger, M. (1962). *Being and time*. San Francisco: Harper. (Original work published 1927)

Helman, C. (1985). *Culture, health, and illness*. Bristol, England: Wright.

Hendin, H. (1997). *Seduced by death: Doctors, patients and the Dutch cure*. New York: W. W. Norton & Co.

Hertz, R. (1960). The collective representation of death. In R. Needham & C. Needham (Trans.), *Death and the right hand* (pp. 84–86). Aberdeen: Cohen and West.

Higginson, I. J. (1999, November/December). Evidence-based palliative care. *European Journal of Palliative Care, 6*, 188–193.

Hill, P. T. (1992). Individual rights vs. state interests: Ethical concerns in thanatology. *Loss, Grief and Care, 6*(1), 51–59.

Hillerman, B. (1980). Chrysalis of gloom: Nineteenth century mourning costume. In M. V. Pike & J. V. Armstrong (Eds.), *A time to mourn: Expressions of grief in the nineteenth century America*. Stony Brook, NY: The Museums at Stony Brook.

Hinker, F. (1985, December 13). Death films popular. *El Dorado (AR) Times*, p. 16.

Hinton, J. (1979). Comparison of places and policies for terminal care. *Lancet, 1*, 29–32.

Hirschfelder, A., & Molin, P. (1992). *The encyclopedia of Native American religions*. New York: Facts on File.

Hitchcock, J. T. (1966). *The Magars of Manyan Hill*. New York: Holt, Rinehart and Winston.

Hocker, W. V. (1987). *Financial and psychosocial aspects of planning and funding funeral services in advance as related to estate planning and life-threatening illness*. Unpublished article distributed by the National Funeral Directors Association.

Hoebel, E. A. (1960). *The Cheyennes: Indians of the Great Plains*. New York: Holt, Rinehart and Winston.

Hoefler, J. M. (2000). Making decisions about tube feeding for severely demented patients at the end of life: Clinical, legal, and ethical considerations. *Death Studies, 24*, 233–254.

Hohenschuh, W. P. (1921). *The modern funeral: Its management*. Chicago: Trade Periodical Company.

Hohmann, A. A., & Larson, D. B. (1993). Psychiatric factors predicting use of clergy. In E. L. Worthington, Jr. (Ed.), *Psychotherapy and religious values* (pp. 71–84). Grand Rapids, MI: Baker Book House.

Holland, J. C. (1994). Now we tell—But how well? *Journal of Clinical Oncology, 7*, 557–559.

Holy Bible (Revised Standard Version). (1962). New York: Oxford University Press.

Homans, G. C. (1965). Anxiety and ritual: The theories of B. Malinowski and R. A. Radcliffe-Brown. In W. A. Lessa & E. Z. Vogt (Eds.), *Reader in comparative religion: An anthropological approach*. New York: Harper and Row.

Hooyman, N. R., & Kiyak, H. A. (1988). *Social gerontology: A multidisciplinary perspective*. Boston: Allyn and Bacon.

Horowitz, M. M. (1967). *Morne-Paysan: Peasant village in Martinique.* New York: Holt, Rinehart and Winston.

Hostler, Sharon L. (1978). The development of the child's concept of death. In O. J. Z. Sahler (Ed.), *The child and death* (pp. 1–25), St. Louis: C. V. Mosby.

Howarth, G. (2000). Dismantling the boundaries between life and death. *Mortality, 5*(2), 127–138.

Howe, D. W. (1970). *The Unitarian conscience: Harvard University Pres, 1805–1861.* Cambridge: Harvard University Press.

Hu, T. W. (1984). Health services in the People's Republic of China. In M. W. Raffel (Ed.), *Comparative health systems: Descriptive analysis of fourteen national health systems* (pp. 133–152). University Park, PA: The Pennsylvania State University Press.

Huang, Z., & Ahronheim, J. C. (2000, May). Nutrition and hydration in terminally ill patients. *Clinics in Geriatric Medicine. 16*(2), 313–325.

Huber, R., Meade-Cos, V., & Edelen, W. B. (1992). Right to die responses from a random sample of 200. *Hospice Journal, 8*(3), 1–19.

Hughes, J. A., Martin, P. J., & Sharrock, W. W. (1995). *Understanding classical sociology: Marx, Weber, and Durkheim.* London, Sage Publications.

Humphry, D. (1981). *Let me die before I wake: Hemlock's book of self-deliverance for the dying.* Los Angeles: The Hemlock Society.

Humphry, D. (1991). *Final exit: The practicalities of self-deliverance and assisted suicide for the dying.* Eugene, OR: The Hemlock Society.

Hunfield, J. A. M., Wladimiroff, J. W., Verhage, F., & Passchier, J. (1995). Previous stress and acute psychological defense as predictors of perinatal grief—An exploratory study. *Social Science and Medicine, 40*, 829–835.

Huntington, R., & Metcalf, P. (1979). *Celebrations of death: The anthropology of mortuary ritual.* Cambridge, UK: Cambridge University Press.

Hutchins, S. H. (1986). Stillbirth. In T. A. Rando (Ed.), *Parental loss of a child* (pp. 129–144). Champaign, IL: Research Press Company.

Hyner, G. C., & Melby, C. L. (1987). *Priorities for health promotion and disease prevention.* Dubuque, IA: Eddie Bowers.

Idler, E. L., & Kasl, S. V. (1992). Religion, disability, depression, and the timing of death. *American Journal of Sociology, 97*, 1052–1079.

Illich, I. (1976). *Medical nemesis: The expropriation of health.* New York: Pantheon.

Ingram, E., & Ellis, J. B. (1992). Attitudes toward suicidal behavior: A review of the literature. *Death Studies, 16*, 31–43.

Institute of Medicine (1997). *Approaching death: Improving care at the end of life.* Washington, DC: National Academy Press.

Irion, P. E. (1956). *The funeral: An experience of value.* Milwaukee: National Funeral Directors Association.

Is life expectancy now stretched to its limit? (1990, November 2). *New York Times*, p. A13.

Jackson, E. N. (1963). *For the living.* Des Moines, IA: Channel Press.

Jacques, E. (1965). Death and the mid-life crisis. *International Journal of Psychoanalysis, 46*, 502–514.

Jankel, C. A., Wolfgang, A. P., & Hoff, M. J. (1994, March–April). Health care priorities: Opinions of one state's citizens and legislators. *Health Values, 18*(2), 7–14.

Jasper, J. H., Lee, B. C., & Miller, K. E. (1991). Organ donation terminology: Are we communicating life or death? *Health Psychology, 10*, 34–41.

Jauhar, S. (2000, January 4). When decisions can mean life or death. *New York Times*, D-8.

Jecker, N. S. (1994). Physician-assisted death in the Netherlands and the United States: Ethical and cultural aspects of health policy development. *Journal of the American Geriatric Society, 42*, 672–678.

Jeffrey, D. (1993). *"There is nothing more I can do!": An introduction to the ethics of palliative care.* Cornwell, UK: The Patten Press.

Jenkins, R. (1999, February 18). Expert wants trade in live body parts. *The Times*, London, p. 8.

Johnson, C. B., and Slaninka, S. C. (1999). Barriers to accessing hospice services before a late terminal stage. *Death Studies, 23*, 225–238.

Johnson, G. R., Krug, E. G., & Potter, L. B. (2000, Spring). Suicide among adolescents and young adults: A cross-national comparison of 34 countries. *Suicide and Life-Threatening Behavior, 30* (1), 74–82.

Johnson, P. V. (1997, April). Creating meaningful events that celebrate life. *Bradshaw Quarterly.*

Johnson, S. (1990, August 26). Near-death experiences almost always change lives. Charleston, SC, *News and Courier/The Evening Post*, p. 13I.

Joralemon, D. (1995). Organ wars: The battle for body parts. *Medical Anthropology Quarterly, 9*(3), 335–356.

Joralemon, D. (1999). *Exploring medical anthropology.* Boston: Allyn & Bacon.

Jung, C. (1923). *Psychological types.* London: Pantheon Books.

Jung, C. (1933). *Modern man in search of a soul.* New York: Harcourt and Brace.

Jung, C. (1971). The stages of life. In J. Campbell (Ed.), *The portable Jung.* New York: Viking Press.

Kail, R. V., & Cavanaugh, J. C. (1996). *Human development.* Pacific Grove, CA: Brooks/Cole Publishing.

Kaldjian, L. C., Wu, B. J., Jekel, J. F., Kaldjian, E. P., & Duffy, T. P. (1999, December 30). Insertion of femoral-vein catheters for practice by medical house officers during cardiopulmonary resuscitation. *The New England Journal of Medicine, 341,* 2088–2091.

Kalish, R. A. (1965). The aged and the dying process: The inevitable decisions. *Journal of Social Issues, 21,* 87–96.

Kalish, R. A. (1966). Social distance and the dying. *Community Mental Health, 11,* 152–155.

Kalish, R., & Reynolds, D. (1976). *Death and ethnicity: A psychocultural study.* Los Angeles: University of Southern California.

Kalish, R. A., & Reynolds, D. K. (1981). *Death and ethnicity: A psychocultural study.* Farmingdale, NY: Baywood.

Kamerman, J. B. (1988). *Death in the midst of life: Social and cultural influences on death, grief and mourning.* Englewood Cliffs, NJ: Prentice-Hall.

Kane, A. C., & Hogan, J. D. (1986). Death anxiety in physicians: Defensive style, medical specialty, and exposure to death. *Omega, 16,* 11–22.

Kass, L. R. (1971). Death as an event: A commentary on Robert Morison. *Science, 173,* 698–702.

Kastenbaum, R. (1992, Spring). Death, suicide and the older adult. *Suicide and Life-Threatening Behavior, 22,* 1–14.

Kastenbaum, R. J. (1986). Death in the world of adolescence. In C. A. Corr & J. N. McNeil (Eds.), *Adolescence and death* (pp. 4–15). New York: Springer Publishing.

Kastenbaum, R. J. (1991). *Death, society, and human experience* (4th ed.). New York: Macmillan.

Kastenbaum, R. J. (1992a, Spring). Death, suicide and the older adult. *Suicide and Life-Threatening Behavior, 22,* 1–14.

Kastenbaum, R. J. (1992b). *The psychology of death* (2nd ed.). New York: Springer Publishing.

Kastenbaum, R. J. (1995). *Death, society and human experience* (5th ed.). Columbus, OH: Charles E. Merrill.

Kastenbaum, R. J. (1998). *Death, society, and human experience* (6th ed.). Boston: Allyn and Bacon.

Kaufman, S. R. (2000, March). In the shadow of death and dying: Medicine and cultural quandaries of the vegetative state. *American Anthropologist, 102,* 69–83.

Kaufmann, W. (1976). *Existentialism, religion and death.* London: New English Library.

Kavanaugh, R. E. (1972). *Facing death.* Baltimore: Penguin Books.

Kellehear, A. (1990a). *Dying of cancer: The final year of life.* Chur, Switzerland: Harwood Academic Publishers.

Kellehear, A. (1990b). *Dying of cancer.* London: Harwood.

Kellehear, A. (1996). *Experiences near death: Beyond medicine and religion.* New York: Oxford University Press.

Kelly, T. E. (1987, February). Predict preneed vital to financial future. *The American Funeral Director,* 31–69.

Kelner, N. J., & Bourgeault, I. L. (1993). Patient control over dying: Responses of health care professionals. *Social Science and Medicine, 36,* 757–765.

Kennell, J. H., Slyter, H., & Klaus, M. H. (1970). The mourning response of parents to the death of a newborn infant. *The New England Journal of Medicine, 283,* 344–349.

Kesselman, I. (1990). Grief and loss: Issues for abortion. *Omega, 21*(3), 241–247.

Kircher, L. T. (1992). Autopsy and mortality statistics: Making a difference. *Journal of the American Medical Association, 267,* 1264–1270.

Klass, D. (1987). Marriage and divorce among bereaved parents in a self-help group. *Omega, 17,* 237–249.

Klima, G. J. (1970). *The Barabaig: East African cattle-herders.* New York: Holt, Rinehart and Winston.

Kloeppel, D. A., & Hollins, S. (1989). Double handicap: Mental retardation and death in the family. *Death Studies, 13,* 31–38.

Klonoff-Cohen, H., Edelstein, S. L., Leftkowitz, E. S., Srinivasan, I. P., Kaegi, D., Chang., J., & Wiley, K. J. (1996, March 8). The effect of smoking and tobacco exposure through breast milk on sudden infant death syndrome. *Journal of the American Medical Association, 273,* 795–798.

Knapp, R. J. (1986). *Beyond endurance: When a child dies.* New York: Schocken Books.

Knobel, P. S. (1987). Rites of passage: Jewish rites. In M. Eliade (Ed.), *The encyclopedia of religion* (12th ed.). New York: Macmillan.

Knox, R. A. (1997, November 26). AIDS scourge is growing, UN reports. *Boston Globe,* pp. A1, A8.

Koenig, B. A. (1988). The technological imperative in medical practice: The social creation of a 'routine' practice. In M. Lock & D. Gordon (Eds.), *Biomedicine examined* (pp. 465–496), Boston: Kluwer Academic Publishers.

Koenig, H. G. (1993). Legalizing physician-assisted suicide: Some thoughts and concerns. *Journal of Family Practice, 37,* 171–179.

Koenig, H. G. (1994). *Aging and god.* New York: Haworth Press.

Kolata, G. (1995, November 2). After 80, Americans live longer than others. *New York Times,* p. A13.

Koocher, G. P., O'Malley, J. E., Foster, D., & Gogan, J. L. (1976). Death anxiety in normal children and adolescents. *Psychiatric Clinics, 9,* 220–229.

Koshal, A. (1994, July). Ethical issues in xenotransplantation. *Bioethics Bulletin, 6*(3).

Kosky, R. (2000, Spring). Perspectives in suicidology: Families, mental illness, and suicide. *Suicide and Life-Threatening Behavior, 30*(1), 1–7.

Kotarba, J. A. (1983). Perceptions of death, belief systems and the process of coping with chronic pain. *Social Science and Medicine, 17,* 681–689.

Kotch, J. B., & Cohen, S. R. (1985). SIDS counselors' reports of own and parents' reactions to reviewing the autopsy report. *Omega, 16,* 129–139.

Kowalczyk, L., & Heisel, W. (1999, September 26). Cadavers becoming commodities? *The Orange County [CA] Register*.

Krauthammer, C. (1999, May 17). Yes, let's pay for organs. *Time*, p. 100.

Kristof, N. D. (1996, September 29). For rural Japanese, death doesn't break family ties. *New York Times*, p. 10.

Kübler-Ross, E. (1969). *On death and dying*. New York: Macmillan.

Kübler-Ross, E. (1975). *Death: The final stage of growth*. Englewood Cliffs, NJ: Prentice-Hall.

Kübler-Ross, E. (1987). *AIDS: The ultimate challenge*. New York: Macmillan.

Kuper, H. (1963). *The Swazi: A south African kingdom*. New York: Holt, Rinehart and Winston.

Kushner, H. S. (1981). *When bad things happen to good people*. New York: Schocken Books.

Kushner, H. S. (1985, October). Lecture given in Charleston, SC.

Kutner, L. (1990, December 6). The death of a parent can profoundly alter the relationships of adult siblings. *New York Times*, p. B7.

Kutner, N. G. (1990). Issues in the application of high cost medical technology: The case of organ transplantation. In P. Conrad & R. Kern (Eds.), *The sociology of health and illness: Critical perspectives* (pp. 356–372). New York: St. Martin's Press.

Labus, J. G., & Dambrot, F. H. (1985–1986). Comparative study of terminally ill hospice and hospital patients. *Omega, 16*, 225–232.

Lack, S. A. (1979). Hospice: a concept of care in the final stage of life. *Connecticut Medicine*, 43(6):369-370.

Lack, S. A., & Buckingham, R. (1978). *First American hospice: Three years of care*. New Haven: Hospice, Inc.

LaFarge, O. (1956). *A pictorial history of the American Indian*. New York: Crown.

LaGrand, L. E. (1991). United we cope: Support groups for the dying and bereaved. *Death Studies, 15*, 207–230.

Lantos, J. (2000, September 8). How to live as we are dying. *The Chronicle of Higher Education, 47*, B18–B19.

Larson-Kurz, C., Wheeler, A. L., & Dickinson, G. E. (1979). *Dying as deviance: A reinforcement of society's norms in the medical profession*. Unpublished paper presented at the Midwest Sociological Society's Annual Meeting in Minneapolis, MN, April 25–28.

Lattanzi-Licht, M., & Connor, S. (1995). Care of the dying: The hospice approach. In H. Wass & R. A. Neimeyer. *Dying: Facing the facts* (pp. 143–162). Washington, DC: Taylor & Francis 143–162.

Lee, P. C. (1995). Understanding death, dying, and religion: A Chinese perspective. In J. K. Parry & A. S. Ryan (Eds.), *A cross-cultural look at death, dying, and religion* (pp. 172–182). Chicago: Nelson-Hall Publishers.

Lemelle, A. J., Harrington, C., & Leblanc, A. J. (2000). *Readings in the sociology of AIDS*. Upper Saddle River, NJ: Prentice Hall.

Leming, M. R. (1980). Religion and death: A test of Homans's thesis. *Omega, 10,* 347–364.

Leming, M. R., & Premchit, S. (1992). Funeral customs in Thailand. In J. Morgan (Ed.), *Personal care in an impersonal world*. Amityville, NY: Baywood Publishing.

Leming, M. R., Vernon, G. M., & Gray, R. M. (1977, July). The dying patient: A symbolic analysis. *International Journal of Symbology, 8*, 77–86.

Lerner, M. (1970). When, why and where people die. In O. G. Brim, H. E. Freeman, and N. A. Scotch (Eds.), *The dying patient* (p. 14). New York: Russel Sage Foundation.

Lessa, W. A. (1966). *Ulithi: A Micronesian design for living*. New York: Holt, Rinehart and Winston.

Levinson, D. J., Darrow, C. N., Klein, E. B., Levinson, M. H., & McKee, B. (1978). *The seasons of a man's life*. New York: Alfred A. Knopf.

Leviton, D. (1977). The scope of death education. *Death Education, 1*, 41–56.

Lewis, C. S. (1961). *A grief observed*. London: Faber and Faber.

Lifton, R. J. (1976). The sense of immortality: On death and the continuity of life. In R. Fulton & R. Bendiksen (Eds.), *Death and identity* (Rev. ed.). Bowie, MD: Charles Press Publishers.

Lifton, R. J., & Olson, E. (1974). *Living and dying*. New York: Praeger.

Lin, D. (1992). Hospice helps make death a time of dignity. *Free China Journal*, 5.

Lindemann, E. (1944, September). Symptomatology and management of acute grief. *American Journal of Psychiatry, 101*, 141–148.

Lino, M. (1990, July). The $3,800 farewell. *American Demographics*, 8.

Lloyd, P. (1980). Posthumous mourning portraiture. In M. V. Pike & J. G. Armstrong (Eds.), *A time to mourn: Expressions of grief in nineteenth century America*. Stony Brook, NY: The Museums at Stony Brook.

Lock, M. (1995). Contesting the natural in Japan: Moral dilemmas and technologies of dying. *Culture, Medicine and Psychiatry, 19*, 1–38.

Logue, B. J. (1994). When hospice fails: The limits of palliative care. *Omega, 29*, 291–301.

Lonetto, Richard. (1980). *Children's conceptions of death*. New York: Springer Publishing.

Long, J. B. (1987). Underworld. In M. Eliade (Ed.), *The encyclopedia of religion* (12th ed.). New York: Macmillan.

Lowenthal, M. F., Thurnher, M., & Chiriboga, D. (1975). *Four stages of life*. San Francisco: Jossey-Bass.

Lowis, M. J & Hughes, J. (1997). A comparison of the effects of sacred and secular music on elderly people. *Journal of Psychology, 131*(1), 45–55.

Ludwig, A. (1966). *Graven images: New England stonecarving and its symbols.* Middletown, CT: Wesleyan University Press.

Lund, D. A., & Leming, M. R. (1975, October). *Relationship between age and fear of death: A study of cancer patients.* Paper presented at the Annual Scientific Meeting of the Gerontological Society, Louisville, KY.

Lynn, J., Teno, J. M., Phillips, R. S., Wu, A. W., Desbiens, N., Harrold, J., Claessens, M. T., Wenger, N., Kreling, B., & Connors, A. F. (1997) Perceptions by family members of the dying experience of older and seriously ill patients. *Annals of Internal Medicine, 126,* 97–106.

Maas, P. J., Pijnenborg, L., & Delden, J. J. M. (1995). Changes in Dutch opinions on active euthanasia, 1966 through 1991. *Journal of the American Medical Association, 273,* 1411–1414.

Magno, J. B. (1990). The hospice concept of care: Facing the 1990s. *Death Studies, 14,* 3109–3119.

Maguire, P., & Faulkner, A. (1992). Communicating with cancer patients: Handling bad news and difficult questions. *British Medical Journal, 297,* 907–909.

Mails, T. E. (1991). *The people called Apache.* Englewood Cliffs, NJ: Prentice Hall.

Malan, V. D. (1958). *The Dakota Indian family.* Bulletin No. 470, Brookings: South Dakota State College.

Maldonado, G., & Kraus, J. F. (1991). Variation in suicide occurrence by time of day, day of the week, month, and lunar phase. *Suicide and Life-Threatening Behavior, 21,* 174–187.

Malinowski, B. (1925/1948). *Magic, science and religion, and other essays.* Boston: Beacon Press.

Malinowski, B. (1929). *The sexual life of savages.* New York: Harcourt, Brace and World.

Mandelbaum, D. (1959). Social uses of funeral rites. In H. Feifel (Ed.), *The meaning of death.* New York: McGraw-Hill.

Mandell, F., McClain, M., & Reece, R. M. (1987, July). Sudden and unexpected death. *American Journal of Diseases of Children, 141,* 748–750.

Mander, J. (1991). *Absence of the sacred: Failure of technology and the survival of the Indian Nations.* San Francisco: Sierra Club Books.

Mangan, K. (1998, February 20). Extending the physician's role to help people who need "palliative care." *The Chronicle of Higher Education,* A12–A13.

Mansnerus, L. (1995, November 26). Dying writer Leary wants creative ending to his story. Charleston, SC, *Post and Courier,* p. 8A.

Marks, S. C., & Bertman, S. L. (1980). Experiences with learning about death and dying in the undergraduate anatomy curriculum. *Journal of Medical Education, 55,* 844–850.

Markson, L., Clark, J., Glantz, L., Lamberton, V., Kern, D., & Stollerman, G. (1997). The doctor's role in discussing advance preferences for end-of-life care: Perceptions of physicians practicing in the VA. *Journal of the American Geriatrics Society, 45,* 399–406.

Marquis, J. (2000, March 14). Study: Men resist medical care. *Los Angeles Times* wire release.

Marrone, R. (1997). *Death, mourning, and caring.* Pacific Grove, CA: Brooks/Cole.

Marshall, V. W. (1975). Socialization for impending death in a retirement village. *American Journal of Sociology, 80,* 1124–1144.

Martin, J. P. (1988). Hospice and home care for persons with AIDS/ARC: Meeting the challenges and ensuring quality. *Death Studies, 12,* 463–480.

Martin, S. C., Arnold, R. M., & Parker, R. M. (1988, December). Gender and socialization. *Journal of Health and Social Behavior, 29,* 333–343.

Martinson, I. M., & Campos, R. G. (1991). Long-term responses to a sibling's death from cancer. *Journal of Adolescent Research, 6,* 54–69.

Matcha, D. A. (2000). *Medical sociology.* Boston: Allyn and Bacon.

Matousek, M. (2000, September–October). The last taboo. *Modern Maturity, 43,* p. 57.

May, P. A. (1990). A bibliography on suicide and suicide attempts among American Indians and Alaska natives. *Omega, 21,* 199–214.

Mayo, W. R. (1916). Address. *Association of American Cemetery Superintendents, 2,* 51.

McBrien, R. P. (1987). Roman Catholicism. In M. Eliade (Ed.), *The encyclopedia of religion* (12th ed.). New York: Macmillan.

McCall, P. L. (1991). Adolescent and elderly white male suicide trends: Evidence of changing well-being? *Journal of Gerontology, 46*(1), 43–51.

McCarthy, A. (1991, September 13). The country of the old. *Commonweal, 118,* 505–506.

McClowry, S. G., Davies, E. B., May, K. A., Kulenkamp, E. J., & Martinson, I. M. (1987). The empty space phenomenon: The process of grief in the bereaved family. *Death Studies, 11,* 361–374.

McCown, D. E., & Pratt, C. (1985). Impact of sibling death on children's behavior. *Death Studies, 9,* 323–335.

McGuire, D. (1988). Medical student, fourth year. In I. Yalof (Ed.), *Life and death: The story of a hospital* (pp. 339–344). New York: Random House.

McGuire, D. J., & Ely, M. (1984, January-February). Childhood suicide. *Child Welfare, 63,* 17–26.

McIntosh, J. (1995). Epidemiology of suicide in the United States. In J. B. Williamson & E. S. Shneidman (Eds.),

Death: Current perspectives (4th ed., pp. 330–344). Mountain View, CA: Mayfield.

McKenzie, S. C. (1980). *Aging and old age.* Glenview, IL: Scott, Foresman.

McNamara, B. (1997, April 4–6). *A good enough death?* Unpublished paper presented at The Social Context of Dying, Death and Disposal Third International Conference, Cardiff University, Wales.

McNamara, B., Waddell, C., & Colvin, M. (1994). The institutionalization of the good death, *Social Science and Medicine, 39,* 1501–1508.

McNeil, J. N. (1986). In talking about death: Adolescents, parents, and peers. In C. A. Corr & J. N. McNeil (Eds.), *Adolescence and death* (pp. 185–199). New York: Springer Publishing.

Mead, H. M. (1991). Sleep, sleep, sleep; farewell, farewell, farewell. Maori ideas about death. In D. R. Counts & D. A. Counts (Eds.), *Coping with the final tragedy: Cultural variation in dying and grieving* (pp. 43–51). Amityville, NY: Baywood Publishing.

Mebane, E. W., Oman, R. F., Kroonen, L. T., & Goldstein, M. K. (1999, May). The influence of physician race, age, and gender on physician attitudes toward advance care directives and preferences for end-of-life decision-making. *Journal of the American Geriatrics Society, 47,* 579–591.

Medhanandi, Sister (1998). The joy hidden in sorrow [Online]. Available: http://www.buddhanet.net/joydeath.htm

Medhanandi, Sister (1998). The way of the mystic [Online]. Available: http://www.buddhanet.net/mystic.htm

Meier, D. E., Emmons, C. A., Wallenstein, S., Quill, T., Morrison, R. S., and Casel, C. K. (1998). A national survey of physician-assisted suicide and euthanasia in the United States. *The New England Journal of Medicine, 338,* 1193–1201.

Mermann, A. C. (1997, July 11). Preparing medical students to provide care for patients at the end of life. *Chronicle of Higher Education,* p. B-3.

Mermann, A. C., Gunn, D. B., & Dickinson, G. E. (1991, January). Learning to care for the dying: A survey of medical schools and a model course. *Academic Medicine, 66,* 35–38.

Merrin, W. (1999). Crash, bang, wallop! What a picture! The death of Diana and the media. *Mortality, 4*(1), 41–62.

Merton, R. K. (1949). The bearing of sociological theory on empirical research. In *Social structure and social theory.* New York: The Free Press.

Merton, R. K. (1968). The self-fulfilling prophecy. In *Social Theory and Social Structure.* New York: Free Press.

Merton, R. K., Reader, G. G., & Kendall, P. L. (Eds.), (1957). *The student-physician: Introductory studies in the sociology of medical education.* Cambridge, MA: Harvard University Press.

Metress, E. (1990). The American wake of Ireland: Symbolic death ritual. *Omega, 21,* 147–153.

Middleton, J. (1965). *The Lugbara of Uganda.* New York: Holt, Rinehart and Winston.

Miles, S. H., & August, A. (1990, Spring–Summer), Courts, gender and "the right to die." *Law, Medicine and Health Care, 18*(1, 2), 85–95.

Miller, M., Hemenway, D., & Rimm, E. (2000, May). Cigarettes and suicide: A prospective study of 50,000 men. *American Journal of Public Health, 90*(5), 768–773.

Mims, C. (1999). *When we die: The science, culture, and rituals of death.* New York: St. Martin's Press.

Mino, J. (1999). Assessing the outcome of palliative care. *European Journal of Palliative Care, 6,* 203–205.

Moller, D. W. (1996). *Confronting death: Values, institutions, and human mortality.* New York: Oxford University Press.

Moller, D. W. (2000). *Life's end: technocratic dying in an age of spiritual yearning.* Amityville, NY: Baywood Publishing.

Montagu, A. (1968). *The natural superiority of women* (Rev. ed.). New York: Collier Books.

Montgomery, L. (1996, December 12). AMA to teach physicians how to aid the dying patient. Charleston, SC, *Post and Courier,* p. 4A.

Moody, R. A., Jr. (1975). *Life after life: The investigation of a phenomenon—Survival of bodily death.* Boston: G. K. Hall.

Morgan, J. D. (1995). Living our dying and our grieving: Historical and cultural attitudes. In H. Wass & R. A. Neimeyer (Eds.), *Dying: Facing the facts* (3rd ed.). Washington, DC: Taylor and Francis.

Morris, R. A. (1991). Po starykovsky (the old people's way): End of life attitudes and customs in two traditional Russian communities. In D. R. Counts & D. A. Counts (Eds.), *Coping with the final tragedy: Cultural variation in dying and grieving* (pp. 91–112). Amityville, NY: Baywood Publishing.

Morriss, F. (1987). Euthanasia is never justified. In J. Rohr (Ed.), *Death and dying: Opposing viewpoints.* St. Paul, MN: Greenhaven Press.

Moscicki, E. K. (1995, Summer). Epidemiology of suicide. *International Psychogeriatrics, 7,* 137–148.

Moss, M. S., & Moss, S. Z. (1989). Death of the very old. In K. J. Doka (Ed.), *Disenfranchised grief: Recognizing hidden sorrow* (pp. 213–227). Lexington, MA: Lexington Books.

Mulkay, M. (1993). Social death in Britain. In D. Clark (Ed.), *The sociology of death: Theory, culture, practice* (pp. 31–49). Oxford, UK: Blackwell Publishers.

Mumford, L. (1947, March). Atom bomb: Social effects. *Air Affairs, 1,* 370–382. Reprinted as Mumford, L. (1954). *Assumptions and predictions. In the name of sanity* (pp. 10–33). New York: Norton.

Munnichs, J. A. A. (1966). *Old age and finitude.* New York: S. Karger.

Murphy, P., & Perry, K. (1988). Hidden grievers. *Death Studies, 12,* 451–462.

Murray, J., & Callan, V. J. (1988). Predicting adjustment to perinatal death. *British Journal of Medical Psychology, 61,* 237–244.

Nagamine, T. (1988). Attitudes toward death in rural areas of Japan. *Death Studies, 12,* 61–68.

Nagi, M. H., Puch, M. D., & Lazerine, N. G. (1978). Attitudes of Catholic and Protestant clergy toward euthanasia. *Omega, 8,* 153–164.

Nagy, M. (1948). The child's theories concerning death. *Journal of Genetic Psychology, 73,* 3– 27.

Nardi, P. (1990). AIDS and obituaries: The perpetuation of stigma in the press. In D. A. Feldman (Ed.), *Culture and AIDS* (pp. 159–168). New York: Praeger.

National Association of Medical Examiners. (1996). *So you want to be a medical detective?* St. Louis: Author.

National Cancer Institute (2000, May 14). *Annual report to the nation on the status of cancer, 1973–1997.* Bethesda, MD: Author.

National Center for Health Statistics. (1995). *Health risk behaviors among our nation's youth* (PHS Publication No. 95–1520). Hyattsville, MD: Author.

National Center for Health Statistics (1997, January). *Possible solutions to common problems in death certification.* Washington, DC: Author.

National Funeral Directors Association. (1997). *Fact sheet for 1996.* Milwaukee: Author.

National Hospice and Palliative Care Organization. (2000, February 10). News release.

National Hospice Organization. (1988). *The basics of hospice.* Arlington, VA: Author.

National Hospice Organization. (1993). *1992 annual report.* Arlington, VA: Author.

National Hospice Organization. (1997). *Hospice fact sheet.* Arlington, VA: Author.

National Hospice Organization. (1999). *Hospice fact sheet.* Arlington, VA: Author.

NCI fact sheet (1999). *Marijuana use in supportive care for cancer patients.* Bethesda, MD: National Cancer Institute.

Neale, R. E. (1973). *The art of dying.* New York: Harper and Row.

Neimeyer, R. A., & Brunt, D. V. (1995). Death anxiety. In H. Wass & R. A. Neimeyer (Eds.), *Dying: Facing the facts* (3rd ed., pp. 49–88). Washington, DC: Taylor and Francis.

Nelson, B. (2000, June 11). Research use of embryos sets off clash. Charleston, SC: *The Post and Courier,* 15-A [Newsday release].

Nelson, W. A. (1980). Clinical teaching of care for terminally ill in a psychiatry clerkship. *Journal of Medical Education, 55,* 610–615.

Ness, D. E., & Pheffer, C. R. (1990, March). Sequelae of bereavement resulting from suicide. *American Journal of Psychiatry, 147*(3), 279–285.

Neugarten, B. L. (1968). *Middle age and aging: A reader in social psychology.* Chicago: University of Chicago Press.

Neugarten, B. L. (1974). Age groups in American society and the rise of the young-old. *Annals of the American Academy of Political and Social Science, 415,* 187–198.

NFDA. (1981). *Body donation: A compendium of facts compiled as an interprofessional source book.* Produced by the College of Health Sciences (University of Minnesota) and the National Funeral Directors Association.

NFDA. (1986). *Financial operations survey for fiscal 1985.* Milwaukee: Author.

NFDA. (1999). *Fact sheet for 1998.* Milwaukee: Author.

Nichols, J. (1984, November). Illegitimate mourners. In *Children and death: Perspectives and challenges.* Symposium sponsored by Children's Hospital Medical center of Akron, Akron, OH.

Nicol, M. T., Tompkins, J. R., Campbell, N. A., & Syme, G. J. (1986). Maternal grieving response after perinatal death. *The Medical Journal of Australia, 144,* 287–289.

Nir, Y. (1987) Post-traumatic stress disorder in children with cancer. In J. E. Schowalter et al. (Eds.), *Children and death: Perspectives from birth through adolescence* (pp. 63–72). New York: Praeger.

Nordheim, D. V. (1993). Vision of death in rock music and musicians. *Popular Music and Society, 17,* 21–31.

Norman, W. H., & Scaramelli, T. J. (1980). *Mid-life: Developmental and clinical issues.* New York: Brunner/Mazel Publishers.

Norris-Shortle, C., Young, P. A., Williams, M. A. (1993). Understanding death and grief for children three and younger. *Social Work, 38,* 736–742.

Novins, D. K., Beals, J., Roberts, R. E., & Manson, S. M. (1999, Winter). Factors associated with suicide ideation among American Indian adolescents: Does culture matter? *Suicide and Life-Threatening Behavior, 29*(4), 332–346.

Nuland, S. B. (1994). *How we die: Reflections on life's final chapter.* New York: Alfred A. Knopf.

Nuland, S. B. (1998, November 2). The right to die. *The New Republic, 219*(18), 29–35.

O'Brien, J. M., Goodenow, C., & Espin, O. (1991, Summer). Adolescents' reactions to the death of a peer. *Adolescence, 26,* 431–440.

O'Callaghan, C. C. (1996). Complementary therapies in terminal care: Pain, music creativity and music therapy in palliative care. *The American Journal of Hospice & Palliative Care, 2,* 43–49.

O'Dea, T. (1966). *The sociology of religion.* Englewood Cliffs, NJ: Prentice-Hall.

O'Halloran, C. M., & Altmaier, E. M. (1996). Awareness of death among children: Does a life-threatening illness alter the process of discovery? *Journal of Counseling and Development, 74,* 259–262.

O'Hara, J. (1996, July 1). The Baby Boomers confront mortality. *Maclean's, 109,* 64.

O'Mathuna, D. P. (1996). Medical ethics and what it means to be human. *Irish Bible School Journal,* 12–19.

O'Meara, K. P. (1999). Harvesting fetal body parts. *Insight,* New World Communications.

Oaks, J., & Ezell, G. (1993). *Dying and death: Coping, caring, understanding* (2nd ed.). Scottsdale, AZ: Gorsuch Scarisbrick Publishers.

Ogden, R. D. (1995, December). The right to die: A rejoinder to Bruce Wilkinson's critique. *Canadian Public Policy, 21*(4), 456–460.

Oken, D. (1961). What to tell cancer patients. *Journal of the American Medical Association, 175,* 1120–1128.

Oliver, L. E. (1999). Effects of a child's death on the marital relationship: a review. *Omega, 39,* 197–227.

Olson, M. (1997). *Healing the dying.* Albany, NY: Delmar Publishers.

On death as a constant companion. (1965, November 12). *Time Magazine.*

Onstad, E. (2000, November 28). Dutch approve law on mercy killings, protests start. Associated Press release.

Orbach, I. (1988). *Children who don't want to live: Understanding and treating the suicidal child.* San Francisco: Jossey-Bass.

Orbach, I., Gross, Y., Glaubman, H., & Berman, D. (1985). Children's perception of death in humans and animals as a function of age, anxiety and cognitive ability. *Journal of Child Psychology and Psychiatry, 26,* 453–463.

Osgood, N. J., & McIntosh, J. L. (1986). *Suicide and the elderly: An annotated bibliography and review.* New York: Greenwood Press.

Osgood, N. J., Brant, B. A., & Lipman, A. (1991). *Suicide among the elderly in long-term care facilities.* New York: Greenwood Press.

Oswalt, W. H. (1986). *Life cycles and lifeways.* Palo Alto, CA: Mayfield Publishing.

Oxley, J. M. (1887, February). The reproach of mourning. *Forum, 2,* 608–614.

Palgi, P., & Abramovitch, H. (1984). Death: A cross-cultural perspective. *Annual Review of Anthropology, 13,* 385–417.

Pardue, P. (1968). Buddhism. In *International encyclopedia of the social sciences* (2nd ed., pp. 165–184). New York: Macmillan.

Parkes, C. M. (1972). *Bereavement: Studies of grief in adult life.* New York: International Universities Press.

Parkes, C. M. (1978). Home or hospital? Terminal care as seen by surviving spouses. *Journal of the Royal College of General Practice, 28,* 19–30.

Parsons, T. (1951). *The social system.* New York: Free Press.

Parsons, T. (1958). The definitions of health and illness in light of American values and social structure. In E. E. Jaco (Ed.), *Patients, physicians, and illness.* New York: Free Press.

Parsons, T., & Fox, R. (1952). Illness, therapy and the modern urban American family. *Journal of Social Issues, 8,* 31–44.

Patchner, M., & Finn, M. (1987–1988). Volunteers: The life-line of hospice. *Omega, 18,* 135–143.

Pattison, E. M. (1977). *The experience of dying.* Englewood Cliffs, NJ: Prentice-Hall.

Peck, M. (1980). *Recent trends in suicide among young people.* Los Angeles: Institute for Studies of Destructive Behaviors.

Perry, P. (1988, September). Brushes with death. *Psychology Today, 22,* 14–17.

Phillips, D. P., & Carstensen, L. L. (1986, September 11). Clustering of teenage suicides after television news stories about suicide. *New England Journal of Medicine, 315,* 685–689.

Piaget, J. (1958). *The growth of logical thinking from childhood to adolescence.* New York: Basic Books.

Pike, M. V., & Armstrong, J. V. (Eds.), *A time to mourn: Expressions of grief in the nineteenth century America.* Stony Brook, NY: The Museums at Stony Brook.

Pine, V. R. (1971, June). *Findings of the professional census.* Milwaukee: National Funeral Directors Association.

Pine, V. R. (1986). The age of maturity for death education: A socio-historical portrait of the era 1976–1985. *Death Studies, 10,* 209–231.

Pine, V. R., & Brauer, C. (1986). Parental grief: A synthesis of theory, research, and intervention (pp. 59–96). In T. A. Rando (Ed.), *Parental loss of a child.* Champaign, IL: Research Press Company.

Pirkis, J., Burgess, P., & Dunt, D. (2000). Suicidal ideation and suicide attempts among Australian adults. *Crisis, 21*(1), 16–25.

Plech, E. H. (2000). *Celebrating the family: Ethnicity, consumer culture, and family rituals.* Cambridge, MA: Harvard University Press.

Plopper, B., & Ness, E. (1993, Winter). Death as portrayed to adolescents through top 40 rock and roll music. *Journal on Adolescence, 28,* 793–807.

Pollak, J. M. (1979). Correlates of death anxiety: A review of empirical studies. *Omega, 10,* 97–121.

Poor man's plague. (1991, September 21). *The Economist, 320,* 21–23.

Pospisil, L. (1963). *The Kapauku Papuans of west New Guinea.* New York: Holt, Rinehart and Winston.

Potter, A. C. (1993). Will the 'right to die' become a license to kill? The growth of euthanasia in America. *Journal of Legislation, 19*(1), 31–62.

Powers, W. K. (1977). *Oglala religion.* Lincoln, NE: University of Nebraska Press.

Powner, D. J., Ackerman, B. M., & Grevnik, A. (1996, November 2). Medical diagnosis of death in adults: Historical contributions to current controversies. *The Lancet, 348,* 1219–1224.

Presbyterian Ministers' Fund (1938). *Presbyterian Synod minutes.* Philadelphia: Author.

Prior, L. (1989). *The social organization of death: Medical discourse and social practices in Belfast.* New York: St. Martin's Press.

Probate FAQ. (1999). *Nolo's legal encyclopedia,* Berkeley, CA: Nolo Press.

Proctor, R. N. (1995). *Cancer wars: How politics shapes what we know and don't know about cancer.* New York: Basic Books.

Purtilo, R. (1990). *Health professional and patient interaction* (4th ed.). Philadelphia: W. B. Saunders.

Quill, T. E. (1991). Bad news: Delivery, dialogue, and dilemmas. *Archives of Internal Medicine, 151,* 463–468.

Rachels, J. (1975). Active and passive euthanasia. *New England Journal of Medicine, 292,* 78–80.

Radcliffe-Brown, A. R. (1964). *The Andaman Islanders.* New York: The Free Press.

Raether, H. C., & Slater, R. C. (1974). *Facing death as an experience of life.* Milwaukee: National Funeral Directors Association.

Rahman, F. (1987). *Health and medicine in the Islamic tradition.* New York: Crossroad Press.

Rahman, F. (1989, February 27). Routine CPR can abuse the old and sick. *Minneapolis Star and Tribune,* 9A.

Ramsden, P. G. (1991). Alice in the afterlife: A glimpse in the mirror. In D. R. Counts & D. A Counts (Eds.), *Coping with the final tragedy: Cultural variation in dying and grieving* (pp. 27–41). Amityville, NY: Baywood Publishing.

Ramsey, P. (1970). *The patient as person: Explorations in medical ethics.* New Haven: Yale University Press.

Rando, T. A. (1993). *Treatment of complicated mourning.* Champaign, IL: Research Press.

Randolph, M. (1999). *Eight ways to avoid probate.* Berkeley, CA: Nolo Press.

Range, L. M., & Goggin, W. C. (1990). Reactions to suicide: Does age of the victim make a difference? *Death Studies, 14,* 269–275.

Raphael, B. (1983). *The anatomy of bereavement.* New York: Basic Books.

Raphael, B. (1986). *When disaster strikes: How individuals and communities cope with catastrophe.* New York: Basic Books.

Rappaport, R. A. (1974). Obvious aspects of ritual. *Cambridge Anthropology, 2,* 2-60.

Rasmussen, C. A., & Brems, C. (1996, March). The relationship of death anxiety with age and psychosocial maturity. *The Journal of Psychology, 130,* 141–144.

Report of the Committee on Physician-Assisted Suicide and Euthanasia. (1996). *Suicide and Life-Threatening Behavior, 26*(Suppl.), 1–19.

Rhymes, J. A. (1996, May-June). Barriers to palliative care. *Cancer Control Journal, 3*(3), 1–9.

Rich, C. L., Warsradt, G. M., Nemiroff, R. A., Fowler, R. C., & Young, D. (1991, April). Suicide, stressors, and the life cycle. *American Journal of Psychiatry, 148,* 524–527.

Richardson, V., & Sands, R. (1987). Death attitudes among mid-life women. *Omega, 17,* 327–341.

Richardson, W. C. (1992, June 3). Educating leaders who can resolve the health-care crisis. *Chronicle of Higher Education, 39,* B1.

Richman, J. (1992, Spring). A rational approach to suicide. *Suicide and Life-Threatening Behavior, 22,* 130–141.

Riley, J. W. (1968). Attitudes toward aging. In M. W. Riley et al. (Eds.), *Aging and society: An inventory of research findings.* New York: Russell Sage Foundation.

Ring, K. (1980). *Life at death: A scientific investigation of the near-death experience.* New York: Coward, McCann and Geoghegan.

Robinson, L., & Mahon, M. M. (1997). Sibling bereavement: A conceptual analysis. *Death Studies, 21,* 477–499.

Robson, J. D. (1977). Sick role and bereavement role: Toward a theoretical synthesis of two ideal types. In G. M. Vernon (Ed.), *A time to die.* Washington, DC: University Press of America.

Rodabough, T. (1985). Near-death experiences: An examination of the supporting data and alternative explanations. *Death Studies, 9,* 95–113.

Rogers, R. G., Humer, R. A., & Nam, C. B. (2000). *Living and dying in the USA.* San Diego, CA: Academic Press.

Rosenberg, C. E. (1973, May). Sexuality, class, and role in nineteenth century America. *American Quarterly, 25,* 137.

Rosenberg, J. (2000, May 15). Art of prognosis becoming an increasingly valued skill. *American Medical News,* 1–4.

Rosenberg, J. F. (1983). *Thinking clearly about death.* Englewood Cliffs, NJ: Prentice-Hall.

Rosenblatt, P. (1983). *Bitter, bitter tears.* Minneapolis: University of Minnesota Press.

Rosenblatt, P. C., Walsh, R., & Jackson, A. (1976). *Grief and mourning in cross cultural perspective.* New Haven: Human Relations Area Files Press.

Rosenthal, N. R. (1980, Fall). Adolescent death anxiety: The effect of death education. *Education, 101,* 95–101.

Rotundo, B. (1973, July). The rural cemetery movement. *Essex Institute Historical Collections, 109,* 231–242.

Rowe, J. Clyman, R., Green, C., Mikkelsen, C., Haight, J., & Ataide, L. (1978) Follow-up of families who experience a perinatal death. *Pediatrics, 62,* 166–170.

Rumbelow, H. (1999, April 8). Transplant boom raises prospect of divorce haggling. London: *The Times,* p. 13.

Ryff, C. D. (1991). Possible selves in adulthood and old age: A tale of shifting horizons. *Psychology and Aging, 6,* 286–295.

Rzhevsky, L. (1976). Attitudes toward death. *Survey, 22,* 38–56.

Sade, R. M. (1999, March 8). Cadaveric organ donation: Rethinking donor motivation. *Archives of Internal Medicine, 159,* 438–442.

Sagan, L. (1987). *The health of nations.* New York: Basic Books.

Sahler, O. J. Z. (1978). *The child and death.* St. Louis: C. V. Mosby.

Sakano, J., & Yamazaki, Y. (1990). *Survey of the role of white-collar workers as family members in the Tokyo area.* Tokyo: University of Tokyo, Department of Health Sociology.

Salamat, A. (1987). A young widow burns in her bridal clothes. *Far Eastern Economic Review, 138,* 54–55.

Sanders, C. M. (1980). A comparison of adult bereavement in the death of a spouse, child, and parent. *Omega, 10,* 303–322.

Sanders, C. M. (1999). *Grief the mourning after: Dealing with adult bereavement.* New York: John Wiley & Sons.

Sandner, D. (1991). *Navaho symbols of healing: A Jungian exploration of ritual, image and medicine.* Rochester, VT: Healing Arts Press.

Sargent, A. H. (1888). Country cemeteries. *Association of American Cemetery Superintendents, 2,* 51.

Saunders, C. (1992, Winter). The evolution of the hospices. *Free Inquiry,* 19–23.

Schachter, S. (1991). Adolescent experiences with the death of a peer. *Omega, 24,* 1–11.

Schaefer, D., & Lyons, C. (1986). *How do we tell the children? A parents' guide to helping children understand and cope when someone dies.* New York: Newmarket Press.

Schatz, H. (1976, March 1). The Ik in America's armchairs. *The Minneapolis Tribune.*

Scheible, S. S. (1988). Death and the law. In H. Wass, F. M. Berarado, & R. A. Neimeyer (Eds.), *Dying: Facing the facts* (pp. 301–319). Washington, DC: Hemisphere Publishing Corporation.

Schiedermayer, D. (1994). *Putting the soul back in medicine: Reflections on compassion and ethics.* Grand Rapids, MI: Baker Books.

Schmidt, J. D. (1986). Murder of a child. In T. A. Rando (Ed.), *Parental loss of a child* (pp. 213–220). Champaign, IL: Research Press Company.

Schneider, J. W. (1993). Family care work and duty in a 'modern' Chinese hospital. In P. Conrad & E. B. Gallagher (Eds.), *Health and health care in developing countries: Sociological perspectives* (pp. 154–179). Philadelphia: Temple University Press.

Schofferman, J. (1988). Care of the AIDS patient, *Death Studies, 12,* 433–449.

Schroepfer, T. (1999). Facilitating perceived control in the dying process. In B. de Vries (Ed.), *End of life issues* (pp. 57–76). New York: Spring Publishing.

Schug, S. A., Zech, D., & Dorr, U. (1990). Cancer pain management according to WHO analgesic guidelines. *Journal of Pain and Symptom Management, 5,* 27–32.

Schwab, R. (1992). Effects of a child's death on the marital relationship: a preliminary study. *Death Studies, 16,* 141–154.

Scott, C. R., & Dolan, C. (1991, April 11). Funeral homes hope to attract business by offering services after the service. *Wall Street Journal,* p. B1.

Seale, C. (1991). Communication and awareness about death: A study of a random sample of dying people. *Social Science and Medicine, 32,* 943–952.

Seale, C. (1997). Social and ethical aspects of euthanasia: A review. *Progress in Palliative Care, 5,* 1–6.

Seale, C. (1998). *Constructing death: The sociology of dying and bereavement.* Cambridge, UK: Cambridge University Press.

Seavoy, M. R. (1906, February). Twentieth century methods. *Park and Cemetery, 15,* 488.

Secomb, L. (1999). Philosophical deaths and feminine finitude. *Mortality, 4*(2), 111–125.

Seibert, D., & Drolet, J. C. (1993). Death themes in literature for children ages 3–8. *Journal of School Health, 63,* 86–90.

Sell, L., Devlin, B., Bourke, S. J., Munro, N. C., Corris, P. A., & Gibson, G. J. (1993). Communicating the diagnosis of lung cancer. *Respiratory Medicine, 87,* 61–63.

Selvin, P. (1990, July 17). Cryonics goes cold: People just aren't dying to be frozen. Charleston, SC, *News and Courier,* p. 5C.

Seravalli, E. P. (1988). The dying patient, the physician, and the fear of death. *The New England Journal of Medicine, 319,* 1728–1730.

Shaffer, D., Gould, M., & Hicks, R. C. (1994). Worsening suicide rate in black teenagers. *American Journal of Psychiatry, 151*(12), 1810–1812.

Shalit, R. (1997). When we were philosopher kings. *The New Republic, 216*(17), 24–28.

Sharlet, J. (2000, January 28). The truth can comfort the dying, a physician argues. *The Chronicle of Higher Education, 47,* A20–A21.

Sheehy, G. (1976). *Passages: Predictable crises of adult life.* New York: E. P. Dutton

Shine, T. M. (2000) *Fathers aren't supposed to die: Five brothers reunite to say good-bye.* New York: Simon and Schuster.

Shneidman, E. (1973). Megadeath: Children of the nuclear family. In *Deaths of man.* Baltimore: Penguin Books.

Shneidman, E. (1980). *Voices of death*. New York: Harper and Row.

Silverman, E., Range, L., & Overholser, J. (1994–1995). Bereavement from suicide as compared to other forms of bereavement. *Omega, 30,* 41–51.

Silverman, P. R. (2000). *Never too young to know: Death in children's lives*. New York: Oxford University Press.

Silverman, P. R., Nickman, S., & Worden, J. W. (1992). Detachment revisited: the child's reconstruction of a dead parent. *American Journal of Orthopsychiatry, 62,* 494–503.

Simonds, O. C. (1919). Review of progress in cemetery design and development with suggestions for the future. *Association of American Cemetery Superintendents, 24,* 41.

Skinner, B. F. (1990, August 7). Skinner to have last word. *Harvest Personnel*.

Sklar, F., & Hartley, S. F. (1990). Close friends and survivors: Bereavement patterns in a "hidden" population. *Omega, 21*(2),103–112.

Slater, P. (1974). *Earthwalk*. Garden City, NY: Doubleday.

Sloane, D. C. (1991). *The last great necessity: Cemeteries in American history*. Baltimore: The Johns Hopkins University Press.

Smith, A. M. (1994). Palliative medicine education for medical students: A survey of British medical schools, 1992. *Medical Education, 28,* 197–199.

Smith, B. (1910, March). An outdoor room on a cemetery lot. *Country Life in America, 17,* 539.

Smith, D. C., & Maher, M. F. (1993). Achieving a healthy death: The dying person's attitudinal contributions. *The Hospice Journal, 9,* 21–32.

Smith, J. M. (1996). *AIDS and society*. Upper Saddle River, NJ: Prentice Hall.

Smith, M. T. (1988, December). Why you might go for a cash-value policy. *Money,* 153–161.

Smith, R. S. (1941, May). Life insurance in fifteenth century Barcelona. *The Journal of Economic History,* New York.

Smith, T. L., & Walz, B. J. (1995). Death education in paramedic programs: A nationwide assessment. *Death Studies, 19,* 257–267.

Smith, W. J. (1985). *Dying in the human life cycle*. New York: Holt, Rinehart and Winston.

Somers, T. S. (1995, June 22–25). *Relict, consort, wife: The use of Connecticut Valley gravestones to understand concepts of gender in the late eighteenth and early nineteenth centuries*. Unpublished paper presented at the Annual Meeting of the Association for Gravestone Studies, Westfield, MA.

Spencer, R. F., & Jennings, J. D. (1965). *The Native Americans*. New York: Harper and Row.

Spiegel, D. (1998). Getting there is half the fun: Relating happiness to health. *Psychological Inquiry, 9*(1), 66–68.

Srinivas, M. N., & Shah, A. M. (1968). Hinduism. In *International encyclopedia of the social sciences* (Vol. 6, pp. 358–366). New York: Macmillan.

Stack, S. (1992, Summer). The effect of the media on suicide: The great depression. *Suicide and Life-Threatening Behavior, 22,* 255–267.

Stalman, S. D. (1996). Children and the death of a sibling. In C. A. Corr & D. M. Corr (Eds.), *Handbook of childhood death and bereavement* (pp. 149–164). New York: Springer Publishing.

Stambrook, M., & Parker, K. C. H. (1987, April). The development of the concept of death in childhood: A review of the literature. *Merrill-Palmer Quarterly, 33,* 133–157.

Steiger, B. (1974). *Medicine power: The American Indian's revival of his spiritual heritage*. Garden City, NY: Doubleday.

Stephenson, J. S. (1985). *Death, grief, and mourning*. New York: The Free Press.

Stevenson, I., Cook, E. W., & McClean-Rice, N. (1989). Are persons reporting "near-death experiences" really near death? A study of medical records. *Omega, 20*(4), 45–54.

Stillion, J. M. (1985). *Death and the sexes*. Washington, DC: Hemisphere/McGraw-Hill.

Stillion, J. M. (1995). Death in the lives of adults: Responding to the tolling of the bell. In Wass, H., & Neimeyer, R. A. (Eds.), *Dying: Facing the facts* (3rd ed., pp. 303–322). Washington, DC: Taylor and Francis.

Stillion, J. M., & McDowell, E. E. (1991). Examining suicide from a life span perspective. *Death Studies, 15,* 327–354.

Stockwell, E. G. (1976). *The methods and materials of demography*. New York: Academic Press.

Stoeckel, H. H. (1993). *Survival of the spirit*. Reno: University of Las Vegas.

Stolberg, S. G. (1997, May 16). Senate tries to define fetal viability. *New York Times,* p. A18.

Strauss, B., & Howe, R. (1991). *Generations: The history of America's future 1584 to 2069*. New York: William Morrow.

Stricherz, M., & Cunnington, L. (1982). Death concerns of students, employed persons, and retired persons. *Omega, 12,* 373–379.

Stryker, J. (1996, March 31). Life after Quinlan: Right to die. *New York Times*.

Sturgill, B. (Producer). (1995). *Death on my terms: Right or privilege* [Videotape]. Chicago: Terra Nova Films.

Sudnow, D. (1967). *Passing on: The social organization of dying*. Englewood Cliffs, NJ: Prentice Hall.

Suicide among older persons in the United States, 1980–1992. (1996, January 12). *Morbidity and Mortality Weekly Report, 45,* 3–6.

Suicide rate among elderly climbs by 9% over 12 years. (1996, January 12). *New York Times,* p. A11.

Sullivan, A. (1990, December 17). Gay life, gay death. *The New Republic, 210,* 19–25.

Sullivan, A. D., Hedberg, K., & Fleming, D. W. (2000, February). Legalized physician-assisted suicide in Oregon—the second year. *The New England Journal of Medicine, 342*(8), 598–604.

Sweeting, H. N., & Gilhooly, M. L. M. (1992). Doctor, am I dead? A review of social death in modern societies. *Omega, 24,* 251–269.

Tamm, M. E., & Granqvist, A. (1995). The meaning of death for children and adolescents: A phenomenographic study of drawings. *Death Studies, 19,* 203–222.

Tanner, J. G. (1995). Death, dying, and grief in the Chinese-American culture. In J. K. Perry & A. S. Ryan (1995). *A cross-cultural look at death, dying, and religion* (pp. 183–192). Chicago: Nelson-Hall Publishers.

Taylor, G. W. (1990, July 16). AIDS mercy killings. *Maclean's, 103,* 14.

Taylor, L. (1980). Symbolic death: An anthropological view of mourning ritual in the nineteenth century. In M. V. Pike & J. V. Armstrong (Eds.), *A time to mourn: Expressions of grief in the nineteenth century America.* Stony Brook, NY: The Museums at Stony Brook.

Taylor, R. B. (1988). *Cultural ways* (3rd ed.). Boston: Allyn and Bacon.

The dying of death. (1899, September). *Review of Reviews, 20,* 354–365.

The ideas of a plain country woman. (1913, April). *Ladies Home Journal, 30,* 42.

The man nobody envies: An account of the experiences of an undertaker. (1914, June). *American Magazine, 77,* 68–71.

Thearle, M. J., Vance, F. C., Najman, J. M., Embelton, G., & Foster, W. J. (1995). Church attendance, religious affiliation and parental responses to sudden infant death, neonatal death and stillbirth. *Omega, 31,* 51–58.

Theodorson, G. A., & Theodorson, A. G. (1969). *Modern dictionary of sociology.* New York: Thomas Y. Crowell.

Thompson, K. E., & Range, L. M. (1992–1993). Bereavement following suicide and other deaths: Why support attempts fail. *Omega, 26,* 61–70.

Thornton, G., Robertson, D. U., & Mlecko, M. L. (1991). Disenfranchised grief and evaluations of social support by college students. *Death Studies, 15,* 355–362.

Thornton, G., Wittemore, K. D., & Robertson, D. U. (1989). Evaluation of people bereaved by suicide. *Death Studies, 13,* 119–126.

Timmermans, S. (1999). *Sudden death and the myth of CPR.* Philadelphia: Temple University Press.

Titus, S. L., Rosenblatt, P. C., & Anderson, R. M. (July, 1979). Family conflict over inheritance of property. *The Family Coordinator,* pp. 337–338.

Tobin, S. (1972). The earliest memory as data for research in aging. In Kent, D. P., Kastenbaum, R., & Sherwood, S. (Eds.), *Research, planning and action for the elderly.* New York, Behavioral Publications.

Tolchin, M. (1989, July 19). When long life is too much: Suicide rises among elderly. *New York Times,* p. 1.

Tonkinson, R. (1978). *The Mardudjara aborigines.* New York: Holt, Rinehart and Winston.

Toynbee, A. (1968). *Man's concern with death.* London: Hodder and Stoughton.

Trammell, R. L. (1975). Saving life and taking life. *Journal of Philosophy, 72,* 131–137.

Trauger-Querry, B., & Haghighi, K. R. (1999). Balancing the focus: Art and music therapy for pain control and symptom management in hospice care. *The Hospice Journal, 14*(1), 25–37.

Tuckerman, H. (1856, November). The law of burial and the sentiment of death. *Christian Examiner, 61,* 338–342.

Tully, M. (1994, September). When body and soul go their separate ways. *Worldwide report: Death.* London: BBC Worldwide.

Turnbull, C. M. (1972). *The mountain people.* New York: Simon and Schuster.

Turnbull, C. M. (1983). *The human cycle.* New York: Simon and Schuster.

Turner, J. H. (1985). *Sociology: A student handbook.* New York: Random House.

Twycross, R. G. (1996, February). Euthanasia: Going Dutch? *Journal of the Royal Society of Medicine, 89,* 61–63.

Tylor, E. B. (1873). *Primitive culture.* London: John Murray.

Tyson, E. (1997, February). Save thousands on wills and trusts with computerized legal programs. *Money,* p. 22.

U. S. Department of Health and Human Services. (1999, October 5). AIDS falls from top 15 causes of death. (National Vital Statistics Reports, Vol. 47, No. 25.48, pp. 99–1120). Washington, DC: Author.

U.S. Bureau of the Census. (1975). *Historical statistics of the United States, colonial times to 1970* (pp. 1050–1059). Washington, DC: Department of Commerce, Bureau of the Census.

U.S. to release rules on use of donor organs (1999, October 19). *New York Times,* Science, p. D9.

Uchendu, V. C. (1965). *The Igbo of southeast Nigeria.* New York: Holt, Rinehart and Winston.

Ufema, Joy. (1996, October). Insights on death and dying. *Nursing, 26,* 26–27.

United Nations. (1953, August). *Principles for a vital statistics system* (Statistical Papers, Series M, No. 19, p. 6).

United Network for Organ Sharing. (1997, March 4). *UNOS facts and statistics about transplantations* [Online]. Available: http://204.127.237.11/stats.htm

van Eys, J. (1988, Summer). In my opinion . . . normalization while dying. *Children's Health Care, 17,* 18–21.

Van Gennep, A. (1960). *The rites of passage* (M. B. Vizedom & G. L. Caffee, Trans.), Chicago: University of Chicago Press. (Original work published 1909)

Vander Veer, J. B. (1999, May). Euthanasia in the Netherlands. *Journal of the American College of Surgeons, 188,* 532–537.

Veatch, R. M. (1995). The definition of death: Problems for public policy. In H. Wass & R. A. Neimeyer (Eds.), *Dying: Facing the facts* (3rd ed., pp. 405–432). Washington, DC: Taylor and Francis.

Verhoef, M. J., & Kinsella, T. D. (1996). Alberta euthanasia survey: Three-year follow-up. *Canadian Medical Association Journal, 155,* 885–890.

Vernon, G. M. (1970). *Sociology of death: An analysis of death-related behavior.* New York: Ronald Press.

Vernon, G. M. (1972). *Human interaction.* New York: Ronald Press.

Vernon, G. M., & Cardwell, J. D. (1981). *Social psychology: shared, symboled, and situated behavior.* Washington, DC: University Press of America.

Viney, L. L., Henry, R. M., Walker, B. M., & Crooks, L. (1992). The psychosocial impact of multiple deaths from AIDS. *Omega, 24,* 151–163.

Vobejda, B. (1997, December 5). Abortion rate in U. S. off sharply. *Washington Post,* p. A1.

Voegelin, E. W. (1944). *Mortuary customs of the Shawnee and other Eastern tribes.* Indianapolis: Indiana Historical Society.

Vogt, E. Z. (1970). *The Zinacantecos of Mexico: A modern Maya way of life.* New York: Holt, Rinehart and Winston.

Vovelle, M. (1976). La redecouverte de la mort. *Pensee, 189,* 3 18.

Waechter, E. H. (1985). Children's awareness of fatal illness. In S. G. Wilcox & M. Sutton (Eds.), *Understanding death and dying* (3rd ed., pp. 299–306). Palo Alto, CA: Mayfield Publishing.

Walker, B. A. (1989). Health care professionals and the near-death experience. *Death Studies, 13,* 63–71.

Walling, A. D. (1996, November 1). Risk factors for sudden infant death syndrome. *American Family Physician, 54,* 2099–2100.

Walsh, K. (1974). *Sometimes I weep.* Valley Forge, PA: Judson Press.

Walter, T. (1995). Natural death and the noble savage. *Omega, 30,* 237–248.

Walter, T. (1999). And the consequence was . . . In T. Walter (Ed.), *The Mourning for Diana* (pp. 271–278). Oxford: Berg.

Walter, T. (1999). The questions people asked. In T. Walter (Ed.), *The mourning for Diana* (pp. 19–47). Oxford: Berg.

Wanzer, S. H., et al. (1989). The physician's responsibility toward hopelessly ill patients. *The New England Journal of Medicine, 320,* 844–849.

Waskel, S. A. (1995, March). Temperament types: Midlife death concerns, demographics and intensity of crisis. *The Journal of Psychology, 129,* 221–234.

Wass, H. (1979). Death and the elderly. In H. Wass (Ed.), *Dying: Facing the facts.* Washington, DC: Hemisphere.

Wass, H. (1995). Death in the lives of children and adolescents (pp. 269–301). In H. Wass & R. A. Neimeyer (Eds.), *Dying: Facing the facts.* Washington, DC: Taylor and Francis.

Wass, H., Berardo, F. M., & Neimeyer, R. A. (Eds.), (1988). *Dying: Facing the facts.* Washington, DC: Hemisphere Publishing Corporation.

Wass, H., Miller, D., & Redditt, C. A. (1991). Adolescents and destructive themes in rock music: A follow-up. *Omega, 23,* 199–206.

Wass, H., & Sisler, S. (1978, January). *Death concern and views on various aspects of dying among elderly persons.* Paper presented at the International Symposium on the Dying Human, Tel Aviv, Israel.

Waters, H. F. (1994, April 18). Teenage suicide: One act not to follow. *Newsweek,* p. 49.

Watts, D. T., Howell, T., & Priefer, B. A. (1992). Geriatricians' attitudes toward assisting suicide of dementia patients. *Journal of the American Geriatrics Society, 40,* 878–885.

Weaver, A. J., & Koenig, H. G. (1996). Elderly suicide, mental health professionals, and the clergy: A need for clinical collaboration, training, and research. *Death Studies, 20,* 495–508.

Weber, L. J. (1981). The case against euthanasia. In D. Bender (Ed.), *Problems of death: Opposing viewpoints.* St. Paul, MN: Greenhaven Press.

Weber, M. (1966). *The theory of social and economic organization.* New York: The Free Press.

Weber, M. (1968). Bureaucracy. In H. Gerth & C. C. Mills (Trans.), *Max Weber.* New York: Free Press.

Weed, H. E. (1912). *Modern park cemeteries.* Chicago: R. J. Haight.

Wehbah-Rashid, J.A.R. (1996). Explaining pregnancy loss in matrilineal southeast Tananzia. In R. Cecil (Ed.), *The anthropology of pregnancy loss: Comparative studies in miscarriage, stillbirth and neonatal death* (pp. 75–93). Oxford: Berg.

Weigand, C. G., & Weigand, P. C. (1991). Death and mourning among the Huicholes of western Mexico. In D. R. Counts & D. A. Counts (Eds.), *Coping with the final tragedy: Cultural variation in dying and grieving* (pp. 53–68). Amityville, NY: Baywood Publishing.

Weijer, C. (1995, Spring). Learning from the Dutch: Physician-assisted death, slippery slopes and the Nazi analogy. *Health Law Review, 4*(1), 23–29.

Weisman, A. D. (1990–1991). Bereavement and companion animals. *Omega, 22,* 241–248.

Weisman, A. D. (1993). *The vulnerable self: confronting the ultimate questions.* New York: Insight Books.

Weisman, A. D., & Kastenbaum, R. (1970). *The psychological autopsy: A study of the terminal phase of life.* New York: Behavioral Publications.

Weiss, G. L., & Lonnquist, L. E. (1994). *The sociology of health, healing, and illness.* Englewood Cliffs, NJ: Prentice-Hall.

Weiss, G. L., & Lonnquist, L. E. (2000). *The sociology of health, healing, and illness.* Upper Saddle River, NJ: Prentice Hall.

Weiss, H. (1995, September–October). Dust to dust: Transforming the American cemetery. *Tikkun Magazine,* 2–25.

Wells, R. A. (1887). *Decorum: A practical treatise on etiquette and dress of the best American society.* Springfield, MA: King, Richardson.

Wells, R. V. (2000). *Facing the "King of Terrors": Death and society in an American community, 1970–1990.* Cambridge, UK: Cambridge University Press.

Wertenbaker, L. T. (1957). *The death of a man.* New York: Random House.

Werth, J. L., Jr. (1995). Rational suicide reconsidered: AIDS as an impetus for change. *Death Studies, 19,* 65–80.

Wheeler, A. L. (1973, April). *The dying person: A deviant in the medical subculture.* Unpublished paper presented at the Southern Sociological Society Annual Meeting, Atlanta, GA.

Wholey, D. R., & Burns, L. R. (2000). Tides of change: The evolution of managed care in the United States. In C. C. Bird, P. Conrad, & A. A. Fremont (Eds.), *Handbook of medical sociology* (pp. 217–237). Upper Saddle River, NJ: Prentice Hall.

Wichstrom, L. (2000, May). Predictors of adolescent suicide attempts: a national representative longitudinal study of Norwegian adolescents. *Journal of the American Academy for Child and Adolescent Psychiatry, 39*(5), 603–610.

Wilkins, R. (1996). *Death: A history of man's obsessions and fears.* New York: Barnes and Noble Books.

Wilkinson, B. W. (1995, December). The 'right to die' by Russell Ogden: A commentary. *Canadian Public Policy, 21*(4), 449–455.

Will, J. (1988). Preneed: The trend toward prearranged funerals. In H. Raether (Ed.), *The funeral director's practice management handbook.* Englewood Cliffs, NJ: Prentice-Hall.

Williams, M. D. (1981). *On the street where I lived.* New York: Holt, Rinehart and Winston.

Williams, T. R. (1965). *The Dunsun: A North Borneo society.* New York: Holt, Rinehart and Winston.

Willinger, M. (1995). Sleep position and sudden infant death syndrome. *Journal of the American Medical Association, 273,* 818–819.

Wolf, S. S. (1995). Legal perspectives on planning for death. In H. Wass & R. A. Neimeyer (Eds.), *Dying: Facing the facts* (pp. 163–184). Washington, DC: Taylor and Francis.

Worden, J. W. (1982). *Grief counseling and grief therapy: A handbook for the mental health practitioner.* New York: Springer Publishing.

World Health Organization. (1986) *Cancer pain relief.* Geneva: Author.

Wortman, C. B., & Silver, R. C. (1992). Reconsidering assumptions about coping with loss: An overview of current research. In L. Montada, S. Filipp, & M. J. Lerner (Eds.), *Life crises and experiences of loss in adulthood* (pp. 341–365). Hillsdale, NJ: Erlbaum.

Wyman, L. C. (1970). *Sandpaintings of the Navaho Shootingway and the Walcott Collection.* Washington: Smithsonian Institution Press.

Yalom, I. (1980). *Existential psychotherapy.* New York: Basic Books.

Yang, B. (1992, January). The economy and suicide: A time-series study of the USA. *American Journal of Economics and Sociology, 51,* 87–99.

Yang, B., Lester, D., & Yang, C. (1992). Sociological and economic theories of suicide: A comparison of the U.S.A. and Taiwan. *Social Science and Medicine, 34,* 333–334.

Yokota, F., & Thompson, K. M. (2000, May 24/31). Violence in G-rated animated films. *Journal of the American Medical Association, 283,* 2716–2720.

Young, E., Bury, M., & Elston, M. A. (1999, November). "Live and/or let die": Modes of social dying among women and their friends. *Mortality, 4,* 269–289.

Youngner, S. J., Landefeld, C. S., Coulton, C. J., Juknialis, B. W., & Leary, M. (1989). Brain death and organ retrieval: A cross-sectional survey of knowledge and concepts among health professionals. *Journal of the American Medical Association, 261,* 2206–2210.

Zucker, A. (2000). Assisted suicide. *Death Studies, 24,* 359–361.

Zunin, L. M., & Zunin, H. S. (1991). *The art of condolence.* New York: HarperCollins.

Zusman, M. E., & Tschetter, P. (1984). Selecting whether to die at home or in a hospital setting. *Death Education, 8,* 365–381.

INDEX

CREDITS

Photo Credits